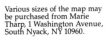

Copyright © by Marie Tharp.

Mercator Projection 1:48,000,000 at the Equator.
Depth and Elevations in Meters.

BASIC PHYSICAL GEOLOGY

EDWIN SIMONS ROBINSON
VIRGINIA POLYTECHNIC INSTITUTE AND STATE UNIVERSITY

BASIC PHYSICAL GEOLOGY

JOHN WILEY & SONS
NEW YORK • CHICHESTER • BRISBANE • TORONTO • SINGAPORE

to Evan, Lindsay, and Valarie

Cover photo: Sandstone laminations near Petra in Jordan. Iron oxides impart the red and brown coloring (Carl Purcell/Photo Researchers).

Copyright © 1982, by John Wiley & Sons, Inc.

All rights reserved. Published simultaneously in Canada.

Reproduction or translation of any part of
this work beyond that permitted by Sections
107 and 108 of the 1976 United States Copyright
Act without the permission of the copyright
owner is unlawful. Requests for permission
or further information should be addressed to
the Permissions Department, John Wiley & Sons.

Library of Congress Cataloging in Publication Data:

Robinson, Edwin S.
   Basic physical geology.

   Includes index.
   1. Physical geology. I. Title.
QE28.2.R6   551   81-16384
ISBN 0-471-72809-8   AACR2

Printed in the United States of America

10 9 8 7 6 5 4 3 2 1

PREFACE

This textbook is concerned with how the earth works and how it is put together. It is an introduction to dynamic geologic processes, minerals and rocks, and the structure and landscape of the earth. The level of presentation is intended for the serious college student, including those majoring in engineering, geology, and other physical sciences. The writing style is for the nineteen- or twenty-year-old student who wants to become involved with the discussions throughout the book.

The opening section, Introduction to the Earth, gives an overview of the main features of the earth and briefly introduces the topics of physical geology. Most of these topics are shown to be interdependent. Chapter 1 provides enough information about all of these topics so that they can be studied separately in later chapters with an understanding of how each one is part of a unified science.

The following chapters are grouped into four sections beginning with The Earth's Size, Shape, and Interior Zones. This section looks at the entire planet, its topography and gravity, its earthquakes and deep interior zones, and the concept of plate tectonics that describes the opening of oceans and the movement of continents over the face of the globe. This is the "big picture" of how the earth works. By presenting a thorough development of the idea of plate tectonics in these early chapters, instead of near the end of the book, this concept can be used effectively as a unifying theme for many of the topics presented in the remaining chapters. The next section, The Rocks of the Earth, examines the earth's interior in more detail. It describes minerals and rocks, and what they reveal about conditions inside the earth. First minerals are discussed, then each of the major rock groups, and finally how these rocks are molded into structures on a grand scale by mountain-building processes. This is followed by The Face of the Earth, a section that concentrates on the landscape and processes acting on or close to the earth's surface. Compared with the previous sections, the discussion here shifts much more toward a world we can feel and touch. This is a world of rivers, glaciers, waves, and wind that all help to carve and to shape the scenery of the land surface. The final section, Other Planets, describes the solar system. It is also a review of what we have learned about the earth, since this is the knowledge we use to determine the extent to which other planets are similar to the earth, or differ from it.

Chapters and topics in the book are arranged to give a comprehensive survey of the earth first, followed by intensive study of earth materials and structure, and finishing with a careful examination of near surface features and processes that perhaps are the most familiar to us. This is a sequence that I have used successfully in the classroom for more than a decade. However, opinions about how the topics of physical geology should be arranged in a book are diverse. Some professors and students may find advantages in following the sequence presented in this book. Others may prefer to study the chapters and sections in some other sequence. Chapter 1 provides a student with enough background to understand most of the material in any subsequent chapter without reference to other chapters. This background allows considerable flexibility for rearranging the sequence of chapters and sections in keeping with individual tastes and practices.

Each chapter in the book presents all of the kinds of information that make up one of the fields of specialization found in the modern community of professional geologists. The two exceptions are Chapters 1 and 5, which present topics with which almost all geologists are familiar. Otherwise, for instance, Chapter 3 discusses most of the things a seismologist should know about whereas Chapter 7 concentrates on the areas of interest to the igneous petrologist. Perhaps this is the most natural way to organize information in a book because it happens to be the way that we geologists have divided that information among ourselves.

This philosophy has also influenced the arrangement of chapters. For example, the chapters on igneous and metamorphic rocks follow one another immediately after the chapter on mineralogy because of the common interests of igneous and metamorphic petrologists in the crystal chemistry of high pressure and temperature environments. These subjects are followed by the chapter on sedimentary rocks, which consist of fragments of igneous and metamorphic rocks as well as older sedimentary rocks. The features of sedimentary rocks motivated geologists to look more carefully at a multitude of surface processes. Therefore, a natural sequence is to discuss these processes in the chapters following the description of sedimentary rocks.

Insofar as possible the information in each chapter is arranged in the sequence of look, describe, and then analyze. This sequence reflects the natural inclination of most of us to become curious enough about many of the things we see to want to learn more about them. But this is certainly not our only motivation to study the earth. There are many practical reasons related to the search for oil, gas, valuable ores, and other commodities of economic importance. Information about them is distributed throughout the book. Various aspects about the search for oil are included in Chapters 2 to 4, 9, and 10. Topics pertaining to ores are found in Chapters 3, 6 to 8, 10, and 12. Other practical information about water resources, industrialization, and urban development is included in Chapters 11 to 15. These applied aspects of geology have more meaning and interest to many students if they learn something of the sources of information for these topics. Therefore, most chapters of the book have descriptions of how geologists conduct field surveys and laboratory experiments.

Much of the book consists of relatively straightforward descriptive material. However, some analytical topics require the student to read and to ponder carefully. Although no college-level science background is required, rudimentary knowledge of chemistry, algebra, and trigonometry is assumed. The book uniformly presents the world from the viewpoint of a geologist. But the geologist needs to use some chemical formulas and mathematical expressions to describe certain relationships. These formulas and expressions are used in various parts of the book where they are appropriate to the discussion.

While writing this book I had numerous discussions with the members of the faculty of the Department of Geological Sciences of Virginia Polytechnic Institute and State University. These discussions have led to a clearer and more complete coverage of the various topics. In particular, I acknowledge with appreciation the efforts of Richard K. Bambach, M. Charles Gilbert, David A. Hewitt, J. Frederick Reed, and Paul H. Ribbe who critically read parts of the manuscript during its initial stages of preparation. Gerald V. Gibbs and Karen Geisinger prepared unique computer drawn crystal structure diagrams. During various stages of revision the entire manuscript was read critically by W. Gary Hooks of the University of Alabama, Winthrop D. Means of the State University of New York–Albany, Hamilton Johnson of Tulane University, J. Allan Cain of the University of Rhode Island, Joseph Lintz of the University of Nevada, Leonard M. Young of Northeast Louisiana University, Gunnar Kullerud of Perdue University, and Rolfe C. Erickson of Sonoma State University in California. Many of their suggestions have been incorporated into the final manuscript. Donald H. Deneck, geology editor at Wiley, provided much of the encouragement and guidance that went into the completion of this project. My wife Valarie helped with various parts of the manuscript and gave me the needed push to get the job done.

Edwin S. Robinson

CONTENTS

SECTION ONE
INTRODUCTION TO THE EARTH

1. ROCKS AND TIME **3**

SECTION TWO
THE EARTH'S SIZE, SHAPE, AND INTERIOR ZONES

2. GRAVITY AND THE SHAPE OF THE EARTH **41**

3. EARTHQUAKES AND SEISMOLOGY **75**

4. THE EARTH'S MAGNETISM **117**

5. GLOBAL TECTONIC PROCESSES **143**

SECTION THREE
THE ROCKS OF THE EARTH

6. MINERALOGY **177**

7. INTERNAL PROCESSES AND IGNEOUS ROCKS **217**

8. METAMORPHISM **280**

9. SEDIMENTARY ROCKS **304**

10. TECTONISM AND THE ARCHITECTURE OF THE EARTH'S CRUST **349**

SECTION FOUR
THE FACE OF THE EARTH

11. GROUND WATER **403**

12. WEATHERING **425**

13. RUNNING WATER, MASS WASTAGE, WIND, AND CONTINENTAL LANDSCAPES **453**

14. GLACIERS AND THE CONTINENTAL LANDSCAPE **505**

15. GEOLOGIC PROCESSES IN THE MARINE ENVIRONMENT **545**

SECTION FIVE
OTHER PLANETS

16. THE EARTH IN THE SOLAR SYSTEM **599**

APPENDIXES 629

GLOSSARY 639

PICTURE CREDITS 665

INDEX 667

BASIC PHYSICAL GEOLOGY

SECTION ONE
INTRODUCTION TO THE EARTH

1

ROCKS AND TIME

Why Study Geology?
The Form of the Earth
Minerals and Rocks
   Minerals
   Igneous Rocks
   Sedimentary Rocks
   Metamorphic Rocks
   Weathered Rock
   Geologic Maps
   Geologic Cross Sections and Columns
Uniformitarianism and Geologic Processes
   Principle of Uniformitarianism
   Volcanic, Plutonic, and Metamorphic Processes
   Erosion and Sedimentation
   Tectonic Processes
Earth History and Geologic Time
   Superposition and Crosscutting Relationships
   Stratigraphic Correlation
   Subdivisions of Geologic Time
   Controversy about Geologic Time
   Radiometric Dating of Rocks
   Calibration of the Geologic Time Scale
The Perspective of Geologic Time

WHY STUDY GEOLOGY?

The world we inhabit is a planet of fascinating complexity. To make sense of this complexity we must describe and explain as objectively as possible the materials, features, and processes that can be observed in our surroundings. The landscape of the earth, the nature and distribution of earth materials, and conditions in the earth's interior are the subjects of principal interest in the science of physical geology.

Much has been learned about the earth in our search for natural resources. All of the petroleum and coal and other natural fuels are discovered in the rocks of the earth. So are the ores from which metals are extracted, the raw materials for concrete, stone, asphalt, and other building supplies, and important sources of fresh water. These precious commodities are not easy to find. There have been many unsuccessful searches and some accidental discoveries of important deposits. In the long history of prospecting winners and losers alike have contributed basic information about how natural resources are associated with various kinds of rock in the earth. From their efforts we have gained a knowledge of geology that we now use to plan much of the exploration for industrial raw materials.

The valleys and plains, rivers and bays, and other features of the landscape have had a profound influence on our culture and civilization. In many important ways these features determine how our communities and transportation routes develop. But the landscape is continually changing, slowly in some places, and elsewhere with unexpected suddenness. Erosion and floods, landslides, earthquakes, and other geologic processes alter the land surface. Knowledge of these natural phenomena and the extent to which they can be predicted is important in the many land-use and building activities of our modern industrialized society. Geologists work with civil engineers in planning flood control, harbor stabilization, soil conservation, and water resource development projects. The strength and durability of rock and soil must be considered when planning the construction of dams, tunnels, and highways. Through these many activities we continue to learn about the nature of the earth.

Advances in our understanding of the earth have resulted from general intellectual interest as well as from attempts to contend with specific problems related to community and industrial development. Some individuals, motivated purely by curiosity, have formulated important geological principles. But whatever the intellectual or practical reason for a particular geological study, it often yields more and different information than was initially expected. Some knowledge of physical geology can help most of us to ask more intelligent questions, to separate fact from fancy in the arguments and exhortations to which we are persistently subjected, and perhaps to find a few answers.

Knowledge of the earth comes from observations made on or near the surface. Ideas about the deeper interior, which is hidden from view, are inferences based upon these surface observations. The different kinds of information pertaining to physical geology can be grouped into topics which are the separate subjects of the chapters in this book. To a certain extent these topics can be studied independently of one another. But there is a degree of interdependence. Some facts about rocks, processes, and geologic time are introduced in this first chapter to provide a background for the more thorough discussions in the chapters that follow.

THE FORM OF THE EARTH

The earth is a nearly spherical planet approximately 12,740 km (7916 mi) in diameter. Its solid surface is formed into **continental platforms** and **ocean basins.** The continental platforms, making up about 30 percent of the surface, stand at an average elevation of 670 m (2200 ft) above sea level. Here we find vast plains and plateaus, rolling countryside of hills and valleys, and high mountain ranges. The highest peak is Mt. Everest reaching 8848 m (29,028 ft) above sea level.

The varied continental landscape has been shaped by several processes. Streams erode sediment from some places and deposit it elsewhere. Glaciers spread slowly over parts of the land, scraping the surface, and later melt away leaving scattered deposits of rock debris. Winds shift the lose sand and dust over the landscape. Other processes originating inside the earth produce volcanoes, earthquakes, and slow warping of rock layers, all of which leave imprints on the landscape. The irregular surface of the continents is constantly changing.

The ocean basins occupy about 70 percent of the surface of the earth. Water overfills these basins, spreading in shallow seas on the margins of the continental platforms so that about three-quarters of the earth's surface is submerged. The floor of the ocean basins is at an average depth below sea level of 3900 m (12,800 ft). The landscape of the ocean floor displays broad, nearly flat abyssal plains, rolling hills, oceanic ridges as rugged as continental mountains, and deep trenches.

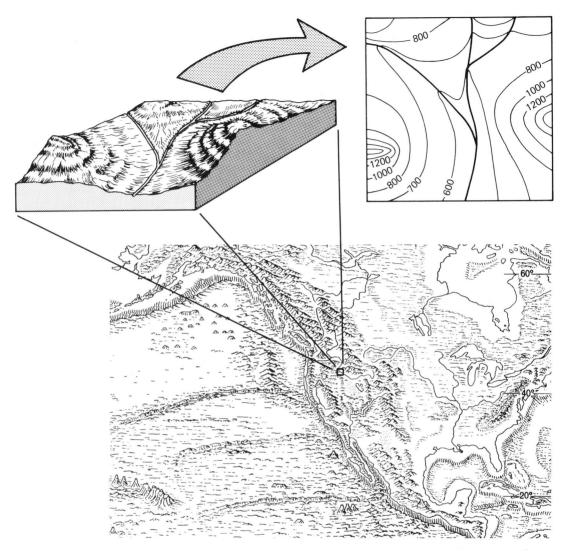

Figure 1-1 Artistic representation of the land surface over North America and the bordering ocean floor (after B. C. Heezen and M. Tharp and other sources), and more detailed representation of the land surface over a small area by means of topographic contour lines, which are lines connecting points of equal elevation above sea level.

The ocean bottom is 11,035 m (36,204 ft) beneath the surface of the sea in the Marianas Trench of the Pacific Ocean Basin. The character of the ocean floor is indeed as varied, but distinctly different, from the continental landscape.

We use topographic maps to represent the form of the earth's surface. This can be accomplished by artistic shading designed to exaggerate certain features. The landforms are depicted in this way in Figure 1-1A. For a more quantitative and detailed display, topographic contour maps are used. The example in Figure 1-1B illustrates how **contour**

6   Introduction to the Earth

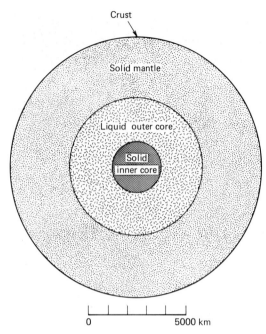

Figure 1-2   Principal interior zones of the earth.

lines, which connect all points of equal elevation, indicate the shape of the land. A topographic profile reveals the variation of the surface elevation along some particular line of direction.

The interior of the earth consists of concentric zones that are illustrated in Figure 1-2. We know this principally from studies of earthquake vibrations which travel at different speeds through these zones. The outermost part is called the **crust.** It consists of rock with a density lower than 3 gm/cm^3 (the **density** of a specimen of rock is found by dividing its mass by its volume). The earth's crust extends to depths of between 30 km and 60 km beneath the continents, and from 5 km to 10 km beneath the ocean basins. Below the crust is the solid **mantle** of the earth. Rock density in this zone is about 3.3 gm/cm^3 just below the crust, and increases to almost 6 gm/cm^3 near its base at 2900 km. Extending from this depth to the center is the **core** where density increases from almost 10 gm/cm^3 to more than 12 gm/cm^3. The outer part of the core reaching to 5100 km is inferred to be liquid from the nature of the vibrations that pass through it. The inner part of the core is believed to be solid.

MINERALS AND ROCKS

The continental platforms and the ocean floor are covered largely by a layer of **regolith** formed from soil, gravel, sand, clay, and other loose rock fragments. Solid **bedrock** exists everywhere beneath the regolith and is exposed in **outcrops** from place to place. Outcrops are abundant in mountainous areas and some arid regions. In some places great thicknesses of bedrock are exposed. But in other areas that are covered by a blanket of regolith ranging in thickness from less than a few meters to more than 100 m, samples of bedrock must be obtained by drilling and from a few small and widely separated outcrops. In recent decades bedrock specimens have been collected from the ocean basins by deep-sea drilling and shipboard dredging operations. By these techniques geologists have collected specimens from almost every region of the world, so that we can study the variety and composition of bedrock that exists near the earth's surface.

From a careful examination of representative samples and the appearance of outcrops and drill cores, it is possible to classify the varieties of bedrock into three general groups. These include **igneous** rocks, **sedimentary** rocks, and **metamorphic** rocks. The distinguishing properties of these different kinds of rock can usually be seen in hand specimens or by examination of polished surfaces and thin sections under a microscope (Figures 1-4 through 1-6). A thin section is made by mounting a chip of rock on a glass slide and grinding it away until only a paper-thin translucent layer remains. All of these rocks consist of grains of different minerals.

Minerals

The word **mineral** has been used in different ways. Coal, petroleum, building stone, ores, and other such commodities are commonly referred to as mineral resources. But for certain scientific purposes where a more restricted and specific definition is needed, we consider a mineral to be a naturally occurring inorganic substance with a definite chemical composition and geometrical arrangement of atoms. According to this definition we can recognize more than 2000 distinctive kinds of minerals. Only a few of these are abundant, and most are very rare.

One of the abundant minerals in the earth's crust is **quartz,** which has the chemical composition SiO_2. It is the principal constituent used in the manufacture of glass. In many places the sand along beaches is principally quartz grains. A crystal of quartz showing well-formed faces and a diagram of the internal arrangement of silicon and oxygen atoms illustrate (Figure 1-3) the geometrical properties of this mineral.

Quartz belongs to a group of minerals in which silicon and oxygen are important constituents. These are known as the **silicate** minerals, and they account for more than 98 percent of the mass and volume of the earth's crust. Oxygen and silicon alone make up about 75 percent of the weight and 93 percent of the volume of the crust. A few other elements including aluminum and iron are also important. Because we do not have the technical capability to separate economically these metallic elements from the other constituents, the silicate minerals are not useful as ores. They are used to a certain extent, however, in the manufacture of ceramics and building supplies.

Some minerals can be processed to extract metals. These minerals are recognized as ores where they have become concentrated in zones which can be mined. An example of an ore mineral is **galena** (PbS), a silvery metallic-appearing mineral that is the principal

Figure 1-3 A cluster of quartz crystals, and a model showing how the silicon and oxygen atoms are geometrically arranged.

source of lead. It is a member of the sulfide mineral group in which metallic elements are combined with sulfur. The sulfide minerals occur in rare but economically important concentrations of ore in the earth's crust. There are many other important groups of ore minerals.

Minerals are useful indicators of conditions in the earth's interior. The temperature and pressure environment as well as the available chemical elements influence the kinds of minerals which can be synthesized in the earth. **Diamond** is a mineral variety of carbon that is produced under high pressure. Rare bits of diamond formed in the mantle are brought to the earth's surface by volcanic eruption processes. Another mineral variety

8 Introduction to the Earth

of carbon is **graphite** which is used in pencils. It is produced under low pressure in the crust, and the atoms of carbon are grouped in a geometrical arrangement completely different from that of diamond.

Igneous Rocks

From time to time melting occurs in parts of the upper mantle and perhaps in deeper parts of the crust. We see evidence of this in the lavas erupted from volcanoes. The rock that is formed directly from solidification by cooling such molten material is defined as **igneous** rock. As this rock-forming liquid cools, some minerals crystallize before others, but eventually all of the elements present become incorporated into the solid rock.

The abundant igneous rocks in the earth's crust consist almost entirely of silicate mineral grains. When we examine a specimen of this kind of rock, we find a mass of irregular grains of individual minerals interlocked with one another. Different varieties of igneous rock are distinguished by the size of the grains and the kinds of minerals present.

As lavas spill from volcanoes onto the earth's surface, they cool and solidify rapidly. There is too little time for large mineral grains to develop. As a result the rocks formed from these lavas consist of grains so small that they can be seen only under a microscope. The most abundant of these fine-grained igneous rocks is **basalt,** a dark-colored rock. In some places where basaltic lavas from a succession of eruptions have spread one upon another over the countryside, we now find layers of basalt. Some features of this kind of rock are illustrated in Figure 1-4. Basalt is a silicate rock completely lacking in quartz.

Some igneous rocks are coarse grained,

Figure 1-4 Basalt layers are exposed in this cliff near Christchurch, New Zealand. A thin section prepared from a specimen of this rock reveals microscopic interlocking mineral grains.

and the individual mineral constituents can be seen without magnification. Growth of these large crystals requires slow cooling and crystallization of a rock-forming fluid. Although no one has actually observed the synthesis of such rocks, it is clear that solidification must take place beneath the earth's surface. The most abundant coarse-grained igneous rock found near the earth's surface is **granite,** which consists of quartz and other silicate minerals. Figure 1-5 displays some characteristics of this variety of rock. The assorted light- and dark-colored grains give it an attractive appearance which for many centuries has enhanced its value as a building stone. Granite is widely distributed in large, somewhat irregular masses in the upper part of the earth's crust.

Sedimentary Rocks

Sediment is accumulating in many places on the earth's surface. Sand collects along beaches, pebbles and cobbles are washed along by streams, silt and clay accumulate on the deltas of large rivers, and the shells of a myriad of organisms collect on the sea bottom. Water percolates through the small openings between these grains of sediment dissolving some substances while others precipitate. Slowly these fragments are compressed under the weight of overlying sediment. Precipitates form on their surfaces binding them together more firmly. Eventually a solid mass of **sedimentary** rock is formed.

Exposures of sedimentary rock usually reveal a more or less distinct layering. In a series of layers you may be able to distinguish different kinds of sedimentary rock according to the composition and size of the constituent fragments. Some important examples are illustrated in Figure 1-6. Features indicative of the environment in which the sediment accumulated, such as a rippled surface reminiscent of a shallow sea floor, are commonly preserved in individual layers.

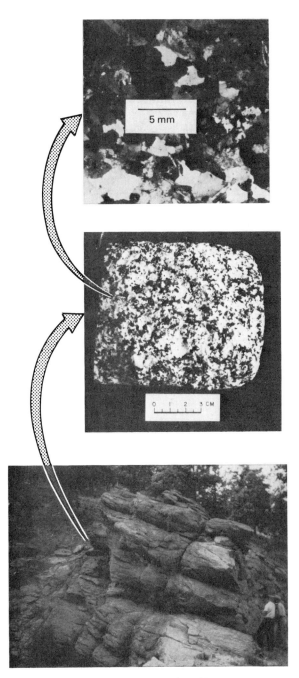

Figure 1-5 Granite is exposed in this outcrop near Boulder, Colorado. The different mineral grains can be seen in the specimen of this rock. The thin section reveals more clearly how the mineral grains are fitted together.

Figure 1-6 Alternating layers of sandstone and shale are exposed near Parkersburg, West Virginia. Constituent grains in the sandstone and thin laminations in the shale can be seen on closer examination.

The most abundant kind of sedimentary rock is **shale**. It is formed mostly from microscopic silt and clay particles. Shale layers consist of thin laminations and may split easily along closely spaced surfaces. Depending upon the content of organic material, iron oxides, and small amounts of other elements, shales display a variety of colors including black, purple, red, brown, green, and gray. It is not a durable rock, and little of value has yet to be extracted from it. This may change with the development of an economical process for removing petroleum now locked in the interstices of some shales that are rich in organic constituents. In this event vast accumulations of oil shale in western North America may become an important natural resource.

Sandstone is another common type of sedimentary rock. The constituent particles are visible without magnification, but they are mostly smaller than 2 mm in diameter. They are usually cemented together by calcium carbonate or silica and minor amounts of iron oxides which may impart a red, brown, or yellow color to the rock. Some kinds of sandstone are very durable and have been widely used for building stone.

Shell fragments and particles of calcium carbonate from decomposed shells are the basic constituents of **limestone,** another abundant sedimentary rock. Varieties of limestone with only small amounts of certain impurities are the source of lime used in cement. Beds of durable limestone are also quarried for building stone.

Metamorphic Rocks

Once formed, rocks may again be subjected to conditions that drastically alter their character. Some of the original constituents decompose or partially melt, and new and different minerals are produced. **Metamorphic rocks** are formed by these kinds of transformations which usually involve an increase in either or both pressure and temperature. They can be made from igneous, sedimentary, or other more ancient metamorphic parent rocks.

Metamorphism of a rock involves some recrystallization of the original minerals. Sometimes the same kinds of minerals are reformed into new interlocking grains. **Marble** is the metamorphic rock formed by recrystallization of limestone. By raising the temperature and pressure on the limestone, new interlocking grains of **calcite,** a calcium carbonate mineral, can be produced from the original calcite grains that made up the former shell fragments. In other kinds of metamorphic rock some or all of the original silicate minerals become altered to different mineral varieties. Depending upon the extent of recrystallization, it may or may not be possible to recognize the kind of rock that existed prior to metamorphism.

Some varieties of metamorphic rock display a distinct zoning of different minerals. The coarse-grained rock illustrated in Figure 1-7, which we call **gneiss,** consists of layered zones of alternating light- and dark-colored minerals. **Slate** is a very fine-grained metamorphic rock which is easily split along very smooth parallel surfaces because of the alignment and zonation of minerals. This property has traditionally made slate useful for chalkboards and roofing tile.

Weathered Rock

When bedrock is exposed at the earth's surface, it reacts chemically with the atmosphere and the water that seeps into cracks and interstices. Some of the constituent silicate minerals decompose, iron and aluminum may become oxidized, and clays form. The appearance and durability of the rock is changed. In some places the effects of weathering form a coating only a few millimeters thick. Elsewhere the weathered rock may reach depths of several tens of meters.

12 **Introduction to the Earth**

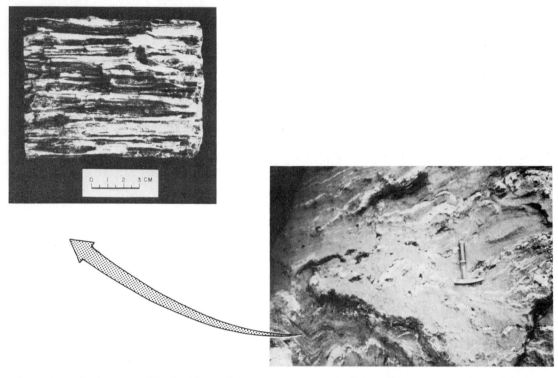

Figure 1-7 Gneiss exposed in the Blue Ridge Mountains near Charlottesville, Virginia. Light- and dark-colored minerals have been segregated into parallel bands in this rock.

Geologic Maps

We can represent the distribution of different kinds of rock and patterns of fractures and deformation on **bedrock geologic maps.** On such a map the bedrock surface is portrayed as it would appear without the natural cover of regolith and vegetation. The example in Figure 1-8 shows how the different kinds of rock making up the bedrock surface can be represented by symbolic patterns or colors. Depending upon the purpose for which the map was prepared, these symbols may distinguish igneous from sedimentary rock or different varieties of these principal kinds of rock. It is also common to group rocks according to age. Older and younger layers and masses can be distinguished by different symbolic patterns.

The geologist pieces together this kind of map from information obtained at outcrops and perhaps from holes drilled through the regolith into the buried bedrock. Other survey methods including measurement of gravity and magnetic field changes, and vibrations from earthquakes or explosions also provide information about the hidden bedrock surface.

Geologic Cross-Sections and Columns

The rocks and features which would be exposed along the surface of a vertical cut in the earth can be portrayed diagrammically on a vertical cross section. The edges of the

Rocks and Time   13

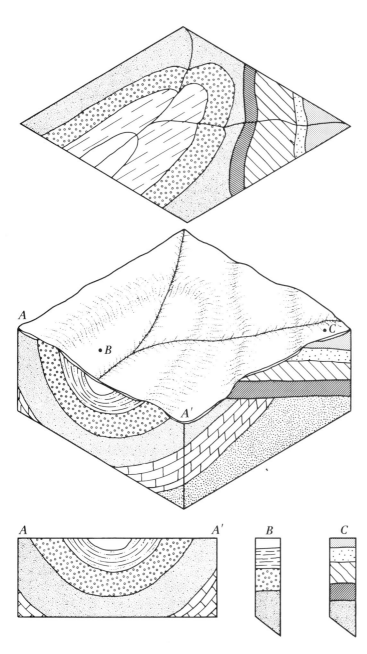

Figure 1-8 The block diagram illustrates a landscape and underlying rock layers. Above is a bedrock geologic map of the surface of the block diagram. Below is a vertical cross section showing subsurface distribution of rock layers beneath the line A-A' on the geologic map, and geologic columns showing sequences of rock directly beneath points B and C on the geologic map.

block diagram in Figure 1-8 are examples, and one in particular taken along the line AA' is shown separately. A modification of the cross section is the **geologic column,** also illustrated in Figure 1-8. This kind of diagram reveals the vertical distribution of different kinds and ages of bedrock beneath a point on the earth's surface.

Geologic cross sections and columns represent the geologist's best estimate of what lies underneath the landscape. Except where we have obtained information directly from a few costly holes drilled into the earth, all that we know about rocks in the interior has been pieced together from data obtained on the land surface.

UNIFORMITARIANISM AND GEOLOGIC PROCESSES

The Principle of Uniformitarianism

Rocks and landforms display clues of the geologic processes that formed them. Some of these clues are quite obvious. For example, you can compare (Figure 1-9) features of an ancient sandstone layer with similar features now being formed in loose sand along a seacoast. Then you can watch how the action of waves and tides produces these sand ripples and alternating thin layers of coarse and fine particles. You can observe how shells become imbedded in the sand. Therefore, you have reason to believe that the ancient layer of sandstone, now far inland and high above the sea, could have been formed long ago by similar processes. Elsewhere we find shale layers with features much like those that we can now see forming in the mud and clay on lake bottoms. These and many other comparisons illustrate how we can associate features in rocks with different processes presently acting on the earth.

The idea of explaining the origin of rocks and landscape in terms of present-day processes was first proposed by the Scottish naturalist James Hutton (1726-1797). In a time when untested assertions and supernatural events were invoked to answer many questions, he discovered natural explanations. Hutton recognized that sedimentary rock played an important role in a cycle of geologic processes. He saw that it was formed from the fragments of more ancient rock. But he also saw how it was being weathered and eroded to provide the debris for rock yet to be formed. He found other evidence of ancient rock which had become molten, then crystallized again to produce newer igneous rock. Everywhere he looked the bedrock and landscape indicated continuing repetition of weathering, erosion, melting, crystallization, and many other processes which we know are active at the present time. Hutton could find no evidence of a beginning or an ending to this succession of events. Given sufficient time, all of the rock and landscape features he examined could have been produced by these natural processes. He concluded that in the past the earth had been shaped by the same kinds of processes and events that are important at the present time.

We now define as the **Principle of Uniformitarianism** the philosophy of explaining earth history and the origin of ancient rock in terms of the same kinds of natural processes that we know are active at the present time. Briefly summarized, "the present is the key to the past." As the science of geology developed, this proposition was rigidly interpreted by some scientists who refused to admit any explanation of a geological problem not documented by evidence of the present balance of processes. We now realize that far in the geologic past some conditions were substantially different from the present, and strict adherence to the narrow interpretation of uniformitarianism is not always justified. As we continue to explore the earth, some questions certainly will be answered in terms of processes we have not yet recognized. The Principle of Uniformitarianism urges a thorough examination of the world around us rather than acceptance of answers that are beyond our powers to verify. Let us now consider, briefly, some of the important geologic processes that this philosophy has helped us to discover.

Volcanic, Plutonic, and Metamorphic Processes

Iceland is one of the places where volcanoes are a part of everyday life. Buildings are heated and power is generated from volcanic steam and hot water. Most of the time the people there can safely go about their business. But occasionally rumblings in the earth warn them to leave for safer surroundings. Then, in a fiery display, glowing ash, hot gases, and lava erupt and spread over the countryside. But despite obvious risks, some

Figure 1-9 Ancient sandstone layers (A and B) in the Appalachian Mountains display a rippled surface similar to a modern sand surface (C) along the eastern coast of Scotland, and alternating thin laminations of coarse and fine particles similar to (D) the layer of beach sand on the coast of South Carolina.

geologists remain to watch the eruption and to try to collect samples of the lava, ash, and gas (Figure 1-10). They do this to learn first hand about **volcanic** processes and to find out about the igneous rock produced by these processes.

Igneous rock is produced from molten material called **magma.** The lava from volcanic eruptions is magma that has spilled out on the earth's surface. To understand the details of how igneous rock forms, it is important to learn as much as possible about other condi-

Figure 1-10 Geologist collecting samples of lava from a pool on Kilauea Volcano on the island of Hawaii. (D. A. Swanson, U.S. Geological Survey.)

tions such as what gases are present when the lava solidifies and what minerals crystallize as the lava cools through different temperatures. Much of this information can be found only from observations made during an eruption. These studies help us to understand different kinds of volcanic processes and to see how these processes leave their imprints on igneous rock such as basalt.

But most magma that is melted from the mantle or deeper parts of the crust never reaches the earth's surface. It stays concealed and eventually solidifies to produce various kinds of igneous rock, such as granite. These rocks are the products of **plutonic** processes. This name is taken from Pluto, the ancient Roman god of the lower world. How do these underground processes work? Because they remain hidden, everything that we know about them comes from circumstantial evidence. Sometimes we can detect vibrations of small earthquakes caused by movement of magma in zones several tens of kilometers beneath active volcanoes. However, it is very difficult if not impossible for us to determine how the composition of the magma may have changed as it moved upward through fractures and by melting the overlying rock. It is also difficult to know the details about how a mass of granite was produced, because the particular plutonic processes have long since ceased to be active by the time the rock becomes exposed by erosion at the surface of the earth. The same is true of **metamorphic** processes which cannot be directly observed. However, we can learn something about these hidden processes from laboratory experiments that test chemical reactions between different mineral combinations at temperatures of 1500° C or more, and pressure which can be 50 thousand times greater than atmospheric pressure on the surface of the earth. Under these conditions we can synthesize groups of minerals similar to those in igneous and metamorphic rock specimens. In this way we can figure out the probable conditions of pressure, temperature, and chemical environment that are related to plutonic and metamorphic processes.

After we have discovered the particular pressure and temperature conditions that are needed to produce igneous and metamorphic rocks, we must then figure out where these conditions exist in the earth. We know that heat released by radioactivity has an important effect on temperature. We know that pressure increases with depth because of the weight of overlying rock, which depends on rock density. Rocks and lava erupted from volcanoes give us some clues about density and radioactivity deep in the crust and in the upper mantle. We get other clues from measurements of earthquakes and the earth's gravity and magnetism.

From experiments we have learned that common igneous rocks become molten at temperatures between approximately 600° C and 1300° C. These experiments also show that the melting point depends upon the relative concentrations of different minerals. But what causes heat to build up in certain zones so that magma is produced? We still cannot answer that question.

Another important consideration is the way that magma changes as solidification progresses. We know that certain minerals will crystallize at higher temperatures than others. As the magma cools, minerals that crystallize first may settle to the bottom of the magma zone. This alters the composition of the remaining fluid so that different minerals crystallize later on. By learning about the order in which different minerals crystallize from magma we can better understand the origin of the typical combinations of minerals that make up different kinds of igneous rock. We can also get a clearer idea about how nature produces important ores of lead, zinc, copper, gold, silver, and nickel which are associated with igneous masses.

Erosion and Sedimentation

Most of us have never watched a rampaging river. But people living near one of the world's great rivers cannot fail to see its tremendous power. The mightly Brahmaputra River winds its way across northeastern India and Bangladesh. Almost every year heavy rains cause the river to flood. This is a mixed blessing. Some destruction is certain, but the land is renewed by deposits of fertile mud that are left behind when flood waters recede. The power of the Brahmaputra is evident in other ways. In places it is more than 5 km wide and 200 m to 300 m deep. The swirling water undercuts the river banks, and from time to time slices of land several hundred meters wide collapse into the torrent, often destroying parts of villages. The Brahmaputra, like the other mighty rivers, has a personality that is woven into the history of civilization.

But long before we were around to make our history, ancient rivers left their imprints. They carved most of the important features of the landscape, and they deposited layers of sediment that are now the pages in a much older history of the earth. Rivers then and now have played a very important role in the geologic processes of erosion and sedimentation that act on the surface of the earth. We can get first-hand information about how many of these processes work. We can observe grains of sediment being produced by weathering of bedrock. For example, in Zion Canyon in southwestern Utah water percolating through the cracks and other small openings near the base of sandstone cliffs dissolves the calcium carbonate that binds sand grains together (Figure 1-11). Gradually a cliff is undercut as the loosened sand grains fall away. Eventually this causes rock on the cliff face to collapse. Falling blocks of rock hit the ground and are further broken apart by the impact.

We can observe many other natural processes that break up bedrock to form sediment. Waves pounding on coasts and glaciers spreading slowly over the land can grind bed-

18 Introduction to the Earth

(A)

(B)

Figure 1-11 (A) Near the base of a sandstone cliff in Zion Canyon, Utah, water seeps out from the rock dissolving some of the calcium carbonate that binds together the constituent grains of sand. Eventually (B) the cliff is undercut to the extent that overlying rock collapses.

rock into particles of loose sediment. We can also observe how this loose sediment is carried away in various ways and then is deposited. A large part of it is washed into rivers by rain and melting snow. When it reaches the river, the turbulent water keeps much of it suspended and it is borne along with the current. Some sediment becomes embedded in glaciers and is slowly transported by the flowing ice. Wind and currents produced by waves along coasts also help to move sediment. But eventually the water becomes less turbulent or the ice melts, and the sediment is deposited. This happens on deltas where rivers flow into the sea or where rushing mountain streams reach broad valleys or plains (Figure 1-12). Broad areas of North America and northern Eurasia are covered by the sediment from huge glaciers that melted between 10,000 and 20,000 years ago. The processes that erode, transport, and deposit sediment all leave their marks. The rippled surfaces and laminations that you see in Figure 1-9 are typical of these marks.

A recent accumulation of sedimentary debris is usually quite porous. Gradually it compacts under its own weight. Then it becomes solidified when the grains of sediment are cemented together by compounds such as silica (SiO_2), calcium carbonate ($CaCO_3$), and iron oxides that precipitate from the water percolating through the pore openings. In this way nature produces the layers of sedimentary rock that you see along roadsides, in mountains, and in so many other places.

Tectonic Processes

Have you ever felt an earthquake? Many people who have felt one find the experience to be quite unnerving. There can be a sudden jolt or series of jolts, or a slower and more rhythmic shaking of the ground. Most earthquakes happen when there is sudden movement of rock along a large fracture in the earth's crust or in the upper mantle. Such movement is caused by **tectonic** processes.

Rocks and Time 19

Figure 1-12 Fan-shaped deposits of sediment transported by mountain streams onto the floor of Death Valley in California. (H. Drewes, U.S. Geological Survey.)

Figure 1-13 Folded and fractured rock layers in California. (U.S. Geological Survey.)

Rocks become folded and fractured by these processes. Some folds are quite small (Figure 1-13), but others are several tens of kilometers across and reach deep in the crust. Besides being crumpled into folds, rock can be cut by **faults**. A fault is a fracture along which the rock on one side has shifted relative to the rock on the other side (Figure 1-13). On a much larger scale tectonic processes build mountain ranges, uplift parts of the crust into high plateaus, and even alter the shapes and the arrangements of the continental platforms and ocean basins.

The folding that we observe and the displacement of rock that we see along faults have occurred over long periods of time. Slowly the rock bends, and little by little it is shifted along a fault. But each small shift can produce a sudden jolt when the rock masses momentarily grind past one another. The result is an earthquake. The energy generated by one mass of rock grinding against another causes vibrations that we call **seismic waves** to propagate away from the fault. When these waves reach the earth's surface, they produce the shaking that is sometimes so destructive. From careful study of seismic waves that have traveled through the earth's interior we have discovered the zones illustrated in Figure 1-1, and we have learned about some of the properties of materials in these zones. Seismic waves tell us that the outer part of the core is liquid. This interesting fact helps us to understand why the earth has the properties of a giant magnet. If the liquid part of the core contains a hot ionized fluid, then currents produced by temperature differences could generate a magnetic field.

Heat energy and gravity play very important roles in the tectonism of the earth. The transfer of heat and related temperature variations affect the plasticity and density of rocks. When these effects are combined with gravity, some kind of deep-seated plastic flow is possible. This would produce movement and deformation in the overlying rock masses. Global patterns of earthquake occurrence and the directions of faulting related to these shocks tell us that the outer shell of the earth is broken by several large fractures. This outer shell consists of the crust and the upper part of the mantle and is about 150 km thick. It is separated by the large fractures into several adjoining plates (Figure 1-14). Not only do most earthquakes occur near the edges of these plates, but most active volcanoes are situated there, too. These plates appear to be drawing apart along some fractures. Else-

Figure 1-14 *(opposite)* Outline of the major plates making up the outer shell of the earth. Principal zones of earthquakes and volcanic activity are located along the borders of these plates. (From "Earthquakes and Continental Drift" by P. J. Wyllie, reprinted from *Univ. of Chicago Magazine,* Jan.–Feb. 1972, copyright © of the Univ. of Chicago.)

where they are overriding or moving sideways relative to one another (Figure 1-15). Mountain building and other large-scale tectonic processes are related to these plate movements. As they move the continents embedded in them slowly shift their positions on the face of the earth.

EARTH HISTORY AND GEOLOGIC TIME

Humans have not been around long enough to have watched much earth history unfold. How do we know what happened before we came on the scene? The geologic history of the earth is interpreted from evidence preserved in rock that was formed at different times and in different locations. A small part of this history is retained in the sandstone layer shown in Figue 1-9. Although it is now exposed high in the Appalachian Mountains, it contains evidence of a shallow sea that formerly covered this area. Fossils indicate the kinds of creatures that lived in this ancient environment. Somewhere else a record of earlier volcanic activity is preserved in layers of ash and basalt. The earth's geologic history is pieced together by using this kind of information to answer some important questions. What was the succession of events, and how did conditions change in some particular place during a long interval of time? What events were occurring, and what conditions prevailed in different places at some particular moment in time? How much time is involved? We will consider these questions separately in an effort to comprehend more clearly the immensity of geologic time. With

Rocks and Time 21

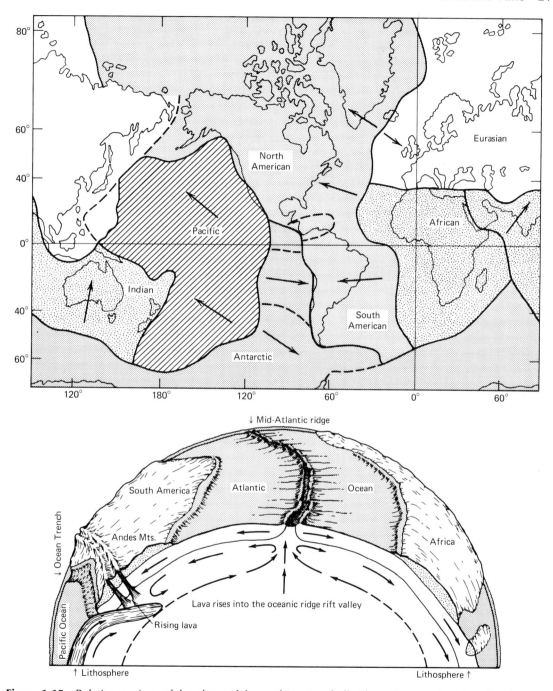

Figure 1-15 Relative motions of the plates of the earth's outer shell. Along the oceanic ridges the plates are drawing apart, and upwelling material is added to their margins. Elsewhere near submarine trenches bordering some continents or arcuate groups of islands the plates are colliding, and the edge of one descends below the other. The plates are approximately 150 km thick. (From "Earthquakes and Continental Drift" by P. J. Wyllie, reprinted from *Univ. of Chicago Magazine,* Jan.–Feb. 1972, copyright © by the Univ. of Chicago.)

this historical perspective we can better appreciate the effects of various geologic processes which may seem almost imperceptible in terms of normal human experience.

Superposition and Cross-cutting Relationships

We can learn about the succession of events and conditions revealed by the rocks of some particular location by finding out the relative ages of the different layers and masses of rock present. In specifying the relative age of a rock we are stating simply that it is older than some rocks and younger than others. We are not attempting to say exactly how long it has existed.

Where parallel layers of sedimentary rock are found, we can assume that each layer was originally deposited in a nearly horizontal position on top of other older layers. This assumption is the basis for the **Principle of Superposition,** which states that a layer of sedimentary rock is younger than layers beneath it, and older than layers above it except if the sequence has been subsequently overturned by tectonic activity. In some particular location, then, we can use this principle to determine relative ages of sedimentary rocks.

In the course of plutonic activity, magma can be forced into layers and masses of older rock. In some places the magma migrates by melting into the rock, and elsewhere it is injected into fractures where it eventually solidifies into igneous rock. Wherever we find igneous rock cutting across other layers and masses of rock, we can assume that it is younger than the rock it transects. An example of such a cross-cutting relationship is seen in Figure 1-16.

Another important kind of cross-cutting feature is the **angular unconformity** which is formed where layers of sedimentary rock have been deposited on the edges of older tilted rock layers so that the younger layers are inclined differently than the older layers.

Figure 1-16 An inclined zone of igneous rock cuts through more ancient rock in central Spain. Magma was injected into a fracture in the older rock and crystallized to form the cross-cutting igneous mass.

The angular unconformity pictured in Figure 1-17 was first described by James Hutton. He recognized that a sequence of events was represented here which began with the deposition of a series of sedimentary layers in a shallow sea. Then deposition ceased, and tectonic stresses caused the layers to be uplifted and tilted into the nearly vertical position in which we now see them. These rocks were eroded, and eventually more sediment was deposited in a horizontal layer across their upturned edges to form the angular unconformity.

Where the continuity of layers or masses of rock is interrupted by injected igneous rock, angular unconformities, faults, and other indications of discontinuity, relative ages can be determined according to the **Principle of Cross-cutting Relationships.** This proposition states that where discontinuities exist, younger features cut across older ones.

Superposition and cross-cutting relationships can be used together to establish the relative ages of rocks in some particular location. Look at the simple example in Figure 1-18 which illustrates both superposed parallel sedimentary layers and cross-cutting features. You can work out the relative ages of

Figure 1-17 Angular unconformity exposed on Siccar Point near North Berwick, Scotland. This outcrop was discovered by James Hutton who used it to illustrate his arguments about uniformitarianism. Gently inclined layers lie on top of the upturned edges of nearly vertical layers.

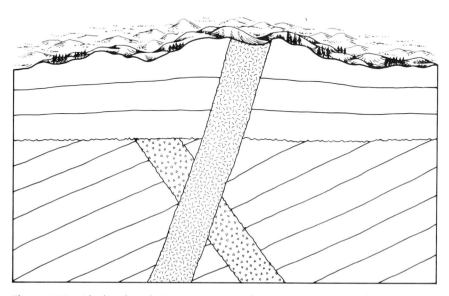

Figure 1-18 Idealized geologic cross section showing superposed sedimentary layers, cross-cutting igneous rock units, and an angular unconformity.

these rock units by applying the two principles we have discussed. The oldest rocks form the lower series of parallel sedimentary layers. These are cut by an igneous mass, and all together these rocks are truncated at the unconformity. Younger sedimentary beds lie on top of the unconformity. These in turn are cut by another igneous mass which also transects the unconformity and the older rocks. The youngest feature is the topography on the upper surface that must have been produced by erosion after all of the rocks had been formed. Once these age relationships are known we recognize the succession of geologic events and conditions which make up the history of this location.

This record of local geologic history is, of course, incomplete. The unconformity indicates an interval of erosion during which time the evidence of some past events and conditions was destroyed. Even when rocks are being formed only some features are preserved, whereas others become obliterated. The geologist is presented with a fragmentary record from which to reconstruct the history of any particular location.

Stratigraphic Correlation

Let us next consider how to recognize the events that occurred and the conditions that prevailed in different places at some particular moment in time. To accomplish this we must again look at the layers, or **strata** as they are commonly called, of sedimentary rock. Suppose that we can examine several layers in each of many locations. It is necessary to determine the particular layers in these different locations which were formed during the same brief interval of time. This task is called **stratigraphic correlation.** It cannot be done simply by identifying the same kinds of rock in different locations. One layer of shale looks very much like another which may have been formed at a different time. Furthermore, during some particular interval of time a layer of shale could have been forming in one location while sandstone was accumulating in another place.

The fossils embedded in sedimentary layers have proved to be the most reliable indicators for stratigraphic correlation. Their value for this purpose was first recognized during the early nineteenth century. While surveying the bedrock of the English countryside for the purpose of planning canal routes, a civil engineer named William Smith (1769-1839) discovered that sedimentary strata extending over large areas could be identified by distinctive groups of fossils. Independently the naturalist Georges Cuvier (1769-1832) recognized that different sedimentary strata in France contained distinctive fossil assemblages.

Following the pioneering efforts of Smith and Cuvier, geologists have examined many sequences of fossiliferous sedimentary strata from most regions of the world. They have determined, using the Principle of Superposition, the relative ages of thousands of guide fossils. A good **guide fossil** represents a variety of animal or plant that was widely distributed over the earth during a short interval of geologic time. The example in Figure 1-19 illustrates in a simple way how combinations of guide fossils are used for stratigraphic correlation. Similar fossils indicate that the sandstone layer at Site 2 is the same age as the upper sandstone layer at Site 1. The shale layer at Site 2 correlates with the bottom sandstone layer at Site 1. This indicates that different kinds of rock were being deposited at the same time at these two locations. The middle limestone layer at Site 1 has no corresponding layer at Site 2. This could mean that nothing was being deposited then at Site 2, or that the deposits were destroyed by erosion before deposition of the top layer. In either case there is an interval of earth history that has left no record at Site 2.

The system of guide fossils which has been pieced together since the beginning efforts of William Smith and Georges Cuvier now pro-

Rocks and Time   25

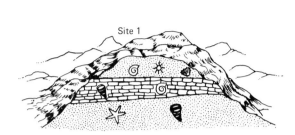
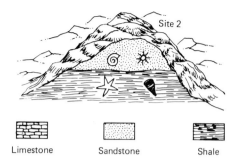

Figure 1-19 Fossils can be used to match up rock layers from different places. This matching is called stratigraphic correlation.

vides us with the most reliable way for determining which sedimentary layers from different locations were formed during some particular interval of time. Some of these fossils are microscopic and others can be seen without magnification. Examples seen in Figure 1-20 illustrate the elaborate detail that can be preserved. As more is learned about the nature of these former organisms, they are not only useful for stratigraphic correlation, but also provide information about environmental conditions during the time of their existence. In these ways fossils contribute to the record of earth history.

Subdivisions of Geologic Time

Earth history spans four long **eras** of time. The **Precambrian** era was the earliest and began with the origin of the earth. The ancient rocks formed during this era preserve evidence of many kinds of events and conditions similar to those of the present. Continents were shaped by running water and glaciers. Mountains raised by tectonic and volcanic processes were eventually reduced by erosion. Oceans covered a large part of the earth's surface. Some conditions were different. Iron compounds in some strata suggest the absence of oxygen in the atmosphere during a long interval of Precambrian time. The first forms of life developed during this era. The fossil impressions illustrated in Figure 1-21

Figure 1-20 Examples of fossils of Paleozoic marine brachiopods illustrating the variety and detail of shells and ornamentation which can be described for purposes of identification and classification.

Figure 1-21 Fossil impressions of wormlike creatures discovered in slightly metamorphosed sedimentary rock of late Precambrian age located near Durham, North Carolina. These are believed to be the oldest fossils of multicelled animals to have been found in North America.

are part of the sparse evidence of Precambrian life in North America. Fossils such as these are very rare, and almost all Precambrian rock is devoid of these kinds of features. There are no useful guide fossils for correlating the sedimentary rocks formed during this era.

The **Paleozoic** era follows, and sedimentary strata from this interval of time display an abundant record of fossil marine organisms. These strata are now observed to extend far inland indicating that shallow Paleozoic seas from time to time spread over much of the area of the continental platforms. Toward the end of this era some large amphibians developed and became the first creatures to walk on the land. But most of the organisms continued to dwell in the sea. A great many Paleozoic marine animal forms have been preserved and are useful as guide fossils. A few examples seen in Figure 1-20 display elaborate shell ornamentation.

Reptiles developed as the most imposing creatures of the **Mesozoic** era. During this interval of time huge dinasaurs inhabited the continental swamps and plains. The variety of marine organisms also increased, and a large number of reliable Mesozoic guide fossils have been classified. Again there is evidence of the advance of shallow seas far inland during different intervals of this era.

The modern **Cenozoic** era is the span of geologic time that includes the present. Mammals emerge as the dominant animals of this interval of history, eclipsing the reptiles and amphibians of earlier eras. All of the principal kinds of rock, igneous and metamorphic as well as sedimentary, continue to be regenerated from older rocks during Cenozoic time.

The geologic eras are subdivided into a succession of shorter time intervals called **periods.** These include the seven Paleozoic periods, three Mesozoic periods, and two Cenozoic periods arranged in chronological order in Figure 1-22. Sedimentary strata formed during one or another of these periods are recognized by distinctive groups of guide fossils, which also provide a basis for separating the periods into still shorter increments of time defined as **epochs.** The epochs of the Cenozoic periods are included in Figure 1-22.

This chronological system of eras, periods, and epochs is called the **geologic time scale.** It is the system that geologists have come to accept when referring to different intervals of earth history. For example, if certain layers of limestone and shale are referred to as Devonian strata, we know that they must be older than some other layers which have been identified as Permian sediments. Cambrian strata are the oldest rocks containing an abundant assortment of fossils. Rocks more ancient than these would have to be of Precambrian age, and very few fossils are preserved from this era.

The geologic time scale evolved in a rather haphazard way during the nineteenth century. It emerged from the descriptions and correlation of many sedimentary assemblages from different locations. Strata exposed in the Devonshire district of Great Britain were named Devonian rocks. Other layers located in Wales were called Cambrian rocks. Later,

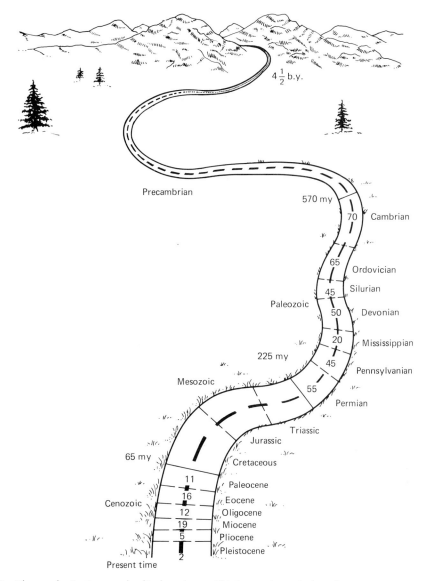

Figure 1-22 The geologic time scale displayed as a "highway through time."

the Devonian assemblage was found to be younger than the Cambrian assemblage, so the same names became designations for the time intervals when they were formed. Eventually rocks from many other regions were correlated with those exposed in Devonshire, and all were recognized as having formed during the same Devonian period. Similarly, other names used originally to describe rocks of a particular location came to represent the time intervals when these and other strata that correlated with them had been deposited.

28 Introduction to the Earth

The geologic time scale finally adopted by common agreement among geologists turned out to be a somewhat arbitrary system. The different periods do not necessarily correspond to any particularly significant worldwide events or conditions. They are all different in the lengths of time represented. However, each period and epoch has associated with it a distinct assemblage of organisms. This system of eras, periods, and epochs developed without plan or intention, but continues by common consent to be the universally accepted subdivision of geologic time.

Controversy about Geologic Time

Most nineteenth-century geologists realized that if uniformitarianism provided a meaningful basis for examining geologic history, then vast lengths of time would be required. Otherwise how could the almost imperceptibly slow processes have produced the variety of features that we now see in the rock record? James Hutton argued that the sedimentary rocks which he observed had formed from the debris of older rocks, and these older rocks in turn had consisted of the weathered products of still more ancient rocks. He could find no evidence of a beginning and saw no end to the ongoing cycle of formation, disintegration, and reconstitution of rock materials. How then could we discover the age of the earth?

In 1859 Charles Darwin published **Origin of Species.** In this remarkable book he argued that living creatures on the earth evolved by a process of natural selection and adaptation to environment. But nature would require a vast length of time to produce by this slow process the variety of creatures that have lived on the earth. To find support for his ideas Darwin turned to geologists who knew that the rock record could be explained by uniformitarianism only if the earth proved to be very old.

Their early attempts to discover the age of the earth were based mostly upon estimates of the total thickness of sedimentary rocks and assumptions about accumulation rates for sedimentary material. Several geologists compiled measurements of the thicknesses of strata produced during the different geologic periods. They then attempted to determine average rates of sedimentation by observing how fast material accumulated on major river deltas, along coastlines, and as a result of floods. Considerable variability was found for accumulation rates of different kinds of sedimentary debris. Rates were different depending on the conditions of deposition for a particular kind of material. Also, the estimates of the total thickness of sedimentary rocks produced during geologic time varied greatly depending upon the locations of strata chosen to represent the different geologic periods. The existence of unconformities added further uncertainty in that they indicated the passing of unknown lengths of time for which the rock record was missing. All of these uncertainties contributed to the wide range of estimates of the age of the earth which are presented in Table 1-1.

The estimates of geologic time given in Table 1-1 are consistent in one important aspect: They suggest considerably greater lengths of time than the calculations of the eminent nineteenth-century physicist Lord Kelvin. He based his arguments upon measurements of temperatures near the surface of

Table 1-1
Some Nineteenth-Century Estimates of the Age of the Earth

Age	Scientist	Date Proposed
100,000,000 years	Huxley	1869
200,000,000 years	Haughton	1878
45,000,000 to 70,000,000 years	Walcott	1893
100,000,000 years	Geikie	1899

the earth and speculations about the heat loss of the earth and the sun. Observing that temperature increased with depth in deep mine shafts, Kelvin reasoned that the earth was cooling, probably from an originally molten state. From experimental data on the heat-conducting properties of common rocks and temperature gradients in the earth, he was able to calculate estimates of the time of crystallization of the outer crust of the earth. Only after this crystallization occurred could the processes of erosion and sedimentation commence. Lord Kelvin also considered the effect of the heat from the sun on the earth. Using plausible arguments of classical physics, he concluded that the sun must be cooling, and formerly contributed to a much higher surface temperature on earth than now exists. He reasoned that at most 20 to 40 million years had elapsed since the formation of a solid outer crust on the earth. This conclusion contradicted the geological evidence. So the scientific controversy raged during the closing decades of the nineteenth century. But one essential kind of information was lacking. No one then knew about radioactivity. This proved to be the key to finding the age of the earth.

Radiometric Dating of Rocks

Most of us who have grown up in the atomic age began learning about radioactivity during our early school years. We know that certain chemical elements that we call radioactive elements eventually become converted into other elements. The conversion process is called radioactive decay. When this happens, heat energy is released. It is this energy that is harnessed in nuclear power plants. When scientists discovered that heat energy was released by the decay of radioactive elements trapped in rocks, they realized that Kelvin's arguments about the age of the earth could not be correct. His calculations were based on the assumption that no heat was being generated inside the earth. The fact that it is being generated by radioactivity means that the earth could have remained relatively cool for an indefinitely long period of time. The temperature conditions cited by Kelvin are simply the result of the outflow of heat that is produced continuously by radioactivity.

Knowledge about radioactivity not only showed Kelvin's arguments to be incorrect, it also provided a way to prove that the earth was, indeed, very old. To explain this we must first discuss what happens to individual atoms that undergo radioactive decay. Perhaps you have already been introduced to the basic ideas about the structure of atoms. If so, some of the following discussion will be a review.

An atom is the smallest unit of a chemical element. It consists of a nucleus made of protons and neutrons, and electrons that revolve in orbits around the nucleus (Figure 1-23). The mass of an atom depends on the number of protons and neutrons. Many chemical elements have two or more varieties of atoms that are distinguished by small differences in mass. These different varieties are called isotopes. For convenience we represent them by the symbol for the element followed by a number that tells how many protons and neutrons are in an atomic nucleus. For example,

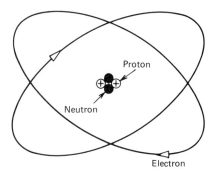

Figure 1-23 Simplified model of a helium atom consisting of two neutrons and two protons in the nucleus and two orbiting electrons.

the four isotopes of lead are Pb^{204}, Pb^{206}, Pb^{207}, and Pb^{208}. Although each isotope has a slightly different mass, they all exhibit essentially the same properties in standard chemical reactions.

The lead isotopes and the isotopes of most other chemical elements are stable. This means that the properties of their atoms do not change. But a few isotopes are unstable. These are the radioactive isotopes. An atom of a radioactive isotope has the capacity to eject bits and pieces of itself in an apparently spontaneous manner. This is the process of radioactive decay. The bits and pieces come in two varieties. Alpha (α) particles consist of two protons and two neutrons. Beta (β) particles are individual electrons. By ejecting these particles an atom of one isotope converts itself into an atom of another isotope. For example, an atom of the unstable uranium isotope U^{238} ejects several α-particles and β-particles to convert itself into Pb^{206}, which is a stable isotope. In this process we call U^{238} the parent, and Pb^{206} is the daughter. In a similar process the other uranium isotope U^{235} is the parent that produces the daughter product Pb^{207}.

We have no way of knowing when some particular unstable atom will begin to undergo radioactive decay. But we do know that where a very large number of atoms are present, radioactive decay at any moment in time takes place at a rate which is proportional to the number of parent atoms existing at that time. This means that the mass of a radioactive substance diminishes each moment at a rate depending on the mass remaining at that moment. Such a process is described by the equation:

$$dm/dt = \lambda m \qquad (1\text{-}1)$$

where m is the mass of a radioactive substance that exists at some moment, dm is the amount of that mass which decays during the brief interval of time (dt) beginning at that moment, and λ is the decay constant. The de-

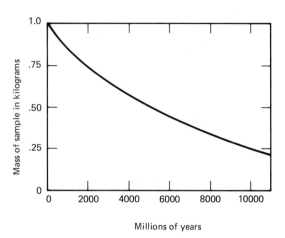

Figure 1-24 Curve showing how the mass of a sample of U^{238} changes with time.

cay constant represents the proportion of mass that will decay during dt. You can tell from Eq. (1-1) that fewer and fewer grams of a radioactive substance decay with each passing moment because there are fewer and fewer remaining grams of the parent. The graph in Figure 1-24 illustrates how the mass of a sample of U^{238} changes with time. We know from laboratory tests that U^{238} has a decay constant of 5×10^{-18} sec^{-1}. But you are probably more familiar with a term called the **half-life** which is the time required for one half of the mass of the sample to decay. Look again at Figure 1-24. You can tell that after about 4.5×10^9 years one-half of the sample would have decayed. If you convert this into seconds, you can then calculate that the half-life ($t_{1/2}$) is related to the decay constant (λ) according to the formula: $t_{1/2} = 0.693/\lambda$. Each radioactive isotope has a distinctive decay constant that can be measured in laboratory tests.

How do we use this knowledge about radioactivity to find out about geologic time? We know that very small concentrations of radioactive isotopes exist between and within some mineral grains in rocks. The isotopes of most interest to geologists are described in

Table 1-2
Some Radioactive Isotopes, Their Sources, and Decay Products

Radioactive Isotope	Stable Decay Product	Decay Constant	Half-Life	Mineral Sources
Uranium 238	Lead 206	$1.54 \times 10^{-10} \text{yr}^{-1}$	$4.5 \times 10^9 \text{yr}$	Uraninite, Pitchblende, Zircon
Uranium 235	Lead 207	$0.97 \times 10^{-9} \text{yr}^{-1}$	$0.7 \times 10^9 \text{yr}$	Uraninite, Pitchblende, Zircon
Thorium 232	Lead 208	$4.96 \times 10^{-11} \text{yr}^{-1}$	$14 \times 10^9 \text{yr}$	Sphene, Apatite, Zircon
Rubidium 87	Strontium 87	$1.47 \times 10^{-11} \text{yr}^{-1}$	$47 \times 10^9 \text{yr}$	Biotite, Muscovite, Microcline
Potassium 40				
11% decays to Argon 40		$0.55 \times 10^{-9} \text{yr}^{-1}$	$1.3 \times 10^9 \text{yr}$	Biotite
89% decays to Calcium 40		$0.46 \times 10^{-9} \text{yr}^{-1}$	$1.5 \times 10^9 \text{yr}$	Muscovite, Hornblende
Carbon 14	Nitrogen 14	$1.2 \times 10^{-4} \text{yr}^{-1}$	5720 yr	Wood, Peat, Bone, Sea water

From Nuclear Geology *(H. Faul, Ed.), John Wiley and Sons, 1954.*

Table 1-2. How did these isotopes become part of a rock? Let us suppose that a radioactive substance became trapped there when nature first put that piece of rock together. Let us suppose further that ongoing radioactive decay of that trapped substance has been producing a daughter substance that also remains trapped in the rock. If these two assumptions are correct, then we can find out the age (t) of that rock. By integrating Eq. (1-1) and rearranging the result we can obtain the equation:

$$t = \frac{1}{\lambda} \ln\left(1 + \frac{D}{P}\right) \quad (1\text{-}2)$$

where P is the mass of the radioactive parent that has a decay constant λ, and D is the mass of the daughter product trapped in a rock specimen.

The practical methods for determining the amounts of various isotopes present in a rock specimen involve careful sample preparation and mass spectrometer analysis. These isotopes are extracted by means of filters and chemical reagents. The mass spectrometer is used to measure relative quantities of different isotopes on the basis of their different atomic masses and electrical charges. The analysis can be carried out with very small samples of an element, often a few micrograms or less. In Table 1-2 several common rocks and minerals are listed which are known to contain elements from one or more of the important radioactive decay series. Ordinarily these materials contain relative quantities of a few parts per million (ppm) of the important elements. Thus, if a sample of muscovite contains 15 ppm of lead, we could hope to extract 15 micrograms of that element from a one-gram sample of the mineral. Generally, only a few grams or less of sample material are needed to extract a sufficient quantity of the radioactive element and its decay product for accurate mass spectrometer analysis. Values of *P* and *D* found by analyzing this material are used in Eq. (1-2) to find the age of the specimen.

The calculation of the age of a rock is based on the assumption that the amounts of parent and daughter isotopes are related only to radioactive decay within the rock. But if certain amounts of either parent or daughter material are lost, then a measurement of the ratio *(D/P)* does not provide an accurate basis for age determination. From our knowledge of the atomic structure of minerals we know that under the proper conditions these isotopes can migrate atom by atom through mineral crystals by a process called molecular diffusion. When the temperature rises above a critical level called the blocking temperature, molecular diffusion of isotopes progresses much more rapidly. The blocking temperature depends on what substances are present. If rock temperatures remain below the blocking temperature, then isotope loss by diffusion usually does not introduce a significant error. If a rock is heated to the blocking temperature at some time long after it was formed, however, then diffusion may introduce significant error in age calculations. However, we have discovered ways of taking advantage of diffusion to learn not only the date when the rock was first formed, but also something of its later history of metamorphism.

In many rock specimens more than one radioactive isotope is found. It is then possible to use each one and its daughter product to make independent age determinations. Insofar as the independent calculations yield approximately the same age for the rock, we gain confidence in the accuracy of that age, and we can assume that diffusion or other causes of isotope loss have been minimal. Because the decay constants of various radioactive isotopes are so different from one another (Table 1-2), we can be quite confident of the accuracy of a rock age based upon two or three separate isotope ratios from a specimen that yield the same age.

The time range over which accurate age determinations can be obtained depends partly on the decay constant of the radioactive isotope. A certain minimum amount of an isotope is needed in order to process it and analyze it by mass spectrometry. And, consequently, a certain minimum time is required before enough of the daughter product is produced for an accurate analysis to be made.

The oldest earth rocks for which isotopic ages have been calculated appear to have formed over 3800 million years ago, which establishes a lower limit for the age of the solid earth. But recent analyses of some rock samples from the moon yield ages of more than 4500 million years. These moon rocks are the same age as all of the meteorites that have fallen to the earth for which we have isotopic dates. There are strong arguments for assuming that the earth is approximately the same age as the moon and these meteorites. So by inference, the age of the earth can be assumed to exceed 4500 million years.

At the present time the most reliable isotopic determinations of age are obtained from igneous rocks. The interlocking mineral grains form a concentrated mass from which it is difficult for the radioactive isotope to escape. It is somewhat more difficult to obtain dependable estimates of age for sedimentary rocks. Most sedimentary rocks which are likely to include radioactive elements in the constituent minerals also have minute openings between grains. Water seeping through these pores may remove or contaminate the isotopes used for dating. Under ideal conditions it is sometimes possible to obtain good age information from sediments, but the likelihood of finding such conditions is still not high.

We need to make some special mention of how radioactive carbon C^{14} is used in isotopic dating of geologic events. It is found in organic material that has become embedded in regolith and rock. The half life of C^{14} is about 5730 years, which is much shorter than the half-life of other isotopes described in Table 1-2. This short half-life restricts the use of C^{14} to dating events that occurred within the past 50,000 years. After that much time has elapsed so much C^{14} has decayed that it is difficult to measure accurately the amount remaining in a specimen.

Where does C^{14} come from? It is produced in the atmosphere when neutrons from cosmic radiation bombard nitrogen atoms. The C^{14} atoms then decay to produce the stable daughter product N^{14}. We believe that the level of cosmic radiation has not changed significantly during the last 50,000 years. We also believe that the amount of nitrogen in the atmosphere has stayed the same during that period of time. This implies that the production and the radioactive decay of C^{14} are in balance. Therefore, the amount of C^{14} in the atmosphere has not been changing. There are two stable carbon isotopes, C^{12} and C^{13}, that exist together with C^{14} in the atmosphere, and the ratios C^{14}/C^{12} and C^{14}/C^{13} remain constant.

Living organisms continually exchange carbon with the atmosphere. As a result, the C^{14}/C^{12} and C^{14}/C^{13} ratios for carbon in the body tissue of living plants and animals are the same as the ratios in the atmosphere. But death stops the exchange of carbon. Then the ratios for carbon trapped in the dead organism begin to change. Because C^{12} is much more abundant than C^{13}, the ratio C^{14}/C^{12} is of greatest interest for dating events in the past. Radioactive decay of C^{14} causes this ratio to change. We know the decay rate, and so we can calculate the time that elapsed since the organism died.

Calibration of the Geologic Time Scale

How can we find out the expanses of time represented by the eras, periods, and epochs of the geologic time scale? These intervals of earth history are established from guide fossils. But these fossils occur only in sedimentary rock, and they do not tell us the length of time the rock has existed. Isotopes of radioactive decay series can be analyzed to determine how long a rock has existed. But uncontaminated isotope samples are found principally in igneous rocks which do not contain the fossils needed to identify the era, period, and epoch when they crystallized. If we want to find the lengths of time represented by these intervals of the geologic time scale, we must study combined sedimentary and igneous rock assemblages.

The idealized geologic cross section in Figure 1-25 illustrates superposed fossil-bearing sedimentary strata and crosscutting igneous masses and unconformities. Here we observe that a 510 million-year-old igneous mass is truncated by an unconformity during the Cambrian period. Younger Cambrian and Ordovician strata overlie this unconformity and are cut by another 470 million-year-old igneous mass. This igneous mass terminates at a second unconformity above which still younger Orodivician sediments have been deposited. From these relationships we recognize that the Ordovician period began sometime between 470 and 510 million years ago. To establish this time more closely, we might estimate rates of sedimentation for the layers between the unconformities. If we can show that the unconformities represent only short gaps in the rock record and that the Cambrian and Ordovician sediments accumulated at about the same rate, we can interpolate a date about 500 million years ago for the end of Cambrian time and the start of the Ordovician period.

In many parts of the world these kinds of relationships have been studied to establish the time span of the geologic eras, periods, and epochs. The dates given in Figure 1-22 have been worked out in this way. The combined expanse of the Paleozoic, Mesozoic, and Cenozoic eras for which we have a usable system of guide fossils makes up only about 13 percent of the length of geologic time. Far longer is the extent of the Precam-

34   Introduction to the Earth

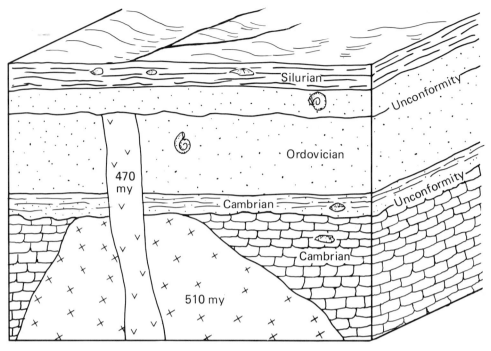

Figure 1-25 Idealized geologic cross section illustrating superposed fossiliferous sedimentary rocks and cross-cutting igneous rocks possessing elements of radioactive decay series. By using the igneous rock ages determined from radioactive isotope ratios together with relative ages determined from sedimentary layers with fossils, the end of the Cambrian period can be interpolated to be 500 m.y. ago.

brian era during which the earth was formed and developed in important ways.

THE PERSPECTIVE OF GEOLOGIC TIME

With the formulation of the Principle of Uniformitarianism, James Hutton provided a rational basis for inquiry about the history of the earth from observations of rocks. In support of his arguments he described rocks which clearly revealed many aspects of how they were produced and how they were later deformed by different geologic processes. In a time when it was not uncommon to invoke supposed supernatural events to explain the features of the earth, Hutton offered the alternative of explaining these things using evidence that could be seen by any interested observer.

As the acceptance of the idea of uniformitarianism grew, so did the subject of geology as a formal science. More observers began to participate in the study of the earth, and deeper insight was gained into the effect of various geological processes and the duration of time involved in the history of the earth. Hutton realized that an immensity of time was involved, even though he was unable to discover a specific beginning. Eventually geologists figured out that there was a beginning for the earth about $4^1/_2$ billion years ago. We have also come to understand that although geologic processes that we can observe during the present were also active in the past, the effects of these processes in altering the earth may not always have been the same. It is difficult to comprehend the vastness of geologic time which scientific investigation leads us to conclude is almost one million times as long as the length of time of recorded human history.

We have considered the importance of the fossil record in the study of various geological problems. Viewed in conjunction with radioactive isotopic calibration of the geologic scale, it becomes clear that the Paleozoic, Mesozoic, and Cenozoic eras together represent only about one-sixth of the length of the Precambrian era. The variety and number of life forms with hard skeletal parts began to increase rapidly with the commencement of the Cambrian period. Prior to this most organisms had soft bodies not durable enough to produce fossils. By middle Paleozoic time we discover that in addition to marine plants and animals, terrestrial forms began to develop. Giant reptiles assumed dominance in the animal kingdom during the Mesozoic era, but yielded to the warm-blooded animals in Cenozoic time. But continuing to develop along with these large animals we find a persistently diversifying marine flora and invertebrate fauna. This continually changing display of life points clearly to the fact that the earth has undergone change in environment in the past and indicates that the balance of various geologic processes has not always been as it is today.

Finding life already somewhat advanced by Cambrian time we have to look into the vast expanse of Precambrian time to explore the origins of life on earth. Because life is associated with the earth's atmosphere, the evolution of this gaseous envelope is an obvious consideration. There is strong evidence that indicates an almost complete lack of oxygen in the early Precambrian atmosphere. Certain chemical elements which readily combine with oxygen have been found unoxidized in Precambrian sediments. Chemical compounds present in these rocks suggest chemical weathering processes different from those active at the present time. To produce the first living organisms in this environment nature may have synthesized carbon compounds from elements in the atmosphere and perhaps using elements found in escaping volcanic gases. We know that some of the simpler forms of complex carbon molecules can be synthesized in the laboratory by electrically shocking gaseous mixtures of the constituent elements. In nature such a synthesis may have occurred during the release of volcanic gases during electrical storms.

Once synthesized, these earliest organisms required food and needed protection from the lethal ultraviolet radiation given off by the sun. These primeval creatures must have been anaerobic to survive without oxygen. Most likely they lived in water that would provide a medium in which they could move and yet be shielded from solar radiation. The environment must have provided some prebiologic carbon compounds as food. Once developed, the early organisms began to diversify and spread through the oceans of the world. It is now quite clear that the development of ecological systems of respiring organisms contributed significantly to the alteration of the atmosphere by releasing oxygen. But at the same time an excess of oxygen would have proved lethal to these primitive creatures. So at that critical time the geological environment played an essential temporary role by providing the chemical elements which could combine with the newly forming oxygen, thus restricting its accumulation in the atmosphere. Eventually these elements became oxidized and could no longer provide a means for preventing oxygen from accumulating in the atmosphere. But enough time had elapsed for evolution of life forms more compatible with an environment containing free oxygen. As the oxygen content of the atmosphere increased we find evidence of the rapid evolution of new varieties of plants and animals.

The interrelationships between biological and geological processes near the surface of the earth cannot be overemphasized. Weathering of rock material, erosion, and transport of sedimentary debris all are closely associated with a variety of organisms and the associated biochemical reactions. It therefore seems reasonable to believe that as life developed near the surface of the earth the effectiveness of surface geological processes in modifying the earth also changed.

When we consider the complex interplay of a large number of important factors necessary for the synthesis and continued proliferation of life, it seems highly improbable that the proper sequence of events could take place. Even if individual steps in the process are more likely to occur, the chances of obtaining the necessary combination of steps is remote. However, by viewing such apparently improbable happenings from the perspective of the immense expanse of Precambrian time, the likelihood of this synthesis increases. From statistical considerations we can predict the probability that a rare event will occur x number of times during n number of years if we can estimate the probability (P_0) of its occurrence during one year. The curves shown in Figure 1-26 indicate these probabilities for different values of x, if an event has only one chance in a million ($P_0 = 10^{-6}$) of occurring in one year. Observe that such an event is almost certain (98 percent probability) to occur at least once in four million years, and we could expect it to occur about 50 times in 65 million years. Thus,

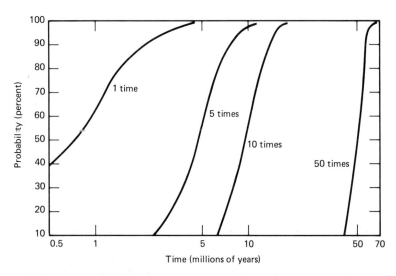

Figure 1-26 Probability of a rare event to occur at least x times in n years if there is only one chance in a million that it would occur during a single year. To find the probability of an event occurring once, look at the curve labeled *1 time*. Note that in one million years there is a 62 percent chance that the event will happen, but there is an 88 percent chance that it will occur once in two million years. Next look at the curve marked *5 times*. It indicates that there is a 54 percent chance that the event will occur five times during a period of five million years. (From "Significance of the Rare Event in Geology" by P. E. Gretener, modified from illustration in *Amer. Assoc. of Petroleum Geologists Bull.*, page 2198, Nov. 1967.)

certain rare events and improbable combinations of events become almost certain to happen when we realize the extent of geologic time available.

We now understand that the Principle of Uniformitarianism must not be too rigidly interpreted if it is to remain a useful maxim for geologists. Biological processes and chemical processes associated with the earth's atmosphere have changed considerably with time. Insofar as they affect geologic processes, these too have changed. We also must understand that rare events may have occasionally contributed significantly to the processes that modify the earth. But perhaps it is most important to recognize that many observable processes which will be discussed in later chapters probably have been active during significant lengths of geologic time.

STUDY EXERCISES

1. Discuss how an igneous rock specimen can be distinguished from a sedimentary rock specimen.

2. Discuss the geologic principles used to find the relative ages of rocks.

3. Examine Figure 1-27 that displays igneous rocks labeled E, F, G, and H, and sedimentary rocks labeled A, B, C, D, and J. The wavy contact between rocks B and J represents an unconformity.

   (a) In which igneous unit would the isotope mass ratio Ar/K have the smallest value?

   (b) Could any of the igneous rocks have been produced by solidification of lava that flowed onto the land surface, or were all of the units produced by solidification of magma deep underground?

   (c) If rock D is of Devonian age and rock J is of Permian age, then rock H might have been produced during which one of the following periods: Pennsylvanian; Triassic; Silurian; or Cambrian?

   (d) In which sedimentary rock might Mississippian fossils be found?

4. Assume that the mass ratio of $Pb208/Th232 = 0.1$ was measured from a specimen of igneous rock.

   (a) Which one of the following ratios would have a value smaller than 0.1 for this particular specimen: $Pb206/U238$; $Pb207/U235$; $Sr87/Rb87$?

   (b) During which era of geologic time was this rock produced?

5. Why is the Principle of Uniformitarianism important in the reasoning that is used to determine the age of the earth?

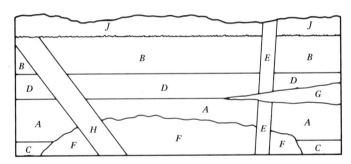

Figure 1-27

SELECTED READINGS

Eicher, Don L., *Geologic Time,* Prentice-Hall, Inc., Englewood Cliffs, N.J., 1968.

Fenton, C. L., and M. A. Fenton, *Giants of Geology,* Doubleday and Co., Garden City, N.Y., 1952.

Faul, Henry, *Ages of Rocks, Planets, and Stars,* McGraw-Hill Book Co., New York, 1966.

Harbaugh, John W., *Stratigraphy and Geologic Time,* Wm. C. Brown Co. Publishers, Dubuque, Ia., 1968.

SECTION TWO
THE EARTH'S SIZE, SHAPE, AND INTERIOR ZONES

2

GRAVITY AND THE SHAPE OF THE EARTH

The Earth's Size and Shape
  Measuring the Size of the Earth
  The Reference Ellipsoid
  Gravity
  Centrifugal Force
  Topography
  The Shape of the Solid Earth
    Continental Platforms
    Ocean Basins
    Comparisons

The Geoid
Topographic Surveying
A Fundamental Level Net
Gravity and Density in the Earth
  Gravity Surveying
  Gravity Anomalies
  Mass Under Continents and Oceans
  Total Mass and Mean Density
  Perspectives

The idea of a spherical earth was introduced at least 2000 years before Columbus. By the time of his westward voyage most educated people believed the earth was spherical. But there was no agreement about its size. Columbus succeeded in promoting his plan for a westward voyage from Spain to Asia in part because he argued the earth was so small. By his estimate Japan lay less than 4500 km west of the Canary Islands, a distance that could be traveled by ship in under three weeks. Others disagreed with this estimate, which eventually proved to be less than one-fourth of the correct distance. Their reasoning, however, was no better than that of Columbus.

Controversy about the size of the earth continued until 1522. In that year **Victoria,** the last of Magellan's five ships, returned to Spain with 18 sailors after a three-year voyage of circumnavigation. After this voyage the true circumference of about 40,000 km (24,000 mi) could be estimated more realistically. Then efforts were begun to measure its size and shape more precisely. Such measurements became increasingly important as more accurate methods of navigation were developed.

During the seventeenth century the earth was discovered to be slightly flattened so that an ellipsoid rather than a sphere came closer to representing its true shape. The polar diameter of this ellipsoid is smaller than the equatorial diameter. Since that time we have been finding out as precisely as possible the dimensions of the ellipsoid that most closely resembles the shape of the earth, and measuring how the actual surface differs from this idealized form. The difference between the idealized form and the actual shape consists of the topography of continents, islands, and the ocean floor. These variations can be described from heights and depths measured above or below the surface of the sea.

The sea level surface itself is not exactly ellipsoidal; rather, it is warped into numerous undulations because of differences in underlying rock density from one place to another. Density is defined as mass per unit volume. It is commonly specified in grams per cubic centimeter, abbreviated as gm/cm^3. As the definition indicates the density of a rock specimen is found by dividing its mass by its volume.

One principle aim of **geodesy,** the branch of geology concerned with the shape and dimensions of the earth, is to determine the form of the ocean surface. This information is crucial to the precise topographic surveying required in many modern engineering activities. Planning the flow of water between distant reservoirs in a large hydroelectric project requires accurate knowledge of elevations above sea level. Similarly, a knowledge of the topography along the route is essential in laying out an oil pipeline. Prediction of the orbits of navigation and communications satellites and the trajectories of rockets is based on precise measurements of the earth's shape, and of the different kinds of terrain in one region or another.

Because of density irregularities in the earth, the strength of gravity is not the same everywhere on the surface. Measurements of gravity from place to place give us a clue to rock density in the crust and mantle. This knowledge of the way mass is distributed in the earth helps us to determine accurately the shape of the sea level surface, as well as to locate different kinds of rock beneath the surface. The mass distribution, shape, and gravity of the earth are interdependent features of fundamental geological interest.

THE EARTH'S SIZE AND SHAPE

Measuring the Size of the Earth

Apollo astronauts viewing the earth from the moon could see directly that it was a globe. Many early philosophers had long ago reached this conclusion from less direct evidence. Because the sun and the moon each had an obviously circular outline, they suggested that the earth might be similar. By combining this philosophical idea with commonplace observations such as a ship disappearing over the horizon they realized that a spherical shape would not only provide a circular outline, but would also explain the disappearance of the ship.

Once its spherical shape was accepted, the next question concerned the earth's dimensions. There are various geometrical methods for finding the size of the earth. One of the most practical involves measurement of shadows cast by objects on its surface.

It has long been known that near the equator there are times when the sun appears to be directly overhead. Because the plane of the earth's equator is inclined at an angle of 23 degrees with the plane of its orbit, the sun is directly overhead twice each year at any point between the Tropic of Cancer and the Tropic of Capricorn. At any given instant the sun is over only one point, and a vertical pole at such a point will cast no shadow. At the same instant similar vertical poles elsewhere will all cast shadows.

Let us consider the relationship between the height of a pole and the length of its shadow. First we assume that light rays traveling from the sun to the earth are parallel. Then, referring to Figure 2-1, we find that the light rays intersect a vertical pole at some angle (α). At point A this angle is 0 degrees, at points B and C the angles are less than 90 degrees, and at point D the angle is 90 degrees. Clearly, the angle can be found from the ratio of the shadow length (D) to the pole height (H):

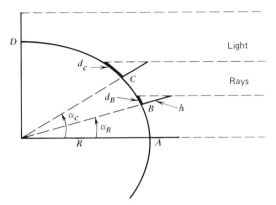

Figure 2-1 The circumference of the earth can be measured by the method of Eratosthenes. A vertical pole of height h situated at point A directly beneath the sun casts no shadow. At the same time similar poles at points B and C cast shadows of lengths d_B and d_C that are used to determine the angles of light rays from the sun and the vertical poles. Ratios of these angles and the distances between poles can be used to calculate the circumference of the earth.

$$\tan \alpha = d/h$$

At point B the angle α and the arc distance AB can be related to the circumference (C):

$$\frac{AB}{\alpha} = \frac{C}{360°}$$

This was apparently known as early as the third century B.C. by the scholar Eratosthenes. From measurements made near the Mediterranean Sea he estimated the earth's circumference to be 252,000 stadia, or about 40,000 km.

Eratosthenes' basic approach is still used. Now the instruments are considerably more refined, however, and corrections are made because the sun's rays are not exactly parallel, and they are distorted by the earth's atmosphere. Modern communications and transportation allow us to make simultaneous measurements at several points. From these measurements we can calculate an approximate circumference (C) of 40030 km. From the familiar formula: $C = 2\pi R$ we find the value of 6371 km for the average radius (R) of the earth.

The Reference Ellipsoid

We know the earth is not a perfect sphere. In fact, it more closely resembles an ellipsoid. This is a geometrical form having a polar radius that differs from its equatorial radius. Actually, the earth's shape is much more irregular than a simple ellipsoid. But we find it convenient to use an ellipsoid as a starting point for describing the earth's shape. The particular ellipsoid that most closely fits the average shape of the earth is called the **reference ellipsoid**. Once we have found its dimensions, we can describe the true shape of the earth by stating how far the numerous irregularities reach above or below this reference surface.

How do we know that the earth is more like an ellipsoid than a sphere? Following the method of Eratosthenes we can make measurements to confirm this fact. In Figure 2-1 let us assume that the points A, B, and C are equally spaced along an arc extending northward from the equator. If the earth were a perfect sphere, we would expect α_C to be exactly twice as large as α_B, because the arc AC is twice the distance of the arc AB. Actually, we would find that α_C is slightly more than twice the size of α_B. This would be expected for an ellipsoid.

Numerous careful measurements along lines in many directions and locations have been made to find the shape of the earth. These measurements indicate a polar radius of 6357 km, and an average equatorial radius of 6378 km. These radii are the dimensions of the reference ellipsoid.

Gravity

There are theoretical reasons for believing that the earth should resemble an ellipsoid. These are based on concepts of gravity and centrifugal force introduced by the British philosopher Isaac Newton (1642-1727). His studies of the motions of planets in the solar system led him to propose a law to describe the mutual attraction of any two bodies of mass. This law has become the well-known Universal Law of Gravitation:

$$F_g = Gm_1m_2/r^2$$

where F_g is called the gravitational force. It is directed along a line between the centers of the two masses m_1 and m_2, which are separated by the distance r. The Universal Gravitational Constant G has been found by experiment to be 6.67×10^{-8} cm³/gram·sec².

What does the Universal Law of Gravitation tell us about the earth's shape? From this law we can find the forces of attraction between particles clustered in a large mass. If these particles were free to move relative to one another, the gravitational forces would arrange the cluster into a sphere. In earlier times this effect was illustrated in the manufacture of shot. A small mass of molten lead dropped from a high tower first forms into a sphere, then solidifies before reaching the ground. Similarly, we might expect the particles making up the earth to become arranged into a spherical mass by gravitational forces.

How strong is the gravitational force near the earth's surface? We can find the strength of F_g between any two bodies by observing how fast they move toward each other. If one mass is a sphere the size of the earth, we can release a small mass (m) from a small height (z) above the surface, and measure the time (t) for it to fall to the surface from a position of rest. The gravitational force on this mass can be found from the formula:

$$F_g = 2mz/t^2$$

Typically, a one gram mass will fall 100 cm to the earth's surface in about 0.452 sec. This indicates a gravitational force of about 980 dynes on the one gram mass (a dyne is the force required to accelerate a one gram mass at the rate of one cm/sec²).

Centrifugal Force

The earth resembles an ellipsoid rather than a sphere. Therefore, something besides gravitational forces must influence its shape. This additional effect is the centrifugal force. It exists because the earth is a rotating mass. Newton proposed that a mass (M) moving in a curved path must experience an outward centrifugal force (F_c) related to its linear velocity (v) and the radius of its curved path (d):

$$F_c = mv^2/d$$

Each particle in the earth follows a circular path around the axis of rotation. Therefore a centrifugal force must act on the particle in an outward direction perpendicular to the axis of rotation. A particle on the earth's surface at the equator travels a distance of 40074 km around its circumference in 24 hours. Here the distance from the rotation axis is 6378 km. Converting these values to units of centimeters and seconds, we calculate that the centrifugal force (F_c) on a one gram particle would be about 3.37 dynes. This force is much smaller than the gravitational force of 980 dynes. Nevertheless, it is strong enough to help to shape the earth into an ellipsoid.

Particles situated away from the equator travel shorter distances in 24 hours. Therefore, they move at slower velocities (v), around paths with smaller radii (d). At the north and south poles the velocity and the radius diminish to zero, and the centrifugal force disappears.

The earth is shaped into an ellipsoid because of the combination of gravitational and

Gravity and the Shape of the Earth

centrifugal forces. The directions of these forces and the resulting net force are indicated in Figure 2-2. The gravitational force acts to form a sphere. The centrifugal force acts to draw out the mass nearer the equator but not at the poles, causing the sphere to be flattened into an ellipsoid.

Topography

If you like to hike in the mountains, you know how spectacular the scenery looks from a steep trail (Figure 2-3). Here the earth does not look like an ellipsoid. The fact that its surface has irregularities like mountains makes it

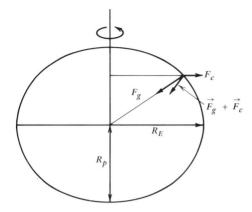

Figure 2-2 Two forces act on a particle on the surface of the earth. F_g is the gravitational force due to the earth's mass, and it is directed toward the earth's center. F_c is a centrifugal force caused by the earth's rotation. F_g and F_c combine to produce a force acting in the direction labeled $F_g + F_c$.

Figure 2-3 Hallett Peak in Rocky Mountain National Park in Colorado viewed from Flattop Trail. (W. T. Lee, U.S. Geological Survey.)

a much more interesting place. Features such as hills and valleys and mountains give form that we call **topography** to the land surface. The earth's topography is the difference between its actual shape and the reference ellipsoid. To describe the shape of the earth accurately we must be familiar with its topography.

Topography is described from knowledge of the height of the land surface above or below sea level. Later we will find out how to measure height. But first we can talk about the earth's principal topographic features. We can use different graphical schemes to display the topography of small areas and large regions. From these displays we can recognize distinctive topographic regions on the continents and in the ocean basins.

The topography of a small area can be represented by contour maps and topographic profiles (Figure 2-4). A contour is a line drawn so that all points along it are at the same height. Where heights are known at numerous points in an area, a contour is drawn so that all points on one side are lower, and all points on the other side are higher than the height represented by the line. Topographic profiles are lines showing the variation in elevation along some particular path.

We must devise other kinds of maps to represent the topography of a large region. Obviously, you cannot depict every hill and gully on a map of North America. Larger features can be represented approximately by artistic shading (Figure 2-5). The topographic character of a region can be described by specifying the mean elevation of the land surface, and the relief. The **relief** is the difference in elevation between the highest and lowest points. These two characteristics of topography can be shown by appropriate colors or symbols on a map.

The Shape of the Solid Earth

We can recognize different regions of the earth's surface by their distinctive topographic features. By describing the typical topographic regions we can form clearer ideas about how the earth's surface actually looks. The best way to proceed is to use a formal classification of different kinds of topographic regions. It is a tedious task to learn a method of classification. Nevertheless a classification scheme is very useful for thinking systematically about the earth's irregular shape, and for remembering how it looks.

We can begin by distinguishing (a) continental platforms from (b) ocean basins. These are the two principal features that mark the earth's solid surface. If we compare the topography typical of the continental platforms and the ocean basins, we find similarities, but interesting and important differences as well. Before making these comparisons we must first describe separately the topographic features of these two regions.

Continental Platforms. We can identify typical topographic regions that are found on all continents. The different kinds of topography were classified in 1967 by the American geographer, R. E. Murphy. They can be used for describing any continent. Murphy's system is not the only one in use, but it is representative of the methods developed for describing a continental surface. The six classes of topography recognized in the Murphy classification are discussed separately. Examples of each of these classes are depicted in Figure 2-5.

Plains. Plains usually are not places of breathtaking scenery, but we find them to be important regions for human activity. Here the landscape has the fewest obstacles to commerce and agriculture. We define plains as gently sloping continental surfaces where local relief is less than 100 m (325 ft). The Atlantic and Gulf Coastal Plains border southeastern North America rising gradually from the seacoast to elevations of a few tens of me-

Gravity and the Shape of the Earth 47

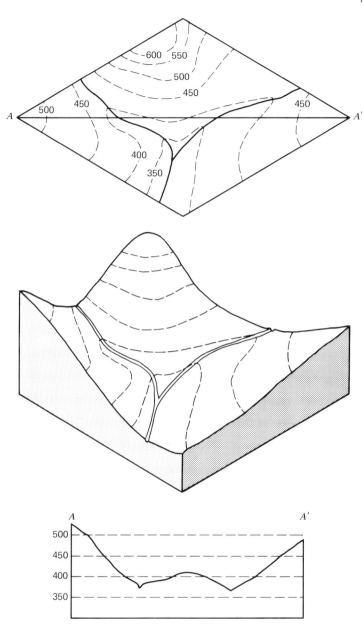

Figure 2-4 The block diagram illustrates a landscape with features that can also be represented by elevation contours, which are lines drawn through points of equal elevation, and by topographic profiles that show elevation along some particular line.

ters about 200 km inland. Low interior plains of the Mississippi Valley rise gradually toward the west from heights of less than 100 m to elevations of more than 1500 m. These plains are marked by shallow river channels, and are almost devoid of high-standing topographic features.

Hills and Low Tablelands. Local relief between 100 m and 600 m is typical of the hill areas of low tablelands where the mean elevation is less than 1500 m (5000 ft). We find that the more varied landscape of a tableland may merge gradationally with a plain, or it may be separated from it by a steep bluff. In

48 The Earth's Size, Shape, and Interior Zones

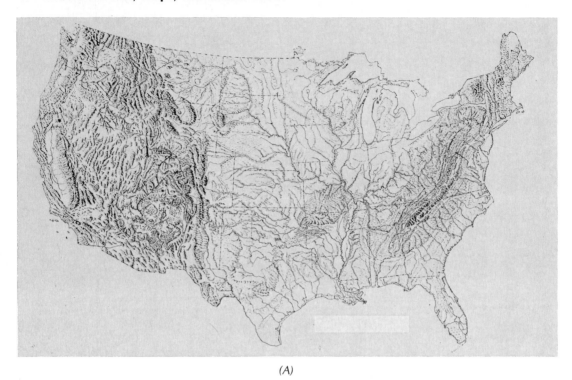

(A)

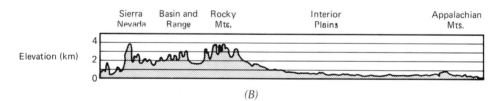

(B)

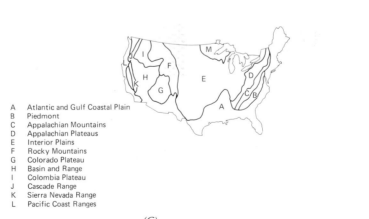

A   Atlantic and Gulf Coastal Plain
B   Piedmont
C   Appalachian Mountains
D   Appalachian Plateaus
E   Interior Plains
F   Rocky Mountains
G   Colorado Plateau
H   Basin and Range
I   Colombia Plateau
J   Cascade Range
K   Sierra Nevada Range
L   Pacific Coast Ranges

(C)

Figure 2-5 opposite *(A) The topographic relief map of the United States (Erwin Raisz, 1954) and (B) the west to east transcontinental topographic profile display the regional landscape features of the principal topographic provinces. (C) Index map shows principal topographic regions.*

southeastern North America this kind of topography extends over the Appalachian Plateaus which separate the interior plains from the Appalachian Mountains. Farther north the hilly terrain of low tablelands reaches to the Atlantic Coast, and extends over much of eastern and central Canada.

High Tablelands. High tablelands stand at mean elevations of more than 1500 m. Local relief is less than 300 m (1000 ft) except in a few places where they are cut by deep canyons. The Colorado Plateau extends over parts of Utah, Arizona, Colorado, and New Mexico. It is the largest high tableland in North America. The magnificent gorges of the Grand Canyon, Zion Canyon, and others cut more than 1500 m into parts of this region. Elsewhere the nearly flat land surface is marked by subdued relief.

Mountains. Most of us find some thrill or challenge in the rugged slopes and majestic heights of mountainous regions. To describe these regions objectively we define mountainous terrain to have steep slopes and local relief of more than 600 m (2000 ft). Mean elevation is not necessarily a distinguishing factor. The peaks and ridge crests of the Appalachian Mountains of southeastern North America rise much higher than the nearby plateau. But the mean elevation of that bordering tableland is higher. Continental mountain regions are commonly elongate chains. Topographic profiles crossing these mountain chains tend to be asymmetrical. Elevation rises more steeply on one side compared with the other.

Widely Spaced Mountains. In some regions isolated mountains rise more than 600 m above the surrounding land surface where the relief is otherwise less than 150 m (500 ft).

Depressions. Some regions are completely surrounded by higher mountains, tablelands, and plains. These regions are topographic basins called depressions. Such a depression is the Basin and Range region of North America. It occupies large areas of Utah, Nevada, Arizona, New Mexico, and southern Idaho. A terrain of widely spaced mountains and a large topographic basin reach over much of this region. It is bordered by the Colorado Plateau and mountainous areas all standing at a higher mean elevation.

North American can be separated into the large topographic regions indicated in Figure 2-5. Other continents can be similarly divided into regions of distinctive topography. Ordinarily a region can be described by one of the six kinds of topography proposed by R. E. Murphy.

Ocean Basins. Compared with knowledge of the continental land surface our comprehension of the ocean floor is rudimentary. Nevertheless, we have measured the depth of the ocean in enough places to begin piecing together a picture of how the submarine landscape looks. These measurements are made with a device called an echo sounder. It records the time required for a sound pulse emitted from a ship to travel to the ocean bottom and echo back to the ship. A more complete description of this device is given in another section of the chapter. The topographic regions of the ocean floor were described by the oceanographer B. C. Heezen and his colleagues after careful analysis of echo sounder data. He separated the submarine landscape into four principal topographic provinces. These include the ocean basin floor, the oceanic ridge system, submarine trench systems, and submerged continental margins. Each of these provinces of the sea bottom has a characteristic topography and depth. Their distribution beneath the North Atlantic Ocean is indicated in Figure 2-6. A collection of echo sounder records in Figure 2-7 illustrates the variety of local topographic features found within these submarine topographic provinces.

The Ocean Basin Floor The largest and most monotonous area of submarine landscape is the floor of the ocean basin. It lies at depths between 4500 m and 5500 m (14,750 to 18,000 ft). **Abyssal plains** extending as almost flat surfaces over large areas make up an important part of the ocean basin floor. These plains are almost devoid of topographic relief (Figure 2-7 A). They may slope gradually over some parts. For example, the southern part of the large Sohm Abyssal Plain (Figure 2-6) lies at a depth of about 5500 m. The surface rises toward the north to a depth of 5000 m. Gradients of between 1 : 1000 and 1 : 5000 have been measured over this surface. This kind of surface slope is typical of other abyssal plains.

Low scarps, which are steep bluffs or cliffs, (Figure 2-7 B) and **midocean canyons** (Figure 2-7 C) interrupt the otherwise featureless surface of an abyssal plain at widely separated intervals. These scarps and canyons may extend several tens or even hundreds of kilometers. Relief is less than 200 m across them, and the canyons are ordinarily less than 10 km wide.

Other regions of the ocean basin floor display **abyssal hill** topography. Here we find low rolling submarine terrain (Figure 2-7 D) where the relief is less than a few hundred meters. These regions merge gradually into bordering abyssal plains as the groups of hills become more and more widely separated, and end with isolated hills rising from the plain.

Elsewhere broad areas of the ocean basin floor rise a few hundred meters higher than the adjacent abyssal plains and hills. These elevated regions are **oceanic rises.** They display greater relief than the surrounding ocean floor. The surface of the large Bermuda Rise (Figure 2-6) varies from gently rolling terrain to groups of rugged hills. The western margin of this rise merges gradually with the bordering Hatteras Abyssal Plain. The eastern margin is more abruptly marked by steep scarps between 100 m and 300 m high, and other rugged relief. These same topographic features are found on other oceanic rises, but are not necessarily arranged in similar patterns.

The monotonous abyssal terrain is interrupted in some places by **seamounts.** These features are submarine mountains, either isolated or in small groups. They rise to heights of more than 1000 m above the surrounding surface. Most seamounts are conical structures. Examples are found in the Kelvin Seamount Group rising abruptly from the Sohm Abyssal Plain (Figure 2-6). The larger individual peaks stand more than 3500 m above the surrounding ocean floor. Seamounts are most abundant in the Pacific. Most rise to sharp peaks. Some, however, are typically steep-sided and flat-topped (Figure 2-7 E). These peculiar seamounts are called **guyots.**

Large and almost straight cliffs or escarpments extend in places along the ocean basin floor. These features are especially prominent in the northeastern Pacific basin where they are from 2400 km to 4800 km in length. The relief exceeds 2000 m in places.

Oceanic Ridges. We find interesting and varied landscape on the oceanic ridges. These are submarine mountain chains that wind through the ocean basins. These submerged mountain ranges are situated in broad sinuous belts.

Oceanic ridges consist of merging and branching regions of rugged relief where the ocean floor is at depths of less than 4500 m. The outer margins of a ridge are transitional into bordering abyssal hills and plains. In

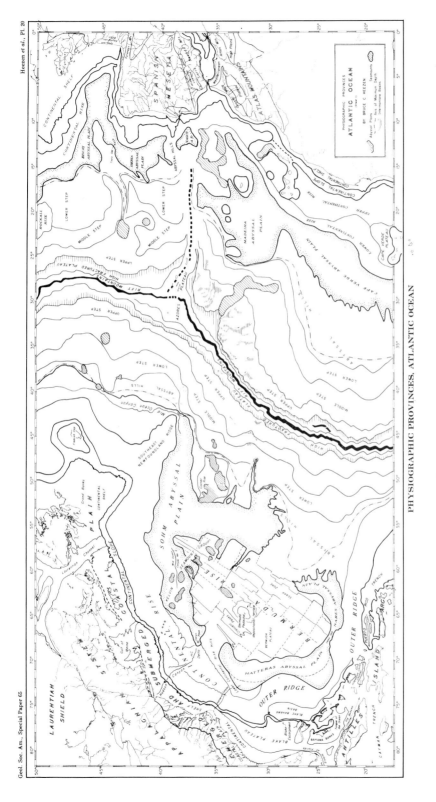

Figure 2-6 Topographic features of the North Atlantic Ocean. (Reprinted from B. C. Heezen, M. Tharp, and M. Ewing, Geological Society of America, *Special Paper 65*, 1959.)

52 The Earth's Size, Shape, and Interior Zones

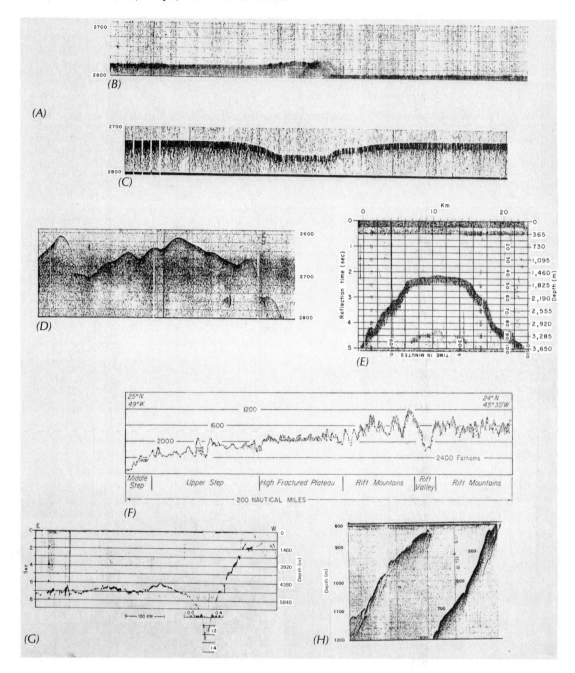

places nearer the center the highest peaks may reach above sea level to form mid-oceanic islands. A topographic feature of most ridges is the narrow and steep-sided trough which extends along the axis. A ridge is more or less bilaterally symmetrical about this axial trough.

The Mid-Atlantic ridge (Figure 2-6) divides the Atlantic Basin. Its rugged topographic regions are evident on a composite echo soun-

Figure 2-7 *(opposite)* Echo sounder profiles across typical topographic features of the ocean floor. Madeara Abyssal Plain *(A)* interrupted by scarp *(B)* and midocean canyon *(C)* Abyssal hills *(D)* indicate rugged relief near the Bermuda Rise. Guyot *(E)* located in the North Pacific Ocean. Composite profile *(F)* crossing the central part of the Mid-Atlantic ridge. The Tonga-Kermadec Trench *(G)* is in the southwestern Pacific Ocean. Continuous profile (note chart resets and scale changes) crossing the continental slope and shelf near New York *(H)*. (Features *A, B, C, D, F* and *H* reprinted from B. C. Heezen, M. Tharp, and M. Ewing, Geological Society of America, *Special Paper 65,* 1959. Profile *E* reprinted from H. W. Menard and R. S. Dietz, *Bulletin of the Geol. Soc. of Amer.,* Oct. 1951. Profile *G* reprinted from D. E. Hayes and M. Ewing in *The Sea,* Vol. IV, Part II, Wiley Interscience, 1971.)

der profile (Figure 2-7 *F*). The axial trough is a deep rift that bisects the ridge. The floor of the axial trough lies more than 1800 m deeper than the bordering peaks that Heezen called the central rift mountains. Flanking these mountains are the high fractured plateaus. These are regions of rugged and irregular landscape, but with less extreme relief than the bordering mountains. Farther out we find the mountainous upper step, then the middle and lower steps where the relief is more subdued.

Perhaps the most striking features of the ridge system are the magnificent fractures cutting across it (Figure 2-9). The ridge appears to be broken into segments by these fractures. The segments are offset from one another. Dynamic processes associated with these features are discussed in Chapter 5.

The oceanic ridges have a distinctive topography which is different than the typical continental mountain chain topography. The symmetry relative to a trough along the ridge axis is not found on the continental landscape. The high fractures which cut through the ridges have no obvious continental counterpart. Although oceanic ridges and continental mountain ranges display rugged elevated landscape, they are distinctly different topographic provinces.

Submarine Trenches. The largest extremes in relief on the surface of the solid earth are found in the oceanic regions. The most tremendous depths are in the submarine trenches. Imagine descending 11,000 m (more than 36,000 ft) into the Marianas Trench. This depth is much greater than the 8848 m (29028 ft) height of Mt. Everest. Submarine trenches are narrow linear zones where depths exceed 5500 m. Most are found in the Pacific Basin. Some facts about the largest trenches are given in Table 2-1.

Table 2-1
Some Oceanic Trenches

Name	Location	Maximum Depth (km)	Length (km)
Marianas	NW Pacific	11.0	2550
Tonga	SW Pacific	10.9	1400
Kurile-Kamchatka	NW Pacific	10.6	2200
Philippine	NW Pacific	10.5	1400
Kermadec	SW Pacific	10.0	1500

From P. K. Weyl, Oceanography, John Wiley and Sons (1970), p. 96.

Submarine trenches are located near continental margins or along the seaward borders of island arcs. Linear chains of islands are called island arcs. The typical trench cross section is evident from the echo sounder profiles in Figure 2-7 *G*. The trenches are asymmetrical, rising more steeply toward the bordering continent or island arc. Their relationship to global dynamic processes is discussed in Chapter 5.

Continental Margins. The global mass of sea water overfills the ocean basins and spills onto the margins of the continental platforms.

The region between ocean basin and the seacoast is a submarine topographic province. This province may be broad, as in the North Atlantic region (Figure 2-6). It is narrow where a continent is bordered by a trench. Topography across a continental margin province not bordered by a trench is depicted in Figure 2-7H.

We can find the important features of the North Atlantic continental margin in Figure 2-8. Immediately bordering the coast is a relatively smooth and gently sloping surface, the **continental shelf,** which lies at depths ordinarily less than 200 m. This shelf is the submerged extension of the continental platform. In places it is depressed to depths of more than 200 m where it is covered by seas such as the Gulf of Maine. These shallow seas are called epicontinental seas. They may reach depths of more than 1000 m, but their basins are almost surrounded by continental shelf.

Bordering the seaward edge of the shelf is the **continental slope** zone. Here gradients of more than 1:40 characterize the steeply inclined ocean floor. The slope may have rugged relief, and bluffs where gradients are more than 1:10. More common, however, is the relatively smooth inclined surface seen in Figure 2-7H.

The **continental rise** zone extends from the base of the continental slope down to the ocean basin floor. Here the surface is more gently inclined at gradients of less than 1:100. Local relief is subdued except for sparsely distributed seamounts and submarine canyons.

Submarine canyons cut into the continental margin in many places (Figure 2-8). They appear as notches in the steep continental slope. They range from less than 1 km to more than 15 km in width, and some are more than 2000 m deep. The Hudson Submarine Canyon is the largest on the North American margin of the Atlantic Ocean. It extends over part of the continental shelf, down the steep slope, and across the broad rise almost to the ocean basin floor.

Comparisons. We can now visualize the shape of the earth from these descriptions of topography. About one-fourth of the earth's solid surface rises above sea level, and three-fourths is submerged. Maximum submarine relief of 11035 m (36,204 ft) is measured between sea level, where the highest ridges rise to form islands, and the depths of the Marianas Trench. This exceeds the maximum continental relief of 8848 m (29,028 ft) from sea level to the peak of Mt. Everest.

The highest oceanic ridges reach heights of 4500 m above the surrounding ocean basin floor. Some continental mountain ranges, the Himalayas of Asia and the Andes of South America, for example, rise higher from the bordering continental terrain. We find equally rugged landscapes in continental mountain chains and oceanic ridges. However, there are no continental counterparts for the deep trough that marks the axis of a ridge, and the great fractures that cut across it. Also, a ridge is approximately symmetrical about its axial trough, whereas continental mountain chains are asymmetrical. These dissimilarities are evident from the profiles crossing the Rocky Mountains (Figure 2-5) and the Mid-Atlantic Ridge (Figures 2-6 and 2-9).

Except for rare scarps and mid-ocean canyons, the abyssal plains are devoid of even the shallow stream channels which relieve the monotony of continental plains. Nowhere on the continents is there featureless terrain comparable to these abyssal plain surfaces. The deep submarine trenches also have no comparable continental features. More similar are the hilly continental tablelands and the rolling abyssal hill terrain and elevated rises of the ocean basin floor. Also the continental landscape of widely spaced mountains is similar in some ways to scattered seamounts on the ocean basin floor.

Topography is varied on the continents and on the ocean floor. However, important differences between continental and ocean basin terrain can be recognized. We will learn more about these differences in later

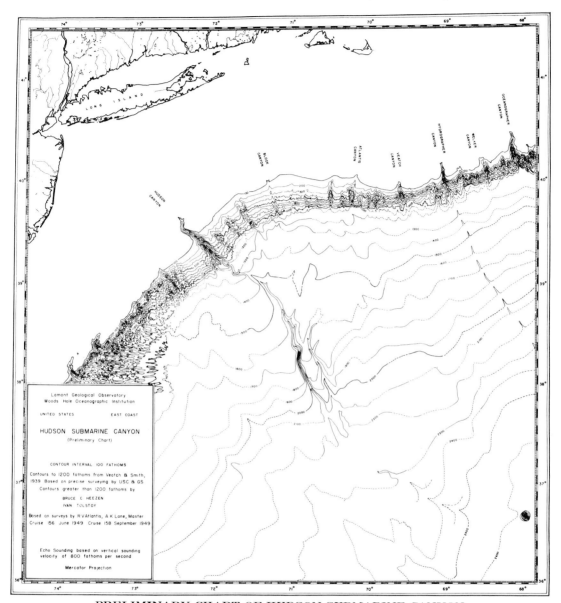

PRELIMINARY CHART OF HUDSON SUBMARINE CANYON
Based on nonprecision soundings taken 1949–1950

Figure 2-8 Bathymetric chart, which is a submarine topographic map, of part of the Atlantic continental margin bordering North America. (Reprinted from B. C. Heezen, M. Tharp, and M. Ewing, Geological Society of America, *Special Paper 65*, 1959.)

Figure 2-9 Bathymetric chart of part of the Mid-Atlantic ridge that illustrates two major transverse fracture zones, the Romanche Fracture Zone and the Chain Fracture Zone. Water depths are between 1400 m and 2000 m over the regions indicated in black. Contour lines indicate water depth in meters. (Reprinted with permission from *Deep Sea Research*, Vol. 11, B. C. Heezen, E. T. Bunce, J. B. Hersey, and M. Tharp, Fig. 2, Jan.–Feb. 1964).

chapters where effects of geologic processes on topography are examined.

The Geoid

Height measurements give us the basic information we need to describe the earth's topography. The height of the land surface is its distance above or below some horizontal surface, usually the surface of the sea. Before we discuss how height is measured we should consider just what sea level is and how closely it corresponds to the reference ellipsoid.

Imagine that the oceans were not disturbed by waves and tides and currents of circulating water. Then the surface of the sea would be an ideal horizontal or level surface. We call this ideal sea level surface the **geoid.** A string fastened to this surface with a small weight hanging from its other end would be exactly perpendicular to the surface. Such a string with a weight on one end is called a **plumb line.** The direction of a plumb line is, by definition, the vertical direction at the particular point where it is hanging. A surface perpendicular to the plumb line is defined as a horizontal or level surface.

A plumb line is free to swing into the direction dictated by the gravitational and centrifugal forces acting on it. This will be the direction of the net force indicated in Figure 2-2 for an idealized ellipsoid. Similarly, the particles of a fluid standing in a basin are free to flow in response to these forces. Consequently, the fluid will form a horizontal surface that is adjusted to the gravitational and centrifugal forces.

To say that a surface is horizontal does not imply that it is a flat plane surface. Indeed, for a rotating globe the ideal horizontal surface will be an ellipsoid. In fact, if the rocks of the earth could flow easily in response to the gravitational and centrifugal forces, the different materials would quickly shift into concentric ellipsoidal shells with the more dense material nearer the center. Rock masses in the earth can flow plastically under proper conditions of pressure and temperature, and we do find concentric zones of the crust, mantle, and core. These masses of different density are sufficiently rigid, however, so that the gravitational and centrifugal forces cannot completely smooth out all irregularities in their arrangement.

How can irregular masses of high or low density rock affect the sea level surface? We can see that plumb lines will be deflected slightly by a nearby irregular mass of high density rock. Because its density is higher than normal, the rock mass will exert an additional local gravitational attraction on the plumb line. This effect is illustrated in Figure 2-10. If the plumb line is deflected, then the level surface will also be changed. This surface must remain perpendicular to the plumb line.

To find the shape of the sea level surface let us move point by point along the profile shown in Figure 2-10. At each point we observe the direction of a plumb line. Then we can connect the plumb line directions by a curve that is everywhere perpendicular to them. In this way we could construct the shape of the actual sea level surface. This surface is defined as the **geoid.** The geoid can then be described by stating its height above or below the reference ellipsoid.

It is not practical to attempt to measure plumb line directions at enough sites to map the geoid. But it can be mapped in other ways. One method for determining the form of the geoid makes use of orbiting satellites. Orbital paths can be predicted very precisely on the basis of the reference ellipsoid. Then the satellite is carefully tracked to detect differences between the actual orbit and the pre-

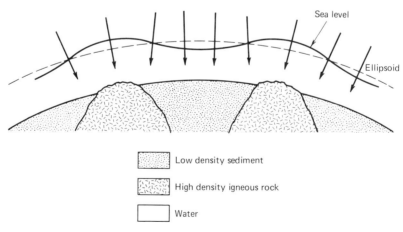

Figure 2-10 Adjustment of the shape of the sea level surface to density irregularities. Note how the plumb lines are deflected toward more dense rock masses. Because the level surface must remain perpendicular to these plumb lines, sea level is seen to be warped upward above the high density rock masses.

58 The Earth's Size, Shape, and Interior Zones

dicted one. By assuming slightly irregular shapes for the earth, revised orbits can be computed until a shape is found which gives the same orbit as is observed. This must be done for many satellites following orbits inclined at a variety of angles with the plane of the equator. Another method of mapping the geoid is done by analyzing measurements of the gravitational force at many different places on the earth's surface. Later we will learn how to make these measurements.

The map in Figure 2-11 illustrates the general form of the geoid. We see that the small differences between the geoid and the reference ellipsoid are at most about 70 m. This indicates how closely the ellipsoid represents the irregular and undulating shape of sea level. If the continental areas of the earth were criscrossed by narrow canals connected to the oceans, the map of the geoid indicates the water level in such canals.

Topographic Surveying

Features such as hills and valleys give form that we call **topography** to the land surface. The topography of continents and islands is mapped by measuring the elevation of the land above sea level. Now that we know about the shape of the sea level surface itself, we can find out about measuring heights above this surface.

First we must define what is meant by the elevation or the height above sea level of a point on the land surface. The correct elevation corresponds to the length of a plumb line reaching from that point down to the geoid. Because a plumb line cannot be lowered to sea level through the solid earth, the method of **spirit leveling** is used to find inland elevations. This method is illustrated in Figure 2-12. A hairline mounted in a horizontal telescope is seen by an observer to intersect different heights on vertical rods, indicating the differences in elevations between points where these rods are situated. By successively relocating rod and telescope it is possible to proceed from a coastal sea level point and estimate inland elevations by summing height increments between rod positions.

The summation of height increments in this manner provides only an estimate of inland elevation. To understand why, suppose that water was added to the ocean, causing sea level to rise from point A to point B, and then to points C and D in Figure 2-12. We would find that these different sea level surfaces are not parallel to one another. Figure 2-13 illustrates why this might be so. The surface close to a rock mass of higher than normal density would be more severely warped than another surface farther above, where the disturbing gravitational attraction of the heavy mass is not so strong. Therefore, the height between these horizontal surfaces changes from one place to another. The heights h_a and h_b in Figure 2-13 are related to values of the gravitational force at these places in the following way: $h_a F_{ga} = h_b F_{gb}$. Small but important differences in the gravitational force between different points on the earth are well known. We will find out more about them later.

Now, look again at Figure 2-12. We see that the correct elevation at D is the sum of the height increments h_1, h_2, and h_3 directly beneath it. By spirit leveling we would measure different height increments h'_1, h'_2, and h'_3 at intermediate positions. To find the height of D, the following gravitational corrections have to be made:

$$h_D = h_1 + h_2 + h_3 = h'_1 \frac{F'_{g1}}{F_{g1}} + h'_2 \frac{F'_{g2}}{F_{g2}} + h'_3 \frac{F'_{g3}}{F_{g3}}$$

The gravitational forces F'_{g1}, F'_{g1}, and F'_{g3} can be measured at the intermediate positions on the land surface. Values of F_{g1}, F_{g2}, and F_{g3} at depths beneath D must be calculated from measured values on the surface and knowledge about the way the gravitational force changes with depth. Later we will discuss how the gravitational force is measured or calculated at different depths.

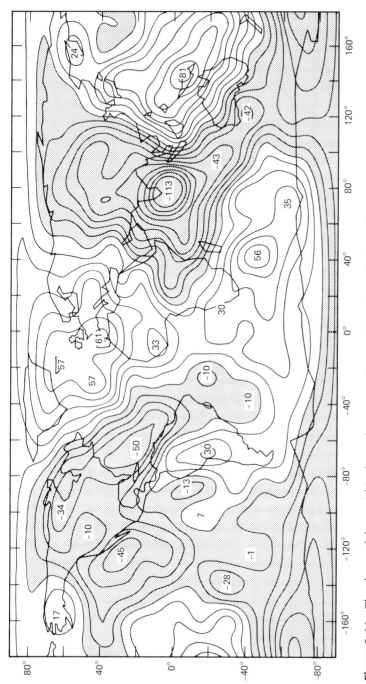

Figure 2-11 The shape of the geoid is shown by contour lines at 10 m intervals that indicate geoid height above (unshaded) or below (shaded) the reference ellipsoid that has a flattening value of 1/298.25. (Reprinted with permission of E. M. Gaposchkin, Center for Astrophysics, Harvard University.)

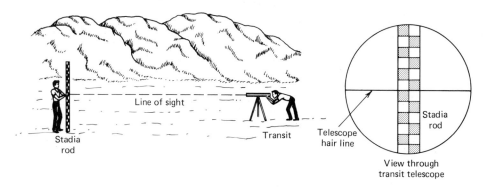

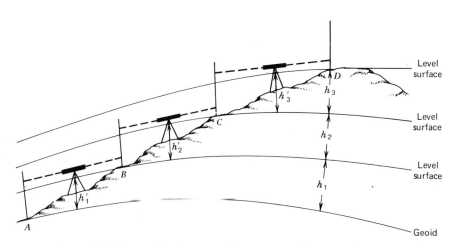

Figure 2-12 The method of spirit leveling is used to measure elevations above sea level. Measurements are made by viewing a vertical stadia rod through a transit telescope, which is a telescope mounted on a tripod. A horizontal line placed in the telescope tube crosses the center of the field of view seen through the telescope. By leveling the telescope, and viewing a stadia rod some distance away, the horizontal line is seen to intersect one of the elevation marks on the rod. This elevation mark is used to determine the difference in elevation between the rod and the transit. By alternately moving the rod and the transit to intermediate points along a profile a series of elevation differences is measured which adds up to the difference in elevation between the ends of the profile.

Failure to make gravity corrections can result in elevation errors of several meters or more at locations far from the coast. These errors can be reduced to less than one meter and perhaps a few centimeters by accounting for effects of gravity. Ordinarily these corrections are not needed in topographic surveys in small areas, but they are important in transcontinental surveys.

Accurate elevations have been established by spirit leveling at a large number of locations. Markers called benchmarks are perma-

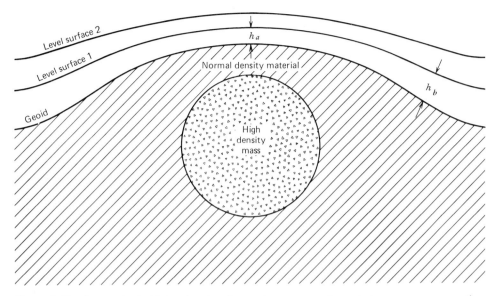

Figure 2-13 Successively higher level surfaces above a mass irregularity are not necessarily parallel because of the way in which the gravitational effect of this mass on a plumb line changes with distance.

nently installed at these places. Topographic survey agencies sponsored by most nations are responsible for this work.

Other methods can be used to find differences in elevation between benchmarks, but they are not as accurate as spirit leveling. A kind of portable barometer, the barometric altimeter, reads in elevation units rather than pressure units. Because atmospheric pressure changes with elevation, heights can be measured with this device. The measurements must be corrected for temperature and humidity because these factors also affect atmospheric pressure.

We can also find differences in elevation by a method called photogrammetry. This involves analysis of vertical aerial photographs of the land surface. Figure 2-14 illustrates how two actual points, a and b, appear at positions a' and b' on photographs. The distance between a' and b' is different on one photograph than on another taken at a distance (d) away from the first. These two photographs are later projected on a screen using two projectors a distance d' apart. Both projectors show point a at the same place on a screen located in position 1. When the screen is moved back a distance (Δx) to position 2, both projectors show point b in the same place. We can then calculate the true difference in elevation (Δz) between points a and b from the formula:

$$\Delta z/d = \Delta x/d'$$

Elaborate laboratory apparatus has been designed for rapid processing of overlapping aerial photographs to produce topographic maps.

The topography of the ocean floor is mapped with a device known as an echo sounder or precision depth recorder (PDR). A sound pulse generator-receiver system is mounted on the hull of a ship which travels along the sea-level geoid. The generator emits 10 to 12 kilohertz sound pulses at regular intervals. Each pulse travels to the ocean floor at a known velocity V close to 1450 m/s. It echoes back and is detected by the pressure-sensitive receiver. The recording ap-

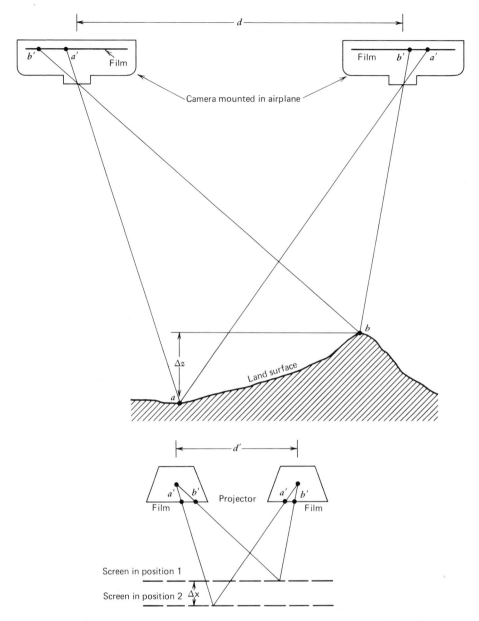

Figure 2-14 The method of photogrammetry can be used to determine elevation differences used for preparing topographic maps. The method utilizes aerial photographs taken of the same area of the land surface from different positions. Points at different elevations appear to be separated by different distances on the photographs. Laboratory projection apparatus is designed to determine elevation differences from the apparent differences in distance that are measured from the photographs.

paratus consists of a pen that begins to move across an advancing paper chart whenever a pulse is emitted. The pen marks the chart when the receiver detects the echo. In this way the travel time t of the pulse to and from the sea bottom is recorded. As the ship moves, travel-time marks for a succession of pulses are displayed on the chart profile, from which depth d can be calculated: $d = Vt/2$.

Reliable echo sounders were developed shortly before 1950. Since then numerous ships equipped with them have recorded a vast amount of information about the topography of the earth's solid surface beneath the oceans.

A Fundamental Level Net

Precise spirit leveling has been done in many parts of the world to establish networks of reference points. Local topographic surveys can be adjusted to these points. In the United States a first-order level net was surveyed by the U.S. Coast and Geodetic Survey (now the National Geodetic Survey). The term **first order** implies the highest standards of accuracy. Benchmarks are situated so that one can be found within 80 km (50 mi) of any place in the country. Differences in elevation are measured to a precision of better than 13 mm over a distance of 100 km (about ³/₄ in. in 100 miles).

The U.S. first-order level net is resurveyed periodically. An interesting fact has emerged from the repeated spirit leveling. Elevations at the benchmarks are changing with time. The land surface is not completely stable. Instead it is rising in some places and subsiding elsewhere. The rates of movement are indicated in Figure 2-15. In later chapters we will look into possible causes of this movement.

GRAVITY AND DENSITY IN THE EARTH

Fortunes are to be made by discovering oil. Our growing industrial economy demands more and more "black gold." The question is how to find it. A century ago the search was guided mostly by schemes based on ignorance if not outright fraud. But the chance discoveries provided valuable information about the various natural settings where oil had accumulated.

Some fabulous oil fields were situated where layers of sedimentary rock included peculiarly deformed beds of salt. Underground pressures caused the salt to flow plastically into domelike structures (Figure 2-16). In Chapter 10 we will discuss these salt structures more carefully and how oil gets concentrated near them. Now we need only remember that they are good places to look for oil. Once this was known, oil drillers began devising methods to locate salt structures buried hundreds or thousands of meters underground. One of the most practical methods was based on minute changes in the gravitational force on the earth's surface.

How can we find salt structures from changes in the gravitational force? The principle is simple. These structures cause density irregularities. We know that the density of salt is about 2.0 gm/cm³. Oil has been found in the places where salt domes protrude into overlying rock having a higher density of about 2.5 gm/cm³. Because of the low density of salt, the gravitational force on the surface directly over a salt dome is not as strong as it is farther away where the underlying rock has higher density.

Geologists are continually asked about the materials that lie hidden underground. These can be fundamental questions about the way the earth is put together. Or we may want to know where some useful raw material might be located. Density is a property of the materials buried in the earth. We already know that irregular variations in density affect the shape of sea level and our measurements of elevation. Now we can look into relationships between density in the earth and the gravitational force on its surface.

64   The Earth's Size, Shape, and Interior Zones

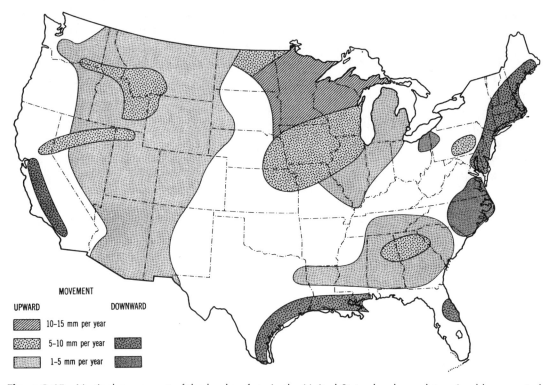

Figure 2-15 Vertical movement of the land surface in the United States has been determined by repeated spirit level surveys. (After S. P. Hand, National Oceanic and Atmospheric Administration, 1972.)

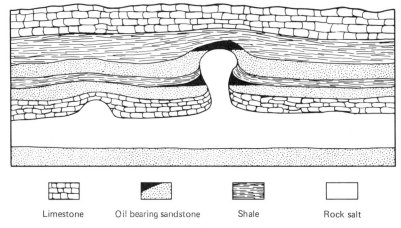

Figure 2-16 Geologic cross section illustrating a salt dome that has penetrated overlying sedimentary layers. Oil and natural gas are known to have accumulated in some of the upturned sedimentary layers near many salt domes.

Gravity Surveying

How can we find out about the density in the earth? We can get some very useful information by knowing the gravitational force at different places on the earth's surface. Before we attempt to interpret variations of the gravitational force we should learn how to detect these variations.

First let us distinguish the gravitational force between two bodies from the gravity of each of these bodies. The term **gravity** refers to the capacity of one body, by virtue of its mass, to attract another body. Gravity is the acceleration that one body imposes on another body if the other body is free to move. To understand this we introduce Newton's Second Law of Motion: that the acceleration (a) of an object of mass (m) is caused by a force (F) according to the expression: $F = ma$. Acceleration is the rate of change in the velocity of the body, that is, the rate at which it speeds up or slows down.

Now, let us write the Universal Law of Gravitation in these different ways:

$$F_g = Gm_1m_2/r^2 = m_1(Gm_2/r^2) = m_1g_2$$

We see that the term $g_2 = Gm_2/r^2$ must be an acceleration. We define g_2 as the gravity of m_2 at the distance r away from its center. At that distance g_2 is constant regardless of the size of m_1. Similarly, we can say that $g_1 = Gm_1/r^2$ is the gravity of m_1 at the distance r. When we talk about the gravity of a body, we are dealing with a property of that particular body. At a certain distance away from that body its own gravity remains unchanged regardless of any other body, large or small, that it may be attracting.

If the earth were a perfect sphere of radius R and mass M, gravity on its surface would be $g = GM/R^2$. The value of g would be the same everywhere on this idealized spherical surface. But we know that the earth is not a sphere. Therefore, we might expect differences in the values of g from one place to another. How can we discover if this is true?

Gravity on the earth's surface can be measured in several different ways. We already know how to find the gravitational force F_g by observing a falling mass. To find the earth's gravity (g) instead, we use the slightly different formula: $g = 2z/t^2$, where t is the time taken by an object to fall a distance z from a position of rest. We noted earlier that near the earth's surface a small mass will fall 100 cm in about 0.452 sec, so that the value of g is about 980 cm/sec^2.

The standard unit of gravity is the **gal** (after Galileo) which is defined as an acceleration of 1 cm/sec^2. For some geological purposes a more convenient unit is the **milligal,** which equals 1/1000 gal. Thus, on the earth's surface, gravity values are close to 980 gals or 980,000 milligals. For modern needs of geodesists and geologists we usually have to measure gravity to an accuracy of 1/10 milligal. This is about 1 part in 10 million of the strength of the earth's gravity on the surface. Some especially precise work requires 1/1000 milligal accuracy, or 1 part in 1 billion of the earth's gravity.

The falling mass experiment is the only completely independent method for measuring gravity to the precision of 0.1 milligal. But these experiments are costly and time consuming. They have been carried out at only a few locations using a mass in a vacuum chamber, and elaborate timing apparatus. The standard value, 980.1024 gals, was measured near Washington, D.C.

A second method of measuring g involves a swinging pendulum. The apparatus consists of a vacuum chamber in which a mass is suspended from a pivot point. When put into motion on the earth's surface, the mass swings back and forth in a period of time T that depends on gravity:

$$g = k/T^2$$

The constant k is related to the shape of the mass and the location of the pivot. It cannot be determined accurately enough to use a pendulum for independent measurement of g. But because k remains the same at different locations we can find the change in gravity between two points from the observed change in the pendulum period. If we know g in one place from a falling mass experiment, we can add or subtract differences found from pendulum measurements to obtain values of g at other places. Pendulum apparatus is portable, and observations at one location require about one hour to complete. The gravity difference between two places can be measured to a precision of approximately 0.1 milligal.

An instrument called a gravimeter consists of a carefully counter-balanced system of masses suspended from springs. As the device is moved from place to place, differences in the earth's gravitational attraction of these masses cause changes in the distension of the springs. The properties of the springs and masses cannot be determined precisely enough to measure gravity independently at a single location. But by making observations at sites where g is already known from falling mass measurements, the response of the gravimeter to changes in gravity is established. Once it has been calibrated in this way, a modern hand carried gravimeter can be used to find other gravity differences in unsurveyed regions. Observations require only a few minutes at a site. Gravity differences as small as 0.02 milligals can be detected with a good instrument. These instruments are so sensitive that if readings are made on each step of a stairway, the changes in gravity caused by moving farther from the center of the earth can be detected.

Although pendulums and gravimeters are simple in principle, great care and expense goes into their construction. The working parts must be sealed in a case where temperature is controlled within 0.1° C and pressure changes are minimized.

These various kinds of instruments have been used to obtain values of gravity at several hundred thousand points on the surface of the earth. From these measurements we find that the value of gravity at the equator is approximately 978 gals, and the value at the poles of the earth is approximately 983.1 gal. This means that an average person weighs about 1/2 kilogram (one pound) more at the poles than on the equator.

There are two main reasons for this difference in the earth's gravity between the equator and the poles. First, gravity-measuring instruments on the equator are farther from the earth's center. We know from the Universal Law of Gravitation that the force diminishes as distance from the center of mass increases. The second reason involves centrifugal force. Earlier we noted that on the equator this force is opposite to the gravitational force. The gravity-measuring instruments actually indicate the net force (Figure 2-2), and cannot distinguish between gravity and centrifugal effects. Therefore, these opposing effects cause a weaker net force at the equator than at the poles where there is no centrifugal force.

This change in gravity between the equator and the poles can be mathematically predicted for a rotating ellipsoid. For the reference ellipsoid we can find the gravity at any latitude (ϕ) from the formula:

$$\gamma = 978.031846(1 + .005278895 \sin^2 \phi + .000023462 \sin^4 \phi)$$

This is called the 1967 Geodetic Reference System Formula (GRS 67). We use the symbol γ to distinguish gravity on this idealized surface from g which represents gravity on the actual earth.

We can compare values of g measured on the earth with values of γ at the same latitude. We find many local differences where g is either higher or lower than γ. These departures range from undetectably small to more than 1000 milligals. They are related to irregularities in the way mass is arranged in the earth.

Gravity Anomalies

Many salt domes have been found by gravimeter surveying in the Gulf Coast region of the United States. Readings at a great many points show that gravity is weaker over some areas than others. Much as we might wish that a salt dome existed everywhere that gravity readings are low, this is not true. To understand this, we must consider some other reasons why gravity varies from one place to another.

We can expect different values of g at two places if one is at a different latitude than the other. This is because gravity on an ellipsoid changes between the equator and the poles. We can correct for such differences by subtracting from each value of g the value of γ computed from the GRS 67 Formula at the latitude of the gravity reading.

If gravity readings are made at different elevations, we might expect a lower value at the higher location. This is because gravity decreases with increasing distance from the earth's center. We call this effect the **free air effect.** Near the earth's surface gravity decreases by 0.3086 milligals for each 1 m increase in elevation. You can check this by calculating the gravity of a sphere the size of the earth on its surface and at a height of 1 m above it.

There is another effect related to the elevations of gravity readings. Where one location is higher than another, there is obviously an additional amount of rock beneath the gravimeter. The attraction of this extra mass adds to the gravity at the higher location. This amounts to approximately $0.0419 \times \rho$ milligals for each 1 m increase in elevation where ρ is the density of the rock. We call this effect the **Bouguer mass effect.** It is named for Pierre Bouguer (pronounced Boo-gay), an eighteenth-century French geodesist.

Gravity readings are routinely corrected for latitude, free air, and Bouguer mass effects. The corrected value is called the **Bouguer gravity anomaly** (Δg_B), and we calculate it from the formula:

$$\Delta g_B = g - \gamma + 0.3086h - 0.0419\rho h$$

where h is the elevation of the reading. Consider the following example. The airport terminal near Houston, Texas, is located 15.2 m (50') above sea level at 29°39' north latitude and 95°17' west longitude. Here gravity was found to be 979293.3 milligals from a gravimeter reading. At that latitude we calculate γ to be 979310.5 milligals. We also calculate a 4.7 milligal free air effect and a 1.7 milligal Bouguer mass effect. From these values we find that the Bouguer gravity anomaly is -14.2 milligals. The word **anomaly** means irregularity. The Bouguer gravity anomaly indicates density irregularity in the earth.

Now we can understand why gravity readings can differ from place to place because of latitude and topographic effects, as well as subsurface density irregularities. Therefore, uncorrected gravity readings do not give reliable indications of the density irregularities such as those caused by salt domes. Bouguer gravity anomalies have been corrected for the latitude and topographic effects. Therefore, these are the values we use to find density irregularities.

Maps and profiles are used to display the variations of Bouguer gravity anomaly values in an area. Contour lines indicate patterns of variation on a map. A change from high to low values indicates a decrease in average rock density. By looking at the map in Figure 2-17 we can quickly see that average density decreases near the center of this gravimeter survey area.

Oil prospectors thought that the low Bouguer gravity anomalies in this area might be caused by a salt dome. A buried mass of salt would cause the average density to be lower than normal. But to test their idea the prospectors had to be more specific. They had to show where such a mass must be located to cause the observed Bouguer anomaly varia-

68 The Earth's Size, Shape, and Interior Zones

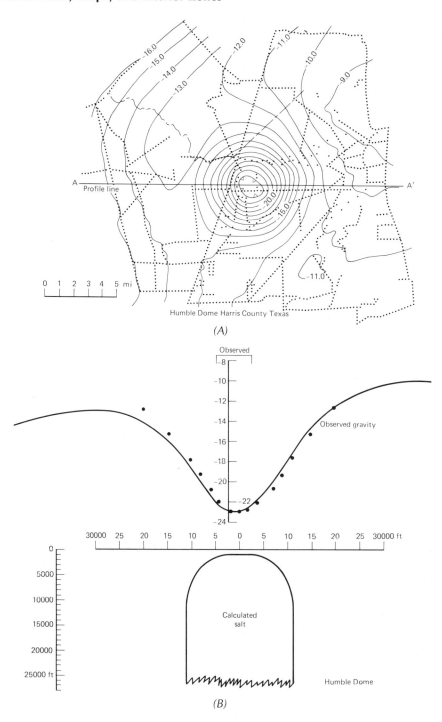

Figure 2-17 *(opposite)* *(A)* Contours on the map show variation of Bouguer gravity anomalies over part of Harris County, Texas, caused by penetration of sedimentary layers by a salt dome. *(B)* Measured Bouguer anomaly variation along the profile compares well with calculated gravity variation over the model of a salt dome shown in the cross section. (From "Elementary Gravity and Magnetics for Geologists and Seismologists," Figures 42 and 43, by L. L. Nettleton, 1971, with permission from The Society of Exploration Geophysicists.)

tion. This is done by mathematical methods derived from the Universal Law of Gravitation. They can be used to calculate the gravity of masses having different shapes.

Gravity variations were computed for several masses with different shapes and in different locations. These variations were compared with the observed Bouguer anomaly pattern. One mass resembling a salt dome was found to produce about the same pattern. This mass and the comparison of gravity variations along a profile crossing the area are shown in Figure 2-17B. A well was drilled to test these results, and a salt dome was found as predicted.

There are many other structures besides salt domes that can be detected from Bouguer gravity anomalies. Wherever these features are formed from rocks of contrasting density, they will cause Bouguer anomaly variations. We have discussed one example of commercial interest. Similar methods of analysis are used to find other kinds of commercial deposits. We can also use these methods to learn more general information about density irregularities that reveal the geologic structure of the earth.

Mass Under Continents and Oceans

The principal zones of the earth, as we know, are the crust, the mantle, and the core. The first clues about the layer we now call the crust came from studies of gravity and the deflection of plumb lines. Later, Bouguer anomalies over the continents were found to be greatly different than Bouguer anomalies in the oceanic regions. These differences are the result of major differences in the thickness and composition of the crust.

During the early decades of the nineteenth century, George Everest, while Surveyor General of India, conducted surveys to find the flattening of the earth. Mt. Everest was named for this well-known British geodesist. He planned to measure the distance of about 600 km between two towns in northern India. An astronomical method and a spirit leveling method were used. The distance measured repeatedly by spirit leveling was found to be 150 m different from the astronomical measurement.

The surveyors were well aware of the fact that the massive Himalayan Mountains caused deflection of plumb lines attached to spirit levels. Knowing the topography of the mountains, they calculated that in the nearest town this deflection should be 16 sec of arc greater than in the more distant town. Later they found from astronomical measurements that the actual difference was only 5 sec of arc. By using the 5 sec value rather than the 16 sec value to correct the spirit leveling data, the 150 m distance error was resolved. But they could not understand why their deflection calculations were wrong.

Explanations of this discrepancy were proposed first by J. H. Pratt in 1854, and shortly afterward in 1855 by G. B. Airy. Both suggested that at some depth, several tens of kilometers below sea level, pressure was the same everywhere around the earth. For this to be true, the mass of rock directly above some point at this depth must be the same as the mass directly above any other point at the same depth.

Pratt and Airy had different conceptions

about the density of rock above the depth of equal pressure. These are illustrated in Figure 2-18. Pratt represented this upper zone by columns reaching from the land surface to this depth. The columns under high mountains were longer than those beneath low valleys. Because all columns had to have the same mass, the density of the longer ones was necessarily lower than the density of the short ones.

Airy proposed two layers of different density above the depth of equal pressure. Density throughout each layer was the same everywhere. But the layer thicknesses changed according to topography. A column reaching from the land surface to the depth of equal pressure included parts of both layers. The lengths of these two parts varied from place to place so that the total mass of one column was the same as any other. Therefore, a column under a high mountain would have a larger proportion of the low density rock than a column beneath a low valley.

Everest's people had not realized that the average density deep beneath mountains could be lower than the average density under lowland regions. This was proposed by both Pratt and Airy. It was then understood that low density rock reaching far under the Himalayas would exert a smaller attraction on a plumb line than formerly expected. This explained why the measured plumb line deflections turned out smaller than values calculated at the time of the survey.

The Pratt and Airy conceptions of density under the land surface each explained Everest's surveying discrepancy. Later, Airy's conception proved to be more geologically realistic. We know this from studies of earthquake vibrations described in Chapter 3. Airy's upper layer is an overly simplified representation of the earth's crust.

Finding plumb line deflection by astronomical measurements is expensive and time consuming. This is not a practical way to look for differences in the crust's thickness predicted by Airy's conception. This can be done much more cheaply and quickly by gravimeter surveying. We will now consider how Bouguer anomalies are related to the thickness of the crust.

First we have to find out about density in the crust, and in the part of the mantle just below it. This part of the mantle corresponds approximately to Airy's lower layer. We know by studying the abundance of different kinds of rock that the average density of the crust is between 2.8 and 2.9 gm/cm^3. In some places rocks from as deep as 300 km have been erupted from volcanoes. These specimens indicate that the density is between 3.3 and 3.4 gm/cm^3 in the upper part of the mantle. So we know that two layers of contrasting density exist in the outer shell of the earth.

Let us next examine the variation of Bouguer anomalies over continents and oceans. Profiles are shown in Figure 2-19. We see that anomalies are positive over much of the ocean and negative over much of the continent. The exceptions in the oceanic regions are over the deep trenches where Bouguer anomalies are strongly negative. Over North America the Bouguer anomalies are more strongly negative in mountainous and high tableland areas than over the lowland regions. Although the Bouguer anomalies remain positive across the Mid-Atlantic ridge, they are relatively lower here than over the adjacent abyssal plains.

Now we can ask if Airy's model (Figure 2-18B) could predict these Bouguer anomaly variations over continents and oceans. For this model, we would expect higher Bouguer anomalies over lowland areas because of the larger proportion of high density rock beneath such regions. Over mountainous areas where the low density layer is relatively thick, we would predict low Bouguer anomaly values. These Bouguer anomaly differences predicted from Airy's model are similar to the observed differences. His model points out a significant difference between continents and ocean basins. The low density layer would be much

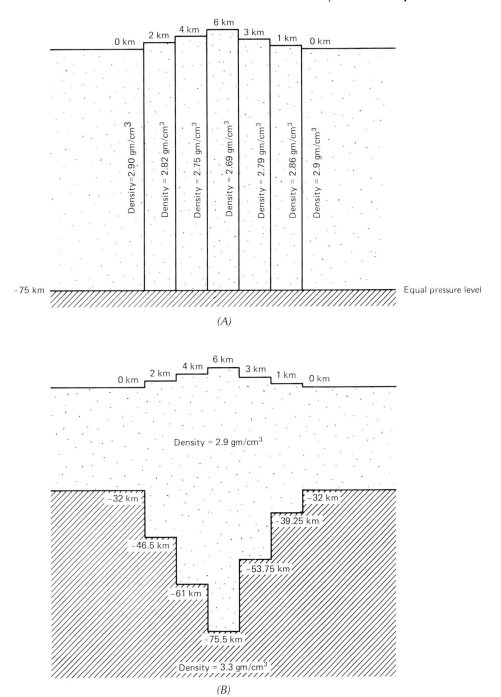

Figure 2-18 Conceptions of density in the earth's outer shell according to (A) J. H. Pratt and (B) G. B. Airy.

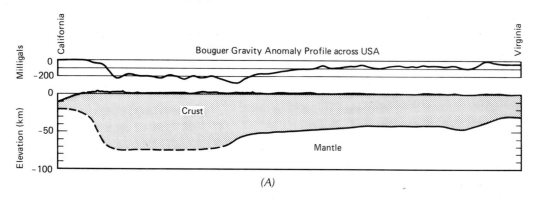

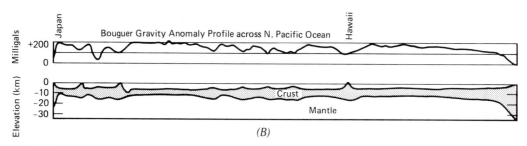

Figure 2-19 Variation of crustal thickness and Bouguer gravity anomalies across (A) North America and (B) the North Pacific Ocean. (Profile A was redrawn from L. C. Pakiser and I. Zietz. p. 513, Fig. 5, *Reviews of Geophysics,* November 1965. Profile B was redrawn from G. P. Woollard and W. Strange, *Monograph 6,* page 72, Figure 8, American Geophysical Union, 1962.)

thicker beneath continents compared with ocean basins. One exception is found over the ocean trenches. Here the solid earth surface is at its lowest elevation, yet the Bouguer anomaly values are strongly negative. We cannot explain this discrepancy from gravity observations alone. We need additional geological information found in later chapters.

Since the time when Airy proposed his simple model, other sources of geological information have confirmed that it is a reasonable, but overly simplified, approximation. We now know that important density irregularities exist in each of the layers we have named the crust and the mantle. The crust is between 30 and 60 km thick in continental regions and extends 5 to 10 km beneath the ocean basin floor. Now we use gravity observations and topographic surveys with other kinds of geological information to find how mass is distributed in the earth.

Total Mass and Mean Density

We want to find how mass is arranged in the deep earth. This will give us a clue about the different kinds of rock beneath the surface. Let us begin by estimating the total mass. We can do this by rearranging the formula for gravity on a sphere: $M = gR^2/G$. We know that g is about 980 gals and R is about 6371

× 10^5 cm. Using these values we estimate that the earth's total mass is 5.96×10^{27} grams.

Next we want to find the volume of the earth. A formula relates the volume (V) and the equatorial (R_e) and polar (R_p) radii: $V = \frac{4}{3} \pi R_p R_e^2$. Using dimensions of the reference ellipsoid we find the earth's volume to be 1.08×10^{27} cm^3.

Density, as we have seen, is mass divided by volume. From our values of M and V we get a figure of 5.52 gm/cm^3 for the earth's mean density. But if we measure density of common rocks near the surface, we find values consistently lower than 3 gm/cm^3. This discrepancy leads us to an important realization: Density must increase with depth in the earth. Materials near the center must have much higher density than the rocks near the surface. Therefore, a general approximation of the arrangement of mass in the earth is that greater mass is concentrated nearer the center.

It is interesting to note from measurements of the moon's mass and volume that its mean density is 3.34 gm/cm^3. This value is much closer to the density of the surface rocks, suggesting that the internal mass distribution of the moon is quite different from that of the earth.

Perspectives

Let us consider a scale model of the earth having an equatorial diameter of 5 m (16.4 ft). Its polar radius would then be about 1.7 cm (0.7 in.) shorter. The highest mountains and deepest submarine trenches would not rise even 0.5 cm (0.16 in.) above or below the sea level surface of the model. Less than 0.03 mm would separate its geoid and reference ellipsoid at places of maximum departure.

Gravity acts to level the earth. If all of the constituent materials could flow like fluids, gravity would cause them to form into homogeneous concentric shells. Because of the strength and rigidity of rock, the earth's solid surface is marked by topographic features. Below this surface variations in crustal thickness and other irregularities as deep as 700 km are indicated by gravity anomalies. This zone where the principal irregularities exist would form only a 30 cm (1 ft) outer shell on our scale model of the earth.

So close does the earth come to being a perfect ellipsoid that we could hardly discern its irregularities on the scale of our model. But we are quite small ourselves compared with the size of the earth. We find that its small imperfections make it a much more fascinating planet.

STUDY EXERCISES

1. Figure 2-20 illustrates a hypothetical distribution of rocks of different densities beneath the ocean along the equator.

    (a) At which location, A, B, or C, is the sea level surface closest to the center of the earth? Explain!

    (b) At which location, A, B, or C, is the observed gravity strongest?

    (c) Between locations A and B is the

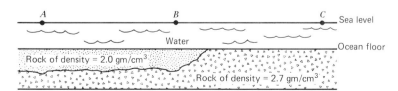

Figure 2-20

shape of the geoid convex upward or concave upward? Explain!

2. Suppose you want to find the difference in elevation (Δh) between locations A and D on Figure 2-21. If gravity corrections described on page 58 were not made, you would get a different value for Δh from measurements made with a spirit level at location B than the value you would get from measurements made with the spirit level at location C.

   (a) From which instrument location, B or C, would the uncorrected value of Δh be largest? Explain!

   (b) At which location, A or D, is the geoid closest to the center of the earth? Explain!

   (c) At which location, A or D, is the Bouguer anomaly most positive?

3. The moon has a radius of 1740 km, and its mean density is 3.34 gm/cm^3. Use these numbers to calculate the mass of the moon and the approximate value of gravity on its surface.

4. Suppose that you used the same pendulum to measure gravity on the earth's surface and on the moon's surface. At which location would the pendulum swing with the shortest period?

5. Explain why a Bouguer anomaly in Denver, Colorado, has a more negative value than a Bouguer anomaly near Norfolk, Virginia.

Figure 2-21

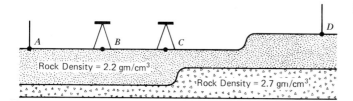

SELECTED READINGS

Heezen, Bruce C., Marie Tharp, and Maurice Ewing, *The Floors of the Oceans, I. The North Atlantic,* Special Paper 65, Geological Society of America, 1959.

King-Hele, Desmond, The Shape of the Earth, *Scientific American,* pp. 67-80, October 1967.

Nettleton, L. L., *Gravity and Magnetics in Oil Prospecting,* McGraw-Hill Book Co., New York, 1976.

Strahler, Arthur N., *The Earth Sciences,* Harper & Row, Publishers, New York, 1971.

3

EARTHQUAKES AND SEISMOLOGY

Earthquakes and Seismic Waves
   Measuring Seismic Waves
   Seismogram Features
   Location of Earthquakes
   Earthquake Magnitude and Energy
   Earthquake Intensity
   Earthquakes and Faults
   Earthquake Prediction and Control

Seismic Waves and Earth Structure
   Elasticity
   Refraction and Reflection of Seismic Waves
   The Earth's Deep Interior
   Seismic Surveying
   The Quest for the Moho
   Density, Pressure, and Elasticity in the Earth

A sudden violent shaking of the ground is called an earthquake. One of the strongest earthquakes in modern time occurred in Alaska on the afternoon of Good Friday in 1964. Robert Atwood, editor of the **Anchorage Daily Times,** was in his home. Here is an excerpt from his eye witness account.

ANCHORAGE

Turnagain Heights—I had just started to practice playing the trumpet when the earthquake occurred. In a few short moments it was obvious that this earthquake was no minor one: the chandelier made from a ship's wheel swayed too much. Things were falling that had never fallen before. I headed for the door. At the door I saw walls weaving. On the driveway I turned and watched my house squirm and groan. Tall trees were falling in our yard. I moved to a spot where I thought it would be safe, but, as I moved, I saw cracks appear in the earth. Pieces of ground in jigsaw-puzzle shapes moved up and down, tilted at all angles. I tried to move away, but more appeared in every direction. I noticed that my house was moving away from me, fast. As I started to climb the fence to my neighbor's yard, the fence disappeared. Trees were falling in crazy patterns. Deep chasms opened up. Table-top pieces of earth moved upward, standing like toadstools with great overhangs, some were turned at crazy angles. A chasm opened beneath me. I tumbled down. I was quickly on the verge of being buried. I ducked pieces of trees, fence posts, mailboxes, and other odds and ends. Then my neighbor's house collapsed and slid *into the chasm. For a time it threatened to come down on top of me, but the earth was still moving, and the chasm opened to receive the house. When the earth movements stopped, I climbed to the top of the chasm. I found angular landscape in every direction. I found my neighbor carrying his young daughter. We found his wife atop one of the high mushroom-like promontories. She was standing alone with her auto, marooned. We climbed up and down chasm walls and under dangerous overhanging pieces of frozen ground to safety. Helicopters were overhead, but they couldn't land near us—the ground was too topsy-turvy. After what seemed to be an endless time, rescuers came and helped the party out of the quagmire that had once been a home.*

Robert Atwood's experience is hardly unique. Devastation by earthquakes is known from thousands of years of history. A year does not pass without severe destruction and loss of lives from a major earthquake. Table 3-1 summarizes the toll of some large shocks.

But these are few compared with the countless smaller tremors reported each day. These earthquakes indicate what a restless and dynamic planet the earth actually is.

Seismology is the science of earthquakes. We get the basic information for this science by measuring ground vibrations with sensitive instruments called **seismographs.** We know that earthquake vibrations are waves spreading away from a place where energy is suddenly released. These are called **seismic waves.** Waves from a strong shock like the 1964 Alaskan earthquake can be detected by seismographs around the world. The energy for most earthquakes is released by the sudden grinding movement between rock masses broken along a fracture. A fracture along which movement has occurred is called a **fault.**

Most earthquakes originate in well-defined regions. We cannot yet predict exactly when and where to expect the next shock. But we can estimate quite accurately the number and size of earthquakes that will occur in some region during a specified interval of time. This information is especially useful for estab-

Table 3-1
Some Destructive Earthquakes During Human History

Date (A.D.)	Location	Approximate Toll in Human Life
7	Hsien, China	Over 100,000
1556	Shensi, China	830,000
November 1755	Lisbon, Portugal	70,000
February 1783	Calabria, Italy	50,000
October 1891	Mino-Owari, Japan	7,000
April 1905	Kangra, India	20,000
December 1908	Messina, Italy	100,000
December 1920	Kansu, China	100,000
September 1923	Kwanto, Japan	27,000
May 1960	Chile	5,700
May 1970	Peru	66,000
December 1972	Managua, Nicaragua	5,000
July 1976	Peking, China	655,000

From Bruce A. Bolt, Earthquakes—A Primer, W. H. Freeman and Co. (1978), Appendix A, Table 1

lishing building codes. In some places law requires that buildings must be constructed to withstand strong shocks.

Earthquake waves convey a wealth of information about the earth's interior. Most of what we know about the deep crust, the mantle, and the core comes from analysis of waves that have passed through these zones. We have learned from theory and experiments how these waves are affected by different materials. Now we can interpret the wave patterns to learn how different materials are distributed in the earth. This knowledge is widely used in the search for oil. Explosives are used to generate waves that are distorted in distinctive ways when they encounter possible oil-bearing structures.

Earthquakes can be devastating events. But we have turned our knowledge of them to advantage in discovering more about how the world is put together.

EARTHQUAKES AND SEISMIC WAVES

Measuring Seismic Waves

When an earthquake occurs, the ground begins to vibrate. We cannot simply watch the earth vibrate from some stationary point far away. We are right there vibrating with it. How can you design a seismograph to record earthquakes while the instrument itself is being shaken about? You can build a seismograph by attaching a mass to a frame in one of the ways pictured in Figure 3-1. When ground vibrations make the frame move, the mass tends to remain stationary, or to move differently because of its inertia. Vertical ground motion can be measured by hanging the mass on a spring. You can use a mass on a beam to detect horizontal motion. The beam attaching the mass to the frame can move sideways. It is inclined slightly so the mass can swing back to the same position after ground motion ceases.

There is a problem with these simple seismographs. The motion between the mass and the frame persists after the ground ceases to vibrate. How can this be controlled? One way is to attach a magnet to the mass so that it moves in a coil fixed to the frame. Then electromagnetic forces set up by the movement of the magnet in the coil oppose this motion. Any movement between the mass and the frame will be halted quickly unless the motion is forced by actual ground vibrations.

Now you need to transcribe the seismograph oscillations on a chart called a **seismogram.** You could simply attach a pen to the mass so that it could mark an advancing chart (Figure 3-1). The difficulty with this method is that we would not see most of the vibrations, so small is the motion between the mass and the frame. Except very near the source of an earthquake, the ground movement is very small. It is important to be able to detect ground movements of less than 1/100 mm.

How can you magnify the seismograph motion? Rather than letting the mass move a pen directly, you can let it generate an electrical signal. This is done by the movement of the magnet in the coil of wire. Whenever a magnet moves in a coil, an electrical current proportional to the motion will be generated in the coil. Then, an electronic amplifier can be used to magnify this current. This magnified current is transmitted to a chart recorder that produces a seismogram (Figure 3-1). The ground vibrations displayed on seismograms have ordinarily been magnified several thousand times. Magnifications of more than one million can be attained.

The seismographs we have described can only respond to movements in some particular vertical or horizontal direction. If you want to measure ground movement completely, you need three seismographs. One detects vertical motion. The others are oriented to detect horizontal motions in, say, the north-south and east-west directions.

We find that a seismograph responds more readily to some kinds of vibration than others

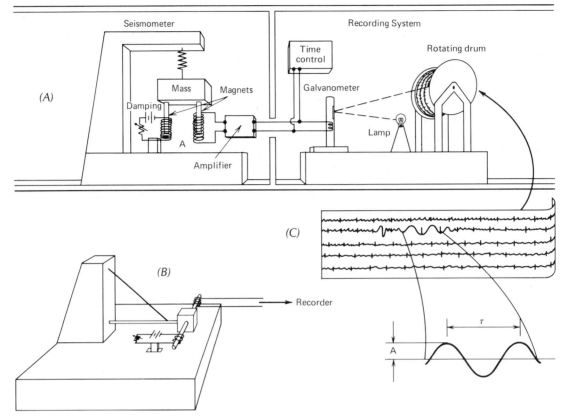

Figure 3-1 Seismographs for recording vertical (A) and horizontal (B) components of ground vibration with an electromechanical and optical system to obtain a seismogram (C) from which the amplitude A and period τ of seismic waves can be measured.

depending on its design. The size of the mass, the length of the beam, and the strength of the spring all affect this response. The important features of a vibration (Figure 3-1) are its frequency (f), its period (τ), and its amplitude (A). The **frequency** is the number of individual oscillations occurring in a given interval of time. In seismology we usually describe frequency as the number of cycles per second (1 cycle/sec = 1 hertz). The period is the time required for one complete oscillation. Therefore, τ = 1/f. The amplitude is the maximum distance that a point on the ground is moved away from its rest position during an oscillation. Usually it takes two or more seismographs of different design to detect all the interesting periods of vibration in one direction. Therefore, seismograph observatories must have at least six instruments to measure the ground motion in the vertical and two horizontal directions.

Much of what we know about earthquake vibrations comes from about 800 permanent seismograph observatories maintained by universities and government agencies of many nations. Here seismographs are in continuous operation. A very important advance in global recording commenced in 1960 when the United States Coast and Geodetic Survey in cooperation with other nations established a worldwide network of 116 observatories equipped with identical instruments. These

Earthquakes and Seismology 79

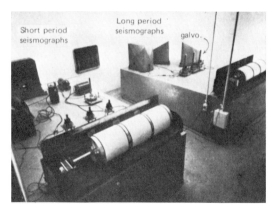

Figure 3-2 A modern seismograph station equipped for continuous recording of vertical, north-south, and east-west components of ground vibration on short period ($1/2 - 2$ sec) and long period (10–100 sec) seismographs. (Earth Sciences Branch, Dept. of Energy, Mines, and Resources, Canada.)

can record vertical and horizontal vibrations in two period ranges: 10 to 100 sec and 0.5 to 2 sec. The interior of one of the observatories is illustrated in Figure 3-2. This group of observatories is called the World Wide Standardized Seismograph Network (WWSSN).

Seismogram Features

Robert Atwood's account of the Alaskan earthquake describes a scene of chaos. Are seismic waves seen on seismograms equally chaotic? Or can we discern patterns of regularity? Seismograms reveal that a sudden disturbance in the earth causes ground vibrations consisting of a series of pulses. Three distinctive groups of pulses are found (Figure 3-3). These are the *P*- or **primary** pulses followed by the *S*- or **secondary** pulses, and then the *L*- or **long** oscillations. Whether the ground motion persists only a few seconds or for several hours, these three groups can always be recognized.

Seismic waves originate at a place called the **focus.** Here energy is released. It radiates outward from the focus causing wave vibrations. These waves travel progressively farther away with increasing time. When they reach a seismograph station, they cause ground motions (Figure 3-3) that can be reconstructed from the vertical and horizontal component seismograms. Ground movement related to P-wave pulses is always in line with the direction that the waves are traveling. Motion caused by S-wave pulses is perpendicular to the direction of travel. The L-waves cause a point on the ground to move in a complexly distorted elliptical path.

Location of Earthquakes

With sudden abruptness the ground heaved and shook and dashed thousands of tons of rock on the quiet campground in southwestern Montana. It was late in the evening of August 17, 1959 when this unexpected earthquake wrought its havoc claiming several human lives. With similar lack of warning the giant earthquake of March 27, 1964 spread devastation along the quiet coastal region of southern Alaska. Can we tell where such shocks might occur next? We can begin to answer this question by finding out where they have occurred in the past.

More than 2000 earthquakes occur every day. Most are too small to feel, or they originate in remote areas. They are detected at seismograph observatories; otherwise we would be unaware of them. Even though these small or remote shocks cause no damage, we are interested in locating them. When the locations of these earthquakes are displayed on maps and cross sections, we can identify distinctive zones where most of them originate. Most destructive quakes also occur in these zones.

The place where an earthquake originates is called the **focus.** We can describe its location by giving the position of a point on the

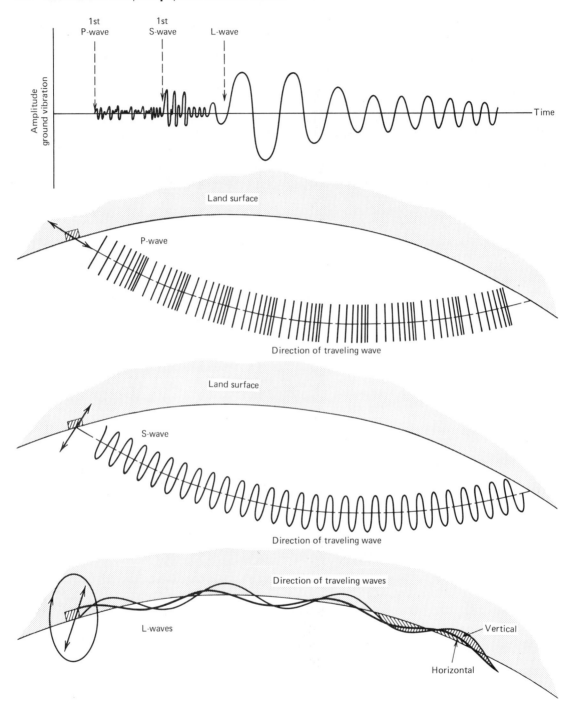

Figure 3-3 Representative seismogram illustrating Primary *(P)* and Secondary *(S)* wave pulses, and a series of Long *(L)* wave oscillations, and diagrams of ground motion associated with these different kinds of seismic waves.

surface directly above, and its depth beneath this point. The surface point is called the **epicenter.** We can find the epicenter and depth to the focus from seismic waves recorded at distant observatories.

Seismologists have devised two methods for locating earthquake epicenters. To use either method we need to know the time required for seismic waves to reach points on the earth's surface at different distances from the epicenter. This information is given on **travel-time curves.** How is a travel-time curve prepared? Suppose you knew the exact location of an earthquake, and the time (t_o) it originated. Then you could find from a seismogram recorded at a known distance (D) from the source, the time (t_p) when the first P-wave pulse arrived there. By subtracting the origin time (t_o) from the arrival time (t_p) you get the travel time (T_p) of a P-wave at the distance D. Travel times can be obtained in this way from the first P-wave pulses detected at many other distances. You could plot these travel times on a graph of time versus distance, and then draw a line through the points to get a P-wave travel time curve. An S-wave travel time curve could be compiled in the same way.

Basically, this is the way seismologists prepared P- and S-wave travel time curves for the earth (Figure 3-4). They started with seismograms from populous areas such as Cali-

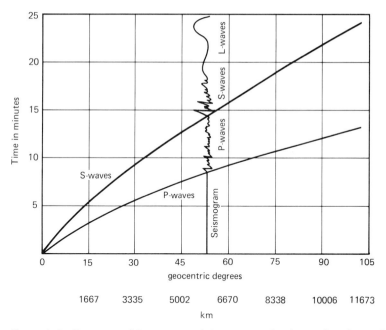

Figure 3-4 P-wave and S-wave travel-time curves for the earth indicate the times required for a P-wave pulse and an S-wave pulse to travel from an earthquake focus to points on the earth's surface situated at different distances from the epicenter. For example, a seismograph situated 6000 km from an earthquake epicenter would detect the first P-wave pulse about 8 minutes after energy was released from the focus, and the first S-wave would arrive about 14 minutes after the time of release of energy from the focus. A seismogram is a record of the variation of ground motion with time. By superposing on the travel time curves a seismogram with the same time scale that was recorded 6000 km from an epicenter, the P-wave pulse is seen to arrive at 8 minutes, and the S-wave pulse is seen to arrive at 14 minutes after the origin time of the earthquake.

fornia and Japan where earthquakes are frequent, and numerous seismographs are operating continuously. Often an earthquake occurred so close to one of these instruments that its location and origin time could be accurately estimated. Waves recorded from these known sources were used to prepare preliminary travel time curves. These were steadily improved by using data from more and more earthquakes. In recent decades seismologists have improved the global travel time curves by recording large nuclear explosions. The location and origin time of such an explosion are known precisely. It generates seismic waves that can be detected around the world. The travel-time curves in Figure 3-4 represent the global average based on information from a wide distribution of earthquakes and nuclear explosions. We see that P-waves will arrive at a place 10,000 km (6000 mi) from the epicenter in about 12 minutes. S-waves arrive $10^1/_2$ minutes later.

Now, how can you use travel-time curves to locate earthquakes that did not originate right under a seismograph? One way is the **S-P epicenter determination method.** This method is illustrated in Figure 3-5. You begin by identifying the first P- and S-wave pulses on seismograms from at least three different locations. Then, find how much later the S-wave comes in after the P-wave. In the example you can see that this time difference is 3.5 minutes at St. Louis. Now, you can see from Figure 3-4 that the time between the arrival of a P-wave and the later arrival of the S-wave is given by the vertical separation of the two curves. Therefore, the distance of the epicenter from a seismograph can be determined from the travel-time curves. To do this you find the distance where the time interval between the two curves is the same as the value measured on the seismogram. Then, taking the seismograph station as the center, draw a circle with this distance as radius on a map. The epicenter should lie somewhere on this circle. By constructing similar circles for other seismograph stations where the quake was recorded, an intersection point can be found that indicates the epicenter location. Usually, a perfect intersection is not found. This is because you have used worldwide average travel-time curves where local travel times may be slightly different. Still more important may be the fact that the earthquake did not occur at the surface, but at some point below.

Ordinarily the time of the first P-wave pulse can be accurately measured on a seismogram. The exact beginning of the S-wave is harder to recognize because it is partly hidden by later P-wave pulses. The **P-O epicenter determination method** uses only the arrival time of the first observable P-wave. This is a method of successive approximations. To begin, you choose an estimate of the epicenter, perhaps by the S-P method. The distances from this position to the various seismograph stations can then be measured. Next, you find P-wave travel times to the seismographs using the travel-time curve (Figure 3-4). By subtracting these predicted travel times from the observed arrival times you can calculate the time of origin of the earthquake. Obviously, the values of origin time calculated from different seismograms must agree, if the estimated epicenter position is correct. Usually the values are different for the first try, and you must test another estimated position. Eventually a location is found which yields a minimum discrepancy in calculated values of origin time. This can often be achieved with relatively few trials. By employing methods of mathematical iteration the computations can be rapidly carried out on an electronic computer. The P-O method is the most precise way to locate an epicenter. Commonly, earthquake epicenters can be determined within 1 or 2 km.

The focal depths of earthquakes are quite difficult to measure unless certain seismic pulses can be clearly recognized. These pulses indicate P-waves that have followed different paths through the earth (Figure 3-5). Later we will talk more about travel paths

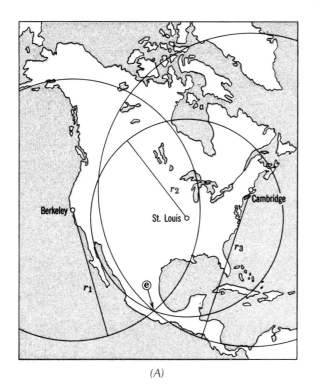

Figure 3-5 An epicenter can be located by finding its distance from each of three seismograph stations. (A) Choosing these distances as radii of circles about the stations, the epicenter lies where these circles intersect. (B) Time differences between arrival of P- and S-waves (S-P time) can be used to find distances on travel-time graphs where P-wave and S-wave curves are separated by the measured amounts. P-wave arrival times alone can be used with estimated epicenter locations and a P-wave travel-time graph in the P-O method to find which location gives the same origin time for data from three or more stations. (C) Focal depths are estimated from differences in arrival times of the phases illustrated on the earth cross section. (Map reprinted from *Physical Geology* by C. R. Longwell, R. F. Flint, and J. E. Sanders, copyright John Wiley & Sons, Inc., 1969.)

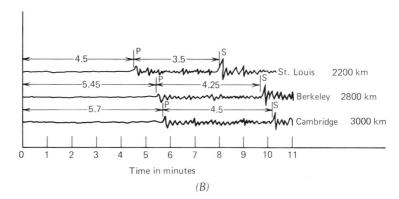

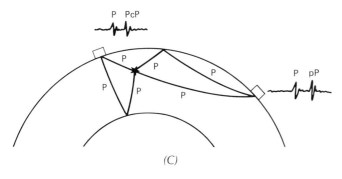

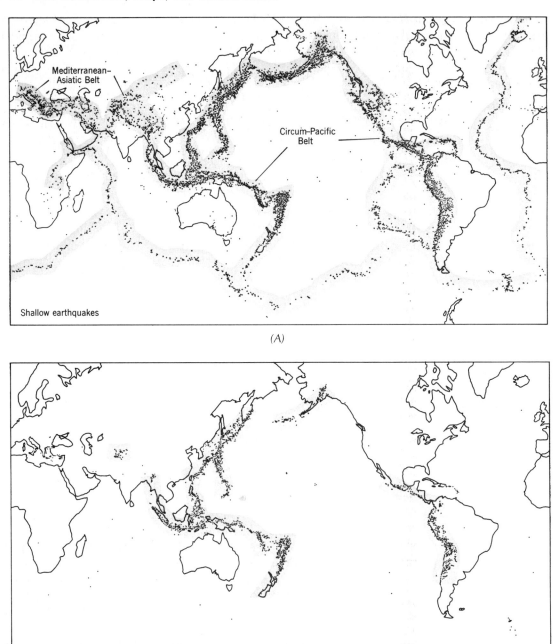

Figure 3-6 Epicenters for earthquakes detected between 1961–1967 at focal depths *(A)* less than and *(B)* greater than 100 km. (From M. Barazangi and M. J. Dorman, *Bulletin of the Seismological Society of America,* Plates 2 and 3, Feb. 1969.)

through the earth, and wave velocities along these paths. For shallow earthquakes occurring less than 70 km beneath the surface, these pulses follow very closely upon each other, and onset times cannot be measured accurately. Because of errors in these travel times, the uncertainty in focal depth calculations may easily be ±15 km or more. For deeper earthquakes this uncertainty becomes less important because it is a smaller part of the total depth.

What have we learned about earthquake zones from epicenter and focal depth measurements? Every year almost 850,000 earthquakes are detected. Seismologists plot their epicenters on charts called **seismicity maps.** The term **seismicity** implies earthquake activity. Typical of these is the world seismicity map (Figure 3-6) showing epicenters for the period of 1961 to 1967. Most earthquakes are seen to occur in well-defined linear zones of seismicity. You can see that earthquake-prone areas such as California and Japan are bordered by zones of intense seismicity. The Pacific Ocean is almost surrounded by these zones. Now, you should compare the world seismicity map (Figure 3-6) with the worldwide topographic map (Figure 2-9). You can see that the principal zones of seismicity are situated beneath regions of the most extreme topographic relief: the submarine trenches, the oceanic ridges, and some continental mountain ranges. This would seem to imply that the earth's topography is affected by the same internal processes that cause earthquakes.

What about the depth of earthquakes? Most shocks originate within 70 km of the surface. These are defined as **shallow** earthquakes. But many deeper earthquakes have been recorded. Foci of the deepest shocks are about 700 km below the surface. Those occurring in the 300 to 700 km range are identified as **deep focus** earthquakes. **Intermediate** earthquake foci are found between 70 km and 300 km deep. Seismologists have found that all intermediate and deep focus quakes originate

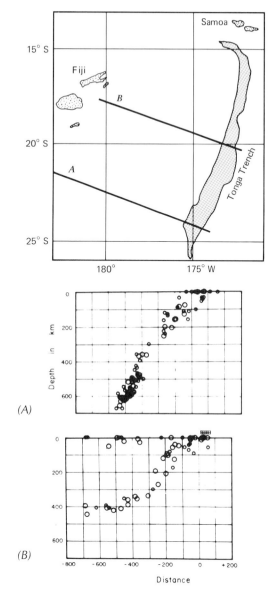

Figure 3-7 Depths of earthquake foci situated near two profiles (A and B) extending from the Tonga Trench to the Fiji Islands in the western Pacific Ocean. (From L. R. Sykes, *Journal of Geophysical Research,* p. 2986 and 2994, June 15, 1966.)

in clearly defined zones parallel to submarine trenches. Cross sections in Figure 3-7 show the typical depth distribution of earthquakes.

These zones where intermediate and deep focus shocks occur are called **inclined seismic zones.** They extend as deep as 700 km into the earth.

Most destructive earthquakes occur in the well-defined zones you can see on the seismicity map. But there are some mavericks. One was the earthquake that destroyed Charleston, South Carolina, in 1886. Seismologists are still baffled about the cause of this shock which occurred unexpectedly in a region far from the zones of intense seismicity.

Earthquake Magnitude and Energy

A major earthquake is a newsworthy event. News commentators described the devastation of Managua, Nicaragua, in 1972. They reported almost 10,000 persons perished in an earthquake that registered 6.8 on the Richter magnitude scale. Some reporters compared this event with the 1964 Alaskan earthquake that cost fewer lives while registering 8.4 on the Richter magnitude scale. What does the Richter magnitude of an earthquake imply, and what does this number indicate about the size of a shock?

Seismologists will never agree completely on the best way to measure the size of an earthquake. But for practical reasons they have accepted a scheme based on the research of Charles F. Richter. He believed that the acceleration of the ground could be measured at a standard distance from the focus to find the size of an earthquake. In 1935 Professor Richter proposed that a particular kind of seismograph, the Wood-Anderson torsion seismograph, be used as the standard instrument for comparing different earthquakes. This device measures ground acceleration directly.

Originally, Richter compared the maximum amplitude (A) on a seismogram recorded 100 km from an earthquake with the corresponding amplitude (A_0) of an idealized standard earthquake. He then calculated a number he called the magnitude (M) from the formula $M = \log (A/A_0)$. The value of A_0 was chosen so that all detectable earthquakes could be described by magnitude numbers between 0 and 10. This is the range of the **Richter magnitude scale.**

Richter found magnitudes for many earthquakes that originated 100 km from one or another of a network of Wood-Anderson seismographs in California. Because each of these shocks was also recorded at other distances, the change in amplitude with distance could be measured. This information was eventually used to modify Richter's original formula. Then the magnitude could be obtained from Wood-Anderson seismogram amplitudes recorded at different distances from the modified formula:

$$M = \log (A/A_0) + 3 \log \Delta - 3.37$$

The **geocentric distance** Δ is the angle between lines reaching from the focus and the seismograph to the earth's center. Seismologists often use this angle rather than distance along the earth's surface.

Wood-Anderson seismographs are no longer the only instruments used to measure magnitude. To use other seismographs you must apply corrections to convert the measurements into amplitudes that are equivalent to Wood-Anderson seismograph results. By common agreement seismologists continue to use Richter's original definition of magnitude as the standard for making these corrections. Ideally, they hope to get the same magnitude number using any seismograph at any distance that Richter would have measured in 1935 with a Wood-Anderson seismograph 100 km away.

Richter magnitudes are simply numbers that seismologists have agreed to use for comparing different earthquakes. Can they state more specifically how these numbers indicate the size of earthquakes? Richter and his colleague, the well-known seismologist Beno

Gutenberg, made theoretical calculations of ground accelerations that would result from releasing energy at points in idealized models of the earth. They compared the theoretical patterns with the natural patterns of amplitude variation observed for a large number of earthquakes. Using these comparisons, they proposed that the energy (E) released at the focus could be estimated from the magnitude (M) according to the formula:

$$\log_{10} E \simeq K + 1.5 M$$

The value of K is between 9 and 12. This formula tells us that the energy release is about 32 times larger for each step of 1 in the Richter magnitude scale. An M7 shock releases 32 times the energy of an M6 shock, and about 1000 times the energy of an M5 shock.

Magnitudes of about 8.5 have been recorded for the largest earthquakes. This indicates an energy release of about 10^{25} ergs, which is equivalent to exploding 31 million tons of TNT! From what we know about the strength of rocks, it is unlikely that an earthquake could ever be larger than M9. Large nuclear explosions have reached magnitudes of about 7. The numbers of earthquakes in different magnitude ranges that occur each year are given in Table 3-2. You can see that over 75 percent of the total energy released comes from the relatively few shocks of magnitude 7.5 or larger. Of the nearly 850,000 quakes that occur each year less than 200 cause any appreciable damage.

Table 3-2
Annual Occurrence of Earthquakes

Magnitude	Number of Earthquakes
8	2
7	20
6	100
5	3,000
4	15,000
3	More than 100,000

From B. A. Bolt, Earthquakes—A Primer, W. H. Freeman and Co. (1978), Appendix A, Table 2.

Earthquake Intensity

The overwhelming nature of the Alaskan earthquake was vividly described by Robert Atwood. Many accounts such as his help us to picture the intensity of earthquake vibrations. Earthquake **intensity** is a measure of the vigor and violence of ground movement.

The conventional ways for estimating earthquake intensity have been based upon levels of human perception and damage. Of the several schemes that have been proposed, the modified Mercalli scale presented in Table 3-3 is now most widely used. It specifies 12 levels of intensity. Information obtained from newspaper reports and interviews with numerous individuals can be used by a seismologist to assign Mercalli intensity values to many places in the vicinity of an earthquake. Variations in intensity are then displayed on **isoseismal** maps (Figure 3-8) by contour lines drawn through points of equal intensity. Old newspaper accounts and historical records have been used to document earthquakes that occurred before seismographs were developed.

Earthquake intensity depends upon the kinds of bedrock and regolith at a location, and the amount of energy received there. The level of energy tends to diminish with distance from the focus. However, the same amount of energy that would cause water-saturated soil and regolith to shake like a bowl of jelly might cause relatively small vibrations in firm bedrock. Geological differences such as these account for the irregularity of isoseismal contours in Figure 3-8. Gutenberg and Richter proposed the following relationship between magnitude (M) and the maximum intensity (I_m) very close to an earthquake epicenter:

$$M \simeq 1.3 + 0.6 \, I_m \qquad (3\text{-}17)$$

Other seismologists have studied relationships between intensity and the amplitudes and periods of actual ground motion. Their

Table 3-3
Modified Mercalli Earthquake Intensity Scale[a]

- I. Not generally felt.
- II. Felt by persons resting, particularly on upper floors of buildings.
- III. Felt indoors, especially on upper floors of buildings. May not be recognized as an earthquake. Vibrations resemble those of a passing light truck.
- IV. Felt by some outdoors, by many indoors. Vibrations resemble those of a passing heavy truck, or a sensation of a heavy object striking the walls. Windows, dishes, and doors rattle or make creaking sounds. Hanging objects swing and stationary cars rock noticeably.
- V. Felt outdoors; sleepers wakened. Liquids disturbed, some spilled. Small unstable objects displaced or upset. Doors swing, close, open. Shutters and pictures move, and pendulum clocks stop, start, or change rate.
- VI. Felt by all, many frightened and run outdoors. People walk unsteadily. Windows, dishes, glassware broken. Objects fall off shelves and pictures fall off walls. Furniture moved or overturned. Weak plaster and poor masonry cracked. Small bells rign, and trees or bushes shaken visibly or are heard to rustle.
- VII. Difficult to stand. Felt by drivers of cars. Hanging objects quiver. Furniture broken. Poor masonry damaged; fall of plaster, loose bricks, stones, tiles, cornices, and architectural ornaments. Weak chimneys broken at roof line. Waves on ponds; water turbid with mud. Sand and gravel banks cave or have small slides. Concrete irrigation ditches damaged. Large bells ring.
- VIII. Difficult to steer motor cars. Ordinary unbraced masonry damaged or partially collapsed; some damage to reinforced masonry but no damage to masonry reinforced against horizontal displacement. Stucco and some masonry walls fall. Twisting or fall of chimneys, factory stacks, monuments, statues, towers, or elevated tanks. Frame houses move on foundations if not bolted in place, and loose panel walls dislocated. Weak piling broken. Branches broken from trees. Springs change flow, wells change level, and temperatures in both may change. Cracking in wet ground and on steep slopes.
- IX. General panic. Poor masonry destroyed, good masonry damaged seriously. Frame structures, not bolted, shift off foundations; foundations generally damaged. Frames crack. Reservoirs seriously damaged. Conspicuous cracks in ground; underground pipes broken. In areas of loose sediment, sand and mud, and water ejected.
- X. Most masonry and frame structures and foundations destroyed. Some well-built wooden structures and bridges destroyed. Serious damage to dams, dikes, embankments. Large landslides occur. Water thrown on banks of canals, rivers, lakes, and reservoirs. Flat areas of sand and mud shifted horizontally. Railroad tracks bent slightly.
- XI. Few, if any masonry structures remain standing. Railroad tracks bent severely, underground pipes completely out of service, many bridges destroyed.
- XII. Damage to man-made structures nearly total. Large rock masses displaced, lines of sight distorted, objects thrown into air.

[a]Modified from Tocher, Don, 1964, Earthquakes and rating scales, *Geotimes*, v. 8, no. 8, p. 19: and Wood, H. O. and F. Neuman, 1931, Modified Mercalli intensity scale of 1931, *Bull: Seismological Society of America*, v. 21, pp. 277-283.

results indicate that vertical ground displacements may be more than 35 cm for earthquake intensities of X or more. Where intensity is smaller than V, vertical ground displacement will be less than 3 cm. When you are in a boat, 35 cm (1 ft) waves on the water are no cause for concern. Waves of this size in the solid earth are very destructive.

Earthquakes and Seismology 89

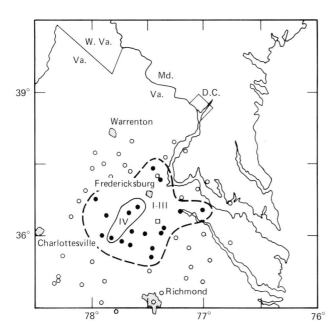

Figure 3-8 Isoseismal map of Mercalli intensities reported for the Fredricksburg, Virginia, earthquake of September 12, 1971. (From G. A. Bollinger, *Earthquake Notes*, page 32, Sept.–Dec. 1971).

Earthquakes and Faults

The earth is broken by fractures where rock masses move and grind together from time to time. Such fractures, as we know, are called faults. Some faults break the land surface, and others are hidden deeper in the crust and upper mantle. The San Andreas fault in southern California (Figure 3-9) can be easily recognized from offset topographic features. This fault can be traced from the Gulf of California northwestward across the state to the Pacific coast near San Francisco.

For many decades surveyors have repeatedly measured the positions of points on both sides of the San Andreas Fault. They have used spirit leveling methods, and more recently laser distance measuring devices and tilt meters that indicate changes in the slope of the land. We have learned from these surveys that a thin sliver of North America consisting of southwestern California and the Baja Peninsula is shifting northwestward at an average rate of 6 cm/yr.

Movement on large faults such as the San Andreas fault is not steady. Rather, it occurs in bumps and jerks (Figure 3-10). First the rock bends and stretches, but movement is resisted by friction along the rough surface of the fault. The breaking point is reached at one place and then another. Over a period of years or centuries these local movements spread along the entire length of the fault.

An earthquake occurs each time there is local movement on a fault (Figure 3-10). Sudden displacement of about 2 m on the San Andreas fault occurred near San Francisco causing the disastrous earthquake that virtually destroyed the city in 1906. Here we have direct evidence of the association of fault movement and earthquakes. Numerous smaller quakes called **aftershocks** commonly follow for a few days or weeks after a large quake. These occur as the two sides of the fault continue grinding together while adjusting to the original large offset.

We can correlate some earthquakes directly with faulting at the earth's surface. Many more come from deep sources. To find out if hidden faulting caused these quakes we must examine their seismic waves. Seismologists have discovered that P-waves traveling

90 The Earth's Size, Shape, and Interior Zones

Figure 3-9 Topographic indication of part of the San Andreas fault in southern California. (R. E. Wallace, U.S. Geological Survey.)

in some directions first push the ground away from the source. P-waves moving in other directions initially draw the ground toward the source. You can find the initial ground movement at an observatory by looking at the seismograms to see if the first P-wave pulse begins upward or downward.

How can you discover from ground movement in different places if hidden faulting might have caused an earthquake? Professor Hugo Benioff explains it in this way (Figure 3-11). The line X-Y indicates a fault. Points A, B, C, and D define a rectangle across the fault. As the rock begins to bend and stretch,

Earthquakes and Seismology 91

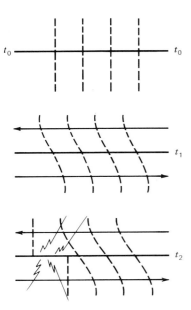

Figure 3-10 These diagrams illustrate the buildup of stress along a fault, and subsequent release of stress by sudden movement along different parts of the fault at different times. The solid line represents the fault. Dashed reference lines indicate the extent to which rock bordering the fault is being strained by a build up of stress. At time t_0 there is no stress along the fault and the dashed lines are straight. At time t_1 arrows show stresses that cause the rock to stretch and bend as indicated by the curved dashed lines. At time t_2 fault movement occurs suddenly at the left-hand end of the fault. This releases the stress as shown by the dashed lines here which are straight again. At time t_3 stress is released by movement in the central part of the fault.

Figure 3-11 Diagrams indicating conditions of strain, fault movement, and directions of first movement of ground vibrations produced by seismic waves generated by fault movement. At a time marked (1) there is no stress along the fault, which is indicated by the line X-Y. At a later time marked (2) stress buildup stretches the rock in some directions (X-B_1 and C_1-Y) and compresses it in other directions (A_1-Y and X-D_1). Fault movement at a time marked (3) releases the stress so that these lines in the rock return to their original lengths. This generates seismic waves that produce ground movement in different initial directions depending upon the path of the wave away from the fault. (Reprinted from Hugo Benioff, *Science*, page 1402, March 1964.)

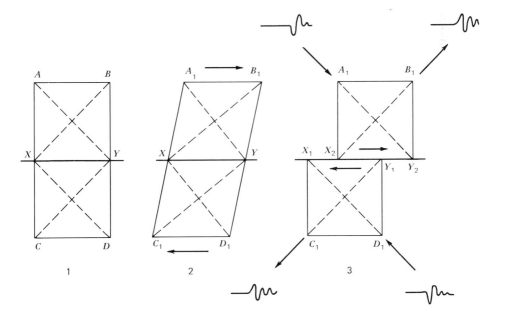

line A-B shifts to A_1-B_1 and line C-D shifts to C_1-D_1. The rectangle is now deformed into the parallelogram $A_1B_1C_1D_1$. The fault line X-Y still remains in the same position. Deforming the rectangle causes the diagonals X-B and Y-C to become extended, while Y-A and X-D are compressed. Then fault movement occurs suddenly. During fault movement you can see that the lines Y-A_1 and X-D_1 are stretched to lengths of Y_2-A_1 and X_1-D_1, and therefore pull the ground inward toward the fault. P-waves that travel outward from these lines first draw ground toward the source. At the same time lines X-B_1 and Y-C_1 are compressed to lengths of X_2-B_1 and Y_1-C_1. The ground is initially pushed outward along these lines. The P-waves in directions of X_2-B_1 and Y_1-C_1 first push the ground away from the source. Because most earthquakes cause this pattern of ground movement, we believe that they result from faulting. By finding the directions of first motion indicated by P-wave pulses from many seismograms surrounding an epicenter, you should be able to find the orientation of a fault surface, and the direction of movement along this surface.

A nuclear explosion does not produce this pattern of movement. Instead, the ground is pushed outward in all directions from the point of detonation. Scientists once thought that nuclear explosions could be distinguished from natural earthquakes by their patterns of ground movement. Unfortunately this pattern is often badly distorted by irregularities in the earth. Thus this is not a reliable means for recognizing explosions.

Earthquake Prediction and Control

The few large earthquakes claim an annual toll of about 10,000 lives. This number is small compared with the loss of life through accidents, war, and other causes over which we have some control. Nevertheless, a large earthquake is one of nature's most forceful phenomena. Fear of such shocks cannot be

Figure 3-12 Earthquake damage to old part of the Veterans Administration Hospital was very severe compared with new earthquake-resistant parts caused by the 1971 San Fernando, California, shock. (Copyright © 1971, *Los Angeles Times*. Reprinted by permission.)

overestimated. Consequently, seismologists are asked to find out about earthquake prediction.

So long as we continue to live and build in earthquake-prone areas we cannot hope to escape damage completely. But lives could certainly be protected by reliable prediction. Structural damage can be reduced by proper construction of buildings in seismically active regions. This is dramatically illustrated by the damage to the Veterans Hospital in San Fernando, California, caused by the 1971 earthquake (Figure 3-12). The older part of the building was destroyed, but newer shock-resistant sections sustained only minor damage.

Seismologists have prepared earthquake risk maps (Figure 3-13) as a first step in predicting seismicity. The probability of a damaging earthquake is estimated for different areas on the map from statistical analysis of previous seismicity. But these maps do not indicate exactly where or when any single shock will occur. Seismic risk maps are used

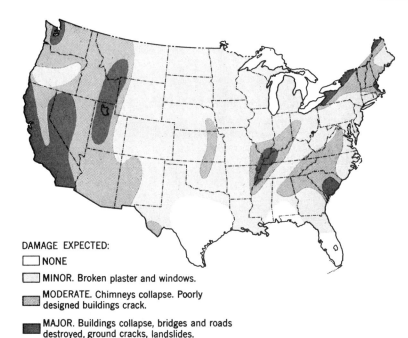

Figure 3-13 This seismic risk map of the United States indicates a statistical estimate of the maximum intensity that could be expected from earthquakes occurring in different regions. This map provides no information about when such quakes might occur. (After S. T. Algermissen, Fourth World Conference on Earthquake Engineering, *Proceedings*, Vol. 1, pp. 14–27, 1969.)

DAMAGE EXPECTED:
- NONE
- MINOR. Broken plaster and windows.
- MODERATE. Chimneys collapse. Poorly designed buildings crack.
- MAJOR. Buildings collapse, bridges and roads destroyed, ground cracks, landslides.

by agencies that recommend design regulations and building codes. Because of this, there can be bitter argument about where the boundaries between high and low risk areas should be located. It is difficult to say exactly where to place a boundary, especially in areas of moderate or low seismicity. However, building regulations may dictate double or triple the cost for a nuclear power plant depending on which side of a boundary it is to be built.

The eventual aim of seismologists is to predict the time and place of an earthquake from events that occur weeks, days, or hours in advance. Several kinds of measurements may be potentially useful for prediction.

1. **Triangulation.** A network of points crossing an active fault is precisely surveyed. The network is periodically resurveyed to detect patterns of movement caused by bending and stretching of the ground. In some places distinctive patterns of movement are known to precede faulting.

2. **Tiltmeter measurements.** The slope of the land surface near a fault is recorded continuously by tiltmeters. Distinctive changes in slope have been found to precede earthquakes at some locations.

3. **Microearthquake recording.** Shocks of Richter magnitude lower than 1 are called microearthquakes. Seismographs close to active faults detect many of these events each day. In some places distinctive changes in the occurrence of microearthquakes precede a large shock.

4. **Water level measurements.** Tide gauges and recorders in wells near active faults sometimes indicate peculiar changes in water level shortly before faulting.

5. **Strain meter measurements.** Extension or compression of the ground is measured by strain meters. These devices measure continuously the distance between two points. Characteristic ground distortion can sometimes be recognized before the onset of a major quake.

6. **Magnetic field changes.** Rocks near the surface distort the earth's magnetic field. When these rocks are bending just prior to faulting, the magnetic field distortion may change. These changes can be detected by sensitive magnetometers that measure the magnetic field strength continuously.

7. **Animal behavior.** Cattle in herds grazing near active faults sometimes become uneasy and skittish shortly before an earthquake. Apparently preliminary ground tremors increase their nervousness.

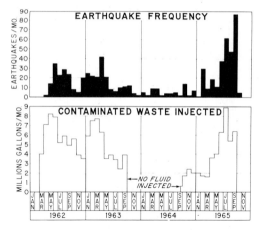

Figure 3-14 Correlation of earthquake occurrence and fluid injection in a 3670 m deep well near Denver, Colorado. (Reprinted from D. M. Evans, *The Mountain Geologist,* page 27, Jan. 1966.)

Seismologists have found no single measurement that reliably indicates the onset of an earthquake. The hope for prediction lies in using combinations of measurements. These may reveal patterns of change before sudden movement on a particular fault. Each fault has its own personality, so prediction patterns for one cannot be used for another. Nevertheless, the patterns that are recognized encourage us to believe that prediction schemes will eventually be developed.

Recently, seismologists have looked into the question of controlling earthquakes. Some ideas about this came from studies of earthquakes that began to occur with alarming regularity near Denver, Colorado, shortly after 1960. After a few years a consulting geologist got access to data on fluid waste disposal in a nearby well 3670 m (12,000 ft) deep. He found a clear correlation between earthquake occurrence and pumping rates (Figure 3-14). Later studies showed beyond any doubt that injection of fluid in the well contributed to the generation of the earthquakes.

Studies of the Denver earthquake magnitudes revealed that much more energy was released from the earth than was put in through pumping. This suggested prior ground distortion that was not quite sufficient to cause faulting. Injection of fluid under pressure provided the additional perturbation needed for fault movement.

Generation of the Denver earthquakes suggests the possibility of dissipating the ground distortion along a fault. If we periodically injected it with water, small magnitude earthquakes could be generated that would cause little or no damage. Otherwise distortion would continue to build up until a large destructive quake occurred. But we can also argue that in an unknown region a large amount of distortion may already exist. In that case fluid injection could trigger a much larger shock than expected. These questions have not been resolved.

SEISMIC WAVES AND EARTH STRUCTURE

Early in the twentieth century many wondrous devices were used to search for oil. They were operated by people called "doodlebuggers," who claimed that the blinking lights and peculiar sounds of their machines were sure indications of oil hidden underground. Of the many doodlebugging schemes, a few were based on scientific principles. Gravity surveying was one of these. Another was called seismic surveying. No one claimed to find oil directly by these methods. Rather, they used seismic and gravity surveying to find the structures most likely to contain oil.

About 1920 oil prospectors began experimenting with seismic waves. Seismologists already knew how to use earthquake waves to study the earth's deep interior. They had discovered the crust, mantle, and core from analysis of seismic waves. Oil prospectors found that these principles could be applied in small-scale experiments. They detonated small explosions to generate seismic waves that echoed from buried rock strata, and were detected with portable seismographs. Features such as salt domes and folded and faulted strata could be found by these echoes. Since 1920, seismic surveying has become the most widely used method of oil prospecting. People doing this work are still called doodlebuggers.

Seismic waves are our best probes of the earth's interior. Oil prospecting is only one of their many uses. Seismic waves have conveyed to us most of what we know about the deeper parts of the crust, the mantle, and the core.

Elasticity

What can seismic waves tell us about the rocks in the earth? We can begin to find out by studying the effects of these waves on different materials. Seismic waves cause temporary distortion of the ground. We define this distortion as **strain.** It is caused by stress in the ground, which implies that stress is associated with seismic waves. We define this **stress** as a force in the ground divided by the area on which it acts (stress = force/area). A substance that is stretched out of shape by stress, but then returns to its original form when stress is removed, is said to have undergone **elastic** strain. One must know about this particular kind of strain to interpret seismic waves.

You can test a substance in several ways to find its elastic properties. In one test the volume (V) of a specimen is reduced by ΔV (Figure 3-15A) when it is squeezed by fluid under pressure ΔP. Here the elastic strain is given by $\Delta V/V$. It is related to the stress (ΔP) by a constant of proportionality (κ) defined as the **bulk modulus:** $\kappa = \Delta P/(\Delta V/V)$. In another experiment we can apply a stress (ΔT) along a surface of a rectangular specimen. This deforms its shape from a rectangle to a parallelogram (Figure 3-15B). Elastic strain is indicated by the angle $\Delta \phi$. It is related to the stress by a constant (μ) called the **shear modulus:** $\mu = \Delta T/\Delta \phi$. The bulk modulus ($\kappa$) and the shear modulus (μ) are called **elasticity coefficients.** They can be measured experimentally, and help to describe the physical properties of a substance. Different kinds of rock have different elasticity coefficients. What do elasticity coefficients have to do with seismic waves? You can find out by measuring the velocities at which seismic waves travel through rock specimens. You would discover that the velocity (V_p) of P-waves, and the velocity (V_s) of S-waves are related to the elasticity coefficients and the density (ρ) of the rock specimen in the following way:

$$V_p = \sqrt{(\kappa + 4/3\mu)/\rho}; \quad V_s = \sqrt{\mu/\rho}$$

It is an important fact that the shear modulus $\mu = 0$ in fluids. This implies that S-waves cannot travel in liquids or gases, but only in solids. Also, we have learned that P-waves always travel faster than S-waves.

Seismic wave velocities, elasticity coefficients, and density are interdependent physical properties of a substance. If we can devise ways to measure the velocities at which P- and S-waves travel in different parts of the earth, we could learn about the elasticity coefficients and density in its interior.

96   The Earth's Size, Shape, and Interior Zones

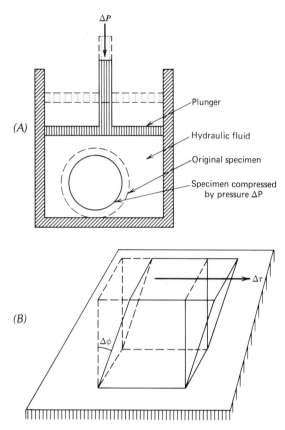

Figure 3-15   Relationships between *(A)* volume strain $\Delta V/V$ caused by change in fluid pressure ΔP on the spherical surface of a specimen, and *(B)* the shear strain $\Delta \phi$ caused by shear stress $\Delta \tau$ exerted on the surface of a specimen.

Refraction and Reflection of Seismic Waves

There are many layers of rock in the earth. How can we distinguish one from another by seismic waves that have traveled through all of them? Before we can understand how this is done, we must see how the seismic waves travel through these layers.

At the source of an earthquake P- and S-waves travel outward in all directions. To simplify things let us begin by looking only at the P-waves. There are many paths that P-waves can follow from the source to a seismograph. The rays in Figure 3-16 show some of these paths.

You can see that a path changes direction at each place it crosses from one layer to another (Figure 3-16A). This effect is called **refraction**. It is like the refraction of light rays which causes a pole sticking up from a clear pool to appear bent at the water surface. We know that the change in direction is related to the wave velocities above and below the boundary. To better understand this, look at the path *ABC* in Figure 3-16. A wave travels toward the boundary at an angle ϑ. There it is refracted, and travels away from the boundary at the angle ϕ. These angles and the velocities V_l and V_j in the upper and lower layers are related by **Snell's law:**

$$\frac{\cos \vartheta}{\cos \phi} = \frac{V_j}{V_l}$$

Each path in Figure 3-16A becomes inclined at smaller angles in deeper layers. Therefore, we can deduce from Snell's law that the velocity must increase in the deeper layers of this particular example.

Eventually, the path is refracted so that the wave travels horizontally along the top of one layer. This is seen at point C in Figure 3-16A. Here we say that the wave has been **critically** refracted. As it follows this part of the path, it continually generates new waves that return upward. These new waves follow paths that are refracted at the same angles as the downgoing path.

Now you can see that P-waves refracted along several paths can be detected at one place. Each of these waves is indicated by a pulse on the seismogram. But these are not the only waves to reach the seismograph. At each place where a wave crosses a boundary some energy is reflected. The reflected energy causes waves to echo back along paths illustrated in Figure 3-16B. The pulses from reflected waves appear mixed with refracted wave pulses on a seismogram. But the seismograph records still more pulses. Each P-wave also generates refracted and reflected S-waves at every boundary it crosses. Similarly, each S-wave generates refracted and reflected P-waves when it crosses a boundary. The directions of these waves are all governed by Snell's law (Figure 3-16C).

You can see that a seismogram is a complicated mixture of pulses that indicates waves from many paths. How can these many waves be sorted out to learn about rock layers in the earth? This is done by preparing travel time curves using seismograms recorded at several distances. Seismograms in Figure 3-16 are drawn on the travel time curves. You can see that the refracted pulses are aligned on one straight line or another. The reflected pulses lie along curved lines that approximate mathematical curves called hyperbolae. Each straight line represents waves that have been critically refracted along the top of one of the layers. Each curved line represents waves reflected from the top of one of the layers.

What information about the earth's interior can you get from a travel time curve? From each straight line (Figure 3-16A) you can find the time (Δt) that a wave takes to travel a distance (Δx) along the top of the layer where it is critically refracted. The wave velocity (V) in that layer is equal to $\Delta x/\Delta t$. But you can see that the slope (S) of the curve is $\Delta t/\Delta x$. Therefore, the slope of the refraction travel time curve is the reciprocal of the velocity in the deepest layer along the path: $V = 1/S$. So, you can find seismic wave velocities in the earth's interior from the slopes of refractions travel time curves. Seismologists can also find these velocities from the shapes of reflection travel time curves (Figure 3-16B), but the analysis is more complicated.

It is not enough to simply find the seismic wave velocity from travel time curves. The next task is to find the thicknesses of the layers. To do this seismologists use the velocities found from the slopes of straight lines, and the distances where these lines intersect. The paths of seismic waves, and the travel time curves in Figure 3-17 can be used to find the thickness (h) of the top layer. Wave velocities V_1 in this top layer and V_2 in the layer below are found from the slopes of the two straight lines. The distance (x) where these lines intersect can also be read from the graph. These values are used in the formula below to find the layer thickness:

$$h = \frac{x}{2}\sqrt{\frac{V_2 - V_1}{V_2 + V_1}}$$

Seismologists derived this formula from Snell's law and the geometry of the refracted wave

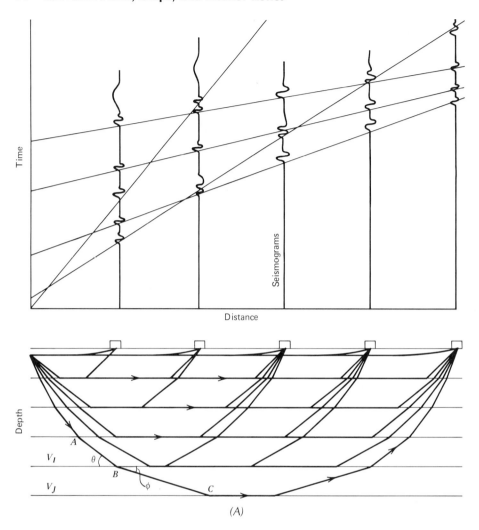

Figure 3-16 Paths of *(A)* refracted and *(B)* reflected P-waves, and corresponding travel time curves for a structure consisting of parallel solid layers. Seismic wave velocity is faster in each deeper layer. Seismograms superposed on the travel time curves show how pulses refracted or reflected from a particular layer all line up along a particular travel time curve. Diagram C shows all of the refracted and reflected P- and S-waves that could be produced by a single P-wave or a single S-wave that strikes the boundary between two solid layers. The path of the incident P- or S- wave that strikes the boundary makes an angle ϑ with the boundary. This is called the angle of incidence. Angles of refraction and reflection of the paths of waves leaving the boundary are indicated by ϕ's. All of these angles are related by Snell's law to the P- and S- wave velocities in the upper layer (V_{P1} and V_{S1}), and in the lower layer (V_{P2} and V_{S2}).

Earthquakes and Seismology 99

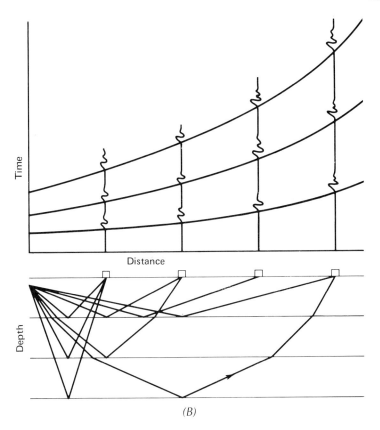

(B)

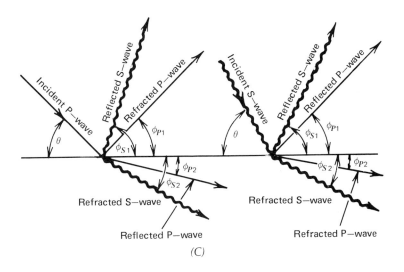

(C)

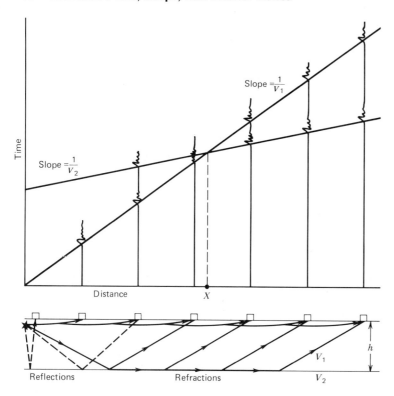

Figure 3-17 Refracted wave paths and travel-time curves, and reflected wave paths between a source and detectors on the surface of a solid layer resting upon another solid. Superposed seismograms show how these straight travel-time curves can be found by the way in which different pulses are aligned.

paths in Figure 3-17. They have derived more complicated equations for finding deeper layer thicknesses. Velocities and intersections obtained from travel time curves are used in all of their calculations. These methods can be used where velocity increases in deeper layers so that wave paths are eventually critically refracted.

You see in Figure 3-16A that the path of a refracted P-wave has several straight segments. Now imagine a great many very thin layers. Let velocity increase with depth. Here the refracted P-wave would follow a path with many straight segments. These are so short that, when joined together, they make a path which is almost a smooth curve. In fact, if the layers are vanishingly thin, the path will be a smooth curve. A travel time curve for seismic waves refracted in this kind of structure will be smoothly curving (Figure 3-18) rather than having several straight line parts. This is the condition in much of the earth's deep interior. By analyzing how slope changes point by point along such a travel time curve, seismologists can calculate the point-by-point increase of velocity with depth.

Seismologists can also find layer thicknesses from reflected waves. The path of a reflected wave to a seismograph close to the source is shown in Figure 3-17. If the velocity in this top layer is known, then the thickness (h) can be found from the travel time (t) of the reflected pulse $h = Vt/2$. If the average velocity above deeper layer boundaries is known, depths to these boundaries can be found in the same way.

The L-waves that follow P- and S-waves on a seismogram also yield information about rock layers in the earth. The periods of these long oscillations change with time in a way that depends on the P- and S-wave velocities, and the layer thicknesses.

It would be ideal if the seismologist could interpret all of the pulses on a seismogram, but usually this is impossible. Smaller pulses are obscured by larger ones, and cannot be

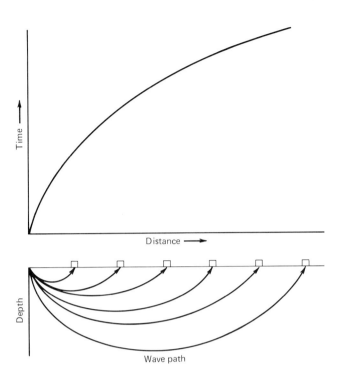

Figure 3-18 Seismic wave paths and travel-time curve where velocity increases continuously with depth.

recognized. Seismologists must use the particular P- and S-wave pulses, and L-waves that can be clearly identified to piece together information about the earth's interior. The ingenious methods of seismology make it possible to obtain this information from measurements made on the earth's surface.

The Earth's Deep Interior

Before the seismograph was invented philosophers imagined that the earth's interior was a blazing inferno. Explosions of its fiery contents were said to cause earthquakes and violent volcanic eruptions. Late in the nineteenth century scientists were learning to decipher the messages carried from the deep interior by seismic waves. These vibrations told a less dramatic story. But seismologists found the scientific challenge of interpreting their meaning fascinating.

First, early seismologists compiled travel time curves for the easily recognized P- and S-waves. In 1897 Professor Wiechert, a German seismologist, described an interesting feature of these travel time curves. He found that refracted P- and S-waves could be clearly identified on seismograms recorded at geocentric distances of less than 103° (about 11,450 km on the earth's surface). Travel times for these waves increased uniformly, as can be seen in Figure 3-4. At geocentric distances greater than 103° these waves could no longer be detected. Then, at geocentric distances beyond 140° (about 15,600 km on the earth's surface), strong P-type pulses could again be recorded. Wiechert surmised from this evidence that a discontinuity exists deep in the earth. Here the nature of the material changes abruptly.

Later, in 1912, Professor Beno Gutenberg argued that the disappearance of refracted S-waves indicated a change from solid to liquid across Wiechert's discontinuity. The disappearance of refracted P-waves at 103° and

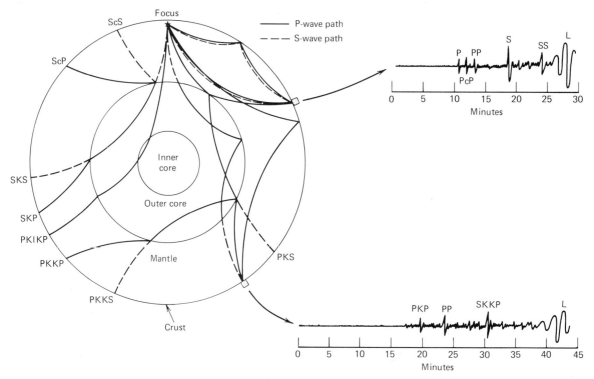

Figure 3-19 First-order seismic discontinuities in the earth, and paths of typical seismic phases that could be detected by seismographs on the surface. Pulses on the seismograms indicate these different phases.

their reappearance beyond 140° was interpreted to mean that P-wave velocity was lower in the liquid. This would cause the paths of P-waves reaching the discontinuity to be refracted more steeply downward rather than being critically refracted back to the surface.

Gutenberg calculated a depth of 2900 km to the discontinuity. He got this figure by analyzing the way that slope changed with distance along the travel time curves. We now call this boundary the **Gutenberg-Wiechert discontinuity.** It separates the solid mantle from the liquid core.

Another important discontinuity was discovered by a Yugoslav seismologist named Mohorovičić. In 1909 he reported that P-wave travel time curves showed a marked change in slope at distances a few hundred kilometers from the source. He interpreted this to mean that two layers could be distinguished. The top layer is about 30 km thick. Here the P-wave velocity is approximately 6 km/sec. In the layer below P-waves travel at an approximate velocity of 8 km/sec. The upper layer is the earth's crust; below it is the mantle. The boundary separating these layers is now called the **Mohorovičić discontinuity,** and nicknamed the "Moho."

Seismologists discovered the discontinuities separating the crust, mantle, and core from refracted P- and S-waves. With this knowledge they could predict other paths for seismic waves (Figure 3-19). You will recall from our earlier discussion that when a seismic wave reaches a boundary, it separates into refracted and reflected P- and S-waves. Some waves reach a seismograph after traveling complicated paths that are refracted at some boundaries and reflected from others. Waves that follow these paths appear as discrete pulses on seismograms. Each pulse is called a seismic **phase.**

A seismic phase is named according to the nature of the wave along different segments of the path. Several phases are identified in Figure 3-19. You can see that the P phase follows a simple curved path. At further distances the path reaches deeper into the mantle before being critically refracted back to the surface. But this phase can only be recorded at distances smaller than 103 degrees. Beyond this distance the path reaches deep enough to impinge on the G-W discontinuity (Gutenberg-Wiechert). Here it separates into the reflected and refracted path segments of other phases.

You can see that the PcP and ScS phases follow paths that reflect from the G-W discontinuity. The PP phase has two identical curved paths that show a reflection from the earth's surface midway between the source and the seismograph.

Only P-type waves can travel in the liquid core. The paths of these waves are identified by the symbol K to distinguish them from P-wave paths in the mantle. So the PKP phase travels downward as a P-wave through the mantle, refracts as a P-wave in the core, then refracts again into the mantle where it continues to the seismograph. The SKS phase consists of an S-wave in the mantle that refracts as a P-wave in the core, then refracts again into the mantle as an S-wave.

The simply refracted P- and S-phases travel only in the mantle and crust. To study the structure of the core seismologists must use other more complicated phases. By analyzing travel time curves (Figure 3-20) and amplitudes of these phases they have discovered yet another major discontinuity in the core. This boundary separates the liquid outer core from a solid inner core at a depth of about 5000 km. The symbol I designates a P-wave path through the solid inner core. Therefore the PKIKP phase indicates a P-wave that has traveled through the central part of the earth. Some seismologists believe that there is evidence of S-wave paths in the inner core. These paths are designated by J.

The Mohorovičić discontinuity and the Gutenberg-Wiechert discontinuity and the boundary between the solid and liquid parts of the core are the three **first-order** discontinuities. These are boundaries where the nature of the material changes abruptly. Elsewhere in the earth seismologists have found **second-order** discontinuities. These are zones where physical properties in the earth change more gradually. The principal discontinuities in the earth are illustrated in Figure 3-19.

Second-order discontinuities are indicated by the way that seismic wave velocities change in the earth. We can find how velocity varies with depth by analyzing the change in slope with distance along travel time curves (Figure 3-20). Two pioneers in this kind of analysis were Beno Gutenberg and Sir Harold Jeffreys. Their calculations of P- and S-wave velocity in the earth are summarized in Figure 3-21. The inflections of their velocity graphs indicate second-order discontinuities near depths of 400 and 700 km. Much new information has been recorded since the 1930s, when Gutenberg and Jeffreys presented their early results. Professor Don Anderson and his colleagues in California have been especially active in this effort. One of their recent velocity graphs is compared with the earlier results in Figure 3-21. Long period L-waves as well as P- and S- waves were analyzed to get these results. The 400 and 700 km discontinuities are more clearly defined on this recent graph. Of even greater interest is the zone between 150 and 250 km where velocities reach their lowest values. This is called the **low-velocity zone.** Except in inclined seismic zones, all earthquakes occur above the low velocity zone. We will discuss this in more detail in later chapters.

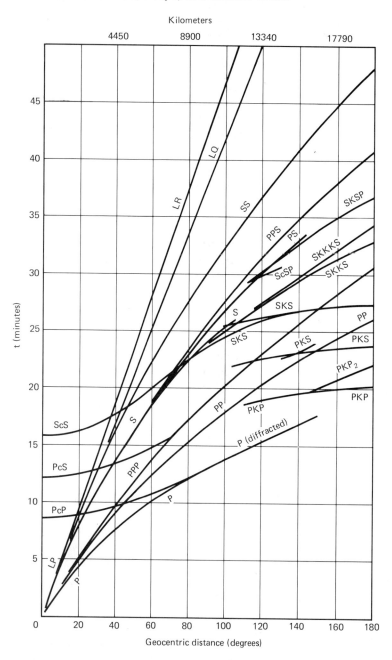

Figure 3-20 Travel-time curves for the most commonly recorded seismic phases. (After Harold Jeffreys, "The Times of the Core Waves," *Monthly Notices of the Royal Astronomical Society,* page 604, Plate 5, Dec. 1939.)

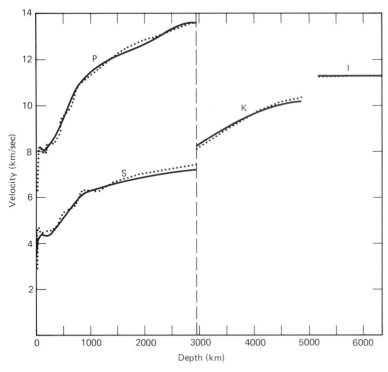

Figure 3-21 Seismic velocities in the earth according to Jeffreys (solid lines) and Anderson and Kovach (dotted lines). After F. D. Stacey, *Physics of the Earth,* Fig. 4.28, John Wiley & Sons, Inc., 1969 and K. E. Bullen, *Physics of the Earth's Interior,* Fig. 2.1, Academic Press, 1959.)

Seismic Surveying

Only earthquakes and large nuclear explosions can produce the strong seismic waves needed to probe the earth's deep interior. But these waves tell us little about the rock layers within the crust. These details can be found by seismic surveying. Basically the procedure is simple. Small seismographs called **geophones** are set out along a line. Then seismic waves are generated by detonating explosives or pounding the ground with a heavy hammer. A typical doodlebugging operation of this kind is illustrated in Figure 3-22. Ordinarily the electrical signals from all of the geophones are transmitted to the instrument truck where they are recorded simultaneously on a single chart. By displaying seismograms from all geophones side by side it is easier to recognize the straight alignments of refracted phases and the curved alignments of reflected phases. After the waves from an explosion are recorded the geophones are moved to another location, and the operation is repeated.

Nature has ways of scrambling seismic phases so that interpreting them is more difficult than one might think. Seismologists have discovered that certain geophone arrangements are more suitable for detecting the strong refracted phases. Other different arrangements enhance the weaker reflected phases. Ordinarily geophones must be placed

Figure 3-22 This diagram illustrates a seismic surveying procedure. Geophones are set out in a line extending away from a hole that is drilled to contain an explosive which is detonated to generate seismic waves. Vibrations received by the geophones are transmitted to the recording truck where they are amplified and inscribed as seismograms on a chart. By displaying these seismograms parallel to one another, alignments of pulses reflected from particular layers can be recognized.

Earthquakes and Seismology 107

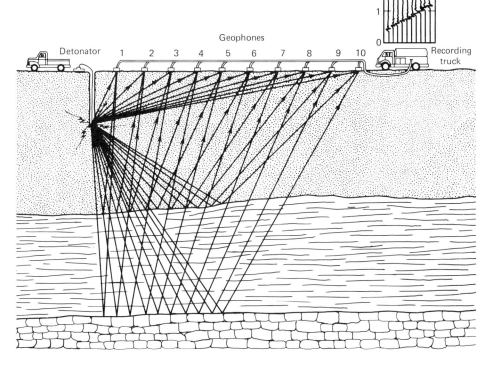

at distances four or five times greater than the depth to which the refracted waves will penetrate. For some practical engineering problems seismologists interpret only the refracted phases. In construction site surveys you can find the thickness of soil and regolith that covers solid bedrock. Where the layer is less than 50 m thick, geophones can be set at 10 m intervals along a 300 m line. Seismic waves are generated by a $1/2$ kg explosive detonated in a shallow hole. An experienced field team of three can complete this experiment in about one-half hour.

At the other extreme is the research experiment designed to probe the crust and upper mantle to depths of approximately 100 km. For this operation several seismic recording trucks, each with an array of geophones, may be located at 50 km intervals along a 1000 km line to record explosions of several tons. In recent years, recordings of nuclear explosions at distances of several thousand kilometers have been made.

Of primary importance in modern oil exploration are reflected seismic waves. From a practical standpoint reflected phases are very difficult to record from layers less than about 100 m deep, and from layers more than about 10 km deep. For the depths between these limits, very clear reflected phases can be recorded from more or less horizontal sedimentary layers. Ordinarily the geophones can be located at intervals of 50 or 100 m along a line approximately 2 km long to record reflections from a depth of several thousand meters.

A common seismic surveying procedure is called reflection profiling. An array of geophones is set out along a 1 or 2 km line, and an explosion is recorded. The array and the explosion shotpoint are then moved 1 or 2 km farther along the line and the experiment is repeated. A professional seismic field party may record 10 to 20 shots during a day of work. The seismograms from many adjacent locations can then be arranged side by side to form a seismic record section (Figure 3-

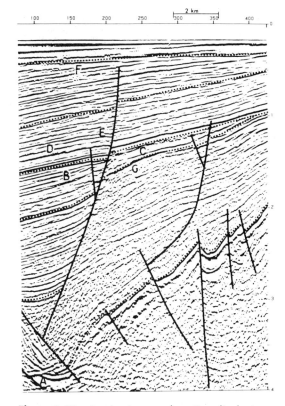

Figure 3-23 A seismic record section displaying pulses that were reflected from folded and faulted sedimentary layers. This should be viewed as a chart on which the horizontal axis is in distance units (kilometers), and the vertical axis is in time units (seconds). Because the seismic waves travel at a velocity (V) of about 5 km/sec, an approximate conversion from time (t) to depth (d) can be made using the formula $d = Vt/2$. Therefore, the deep reflections appearing between 3 and 4 sec down on the record section indicate geologic features between 7.5 and 10 km below the surface. (Reprinted from *Seismic Stratigraphy—Applications to Hydrocarbon Exploration,* Memoir 26, page 309, American Association of Petroleum Geologists, 1977, with permission.)

23). This record section is somewhat similar to a geologic cross section. It reveals numerous reflecting layers, and indicates folds and faults in the sedimentary strata. Such information has much practical importance. It indicates structures where oil might be trapped. There is growing need for similar data in studies of water resources. Seismic surveying can be used to trace subsurface water bearing zones.

The Quest for the Moho

Beginning in the 1950s many earth scientists became interested in the thickness and structure of the earth's crust. Seismic surveying proved to be the best way to find depth to the Mohorovičić Discontinuity and evidence of layering in the crust. These were exciting times for seismologists. Expeditions traveled to many parts of the globe to "shoot for the Moho."

Typical of these ventures was a survey of the crust between Little Rock, Arkansas, and Cape Girardeau, Missouri. Several aspiring seismologists supervised by Professor Robert Meyer set out from the University of Wisconsin in an assortment of trucks carrying explosives, recording apparatus, and short-wave radios. This equipment was operated at the seismic recording stations and shot locations indicated in Figure 3-24. At each station was a recording truck with five geophones set at

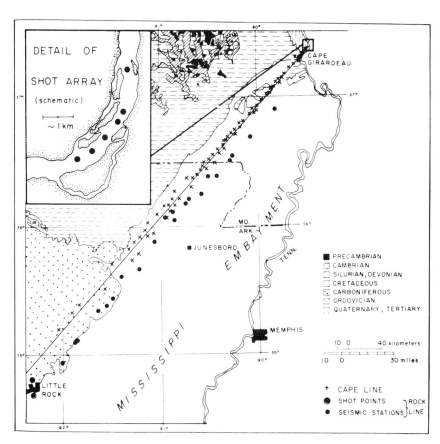

Figure 3-24 Location of seismic refraction survey of crustal structure between Little Rock, Arkansas, and Cape Girardeau, Missouri. (Reprinted from K. McCamy and R. P. Meyer, *Geophysical Monograph No. 10*, page 371, American Geophysical Union, 1966.)

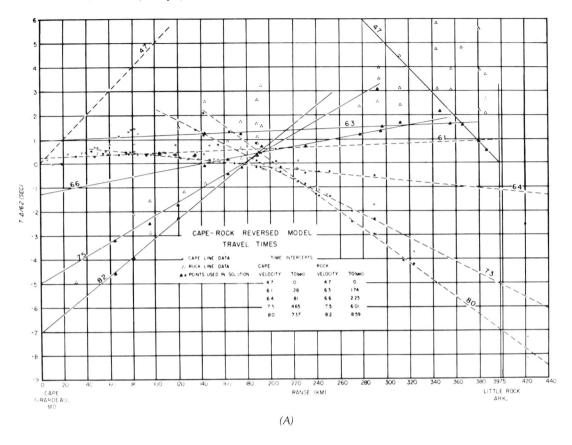

(A)

(B)

0.5 km intervals along a 2 km line. Explosive charges of up to 700 kg were detonated at the shotpoints. The time of detonation was transmitted to the recording trucks by short-wave radio, signaling individual observers to turn on their recording apparatus. Refracted phases from five layers were detected. Figure 3-25 shows the travel time curves and the corresponding interpretation of the layers in the crust.

Numerous seismic experiments similar to the one described above have been carried out in continental areas around the world. Similarly, there have been many measurements in oceanic areas. Marine seismic experiments are accomplished by trailing a floating line of detectors called **hydrophones** from a recording ship. These detect waves generated by detonating depth charges released from another ship. For both marine and continental surveys several seismograms are obtained as the different recording trucks or ships each detect the waves generated at the shotpoint. Typical results of numerous continental and marine seismic surveys are seen in Figure 3-26. These profiles clearly show a difference in continental and oceanic crustal structure. The crust of the earth is characterized by P-wave velocities that are commonly lower than 7.0 km/sec. Velocities in the mantle are generally higher than 7.9 km/sec. The thickness of the continental crust is mostly greater than 30 km, and the oceanic crust is mostly less than 10 km thick. Thus we find seismic evidence of an important difference between continental and oceanic structure. You will recall from Chapter 2 that Bouguer gravity anomalies also indicate this difference.

The seismic results confirm the simple model of G. B. Airy shown in Figure 2-18. In addition, seismic surveys provide considerably more information about layering and irregularities in the crust. By combining Bouguer gravity anomaly data and seismic survey data it is possible to compile diagrams of crustal cross sections showing the density distribution (Figure 3-26) of subsurface rocks.

Density, Pressure, and Elasticity in the Earth

Seismic waves from the earth's deep interior are a means to an end. We can measure the P- and S-wave velocities. This information can then be used to find elasticity coefficients. But you will recall that velocity depends on density as well as elasticity. So we have to find a way to estimate density before we can calculate elasticity coefficients.

You know from Chapter 2 that density in the crust and upper mantle is estimated from rock specimens. Seismic velocities in these specimens can also be measured. Generally we find that faster velocities are recorded in specimens having higher density. Now, look back at the equations that relate velocity, elasticity, and density. You might think that

Figure 3-25 *(opposite)* (A) Travel-time data and (B) calculated crustal structure were obtained from the seismic survey between Little Rock, Arkansas, and Cape Girardeau, Missouri. The graph is not a conventional travel-time curve. Instead, numbers called *reduced travel times* were calculated from the arrival times (T) of pulses on seismograms recorded at distances (Δ) from the shot points where explosive charges were detonated using the expression $(T - \Delta/V)$ where V is an average velocity taken to be 6.2 km/sec. Points on the graph indicate these reduced travel times at different distances. Straight lines drawn through apparent alignments of points are used to identify pulses refracted from a particular layer. The purpose of using reduced travel time is to display alignments of points more clearly than can be done on a conventional travel-time curve. Velocities of P-waves in the different layers are calculated from the slopes of these lines, and are used to determine layer thicknesses. (Reprinted from K. McCamy and R. P. Meyer, *Geophysical Monograph No. 10,* pp. 378 and 380, American Geophysical Union, 1966.)

112 The Earth's Size, Shape, and Interior Zones

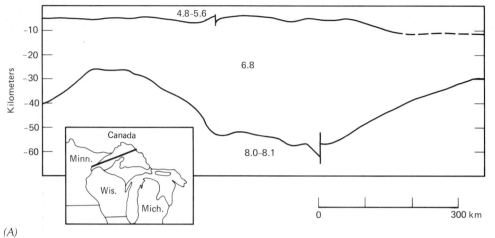

(A)

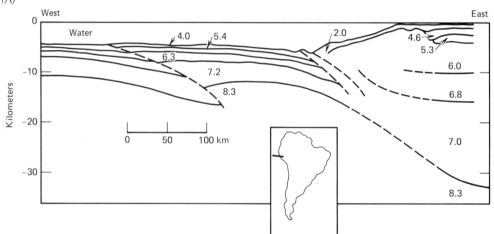

(B)

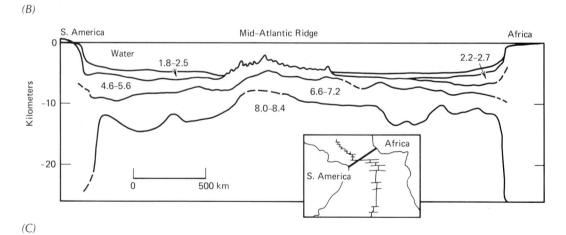

(C)

Figure 3-26 *(opposite)* Cross sections illustrating features of crustal structure determined from seismic surveys *(A)* along Lake Superior, *(B)* near the coast of Peru, and *(C)* across the central Atlantic Ocean. (Top profile was redrawn from T. J. Smith, J. S. Steinhart, and L. T. Aldrich, *Geophysical Monograph No. 10*, page 194, American Geophysical Union, 1966. Middle profile was redrawn from D. M. Hussong, P. B. Edwards, S. H. Johnson, J. F. Campbell, and G. M. Sutton, *Geophysical Monograph No. 19*, page 83, American Geophysical Union, 1976. Bottom profile was redrawn from C. L. Drake, *Geophysical Monograph No. 13*, page 550, American Geophysical Union, 1969.)

velocity should decrease where density increases. But this is contrary to what we observe. Where both velocity and density increase, elasticity coefficients must increase even more.

The earth's mean density of 5.5 gm/cm³ is considerably higher than the values for the crust and upper mantle. Therefore, we know that density in the deep interior must be still higher. Seismic wave velocities also increase with depth in the mantle. This implies the elasticity coefficients, too, are increasing with depth in the mantle.

We have no specimens from deep in the mantle and the core. Therefore, density has to be found indirectly. In 1923 two scientists, E. D. Williamson and L. H. Adams, figured out how this could be done if the same mineral composition existed throughout the mantle. They argued that density increased with depth because the rock was compressed by the weight of the overlying burden. Adams and Williamson knew that this compression and seismic wave velocities could be related to the bulk modulus of the rock. So, they used the equations relating velocity, elasticity, and density to derive the formula that gives the change in density ($\Delta \rho$) occurring over an increment of depth (Δh):

$$\frac{\Delta \rho}{\Delta h} = g\rho(V_p^2 - \frac{4}{3}V_s^2)$$

To use this formula they needed to know the density (ρ) and gravity (g) at some starting depth h near the earth's surface. Then they calculated the density change ($\Delta \rho$) that should occur in the depth increment (Δh). Adding these values to those at the starting point gave them the density ($\rho + \Delta \rho$) and the slightly greater depth ($h + \Delta h$). Calculations were then repeated step by step down through the mantle. This method of Adams and Williamson could not be used to find density in the liquid core because S-waves do not propagate there.

Early attempts at calculating density in the earth were undertaken by Adams and Williamson and later by the Australian seismologist K. E. Bullen. Their results yielded an incorrect value for the mean density. At first this was attributed to imprecise knowledge of the seismic velocities. As better velocity data became available, however, scientists realized that the fundamental assumption of uniform composition in the mantle was not valid. To compute densities that give the correct mean density it is necessary to assume differences in composition. Second-order discontinuities seen on seismic velocity graphs (Figure 3-21) probably indicate compositional changes. Then, the formula of Adams and Williamson can be used to estimate density in different shells, each of which is considered to be of uniform composition. Density in the core is then estimated after calculation of mantle density. Recent results are shown in Figure 3-27. Because of the uncertainties associated with the assumptions about composition, the graph represents approximate density values rather than precise values.

If the density in the earth is known, it then becomes a simple matter to calculate the pressure. Change of pressure (Δp) over a small change in depth (Δh) is easily calculated: $\Delta p = g\rho\Delta h$. Therefore, by starting at the surface where atmospheric pressure is approximately 1 bar (1 bar = 10^6 dynes/cm² ≃

114 The Earth's Size, Shape, and Interior Zones

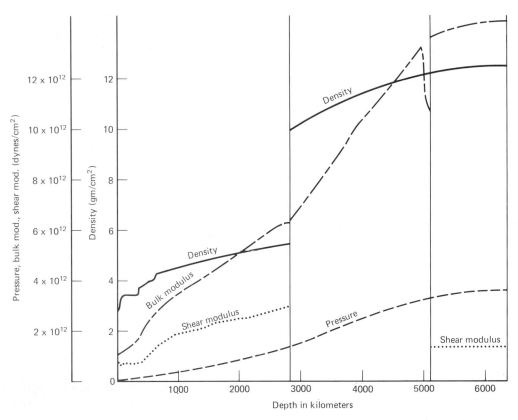

Figure 3-27 Variation of bulk modulus, shear modulus, density, and pressure in the earth. (Prepared from data in *Physics of the Earth* by F. D. Stacey, Table E.1, copyright © John Wiley & Sons, Inc., 1969.)

1 atmos) the pressure at depth is computed step by step from the increments of pressure change. The results are shown in Figure 3-27.

Finally, the elasticity coefficients in the earth can be calculated from the formulas that relate velocity, density, and elasticity. Graphs of elasticity coefficients in the earth are also included in Figure 3-27.

In recent years seismologists have figured out a new and better way to find density and elasticity in the deep interior. They can now measure very long period oscillations that follow strong earthquakes. These oscillations have periods of up to one hour. They are called **free oscillations,** and they indicate that the earth is ringing like a giant bell. Some of the different free oscillations cause the whole earth to alternately expand and contract. Others indicate twisting between the northern and southern hemispheres. Amplitudes and periods of free oscillations depend on the density and elasticity in all of the earth's concentric shells.

Density, pressure, elasticity: these are the things we know about the earth's interior. What kind of material is down there? Is it like anything we find nearer the surface? We know that mixtures of iron and nickel in the core would probably have the right density for the pressure conditions there. Certain compounds of aluminum, magnesium, and silicon could make up the mantle. In Chapter 7 we will speculate more on the earth's internal chemistry.

STUDY EXERCISES

1. Suppose that an earthquake is recorded on a seismograph at location A which is situated 4000 kilometers from the epicenter. The P-wave arrived there at a time of 12 hours and 20 minutes. A seismograph at a different location B recorded the S-wave 7.5 minutes after the P-wave.
    (a) At what time did the S-wave arrive at location A?
    (b) How far is location B from the epicenter?
    (c) At what time did the earthquake occur?
    (d) At which location is earthquake intensity greatest?
    (e) Do you have enough information to find the distance between location A and location B?

   Note: Use Figure 3-4 for travel time information.

2. Identify the seismic phases recorded at locations A, B, C, D, and E in Figure 3-28.

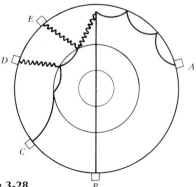

Figure 3-28

3. Suppose that a P-wave was generated by an explosion detonated 5 m beneath the ocean surface. It travels down through the water and hits solid rock on the ocean bottom at a 75 degree angle of incidence. P-wave velocity is 1500 m/sec in the water and 5000 m/sec in the rock.
    (a) When the incident ray hits the rock, how many new refracted and reflected rays are produced? Explain!
    (b) Is the angle of refraction of the P-wave larger or smaller than 75 degrees?

4. Between depths of 1000 and 2000 km S-wave velocity and density both increase.
    (a) Does this imply that the shear modulus increases or decreases with depth? Explain!
    (b) These depths are situated in what zone of the earth?

5. Suppose that a P-wave travels downward through the mantle and hits the core-mantle boundary at a 60 degree angle of incidence. The P-wave angle of refraction is more than 60 degrees, but core density is higher than mantle density.
    (a) Does this imply that near the boundary P-waves in the core travel faster or slower than P-waves in the mantle? Explain!
    (b) Does this imply that the shear modulus in the core is higher or lower than the shear modulus in the mantle? Explain!

6. Referring to Figure 3-29:
   (a) How many layers are indicated?
   (b) What are the seismic wave velocities in the two uppermost layers?
   (c) What is the thickness of the top layer?
   (d) These travel time lines represent refracted waves, reflected waves, or surface waves.

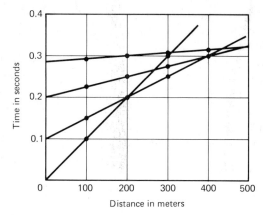

Figure 3-29

SELECTED READINGS

Clark, Sidney P., *Structure of the Earth*, Prentice-Hall, Inc., Englewood Cliffs, New Jersey, 1971.

Dobrin, Milton B., *Introduction to Geophysical Prospecting*, 3rd ed., McGraw-Hill Book Co., New York, 1976.

Jacobs, J. A., R. D. Russell, and J. T. Wilson, *Physics and Geology*, 2nd ed., McGraw-Hill Book Co., New York, 1974.

Richter, Charles F., *Elementary Seismology*, W. H. Freeman and Co., San Francisco, 1958.

4

THE EARTH'S MAGNETISM

Magnetism
The Earth's Magnetic Field Today
   Measuring the Earth's Magnetic Field
   The Main Magnetic Field
   Rock Magnetism
   The Anomalous Magnetic Field

The Ancient Magnetic Field
   Paleomagnetism
   Magnetic Field Reversals
   Paleomagnetic Poles
The Origin of the Main Magnetic Field

The earth is a giant magnet! This astonishing news was published in 1600 by William Gilbert (1544–1603). Trained in medicine, Gilbert was a physician in the court of Queen Elizabeth I of England. But he was truly a renaissance man of science. Gilbert was fascinated by magnetism, a mysterious property of rock called **lodestone.** The term **magnetism** came from the district of Magnesia in ancient Greece where lodestone was mined about 3000 years ago. Because of its magnetism, lodestone attracted iron. Therefore, it also attracted scholars. They carefully studied the effects of lodestone rods called magnets. Eventually people discovered how to make a compass by balancing a magnet on a pivot. But nobody understood why the compass pointed toward north. William Gilbert figured this out be experimenting with a magnetic globe made of lodestone. He discovered that this globe had about the same effect on nearby magnets that the earth has on a compass. So Gilbert concluded that the earth itself must be a giant magnet.

Why is the earth magnetic? William Gilbert proposed that earth magnetism was caused by lodestone in the deep interior. This idea prevailed for more than two centuries. Its eventual undoing came partly from experiments that, interestingly enough, were first done by Gilbert himself. He found that when he heated his lodestone globe until it glowed red, it lost its magnetism. If the deep interior were as hot as many believed, how could lodestone be the cause of the earth's magnetism? Furthermore, slow changes in compass direction indicated that the earth's magnetism was changing with time. How could this happen in solid lodestone? Perhaps a better explanation could be proposed.

During the nineteenth century, physicists learned that lodestone was not the only source of magnetism. Magnetism is a phenomenon inseparably related to the flow of electrical current. But how could currents strong enough to cause the earth's magnetism exist in its interior? Not until seismologists had discovered the liquid outer core could we begin to answer this question. Since then, modern hypotheses have attempted to relate the earth's magnetism to the flow of electrically charged particles in the fluid core. This flow is supposedly caused by thermal currents like those seen in a pot of boiling water. These currents are also affected by the earth's rotation.

If lodestone is not the cause, does it then contribute in any important way to the earth's magnetism? We now know that this rock consists mainly of an iron oxide (Fe_3O_4) mineral called **magnetite.** This mineral is one of a few that can acquire magnetism. Magnetite grains in the crust cause distortions of the earth's magnetism which can be detected with sensitive instruments. We call these distortions **magnetic anomalies.** They are something like gravity anomalies. We can analyze them to find out about geologic features in the crust.

MAGNETISM

Before learning about the earth's magnetism in particular you need to know some facts about how magnets interact with one another. The French physicist Charles Coulomb (1736–1806) discovered that forces between magnetized rods were directed toward points close to the ends of the rods. These points are called **magnetic poles.** Every magnet has two poles, each of which will attract one pole and repel the other pole of a second magnet (Figure 4-1). The two poles of a magnet are called positive and negative because they exert opposite effects on another magnet. The unlike poles of two magnets are mutually attracted, and the like poles repel each other. The force (F) between two poles is given by Coulomb's law:

$$F = \frac{1}{\mu} \frac{P_1 P_2}{r^2}$$

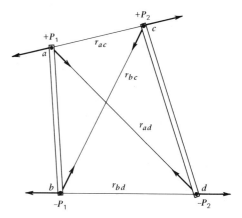

Figure 4-1 The interaction of two long, thin magnets can be explained in terms of forces acting at positive and negative magnetic poles situated very close to the ends of the magnets. Arrows in the diagram represent the direction and strength of forces acting at poles labeled a, b, c, and d. Distances between poles are labeled by r with appropriate subscripts. For example r_{ad} is distance between poles a and d.

where P_1 and P_2 are the **pole strengths** of the two poles, and r is the distance between them. The **magnetic permeability** μ is a property of the medium where the magnets are situated. In a vacuum, and for practical purposes in the earth's atmosphere, $\mu = 1$. The pole strength is a measure of the force that one pole can exert on another. A magnet is said to have one **unit of pole strength** (ups) if one of its poles exerts a force of 1 dyne on another pole 1 cm away. For this magnet $P = 1$ ups. If a magnetic pole exerts a 2 dyne force on another pole 1 cm away, then the magnet has a pole strength $P = 2$ ups.

There is no way to separate the two poles of a magnet. If you break a magnetic rod, you immediately get two magnets. Each will have a positive and a negative pole. For a long, thin rod the poles will be at the ends. But for a magnetic globe like the one William Gilbert used, the two poles will appear to be inside. A thin rod with equivalent magnetism would be considerably shorter than the diameter of

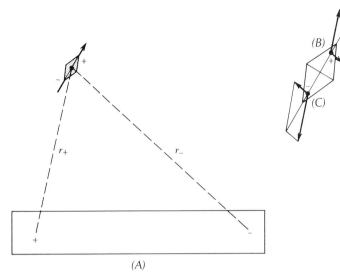

Figure 4-2 Interaction of two dipole magnets: *(A)* orientation of a small magnet in relation to a large dipole magnet; *(B)* forces associated with the + and − poles of the large magnet that act on the + pole of the small magnet; *(C)* corresponding forces that act on the − pole of the small magnet.

the globe. Because a magnet always has two opposite poles, it is commonly called a **dipole.** Pole strength is always the same for both poles.

Now, what can you learn about a large magnet like the earth without first knowing its pole strength and the exact locations of its poles? You know that it can exert force on other magnets some distance away. Therefore we say that a field of force exists in the region surrounding the large magnet. This field of force is called a **magnetic field.** You know from Coulomb's law that the combined force of the poles of the large magnet changes from one place to another. Therefore the magnetic field must differ from place to place. If you can describe this magnetic field, then you can predict the effect of the large magnet on other magnets.

To describe a magnetic field you must know its **intensity** and its **direction** at many different places. These two things can be measured with a test magnet (Figure 4-2). This test magnet should be small enough and weak enough so that it does not significantly change the position of the large magnet. The magnetic field intensity (H) at some point is the combined force (F) exerted by both poles of the large magnet on the positive pole of the test magnet divided by the pole strength (P) of the test magnet: $H = F/P$. Of course, the force (F) is also exerted on the negative pole of the test magnet. Now, suppose that this little magnet can rotate into any alignment. Then the force (F) on its two opposite poles will rotate it into the direction of the magnetic field of the large magnet.

Physicists and electrical engineers commonly use the oersted as a unit of magnetic field intensity. This unit was named for the Danish scientist Hans Christian Oersted (1777–1851). One **oersted** is equal to a force of one dyne divided by one unit of pole strength. But the earth's magnetic field intensity is less than one oersted, and certainly much weaker than the fields used in most physics and engineering operations. Therefore, another unit, the **gamma,** is more convenient for describing the earth's magnetism: 1 gamma = 10^{-5} oersted.

After measuring H and observing the test magnet orientation at many points, you can describe the magnetic field by vectors, or by so called magnetic "lines of force" (Figure 4-3). A vector is an arrow that points in the field direction, and its length is proportional to intensity. "Lines of force" depict a continuous pattern of field direction, and intensity is implied by line spacing. A stronger field is indicated by more closely spaced lines.

You do not necessarily have to use a lode-

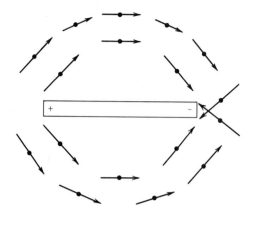

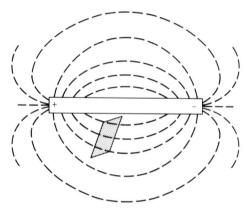

Figure 4-3 Representation of the magnetic field in the space surrounding a dipole magnet: *(A)* at discrete points in space the field direction is shown by a vector arrow whose length is proportional to the field intensity; *(B)* magnetic lines of force that indicate field direction, and where intensity is proportional to the line spacing.

stone test magnet to measure a magnetic field. A coil of wire with electrical current flowing in it will do just as well. The association of magnetism with electrical current was recognized early in the nineteenth century by Hans Christian Oersted. Physicists then went on to discover that a coil with N turns of wire, and cross-sectional area A reacted the same as a test magnet in a magnetic field. When current (i) flows in the coil, it is equivalent to a magnet of pole strength P and length λ according to the formula: $P\lambda = NiA$. It is also true that you can make an electrical current flow by moving a wire in a magnetic field. Once you realize that magnetism is always associated with the flow of electrical current, it is easier to understand how the earth's magnetism might be generated by currents in the outer core.

THE EARTH'S MAGNETIC FIELD TODAY

Measuring the Earth's Magnetic Field

You could account for most of the earth's magnetism by a large magnet placed in the core. You would have to tilt this magnet about $12\frac{1}{2}$ degrees from the axis of rotation to explain why a compass does not point exactly north. But there are still other irregularities in the earth's magnetic field. We know about these irregularities from surveys of the field direction and intensity. Some very ingenious instruments have been invented to survey the earth's magnetic field.

To find the earth's magnetic field direction you have to measure two angles (Figure 4-4). One of these is the declination (d). It is the angle between true north and the compass direction, which we call **magnetic north.** Before declination can be measured you must find true north from astronomical sightings. Then you can compare this line with the compass bearing to get the declination angle.

Inclination (i) is the angle between the magnetic field direction and a horizontal surface. You can measure i with a dip needle, which is a magnet that can rotate vertically. First you have to orient it with a compass so that its vertical plane of rotation is in line with magnetic north. Then the dip needle will rotate into the direction of the earth's magnetic field. You can also use an **earth inductor.**

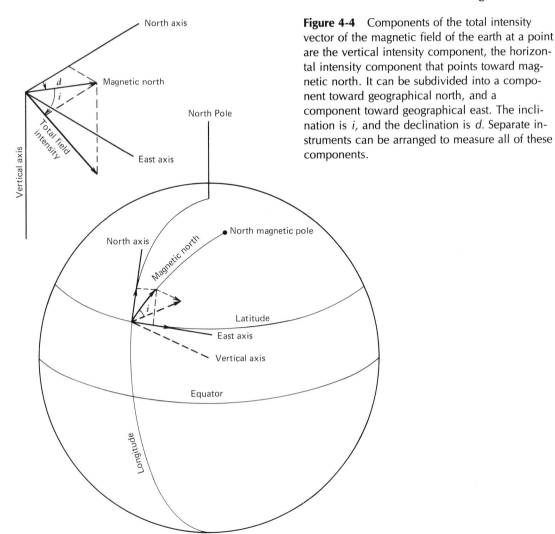

Figure 4-4 Components of the total intensity vector of the magnetic field of the earth at a point are the vertical intensity component, the horizontal intensity component that points toward magnetic north. It can be subdivided into a component toward geographical north, and a component toward geographical east. The inclination is i, and the declination is d. Separate instruments can be arranged to measure all of these components.

This instrument has a coil that spins on an axis. As it spins in the earth's magnetic field, the motion causes electrical current to flow in the coil except when the axis is aligned with the field direction. A good earth inductor is more sensitive than a dip needle, because even a slight misalignment can cause a measurable current in the coil.

Portable instruments called **magnetometers** are used to measure the earth's magnetic field intensity. They use test magnets in interesting ways. One of the simplest to understand is the torsion magnetometer. A test magnet is attached to a horizontal fiber. Rotation of the magnet is counteracted by the strength of the fiber. The device is calibrated so that field intensity can be found from the angle of twist in the fiber.

Another instrument is the nuclear precession magnetometer (Figure 4-5). It finds the field intensity from the motion of nucleii of hydrogen atoms in a bottle of water. These nucleii behave as tiny test magnets. During a cycle lasting a few seconds they can be alter-

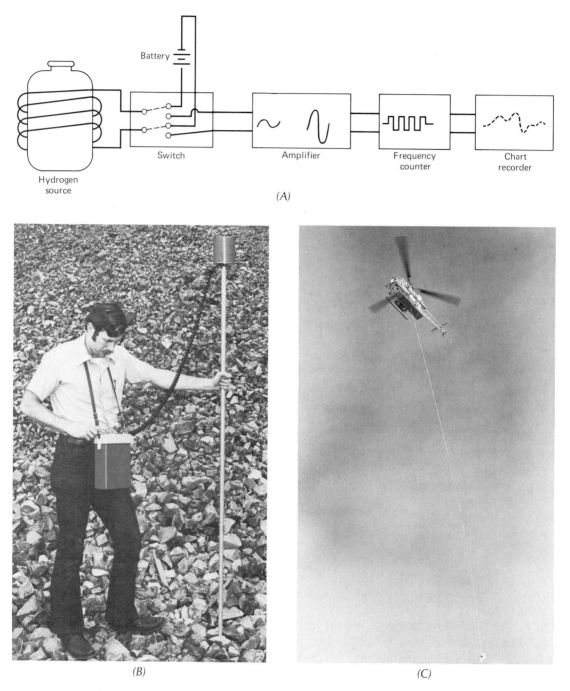

Figure 4-5 Nuclear precession magnetometer: *(A)* schematic diagram of electronic units; *(B)* portable field unit; *(C)* mounted for airborne recording. (Photographs courtesy of E. G. and G. Geometrics, Inc., Sunnyvale, California.)

nately aligned, first with the magnetic field of a coil surrounding the bottle, and then with the earth's field. While changing alignment these nucleii precess like tops. This motion causes an ac electrical current to flow in the coil. The frequency of this current is proportional to the earth's field intensity. It is measured by an electronic frequency meter.

The fluxgate magnetometer is still another device. It was originally designed during World War II to trail from a low-flying airplane to detect submarines. There is enough iron in a submarine to cause distortion of the earth's field which can be detected by a magnetometer several hundred meters away. The fluxgate magnetometer has parallel iron rods inside two oppositely wound coils. The ac electrical currents in these coils generate opposite but equal magnetic fields in the iron rods. If the earth had no magnetism, then these two ac fields would exactly cancel. But the earth's field alternately adds to one and then the other. This unbalances the system so that an ac electrical current is generated in another coil that surrounds both iron rods. This current is proportional to the earth's field intensity.

All of these magnetometers are compact portable devices (Figure 4-5) that can measure field intensity to a precision of one gamma or better. They can be hand carried or trailed from ships and airplanes. They have been used to survey magnetic field intensity over most of the earth's surface.

The Main Magnetic Field

We know from magnetometer surveys that the earth's magnetic field intensity ranges from about 25,000 gammas (0.25 oersted) near the equator to almost 70,000 gammas (0.7 oersted) in the polar regions. You can get a practical idea about how strong this field is by holding a compass near a toy magnet. When the compass is close enough to be deflected, you know that the earth's field is about as strong as the field of the small magnet. In a polar region you would have to hold the compass closer to the toy magnet than you would in an equatorial region to get the same deflection, even though the earth's field is more than 40,000 gammas stronger near the poles. See if you can figure this out from the fact that inclination changes from zero degrees near the equator to 90 degrees near the poles.

Suppose that you have read a magnetometer along many profiles stretching from the equator into the polar regions. What kinds of magnetic intensity features would you observe? The principal feature would be a change in intensity that you would expect from one large magnet in the core which is tilted about $12^1/_2$ degrees from the earth's rotation axis. But you would not find just the simple dipole pattern displayed in Figure 4-3. You would also observe some broad irregularities several thousand kilometers across. Scientists have learned from mathematical analysis of these broad irregularities that they must come from sources of magnetism in the core. You could account for these features with about 11 smaller dipoles tilted at various angles from the axis of rotation. Altogether, you would need 12 magnets in the core to explain the features we have mentioned so far. The large magnet inclined at $12^1/_2$ degrees would have to be about 80 times stronger than all of the smaller ones. The part of the earth's magnetism that we call the **main magnetic field** consists of these features that could result from the 12 fictitious magnets in the core.

We have yet to mention the great many small irregularities in magnetic field intensity that you would also observe in a magnetometer survey. These features are mostly less than a few tens of kilometers wide, and have amplitudes generally smaller than 2000 gammas. They are called **magnetic anomalies.** All of these small irregularities make up the earth's **anomalous magnetic field.** This part of the earth's field comes from magnetized rock masses in the crust. We will talk more about the anomalous field later in this chapter.

When you make a magnetometer survey, you actually measure the earth's **total magnetic field.** This is simply the combination of the main field and the anomalous field. A magnetometer does not know anything about these different parts. It just feels the total magnetic field. To find the particular features of the main field or the anomalous field you have to separate these two parts after a magnetometer survey has been done. Scientists have developed mathematical schemes for doing this separation. These schemes were used to get the main magnetic field maps shown in Figures 4-6 and 4-7.

Now let us look more carefully at the features of the main magnetic field. As we said earlier, most of the main field could be accounted for by 12 magnets in the core, the strongest being about 80 times more powerful than the other 11. If the axis of the one strong magnet were projected to the earth's surface, it would intersect at two places. In 1955 scientists calculated these positions at 78° 34'N–290° 40'E and 78° 34'S–110° 40'E. These two places are called the **geomagnetic poles.** If you think about these pole positions, you will realize that the axis of the strong fictitious magnet does not go exactly through the center of the earth.

If this one strong dipole magnet could completely account for the main magnetic field, then dip needles would show 90-degree inclination at the geomagnetic poles. But this is not observed because of the effect of the other 11 fictitious magnets. So we have to locate two other positions where dip needles indicate vertical inclination. These places are called the north and south magnetic **dip poles.** In 1960 the north magnetic dip pole was at 75° N-101° W, and the south magnetic dip pole was at 67° S-143° E, quite far from the geomagnetic poles.

The dip poles were goals that early polar explorers tried to reach. In journeys full of danger and adventure British expeditions reached the southern dip pole in 1840, and again in 1908. Australians reached this frigid place in 1914. Each time the dip needle surveys revealed a different location for the dip pole. The earth's main magnetic field, then, seemed to be changing with time.

During William Gilbert's lifetime, systematic observations of declination were begun in London. These have been made more or less continuously since that time. Similar measurements were started in Boston in 1723. The slow changes in declination at these places are seen in Figure 4-8. Here is more evidence that the main magnetic field is changing.

With a large increase in the number of magnetic observatories around the world during the twentieth century we now have much more evidence of the time-varying nature of the main magnetic field. These time variations can be classed in three general categories: secular variations, diurnal (daily) variations, and magnetic storms. Secular change consists of the **westward drift,** and an observed decrease in the strength of the main field. The westward drift is an apparent shift of the entire magnetic field relative to fixed points on the earth's surface. All of the large closures indicated by closed contours on the map of the field intensity (Figure 4-6) appear to be shifting westward at the rate of about 0.2 degrees per year. As the westward drift continues, we can see that it would cause field intensity to change at a geographical point, as a high or low magnetic anomaly shifts over the point. In addition to this westward drift we have learned that the actual strength of the earth's magnetic field is decreasing at an approximate annual rate of 1/1500 times its present pole strength. The westward drift and the decreasing magnetism combine to cause a slow change in the field that we call **secular variation.** By using measurements of this variation during the past 50 years or so, scientists can predict the change during the coming decade.

The diurnal variation is a daily cycle. Rec-

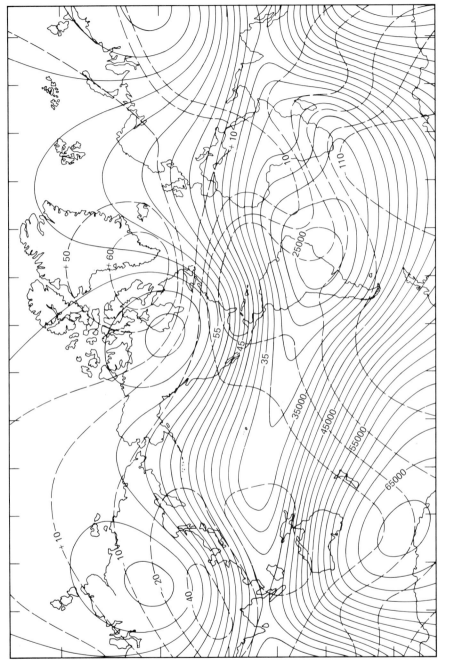

Figure 4-6 Main magnetic field intensity on the surface of the earth contoured at 2000 gamma intervals. Dashed lines indicate annual change in field intensity. (Modified from "Total Intensity of the Earth's Magnetic Field," Epoch 1975.0, chart published by the Defense Mapping Agency Hydrographic Center, Washington, D.C.)

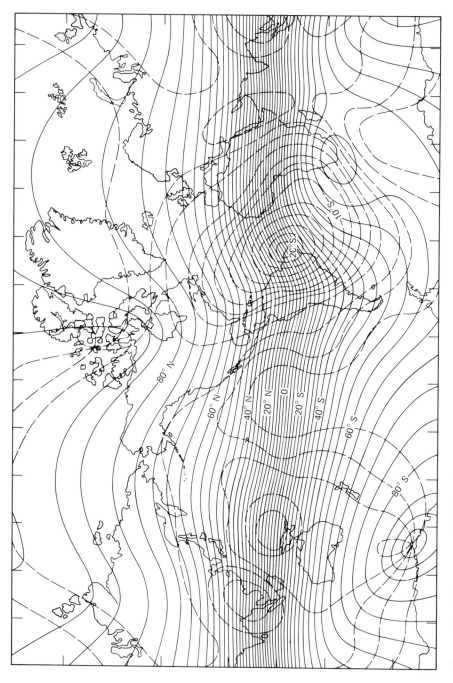

Figure 4-7 Main magnetic field inclination on the surface of the earth contoured at 4 deg intervals. Dashed lines indicate annual change in minutes of arc. (Modified from "Magnetic Inclination or Dip," Epoch 1975.0, chart published by Defense Mapping Agency Hydrographic Center, Washington, D.C.

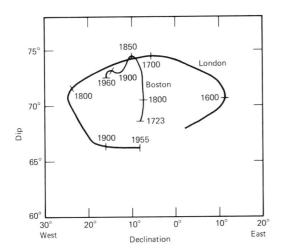

Figure 4-8 Secular change in magnetic declination observed in London and Boston. (Modified from J. H. Nelson, L. Hurwitz, and D. G. Knapp, *Magnetism of the Earth,* Publication 40-1, U.S. Dept. of Commerce, Coast and Geodetic Survey, 1962.)

ords in Figure 4-9 illustrate these field intensity fluctuations. They are generally less than 100 gammas in amplitude. Analysis of these records indicates that the diurnal variation is associated with the gravitational effects of the sun and moon on electrically charged particles surrounding the earth. Therefore, the

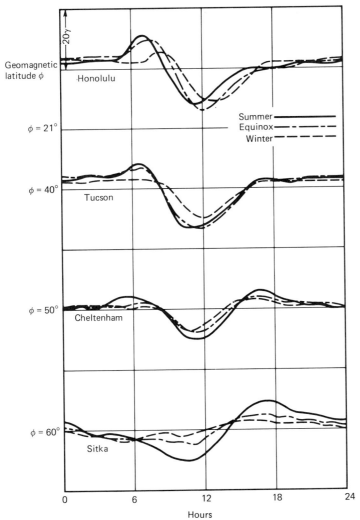

Figure 4-9 Diurnal variation of the magnetic field at different latitudes. (Reprinted from J. H. Nelson, L. Hurwitz, and D. G. Knapp, *Magnetism of the Earth,* Publication 40-1, U.S. Dept. of Commerce, Coast and Geodetic Survey, 1962.)

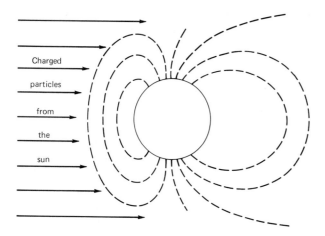

Figure 4-10 Representative lines of force showing the approximate form of the magnetic field in the space surrounding the earth.

diurnal variation is caused by external magnetic field sources.

Magnetic storms are unpredictable phenomena associated with sunspots. The onset of a magnetic storm may be signalled by sudden erratic fluctuations of field intensity sometimes exceeding 1000 gammas. The storm may last only a few minutes or perhaps several days. If you know about short-wave radio broadcasting, you may have heard of radio "blackouts." Magnetic storms can cause this severe interference in short-wave radio communication.

In recent years we have learned from satellite and rocket magnetometers about the magnetic field in the space surrounding the earth. The effect of charged particles moving in space contributes to the external sources of the field that cause a distortion of the lines of force. The form of the magnetic field is shown in Figure 4-10. The field has an important effect upon incoming solar electromagnetic radiation. It traps electrically charged particles in radiation belts that surround the earth. This shields us from intense radiation that could prove lethal to life.

Rock Magnetism

If you like to hike in the wilderness, perhaps you have used a compass to find your way. Occasionally you may come to a place where the compass seems to go haywire. It might be so strongly attracted by nearby outcropping rock that it is useless for finding direction. The reason is rock magnetism. This is caused by magnetic minerals disseminated through the rock. These are called **ferromagnetic** minerals because they are iron compounds that can acquire magnetism. Except in rock such as lodestone, ferromagnetic minerals make up only a small fraction of the mass of most kinds of rock. Ordinarily this is not enough to deflect a compass, but small variations in the content of ferromagnetic minerals can be detected with a sensitive magnetometer.

The most important ferromagnetic minerals are **magnetite** (Fe_3O_4), **hematite** (Fe_2O_3), **ilmenite** ($FeTiO_3$), and **pyrrhotite** ($Fe_{1-x}S$). The atoms in these minerals are geometrically arranged so that magnetic fields caused by individual orbiting and spinning electrons are aligned throughout small discrete zones of a crystal. These zones are called **magnetic domains** (Figure 4-11), and are commonly a few microns in diameter. The electrons in a domain are electrically charged particles. When moving, they amount to an electrical current. You know that there is always a magnetic field associated with an electric current. Because of this electron motion each magnetic domain has a permanent magnetism. Its

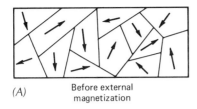

(A) Before external magnetization

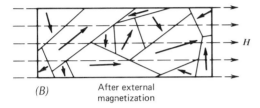

(B) After external magnetization

Figure 4-11 Magnetism of ferromagnetic materials: *(A)* idealized ferromagnetic domains randomly oriented; *(B)* domains that are more closely aligned to an external magnetic field, which is represented by symbol H and the dashed arrows.

strength depends upon the proportion of electron magnetic fields aligned parallel or opposite to one another. Magnetite and pyrrhotite have strong domain magnetism because individual electron fields are principally in the same direction. The domains of hematite and ilmenite have weaker magnetism because adjacent electron fields tend to be oppositely oriented.

The magnetism of a ferromagnetic mineral specimen depends upon the extent that the magnetic fields of its domains are aligned. Some specimens display no outward magnetism because of the random orientation of the internal domain fields. Others containing domains in a preferred orientation have a stronger net magnetism. We describe the strength of magnetism in a specimen by its **magnetic moment per unit volume.** Suppose that a small magnet of pole strength (P) and length (λ) exists in each unit of volume, for example, each cubic centimeter of the specimen. Then, its magnetic moment per unit volume (I) is equal to $P\lambda$.

The magnetic moment per unit volume may exist because of induced magnetism, or it may be remanent magnetism. **Induced** magnetism is acquired by a specimen when it is in another magnetic field. Ferromagnetic minerals possess some induced magnetism because of the earth's main magnetic field. In weak magnetic fields ($H<1$ oersted) the value of I varies in proportion to the intensity H of the field where the specimen is located: $I = kH$. We call k the **magnetic susceptibility** of the material. It represents the capacity to acquire induced magnetism. This property has the same importance in studies of the earth's magnetic field as density has in studies of the earth's gravity field. In weak magnetic fields induced magnetism results from elastic warping or bending of the domains of the specimen. This causes stronger alignment of their individual magnetic fields. In strong magnetic fields ($H>1$ oersted), the value of k for a specimen depends upon whether field intensity has been increasing or decreasing, and the value (H) of this intensity. We can assume k to be constant only where H has low values such as observed near the earth's surface.

A ferromagnetic specimen possesses remanent magnetism when it retains a magnetic field regardless of the existence of other magnetic fields. This implies that there is a preferred orientation of domain magnetism within the specimen. The most important remanent magnetism in the earth's crust is developed when the domains originally form. At temperatures above a critical point between 550° C and 600° C, known as the **Curie temperature,** thermal motions of atoms are so great that magnetic domains cannot exist. Ferromagnetic minerals cannot possess magnetism at temperatures higher than the Curie temperature. This is why William Gilbert's lodestone globe lost its magnetism when he heated it. Domains form when the mineral is cooled through the Curie temperature. When they form in the presence of a magnetic field like the earth's main field, their individual magnetic fields tend to align. The preferred orientation of domain magnetism achieved in this way gives the specimen a remanent magnetic moment per unit volume which is in line with the field where the specimen cooled. Because of its association with temperature, we call this magnetism **thermoremanent magnetism** (TRM).

Temperature increases in the earth to values above the Curie temperature a few tens of kilometers below the surface. Therefore, earth magnetism due to ferromagnetic minerals can exist only in the crust. You can produce remanent magnetism in ferromagnetic materials by placing them in a very strong magnetic field. Here sufficient energy exists to cause growth of favorably oriented domains at the expense of nonaligned domains (Figure 4-11). Because we do not find strong fields in the earth's crust, this way of achieving remanent magnetism is not of geologic importance. Some chemical and physical processes in the crust can alter the magnetism of minerals, but these are largely secondary effects.

Thermoremanent magnetism and induced magnetism in ferromagnetic minerals are the sources of the earth's anomalous magnetic field. Rock masses appear to possess magnetism insofar as they contain disseminated ferromagnetic minerals. The strength of this rock magnetism depends upon the amounts of these minerals present. Some representative values of magnetic susceptability and remanent magnetism for common rocks and ferromagnetic minerals are given in Table 4-1. These values were measured with portable apparatus designed to test the effects of magnetized specimens on specially designed electromagnetic circuits.

The Anomalous Magnetic Field

Take a very close look at the earth's magnetic field, and you will see many irregularities (Figure 4-12A). These are magnetic anomalies, and they are caused by variations in rock magnetism. Magnetic anomalies, as you already know, make up the earth's anomalous magnetic field. Magnetic anomalies, like gravity anomalies, give us clues about rocks in the crust. In some places these anomalies indicate valuable iron ores, or other economic deposits containing ferromagnetic minerals. Elsewhere, magnetic anomalies can give us more general information about geologic features of the crust.

Geophysicists can analyze magnetic anomalies in ways that are not too different from the analysis of Bouguer gravity anomalies. First they make an educated guess about the location of rock masses with contrasting values of magnetic moment per unit volume. Then, using formulas based on Coulomb's Law, they calculate the magnetic field due to

Table 4-1
Some Magnetic Properties of Minerals and Common Rocks

Mineral	Susceptibility	Remnant Magnetic Moment per Unit Volume
Magnetite	0.3 to 0.8 cgs units	0 to .4 cgs units
Ilmenite	0.135	0 to .09
Pyrrhotite	0.125	0 to .08
Rocks		
Basalt	0.00005 to 0.005	0 to .004
Granite	0 to 0.002	0 to .001
Metamorphics	0 to 0.003	0 to .002

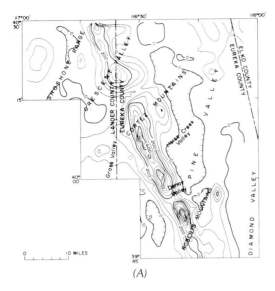

Figure 4-12 Aeromagnetic survey over central Nevada recorded from aircraft flight lines 3000 m above sea level; (A) anomalous magnetic field intensities contoured at 50 gamma intervals; (B) theoretical magnetic field intensity over model consisting of the distribution of magnetized rocks described in (C) and contoured at 50 gamma intervals. (E. S. Robinson, *Geological Society of America Bulletin,* Vol. 81, pp. 2045–2060, 1970.)

this supposed rock assemblage. By trial and error a magnetic field which compares favorably with the measured field is eventually calculated.

Consider the example in Figure 4-12. First, the total magnetic field intensity was measured along parallel profiles 1.5 km apart by trailing a fluxgate magnetometer from a low-flying airplane. Then, the effect of the main magnetic field was subtracted to get the anomalous magnetic field shown by contours (Figure 4-12 A). After several trials the magnetic field (Figure 4-12 B) was calculated for the 18 irregular rock units seen in the illustration (Figure 4-12 C). You can see that the principal features of the anomalous magnetic field are reproduced on this map. The calcu-

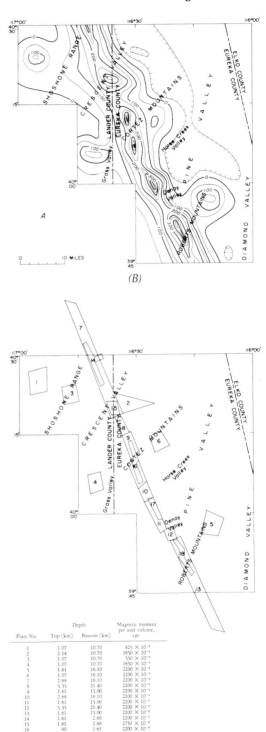

Plate No.	Depth Top (km)	Depth Bottom (km)	Magnetic moment per unit volume, cgs
1	1.07	10.70	825 × 10⁻⁶
2	2.14	10.70	1650 × 10⁻⁶
3	1.07	10.70	550 × 10⁻⁶
4	1.07	10.70	1650 × 10⁻⁶
5	1.61	16.10	2200 × 10⁻⁶
6	1.07	16.10	2200 × 10⁻⁶
7	2.68	16.10	2200 × 10⁻⁶
8	5.35	21.40	2200 × 10⁻⁶
9	1.61	15.00	2200 × 10⁻⁶
10	2.68	16.10	2200 × 10⁻⁶
11	1.61	15.00	2200 × 10⁻⁶
12	5.35	21.40	2200 × 10⁻⁶
13	1.61	15.00	2200 × 10⁻⁶
14	1.61	2.68	2200 × 10⁻⁶
15	1.61	2.68	2750 × 10⁻⁶
16	.80	1.61	2200 × 10⁻⁶
17	.53	1.61	2200 × 10⁻⁶
18	.53	1.61	1650 × 10⁻⁶

(C)

lations were made with an electronic computer so that many trial models with different rock units could be quickly tested.

The anomalous magnetic field over the oceans is especially interesting to geologists. Before 1950 little was known about marine magnetic anomalies. Then geophysicists began using newly invented airborne and shipboard magnetometers to survey the oceans. From these surveys of the earth's total magnetic field they discovered tremendous linear anomalies, some stretching several hundred kilometers. To find the anomalous magnetic field the effect of the main field was subtracted from these surveys of total field intensity. Some zones of the anomalous field have positive intensity values and other zones have negative values. These features are evident in Figure 4-13. Alternating positive and negative magnetic anomalies lay in great parallel strips across the face of the ocean. From place to place these linear anomalies were offset along almost straight lines.

Geologists compared maps of the anomalous field with maps of ocean floor topography. Three things became evident: (1) Linear magnetic anomalies were parallel to oceanic ridges; (2) Patterns of anomalies were symmetrical about ridge axes; (3) Anomaly patterns were offset where the large transverse fractures separate segments of oceanic ridges. Clearly there was some relationship between

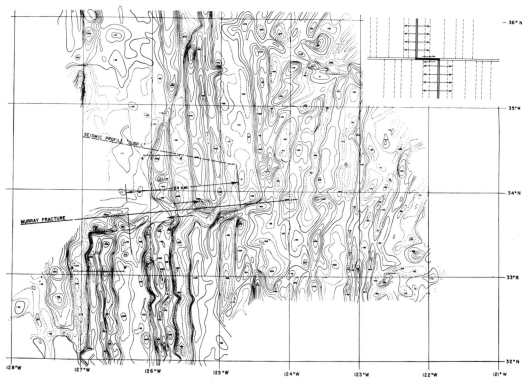

Figure 4-13 Marine magnetic field intensity survey revealing great linear magnetic anomalies over the Murray Escarpment in the eastern Pacific Ocean. Anomalous magnetic field is contoured at 50 gamma intervals with positive values shown by solid contours and negative values shown by dashed contours. (Reprinted from Victor Vacquier in *Continental Drift* [S. K. Runcorn, ed.], Fig. 2, facing page 136, Academic Press, 1962, with permission.)

the anomalous magnetic field and ocean floor topography. Geologists knew that there was an important story in these features. The problem was in decoding the information. The key to the code was found in the newly developing science of paleomagnetism.

THE ANCIENT MAGNETIC FIELD

Paleomagnetism

We all know about fossils. They are the remains of shell and bone that tell us about the earth's ancient creatures. But there are other kinds of fossils. These are ferromagnetic mineral grains. Their remanent magnetism tells us about the earth's ancient magnetic field. **Paleomagnetism** is the science of the earth's ancient magnetic field. We can learn from remanent magnetism in rocks how the earth's field has changed during the past.

We get some of the best evidence of fossil magnetism from volcanic rock such as basalt which forms when lava congeals. Ordinarily, basalt contains some magnetite and hematite. After the lava congeals the rock continues to cool. When its temperature drops below the Curie point, the ferromagnetic mineral grains become magnetized by the earth's main field. These grains acquire thermoremanent magnetism in the direction of the main field at the time when the rock temperature reaches the Curie point. Then, each grain becomes a tiny magnetic vector, pointing in the main field direction. As long as the basalt layer is not moved or tilted by tectonic forces, the thermoremanent magnetism will continue to point in that direction. If the main magnetic field direction changes in the meantime, the rock magnetism preserves a record of the former field direction. By finding a worldwide distribution of rock specimens that was magnetized at the same time you can reconstruct the ancient shape of the main magnetic field.

But how can you tell when the rock was magnetized? Basalt usually contains some radioactive elements such as uranium and thorium. You know that the age of the rock can be found from ratios of the masses of these elements and their radioactive decay products. Usually the temperature in volcanic rock such as basalt drops to the Curie point shortly after the lava congeals. Therefore the age of the thermoremanent rock magnetism is close to the age of the rock itself.

Volcanic rocks are not the only sources of fossil magnetism. Some sedimentary rock layers also preserve evidence of former directions of the main magnetic field. These layers, as you know, are made of the weathered fragments of more ancient rock. Some of the fragments are ferromagnetic mineral grains that were already magnetized. Perhaps they were washed from the land and then carried along by a river until they reached the quiet water of a lake or the sea. Here, with other particles they settled to the bottom, becoming part of a layer of sediment. But while the ferromagnetic grains settled through the water, they were oriented by the main magnetic field. When they reached the bottom and were buried in the sediment, these magnetized grains tended to preserve this alignment. Later, when the sediment was compacted and cemented into rock, evidence of the main field direction was preserved. At the same time that the sediment was accumulating, the remains of organisms were also being fossilized. The ages of guide fossils in these sedimentary layers now tell us the time when the main field was in the direction of the fossil magnetism.

Suppose that you want to collect and analyze rock specimens to find fossil magnetism. What is the procedure? Before you remove a specimen from an outcrop, you first have to mark it with arrows showing its orientation. Then you can return it to a laboratory to test it in a spinner magnetometer. Here it is mounted on a spindle. Magnetic fields from electrical coils near the spindle are then directed to cancel the earth's field completely. This removes any induced magnetism from the specimen. Now, when it is turned on the spindle, its remanent magnetism causes electrical current to flow in a nearby test coil. The direction of this remanent magnetism is found by spinning the specimen in many different orientations and observing how current changes in the test coil.

It is very important to prove that the remanent magnetism of a specimen has remained unchanged over the period of time since it was acquired. There are various effects that could alter the remanent magnetism, thus invalidating it as a reliable sample. The stability of this magnetism depends partly upon the crystalline structure of the ferromagnetic mineral concerned, and partly upon the various geological events that might alter the temperature and the chemical environment of the specimen. Certain tests are used to learn about remanent magnetic stability. As an example, consider the magnetism directions obtained from several samples in a folded sedimentary bed. If the individual samples show the directions seen in Figure 4-14A, we see that they would be parallel were the bed not folded. This implies that the remanent magnetism was acquired prior to folding. Now look at the parallel sample directions shown in Figure 4-14B. They indicate that the rock was remagnetized after folding. In addition to these and other tests, laboratory experiments have been devised to estimate the stability. If you can easily alter the remanent magnetism by placing the specimen in other magnetic fields, you can conclude that various geological events might also have caused alteration. This casts doubt on the long-term stability of the measured direction. Similarly, by raising the temperature of the specimen to near the Curie temperature you can test the effect of thermal events on the permanence of the remanent magnetism in your specimen.

During recent decades geophysicists have measured remanent magnetism in oriented rock specimens spanning most of geological time. Samples come from a worldwide distribution of locations. The stability of this mag-

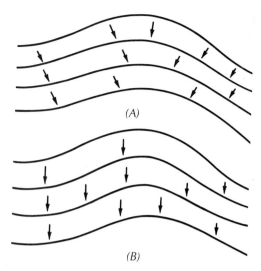

Figure 4-14 Relationships between remanent magnetization directions and bedding of sedimentary strata: (A) angles between magnetization and layering are similar; (B) angles between magnetization and layering are variable, but all magnetization vectors are parallel. (D. W. Strangway, *The History of the Earth's Magnetic Field*,, McGraw-Hill, 1970.)

netism has been proven so that the data can be used to learn about the history of the earth's field. Two important discoveries have emerged: (1) The direction of the earth's main magnetic field has reversed numerous times, and (2) the magnetic pole positions relative to various continents have changed.

Magnetic Field Reversals

What do we mean when we say that the main magnetic field reverses direction? This means that the north and south magnetic poles exchange position. A compass that points north in today's field would point south when the field was reversed. This idea was proposed after geophysicists had measured remanent magnetism in places where several rock layers are exposed. Specimens from some layers were magnetized in the same direction, but interspersed between these are other layers where the magnetism is in the opposite direction. Before these measurements were made no one had supposed that the main magnetic field had reversed itself. At first the measurements were hard to believe, and scientists suspected other causes. Perhaps chemical processes had altered the magnetism in some layers.

Research expeditions traveled to many parts of the world to obtain more measurements of remanent magnetism. With this additional information people began to recognize a pattern. Similar changes in the direction of remanent magnetism were found in many locations. All of the rocks formed during some intervals of time pointed to a main magnetic field much like today's field. But during alternate intervals of time remanent magnetism indicates an oppositely directed main field. This evidence has convinced most geophysicists of the reality of main magnetic field reversals, even though the cause of these events remains unexplained.

We find in the rock record a history of main magnetic field reversals reaching back into Precambrian time. The record is most complete for the Cenozoic era. The sequence of reversals is shown graphically in Figure 4-15. You can see that the field direction has been mostly **normal** for about the past 700,000 years. This means that its direction has been as we find it today. But during the time from 700,000 years to 1.6 million years ago the field was mostly **reverse,** or in the opposite direction. Field reversals seem to occur randomly, and we have no idea when the next one will happen. It takes between 5000 and 10,000 years for the field to reverse itself completely.

Now let us see what magnetic field reversals have to do with magnetic anomalies in the oceanic regions. The connection is found in the concept of **sea floor spreading,** one of the most exciting ideas in modern geology. A large part of Chapter 5 is devoted to this subject. But a brief introduction is useful now so that you can understand how the great linear magnetic anomalies over the oceans (Figure 4-13) are related to magnetic field reversals. Basically, the idea of sea floor spreading is that the oceanic ridges are situated along great rifts where the earth's brittle outer shell is drawing apart. This brittle outer shell consists of the crust and the upper part of the mantle that lies above the seismic low velocity zone. Magma then rises from the mantle to fill the openings beneath the axial troughs. Here it congeals, adding rock to the crust. While cooling, this newly formed rock acquires remanent magnetism. Then it slowly moves away from the ridge axis as the crust continues to draw apart. The remanent magnetism is preserved; it is the source of positive or negative magnetic anomalies depending on whether its direction is normal or reverse. In zones where remanent magnetism and the main field are in the same direction, positive anomalous field values are obtained when the main field is subtracted from the total field. You get negative anomalous field values where remanent magnetism is opposite the main field.

While this is happening, still newer rock is formed along the ridge axis and becomes magnetized by the main field as it cools below the Curie temperature. In the meantime a magnetic field reversal occurs. Therefore, the newer rock is magnetized in the direction opposite the older rock. After several reversals have occurred while rock was continually added along a segment of ridge axis, a series of long, parallel magnetic anomalies develops. These anomalies are alternately positive and negative like the features in Figure 4-13.

Paleomagnetic Poles

We could read a compass at a great many places on the earth to find lines pointing toward magnetic north. If we extended these

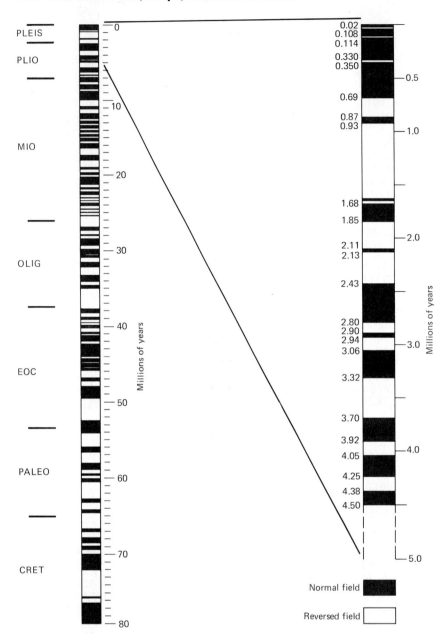

Figure 4-15 A history of the earth's magnetic field reversals during the past 80 million years. Times during which the polarization was the same as at the present time are indicated in black. Times of reversed polarization are shown by white. Reversals during the past five million years are shown in more detail. (From J. R. Heirtzler and others, *Journal of Geophysical Research*, page 2123, March 15, 1968, and F. D. Stacey, *Physics of the Earth*, page 176, John Wiley & Sons, Inc., 1969.)

lines around the earth, they would intersect somewhere near the geomagnetic poles. We could also find the direction of magnetic north from remanent magnetism directions measured in recently formed rocks. Lines extended in these directions, too, would intersect near the geomagnetic poles. But if we tried to use remanent magnetism directions from older rocks, we would have difficulties in locating the geomagnetic poles. The lines found from numerous measurements on one continent would show different pole positions than lines based on the remanent magnetism on other continents. One of the first to detect these discrepancies was the British geophysicist, S. K. Runcorn, a pioneer in the science of paleomagnetism. At first many scientists doubted Runcorn's results. How could there be several different geomagnetic poles? Many argued that the discrepancies in pole positions were simply the result of inaccurate data. But as Runcorn and others improved their experimental methods, the discrepancies did not disappear. If they were real, then there must be an explanation.

Most of the earth's present magnetic field, as we already know, could be explained by one large magnet in the core that is tilted about $12^1/_2$ degrees from the rotation axis. Most scientists agree that the ancient magnetic field, too, could be attributed largely to one large, fictitious magnet. But this magnet would have to be tilted differently, and its pole positions reversed from time to time. Although this assumption has not been proved absolutely, the evidence is in its favor. Studies of remanent magnetism in rocks and archaeological relics spanning the past 3000 years indicates that the geomagnetic pole positions change markedly over short intervals of geologic time, appearing to migrate about the pole of rotation. However, it is found that if these positions are averaged for time periods of about 2000 years, the mean geomagnetic pole lies close to the geographical pole.

Looking at data farther in the past we find evidence of a similar kind of magnetic field in earlier eras. Pole positions determined from fossil magnetism in Triassic rocks from North America are seen in Figure 4-16 A. The positions indicated by Triassic rocks from Europe and Asia are plotted in Figure 4-16 B. The fact that the points are fairly closely grouped suggests a magnetic field that could result mostly from one large magnet in the core. Otherwise, a much greater scatter would be expected. Similar evidence from other periods appears to substantiate the belief that the earth's magnetic field has been similar through much of the geologic past. The geomagnetic pole probably wobbles around the geographical pole, but when averaged over a few thousand years, the two poles are close.

The points in Figure 4-16 reveal an interesting pattern. Data from North America reveal a close grouping in a different location from the closely grouped points obtained from Eurasian data. Taking the mean pole position to be at the center of the group, we see two distinctly different pole positions revealed by two sets of data from different continents. Mean pole positions from other geologic periods, similarly determined from separate North American and Eurasian rocks, are seen in Figure 4-17. The pole positions from the two sets of data are quite close from Cretaceous time up until the present. Prior to Cretaceous time the patterns of change are similar for both groups, but they are offset. You can see that by shifting the North American and Eurasian continents close together like fitting peices of a puzzle, the pre-Cretaceous magnetic pole positions would be much closer. This relationship has been cited as evidence that the two continents have shifted relative to one another, and were much closer prior to the Cretaceous period. It is the explanation proposed by S. K. Runcorn for the discrepancies in geomagnetic pole positions found from remanent magnetism on different continents. A large part of Chapter 5 is devoted to the important subject of shifting continents.

138 The Earth's Size, Shape, and Interior Zones

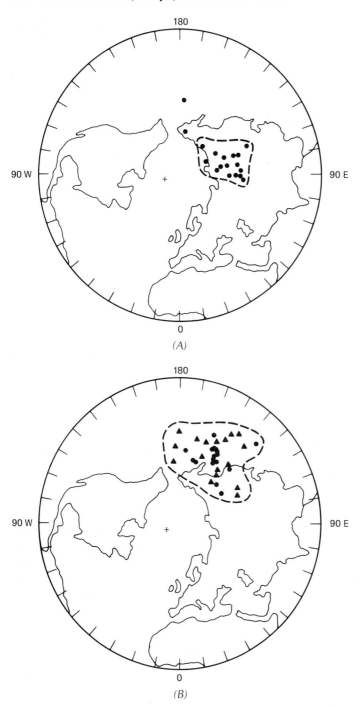

Figure 4-16 Pole positions indicated by remnant magnetization of Triassic rocks: *(A)* North American data; *(B)* Eurasian data. (D. W. Strangway, *The History of the Earth's Magnetic Field,* McGraw-Hill, 1970).

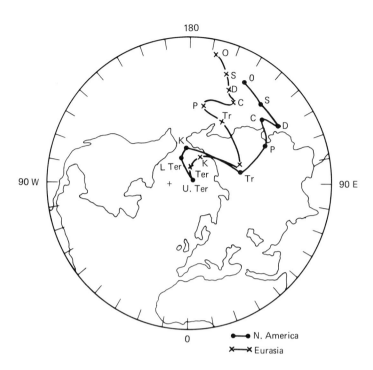

Figure 4-17 Paleomagnetic pole positions indicated by remanent magnetization of North American rocks (•-•) and Eurasian rocks (x-x). Abbreviations: O = Ordovician, S = Silurian, D = Devonian, C = Carboniferous, P = Permian, Tr = Triassic, J = Jurassic, K = Cretaceous, Ter = Tertiary. (D. W. Strangway, *The History of the Earth's Magnetic Field*, McGraw-Hill, 1970.)

THE ORIGIN OF THE MAIN MAGNETIC FIELD

Up to this point we have been concerned with describing the magnetism of the earth. Scientists have measured the field intensity and direction on the surface and in the space surrounding the planet to find the form of the field. Magnetic observatory records and studies of remanent magnetism indicate how the field changes with time. Now we can consider the source of this magnetic field. The attempt to explain the self-reversing magnetic field with the secular and spatial variations we observe on the earth is one of the most intriguing problems of theoretical geophysics.

Early theories attempted to explain the field by a permanently magnetized core, and by undiscovered principles of magnetism that might be associated with large rotating masses. We now know that the high temperature in the core would prevent any permanent magnetism in ferromagnetic materials. Also, the observed magnetic field variations with time are unlikely to be caused by a rigidly magnetized core. No consistent relationship is apparent in the earth or other rotating planets or stars that would point to any undiscovered physical principle. Consequently, these theories have been abandoned. Modern thought is directed toward a mechanism associated with motions of electrically charged fluids in the core. The liquid outer core is probably rich in iron. Consequently, it is probably a mobile fluid with good electrical conductivity properties. Therefore, it seems reasonable to look for possible electromagnetic effects related to fluid movement of the outer core.

First let us consider a simple system that could theoretically produce a self-sustaining magnetic field. The electromechanical dynamo has proved to be very helpful in understanding earth magnetism. The disc dynamo invented by Michael Faraday (1791–1867) has been used to reproduce certain features of the main magnetic field. Two well-known scientists, W. M. Elsasser and E. Bullard,

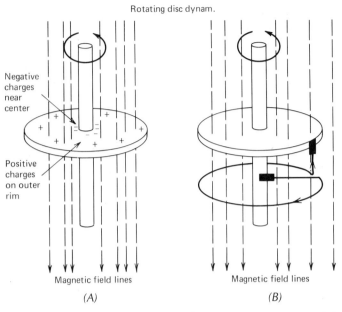

Figure 4-18 The disc dynamo: *(A)* disc rotates in a magnetic field causing separation of electrical charges; *(B)* disc with enclosing coil through which a current flows if the disc rotates. (Redrawn from T. Rikitake, *Electromagnetism and the Earth's Interior,* Elsevier-North-Holland, 1966.)

have developed a **magnetohydrodynamic** theory based upon the principles of fluid mechanics and dynamo theory. You can make a simple disc dynamo with a circular plate that rotates in a magnetic field. This motion will cause a separation of electrical charge, positive charges collecting on the rim and negative charges near the center of the disc, as seen in Figure 4-18A. By connecting an electrical conductor between the rim and the center an electrical current will flow to dissipate the excess charge. If you wrap this conductor into a coil around the disc (Figure 4-18B), the current will generate a magnetic field parallel to the initial field. By turning the disc at a certain critical velocity, the magnetic field generated by the current will be exactly sufficient to maintain the current, and the system becomes a self-sustaining dynamo. If mechanical friction of rotation and electrical resistivity in the conductor could be eliminated, the system would operate indefinitely. Such a self-sustaining dynamo offers intriguing possibilities for explaining the earth's main magnetic field.

One major difficulty with the simple dynamo is that there appears to be no simple way for the direction of the field to reverse. An interesting way to overcome this problem is to couple two dynamos together (Figure 4-19A) placing the conducting coil from one about the other. We know from the work of two scientists, T. Rikitake and D. W. Allen, how the magnetic field of this system can change with time. Some results are shown in Figure 4-19B. Here we see that spontaneous self-reversal is a physical possibility.

Other difficulties arise in explaining some of the irregular features of the earth's main magnetic field. More complicated combina-

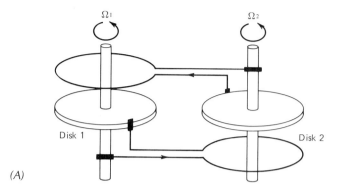

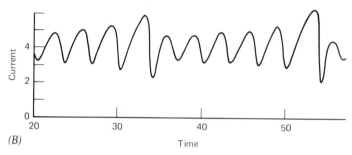

Figure 4-19 (A) A coupled system of two disc dynamos and (B) the corresponding theoretical spontaneous reversals of the associated magnetic field. (Rikitake, *Electromagnetism and the Earth's Interior,* Elsevier-North-Holland 1966.)

tions of disc dynamos have been devised that partly explain these features. The use of electromechanical dynamo models has by no means succeeded in reproducing all of the features of earth magnetism, but this approach to the problem is obviously productive.

The problem of relating a complicated disc dynamo system and the electromagnetism of a rotating conducting fluid sphere is very difficult. Preliminary attempts to solve this problem reveal the possibility of fluid current patterns arising from rotation and thermal convection effects. Thermal convection currents are caused by temperature differences in a fluid. You know how water currents circulate in a pot as you bring it to a boil. These are convection currents. Similar currents in the outer core could generate a magnetic field that would be almost self-sustaining. The patterns in Figure 4-20 illustrate the basic currents that are needed. Beyond this it is difficult for us to predict how successful theoretical geophysicists will be in explaining the details of earth magnetism. Not only are the physical conditions in the core obscure, but the mathematical operations are formidable.

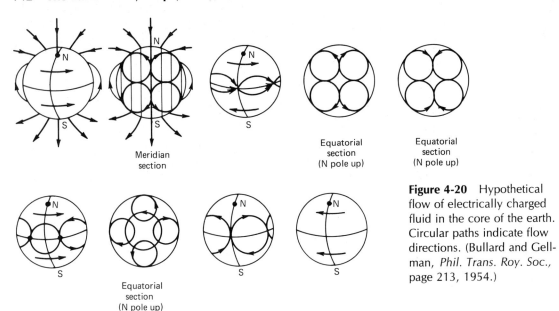

Figure 4-20 Hypothetical flow of electrically charged fluid in the core of the earth. Circular paths indicate flow directions. (Bullard and Gellman, *Phil. Trans. Roy. Soc.*, page 213, 1954.)

STUDY EXERCISES

1. Is the difference between the strongest and weakest values of the main field intensity measured on the earth's surface larger or smaller than the difference between the strongest and weakest values of anomalous field intensity?

2. Examine Figure 4-6 to determine the amount of secular change in the main magnetic field intensity at Anchorage, Alaska, that would be caused by west drift during the next 1000 years.

3. Suppose that a layer of Pliocene age basalt is found lying on a layer of Permian age basalt at a location in Nevada. Explain why the directions of rock magnetism in these two layers would probably be different from one another.

4. What characteristics of the earth's magnetic field can be used as evidence that the earth has a liquid core? Explain!

5. Describe the principal features of marine magnetic anomalies, and explain how these anomalies could be produced.

SELECTED READINGS

Elsasser, W. M., "The Earth as a Dynamo," *Scientific American*, pp. 44–55, May 1958.

Nettleton, L. L., *Gravity and Magnetics in Oil Prospecting*, McGraw-Hill Book Co., New York, 1976.

Strangway, David W., *History of the Earth's Magnetic Field*, McGraw-Hill Book Co., New York, 1970.

5

GLOBAL TECTONIC PROCESSES

The Plate Tectonic Theory
  Hypotheses and Speculation
  Sea Floor Spreading
  Topography and the Plate Tectonic Hypothesis
  Magnetism and the Plate Tectonic Hypothesis
Seismology and the Plate Tectonic Hypothesis
The Age of Ocean Basin Sediment
Hypothesis or Theory
Thermal Convection
Isostasy
The Global Tectonic Scheme

Look closely at a world map. Think about the shapes of the continents. Do they look like pieces of a giant jigsaw puzzle? If you could move these pieces, you could almost fit them together into one large supercontinent (Figure 5-1). Apparently the earliest mention of the interesting similarities of continental borders appears in the writings of the English philosopher Francis Bacon early in the seventeenth century. Later, in 1795, the great German naturalist Alexander von Humbolt commented that the complementary forms of different continents was too exact to be merely the product of chance.

Most geologists like to speculate about how the earth works. Is there some grand process that explains why the continents and ocean basins are shaped as we now see them? You know about the principal features of the earth's topography, seismicity, magnetism, and its interior zones. This information has enabled scientists to recognize some of the major processes that have shaped our planet.

THE PLATE TECTONIC THEORY

Hypothesis and Speculation

Is it reasonable to suggest that the present continents were formerly joined, and have since spread apart? The German scientist Alfred Wegener thought so. In 1912 he published a comprehensive **hypothesis of continental drift.** Wegener proposed that the present continents once formed a single large landmass that he called Pangaea. This supercontinent broke apart during the Mesozoic era and its fragments drifted slowly into the positions where we now see them. Wegener introduced different kinds of geologic information to support his hypothesis. He pointed to rocks of similar age near the matching borders of continents. He argued that zones of deformed rock on different continents lined up to indicate the ancient mountain chains that once reached across Pangaea. His most compelling argument is elegantly simple,

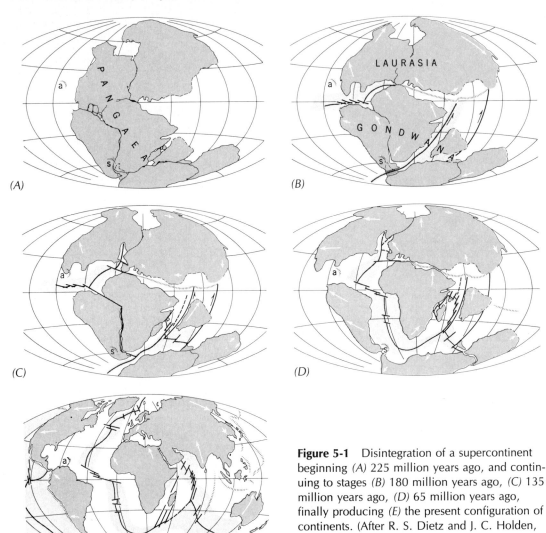

Figure 5-1 Disintegration of a supercontinent beginning (A) 225 million years ago, and continuing to stages (B) 180 million years ago, (C) 135 million years ago, (D) 65 million years ago, finally producing (E) the present configuration of continents. (After R. S. Dietz and J. C. Holden, *Journal of Geophysical Research*, pp. 4939–4956, Sept. 10, 1970.)

however. Wegener showed how the present continents could be fitted together to make one large supercontinent (Figure 5-1). But the idea of continents plowing through solid rock seemed inconceivable to many geologists. They had other ideas about the geologic evidence presented by Wegener, and they maintained that the apparent fit of the continents was meaningless. Wegener's hypothesis became one of the most controversial topics in the science of geology. Have the continents remained in fixed positions, or have they drifted apart as fragments of a former supercontinent?

To move a continent you would need a dynamic process of magnificent proportions. Knowing about such a process would certainly help us to understand some of the earth's grand geologic features. Recall that processes causing physical deformation such as folding and faulting are called tectonic processes. The large-scale processes that could deform whole continents and ocean basins are **global tectonic processes.** Geologists have come up with interesting ideas about these processes which might help us to understand the workings of our planet.

Alfred Wegener thought that continental drift was the result of centrifugal forces related to the earth's rotation, and lunar and solar gravitational forces that tended to pull the continents toward the equator. But opponents of his hypothesis proved that these forces were entirely inadequate to move continents. At the time when Wegener introduced the novel idea of continental drift most geologists believed that folding and faulting in the crust, seismicity, and mountain building were caused by contraction in a cooling earth. Knowing of no global tectonic process that could move continents, these scientists argued that drift was impossible regardless of the shapes of continents.

As the debate about continental drift continued, an interesting idea was proposed to explain the movement of large land masses. Some scientists believed that temperature and pressure in the mantle are sufficient to make the rock ductile like a soft plastic. If the right temperature conditions existed, there might be thermal convection currents flowing slowly in the mantle such as when water boils in a pot. In Figure 5-2 you can see one way that a mantle convection system might work. The rock flows slowly around a so-called **convection cell** because of density differences that result from heating in the lower mantle and cooling in the upper mantle. The well-known English geologist Arthur Holmes was one of the first to propose that continental drift is caused by a global tectonic porcess of mantle convection. In 1928 Professor Holmes published his idea that the continents are rafts of low density rock that float on the more plastic mantle. These huge rafts drift with the convection currents in the upper mantle.

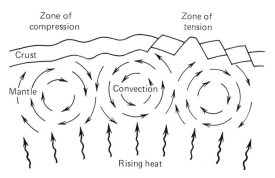

Figure 5-2 Hypothetical thermal convection cells in the ductile rock of the mantle, and resulting deformation of the brittle crust. (Modified from B. F. Howell, *Introduction to Geophysics,* page 327, McGraw-Hill Book Co., 1959.)

But what about the ocean basins? When the ideas of continental drift and mantle convection were first introduced, little was known about almost 75 percent of the earth's surface covered by the sea. Certainly these global tectonic processes must have left an imprint on the ocean floor. The search for this imprint began shortly after 1950, when echo sounders, magnetometers and gravimeters, ocean floor temperature probes, and seismic instruments had been developed for shipboard surveying. During the next decade enough information was gathered for scientists to recognize the principal features of submarine topography and crustal structure. This was the grist for more speculation about global tectonic processes.

Sea Floor Spreading

Sea floor spreading was the most important new hypothesis to come from the recently acquired knowledge about the ocean basins. The basic idea was proposed around 1960 by Harry Hess, a geology professor at Princeton University. We can see in Figure 5-3 how the process of sea floor spreading works. Hess believed that the oceanic ridges are situated

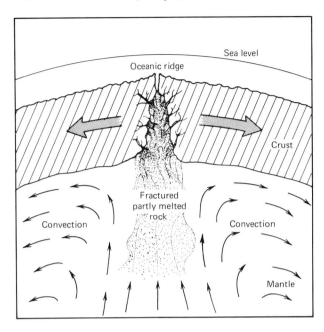

Figure 5-3 The idea of sea floor spreading proposed by Professor Harry Hess was that mantle convection caused the crust to draw apart so that fractures open beneath the axial trough of an oceanic ridge. Magma melted from the mantle rises to fill these openings and congeals in a way that produces new crustal rock.

over rising currents of mantle convection cells. He thought that if convection currents rise and then spread outward horizontally in two opposite directions, the overlying crust would be drawn apart. Then a crack extending through the crust would open beneath the axial trough that marks the center of an oceanic ridge. This crack would be filled by molten rock called **magma** which was melted from the upper mantle by the heat rising by convection. The magma comes from mantle rock that has a density of about 3.4 gm/cm^2. But when it congeals in the crust, rock having a density of 2.9 gm/cm^3 is produced. Harry Hess pointed out that you could expect this density change because of the different conditions in the mantle where the magma forms and in the crust where it congeals. Pressure is lower in the crust, and temperature drops low enough for water to react chemically with magma causing the lower density rock to form. You can see that material is taken from the mantle and added to the crust as a result of sea floor spreading.

Harry Hess viewed sea floor spreading as a continuous process of manufacturing oceanic crust with magma produced by melting in the mantle under the oceanic ridges. Mantle convection continually moves the new increments of crust away from a ridge axis, and opens space for still younger crustal rock to form. But Hess also was convinced that the earth had reached its present size early in Precambrian time, and has remained this size ever since. This meant that if rock were taken from the mantle to make new crust under the ridges, then crustal rock must be returned to the mantle somewhere else. Hess thought that the submarine trenches marked the most likely places where oceanic crust might be broken and drawn down into the mantle. Here the elevated pressure and temperature would cause chemical reactions that could change the broken fragments of crust into higher density mantle rock. Therefore, in the process of sea floor spreading the oceanic crust is a rind on the upper part of a mantle convection cell. It forms beneath the ridges and is carried outward to the trenches where it disintegrates.

How fast does the sea floor move? To answer this question Hess turned to the older idea of continental drift. If Wegener were correct, the Atlantic Ocean began opening during the Mesozoic era about 150 million years ago and has now reached a width of about 3000 km. This implies that the Atlantic has been widening at the rate of 2 cm/yr. But this is the result of sea floor spreading in both directions away from the ridge. Therefore, Hess reasoned that a point on the sea floor moved away from the ridge at the rate of about 1 cm/yr. If this value could be applied elsewhere, then nowhere could the present oceanic crust be more than about 300 million years old. Compared with the ancient Precambrian rock that is found in the continental crust, Hess predicted that the rock of the oceanic crust would prove to be very young.

It seemed obvious to Hess and other scientists that if the sea floor is spreading, then the continents must be drifting. This may not be clear to you if you look only at the Pacific Basin. Here you can see trenches that almost surround this ocean. Supposedly, oceanic crust formed under the Pacific ridges could be completely consumed along these trenches without any movement of the continents. But if you look at the Atlantic Basin, you find very few trenches. Clearly, if the Atlantic sea floor is spreading, then the Americas must be separating from Africa and Europe.

It was logical for scientists to combine the ideas of sea floor spreading and continental drift into a more comprehensive hypothesis about how the earth works. We now call this the **plate tectonic hypothesis.** It describes the movement of rock in the earth's brittle outer shell, which is called the **lithosphere,** and in the deeper more ductile zone that we call the **asthenosphere.** You can see the principal features of the plate tectonic hypothesis in Figure 5-4. The lithosphere varies in thickness from less than 70 km to as much as 200 km, and therefore includes the crust and part of the upper mantle as well. The ductile asthenosphere is a partially molten region where small pockets of magma are interspersed through a matrix of rock. This region is between 200 and 400 km thick. The molten constituents make up as much as 10 percent of the asthenosphere. Beneath it is a completely solid but ductile region called the **mesosphere.**

According to the plate tectonic hypothesis, the lithosphere consists of several large plates. Oceanic ridges, submarine trenches, and some long escarpments crossing the ocean basin floor mark the edges of these plates. The ridges lie along the so-called **constructive plate margins.** Here new lithosphere is produced from magma that rises out of the asthenosphere. Newly formed lithosphere then moves outward in both directions away from a ridge axis. You can see that sea floor spreading is a process that involves not only the crust, but the entire thickness of the lithosphere.

Submarine trenches indicate places where the lithospheric plates are pushing together. Here one plate slides under the other, and is drawn downward into the asthenosphere, as deep as 700 km in some places. This process is called **subduction.** The subducting margin of the lithosphere is eventually changed into the more ductile high density rock of the asthenosphere. This change is stimulated by the increasing pressure and temperature encountered at greater depth in the asthenosphere. Thus you can see that submarine trenches mark the **destructive margins** of lithospheric plates. Here rock from the earth's outer shell is recycled into the asthenosphere.

Some long escarpments across the ocean basin floor exist where two plates are sliding past one another, moving in opposite directions. Here we find the **conservative plate margins** where rock is neither added to nor removed from the lithosphere.

The plate tectonic hypothesis, then, proposes a global tectonic process of moving plates. Borders of the principal plates illustrated in Figure 5-5 are indicated by oceanic

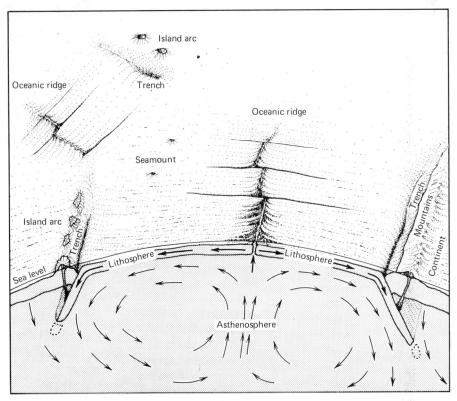

Figure 5-4 This diagram illustrates how the lithosphere forms and spreads away from oceanic ridges, and is conveyed to zones bordering submarine trenches, where it is drawn downward and reassimilated into the asthenosphere accrding to the plate tectonic hypothesis.

ridges, submarine trenches, and some long ocean floor escarpments. The continents are imbedded in these plates. They shift slowly over the face of the earth because the plates are moving. The floor of the ocean spreads outward from the oceanic ridges as a result of the plate motion.

The real challenge of science is to explain the events and features that we observe. A preliminary explanation which is proposed without much substantiating evidence is called an **hypothesis**. A **theory** is an explanation that is supported by an abundance of evidence. You have been introduced to the plate tectonic hypothesis. Do we have enough evidence favoring this idea to call it the plate tectonic theory? You can judge this for yourself after we have looked at the evidence of topography, magnetism, seismicity, and the age of the ocean floor.

Topography and the Plate Tectonic Hypothesis

You can recognize global patterns of topography from maps that emphasize the principal features by artistic exaggeration. These features are clearly evident in Figure 5-6. Here you can easily distinguish the continental platforms from the ocean basins. The shapes of the continental platforms, more than any

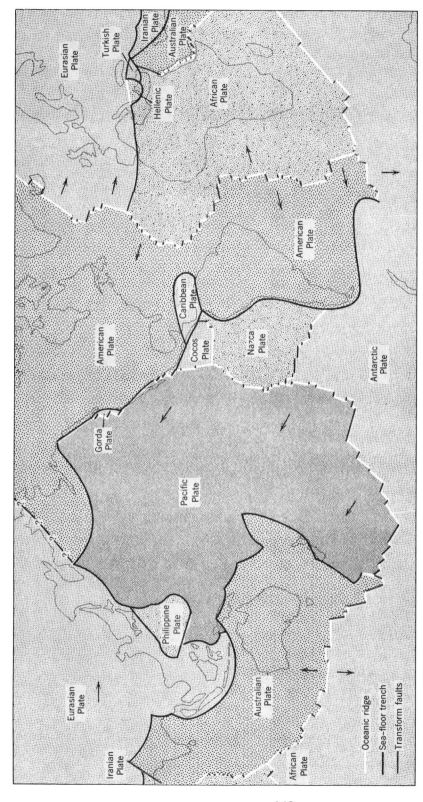

Figure 5-5 The lithosphere can be separated into six large plates and several smaller plate fragments bordered by constructive margins along oceanic ridges, destructive margins marked by submarine trenches, and conservative margins indicated by long ocean floor escarpments. (Reprinted from John Dewey, *Scientific American*, pp. 56–57, May 1972.)

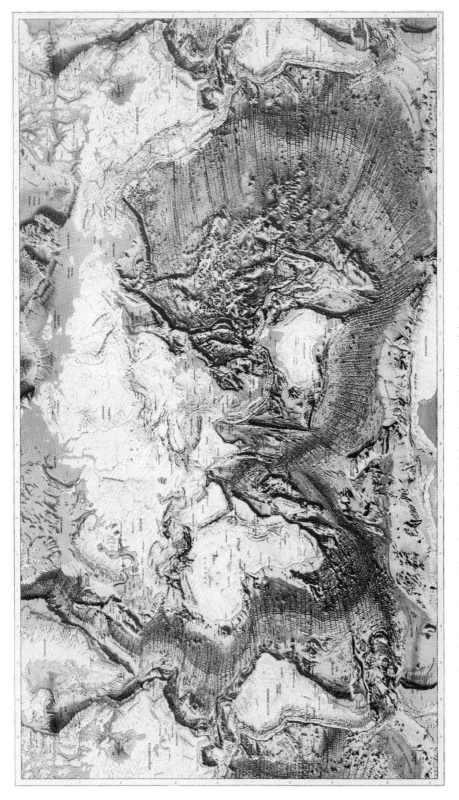

Figure 5-6 Topography of the earth's solid surface. (Reprint of the chart "The Floor of the Oceans" prepared by Tanguy de Remur for the Office of Naval Research, U.S. Navy, from bathymetric studies of B. C. Heezen and M. Tharp.)

other feature of the earth, have stimulated scientists to think seriously about the possibility of continental drift.

Even among proponents of the continental drift hypothesis, there has always been disagreement about exactly how the continents were formerly grouped. Should you use the outer edge of the continental shelf, or the base of the continental slope, or some intermediate contour on the slope when trying to find the correct arrangement? Clearly you can expect some distortion of the original borders after the continents became separated. Erosion has produced submarine canyons that cut deeply into the continental shelf edges. Sediment washed from these canyons has been deposited along the base of the continental slope. The original shapes of the continents are probably least distorted by these processes at depths midway down the continental slope.

A group of English scientists directed by Professor E. C. Bullard proposed a systematic way to arrange the continents in different groupings, and to test these to find the one that produces the closest fit. The method is based on a theorem of spherical geometry that was developed by the eighteenth-century Swiss mathematician Leonhard Euler. According to Euler's theorem, a rigid plate can be moved from one position on a sphere to any other position by a simple rotation about a properly chosen axis that passes through the center of the sphere. Bullard's idea was to find by trial and error the fictitious axis for rotating the outline of one continent against another so that there would be a minimum of gaps and overlaps between them. The fit of Africa and South America was the first to be tested. A trial axis was chosen. Then great circle lines could be drawn from this axis to equidistant points on the borders of the two continents. It is clear from Euler's theorem that angles between great circle lines for all pairs of these matching points should be equal if the axis is correctly chosen. You can see from Figure 5-7 how different trial axes

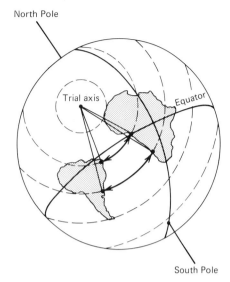

Figure 5-7 Africa and South America are used here to illustrate Euler's theorem, which states that the movement of a rigid plate over the surface of a sphere can be expressed as a simple rotation around some axis. This implies that if Africa and South America were formerly joined, their movement away from one another can be described by a circular path around some axis that can be located by trial and error. Therefore, points along the coasts that were formerly at the same place can be joined by paths lying along circles all having the same center. Two such paths centered on the same trial axis are illustrated.

can be tested until one is found that produces nearly the same angles for all pairs of points.

Bullard and his associates used an electronic computer to test many trial axes. Their best fit was found for an axis at 43.9 degrees N. latitude and 30.3 degrees W. longitude. The slopes of Africa and South America were taken from the 1000-m depth contour on the continental slopes. The angles for all pairs of matching points were within $1^1/_2$ degrees of the average value of 57 degrees. The fit that you see in Figure 5-8 was obtained by rotating the outline of South America by this angle toward the outline of Africa. The gaps and

152   The Earth's Size, Shape, and Interior Zones

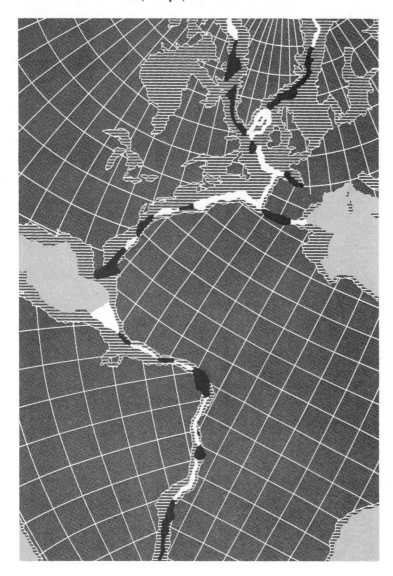

Figure 5-8 The fit of Africa and South America based on rotation according to Euler's theorem through a 57-deg angle about an axis through the earth's center and intersecting its surface at 43.9 deg north latitude and 30.3 deg west longitude. (After E. C. Bullard, J. E. Everett, and A. G. Smith, *Philosophical Transactions*, Royal Society of London, Vol. 258, A1088, Figure 8, 1965.)

overlaps are mostly less than 90 km wide. Bullard believed that this remarkably good fit presents us with a choice of two possible conclusions. Either these two continents were formerly joined or their shapes have no meaning whatsoever in the debate about continental drift. The fit of Africa and South America shown in Figure 5-8 was based solely on shape. Later, the information about rock ages was added to this map. This independent information appears to support the first of Bullard's two choices.

Fitting the outlines of continents is only part of the effort to find topographic evidence that supports the plate tectonic hypothesis. You should also examine the floor of the ocean. Look carefully at the features of the oceanic ridges that are illustrated in Figure 5-6. Observe how the transverse fractures reach across a ridge, separating it into segments.

Global Tectonic Processes 153

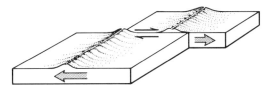

Figure 5-9 Displacement of two segments of a ridge that would result from simple horizontal movement along a transverse fracture in the absence of sea floor spreading. Continuing horizontal movement would increase the offset between the two ridge segments.

When scientists first saw these features, many thought that the transverse fractures looked like faults along which ridge segments had been offset (Figure 5-9). This implied that these ridge segments had been formerly aligned. But if this were true, then there was no obvious relationship between oceanic ridge topography and the shapes of the continents which showed no similar faults crossing their borders.

A new and completely different interpretation of oceanic ridge topography was then proposed by J. Tuzo Wilson, A Canadian geophysicist. Wilson proposed a process called **transform faulting** that was related to the earlier idea of sea floor spreading. You can see how it works from Figure 5-10. Assume that the sea floor is spreading outward from the axes of two adjoining ridge segments. Now look at the part of a transverse fracture that extends between these two ridge axes. Here the sea floor on one side of the fracture must be moving in opposition to the sea floor on the other side. This part of the transverse fracture is called a **transform fault.** But the transverse fracture extends in both directions beyond the ridge axes. Along these extensions of the fracture the sea floor on both sides is moving in the same direction. Small differences in the spreading rates of the adjoining ridge segments could account for the continuation of transverse fractures beyond the ridge axes. These fractures terminate farther out where movement of the sea floor on both sides becomes adjusted to the same rate. The rugged relief of the central rift mountains and high fractured plateaus is produced partly by these adjustments.

Transform faulting is an explanation of the movement of the sea floor relative to transverse fractures and axes of oceanic ridge segments. But it does not explain the origin of these ridge segments, and why they are offset. You can draw a line along the axes of ridge segments and their connecting trans-

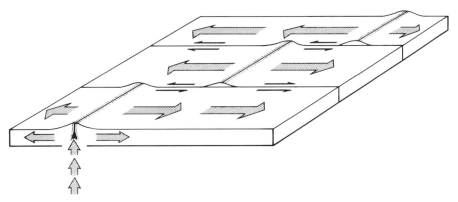

Figure 5-10 Transform faulting is the process of relative movement along transverse fractures related to sea floor spreading and the condition that offsets between ridge segments remain unchanged. Therefore, any relative movement along a transverse fracture is the result of different directions or speeds of sea floor spreading on one side of a fracture compared with the other side.

form faults. Perhaps this indicates the shape of the initial jagged break in the lithosphere when two plates began spreading apart, but we cannot say for sure. If this were true, however, then the offset segments do not represent displacements that have been occurring after the ridge was first formed. Rather, these offset segments are an original feature of the ridge.

Now look again at the transverse fractures. Do they appear to be slightly curved? This possibility is evident in Figure 5-11 for transverse fractures crossing the Mid-Atlantic Ridge between Africa and South America. According to the idea of transform faulting, these fractures should be parallel to the directions of sea floor spreading. This implies that a point on the ocean floor is moving away from the ridge along a curved path. You know from Euler's theorem that this kind of path would be expected for a point on a plate that is moving by simple rotation over a sphere (Figure 5-7).

Do the curved transverse fractures imply that the Atlantic sea floor is spreading by simple rotation? To test this idea you can draw great circle lines perpendicular to the transverse fractures. If simple rotation has occurred, these great circle lines should all intersect at the hypothetical axis of this rotary displacement. You can see in Figure 5-11 that there is a definite zone of intersection. Furthermore, this zone is quite close to the hypothetical axis that Bullard and his associates found by fitting outlines of Africa and South America. This could mean that drifting of these continents, and spreading of the sea floor between them are both controlled by the same process. According to the plate tectonic hypothesis, this process is the movement of two lithospheric plates outward from the Mid-Atlantic ridge (Figure 5-10).

Intersection of great circle lines like those in Figure 5-11 supports the idea of simple rotational movement of rigid plates. Great circle lines perpendicular to transverse fractures crossing other oceanic ridges indicate inter-

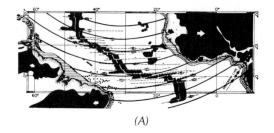

(A)

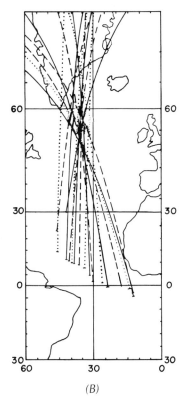

(B)

Figure 5-11 *(A)* Transverse fracture zones crossing the Mid-Atlantic ridge in the central Atlantic Ocean, and *(B)* projections of lines perpendicular to the central Atlantic fracture zones. (Modified from W. Jason Morgan, *Journal of Geophysical Research,* pp. 1965 and 1067, Mar. 15, 1968.)

section zones in different locations. This evidence favors the idea that the lithosphere consists of several rigid plates such as in Figure 5-5.

When you look at global topography (Figure 5-6) the borders between the continental platforms and the ocean basins are the most impressive features. But for purposes of learning how the earth works, the oceanic ridges, the submarine trenches, and some long ocean floor escarpments may prove to be more important topographic features.

Magnetism and the Plate Tectonic Hypothesis

Serious debate about continental drift declined after 1930. Scientists holding opposing views reached a stalemate that could not be resolved with the information available to them. More than 20 years passed before the debate was resumed. New information about rock magnetism stimulated scientists to look again at the possibility of continental drift.

You will recall from Chapter 4 that remanent magnetism in North American rocks pointed toward different paleomagnetic poles than remanent magnetism in Eurasian rocks (Figure 4-16). Similar discrepancies are found when you compare paleomagnetic pole positions determined from remanent magnetism in South American and African rocks. You can see these results in Figure 5-12. Now suppose that Africa and South America have drifted apart by simple rotation about the hypothetical axis found by E. C. Bullard and his colleagues. According to Euler's theorem, the apparent paleomagnetic poles would have shifted from their original positions along circular paths centered about this same axis.

Africa and South America are now separated by a 57-degree angle of rotation relative to this axis. You can see in Figure 5-12 how the paleomagnetic pole positions compare when they are rotated back through this angle. Paleozoic and early Mesozoic pole positions relative to the two continents are almost superposed. Apparent pole positions for late Mesozoic and Cenozoic time are offset. This is the pattern that would be produced if Africa and South America began drifting apart sometime during the Mesozoic era. Similarly, the different Triassic pole positions deter-

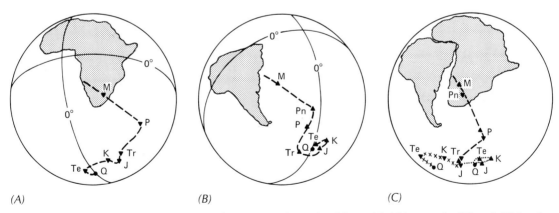

Figure 5-12 Apparent paleomagnetic pole positions determined from (A) African rocks (▼) and (B) South American rocks (▲) for times during the Mississippian (M), Permian (P), Triassic (Tr), Jurassic (J), Cretaceous (K), Tertiary (Te), and Quaternary (Q) Periods. (C) If these two continents and their corresponding apparent paleomagnetic poles are rotated according to Euler's Theorem (Figure 5-7), data from both continents indicate approximately the same pole positions from the Mississippian through the Triassic periods. Thereafter, the data indicate digressing paths of pole migration. This confirms the idea that these continents began to drift apart during Jurassic time. (After K. M. Creer, A Review of Paleomagnetism, *Earth Science Reviews*, Vol. 6, pp. 384, 386, and 394, Elsevier-North-Holland, 1970.)

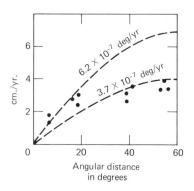

Figure 5-13 Curved lines show the dependence of rate of displacement of a point on a spherical surface upon geocentric distance of the point from the rotational axis for assumed angular displacement rates of 3.7×10^{-7} and 5.6×10^{-7} deg/yr. The points indicate sea floor spreading rates determined from analyses of marine magnetic anomalies at locations shown by geocentric distance from the axis located at 43.9° N, 30.3° W. (Based on data from W. Jason Morgan, *Journal of Geophysical Research,* page 1967, March 15, 1978.)

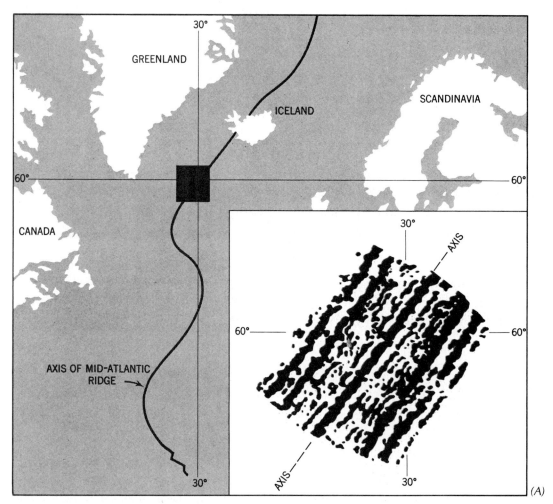

Figure 5-14 Magnetic anomalies over Reykjanes Ridge, a segment of the Mid-Atlantic ridge. *(A)* Map of anomalous field intensity shows positive anomalies in black and negative anomalies in white. *(B) (opposite)* A profile of measured field intensity across the ridge is compared with a simulated profile computed for the model of the crust consisting of alternating zones of positively (black) and negatively (white) polarized rock magnetism. Note that the profiles are more or less symmetrical about the center of the ridge. (After J. R. Heirtzler, X. LePichon, and J. G. Baron, *Deep Sea Research,* pp. 427–443, Vol. 13.)

Global Tectonic Processes 157

mined from North American rocks and from Eurasian rocks (Figure 4-16) suggest that these two continents began drifting apart during Mesozoic time.

We can use this information about apparent paleomagnetic poles to estimate the average rate at which Africa and South America may have moved apart. This information indicates that these two continents first separated sometime between 150 million (Jurassic) and 100 million years ago (Cretaceous).

Assume that they have moved apart by simple rotation around Bullard's axis so that they have now reached an angular separation of 57 degrees. This implies that the average rate of angular displacement has been between 3.7×10^{-7} deg/yr and 5.6×10^{-7} deg/yr.

You can see from Figure 5-7 that when a plate is moved over a sphere by simple rotation, all points on the plate have the same angular displacement. But points near the rotation axis are moved through shorter circular paths over the surface than points farther from the axis. Curves in Figure 5-13 show how the rate of movement over the surface varies with geocentric distance from the rotation axis. For points moving at the same rate of angular displacement those located farther from the axis move at faster rates along paths over the surface.

Again, let us suppose that Africa and South America have been separating by simple rotation around the axis found by Bullard and his associates. According to the information in Figure 5-13, this implies that the South Atlantic Ocean has been widening at a rate that increases toward the south. If the plate tectonic hypothesis is correct, then the rate of sea floor spreading in the ocean between Africa and South America should be faster toward the south. We can use marine magnetic anomalies to test this supposition.

First, you should recall the typical marine magnetic anomaly patterns, and how they can be explained in terms of sea floor spreading and reversals of the main magnetic field. The important features are illustrated in Figure 5-14 for a segment of the Mid-Atlantic ridge that is called Reykjanes Ridge. Here you can see that the anomalous magnetic field consists mainly of long parallel zones. The anomalous magnetic field intensity values are alternately positive and negative in these zones. These features are the great linear marine magnetic anomalies that are found over much of the ocean. Along an oceanic ridge segment these anomalies are parallel to the ridge axis. They make a symmetrical pat-

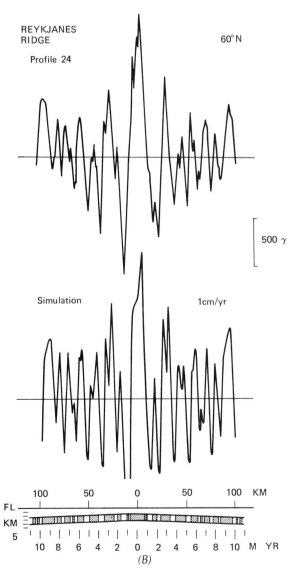
(B)

tern that is centered along the axis. Anomalies in this pattern are continuous between the transverse fractures that mark the ends of the ridge segment. At places where ridge segments are offset along these transverse fractures the symmetrical magnetic anomaly patterns are similarly offset (Figure 4-13).

The marine magnetic anomalies are caused by rock magnetism in the oceanic crust. The anomalies have positive values over the zones where the rock magnetism is in the same direction as the earth's main magnetic field at the present time. Negative anomalies occur over zones where the rock magnetism is opposite the present main field direction. If the idea of sea floor spreading is correct, then remanent magnetism is acquired very near a ridge axis when crustal rock formed from congealing magma cools below the Curie temperature. At some point in the crust this rock magnetism is in the direction of the main field at the time when the Curie temperature was reached. It is preserved as the crust spreads outward from the ridge axis, even though the main field undergoes subsequent reversals. The crust forms, and moves outward from a ridge axis continuously, whereas the main magnetic field undergoes a series of reversals. Therefore, parallel zones in the crust should be formed where the directions of remanent magnetism are alternately normal and reverse. Let us assume that rates of sea floor spreading are equal in opposite directions from a ridge axis. Then, a symmetrical pattern of normally and reversely magnetized zones should have developed. This would produce the pattern of marine magnetic anomalies that your see in Figure 5-14.

Let us continue with this line of reasoning. Assume that sea floor spreading has been a continuous process. Then, extending outward on each side of the ridge axis you should find positive and negative marine magnetic anomalies corresponding to each episode of normal and reverse main magnetic field direction.

You already know the times when the main field was oriented in normal and reverse directions (Figure 4-15). This was learned from the remanent magnetism in continental rock layers. Here is the key to finding rates of sea floor spreading from the spacing of marine magnetic anomalies. You can see from Figure 5-15 how this is done. First you measure the distance (x) from the ridge axis to some particular positive or negative magnetic anomaly. Then refer to the graph that shows the history of main field reversals. On this graph you can find the time (t) since the remanent magnetism was acquired, which is causing the magnetic anomaly. Look at the example in Figure 5-15A where the distance (x) is marked for the fifth positive anomalies outward in both directions from the ridge axis. Then look on the graph (Figure 5-15B) to find how much time (t) has passed since the fifth previous episode of normal main field direction. Now you can calculate the velocity of sea floor spreading (V) from the well-known formula: $V = x/t$.

Scientists have used this method to calculate rates of sea floor spreading in most parts of the ocean. Almost everywhere the rates are found to be a few cm/yr. But at a few locations in the Pacific Ocean rates of as high as 10 cm/yr are known. Results from the Atlantic Ocean between Africa and South America are especially interesting. Some of these values are shown in Figure 5-13. Each point was obtained for a particular segment of the Mid-Atlantic ridge. It indicates the sea floor spreading rate at a geocentric distance measured from Bullard's hypothetical rotation axis. You can tell two things from the points in Figure 5-13. First, the sea floor spreading rates are close to values which you would expect from the estimated rate of continental drift that we discussed earlier. The other important indication is that the sea floor is spreading faster toward the south. This was the supposition that we set out to test by analyzing marine magnetic anomalies.

You can see that we have two independent ways for using the earth's magnetism to esti-

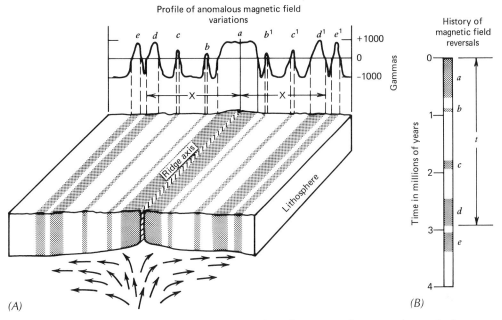

Figure 5-15 Measurement of sea floor spreading rates based upon (A) distances of a particular anomaly from the ridge crest x, and (B) the time since magnetization of the portion of the crust that causes the anomaly t, obtained by correlating the apparent reversals of crustal magnetization and the time scale of paleomagnetic reversals.

mate the rate of widening for the Atlantic Ocean between Africa and South America. In one method we used apparent paleomagnetic pole positions that were determined from continental rock magnetism. In the other method we used marine magnetic anomaly patterns. Approximately the same rates for widening of the ocean were obtained from these two separate sources of information. According to the plate tectonic hypothesis, you should expect these results because sea floor spreading and continental drift are both controlled by the same global tectonic process.

Seismology and the Plate Tectonic Hypothesis

Can we find evidence for distinguishing the lithosphere from the asthenosphere? If so, can we find out if the lithosphere consists of several moving plates? We can learn more about these features of the plate tectonic hypothesis from earthquakes and seismic waves.

Most earthquakes occur in clearly defined zones of seismicity. Look again at Figures 3-6 and 3-7 to see the global distribution of earthquakes. They are concentrated near the oceanic ridges, the submarine trenches, and some ocean floor escarpments. Suppose that these features do mark the constructive, destructive, and conservative borders of the plates that make up the lithosphere (Figure 5-5). If the plates are moving, then you would expect earthquakes along the borders where the edges grind together.

Most earthquakes are caused by movement along fractures in brittle rock. More than 90 percent of these events are shallow focus quakes. The remaining intermediate and deep focus earthquakes are all confined to the inclined seismic zones that extend downward from submarine trenches as far as 700 km. This distribution of earthquakes supports the idea of a brittle outer shell that we call the lithosphere. Absence of earthquakes in the deeper region beneath the lithosphere probably means that the rock is ductile enough to flow in response to stress rather than fracturing. This is the kind of rock we would expect to find in the asthenosphere. Now suppose the lithosphere consists of several plates arranged as shown in Figure 5-4, so that some edges extend downward along the inclined seismic zones. This would explain the presence of brittle rock in the places where intermediate and deep focus earthquakes occur.

How thick is the lithosphere? Seismologists cannot give an exact answer. The earthquakes that occur along oceanic ridge axes and transverse fractures are very shallow. Perhaps the brittle lithosphere reaches only a few tens of kilometers beneath the ridges. On the seaward side of submarine trenches the depth range for shallow quakes is larger. Here the lithosphere could be as much as 100 km thick. The same kind of evidence indicates that brittle rock may extend to depths of 200 km or more in some places under continents. Earthquake activity in the inclined seismic zones is largely confined within a thickness of a few tens of kilometers.

We know that seismic wave velocities increase downward through the crust and into the upper mantle. There a region is reached where P- and S-waves travel more slowly. This is the seismic low velocity zone that is clearly evident in Figure 3-21. Seismologists have analyzed the ways that amplitude and shape of seismic phases are changed when these phases travel through the low velocity zone. The results indicate that it is a relatively soft and ductile zone. This has led scientists to define the asthenosphere to be the region indicated by the seismic low veolcity zone. Seismologists cannot tell for certain if the borders of the asthenosphere are abrupt or gradational upward into the lithosphere and downward into the mesosphere.

We can tell the approximate thickness of the lithosphere from estimates of the depth to the top of the seismic low velocity zone and from the range of earthquake focal depths. Under the oceans it has an average thickness of about 70 km, and in continental areas the average thickness is between 100 and 200 km. Therefore, the lithosphere encompasses the earth's crust and part of the upper mantle. It has the distinguishing property of being brittle enough to produce earthquakes.

Judging from the way that seismic velocities change through the low velocity zone, we estimate that the asthenosphere is between 200 and 400 km thick. But why does this region of ductile rock exist a few hundred kilometers beneath the earth's surface? The most widely accepted explanation is that pressure and temperature create a critically balanced environment in which some of the mineral constituents melt from the rock, whereas others remain solid. In Chapters 6 and 7 we will look more carefully at the conditions that can cause only part of a mineral assemblage to melt. For now it is enough to know that a region which is partly melted in this way would be relatively soft and ductile. Estimates range as high as 10 percent for the proportion of melt interspersed through the otherwise solid asthenosphere.

Under the asthenosphere is the mesosphere. Interpretation of the shapes and amplitudes of seismic phases indicates that this region is completely solid. Absence of earthquakes, however, means that it is probably still ductile enough to flow under stress.

In answer to an earlier question, we see that seismologists do have evidence that a brittle outer shell lies above a more ductile region. Earthquakes within this brittle shell are concentrated in narrow zones such as you would expect for a lithosphere consisting of several individual plates. But what about the movements of these plates? According to the plate tectonic hypothesis they move outward from constructive borders beneath oceanic ridges, and are carried downward by subduction at the destructive boundaries marked by submarine trenches. We can look at the directions of faulting indicated by earthquake waves to determine if these kinds of plate movement are happening at the present time. Look again at our earlier discussion of Figure 3-11 to recall how direction of faulting is determined.

About five years after the World Wide Standardized Seismograph Network (WWSSN) was installed, seismologists completed a fault motion study of more than 100 earthquakes. The quakes were chosen to provide a worldwide distribution of foci located along supposedly typical constructive, destructive, and conservative plate borders. The directions of horizontal movements found in this study are shown in Figure 5-16. You can see that these results are a clear confirmation of the plate tectonic hypothesis. Outward displacement away from ridge crests was found for most

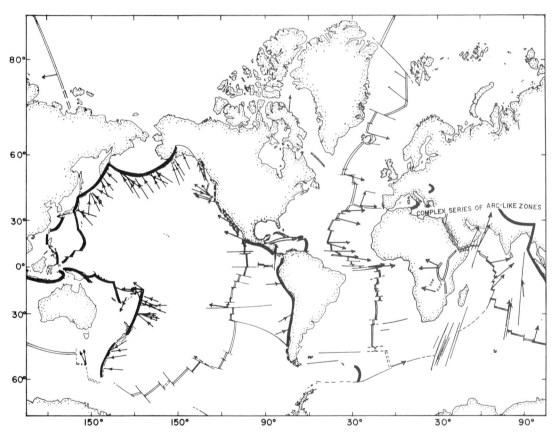

Figure 5-16 Apparent relative movement of lithospheric plates indicated by earthquake focal mechanism studies. Arrows show movement of the plate on which they are drawn. Constructive plate margins are shown by double lines, destructive margins by wide solid lines, and conservative boundaries appear as narrow solid lines. (Reprinted from B. Isacks, J. Oliver, and L. R. Sykes, *Journal of Geophysical Research*, page 5861, Sept. 15, 1968.)

earthquakes located along constructive plate margins. Movements close to submarine trenches were generally in agreement with directions expected near destructive plate borders.

The earthquakes that occurred on oceanic ridges can be separated into a group situated along ridge axes, and another group situated on transverse fractures. Fault movements determined for quakes on ridge axes indicated predominantly vertical displacement. This motion is consistent with the idea of constructive plate margins located above zones where material is rising vertically from the asthenosphere. Fault movements determined for earthquakes located along transverse fractures revealed primarily horizontal displacement. In most cases the relative motion was consistent with directions of transform faulting that you see in Figure 5-10.

The analyses of fault motion for earthquakes situated along destructive margins are more complicated. They indicate that faulting probably occurs within the brittle plate as it is drawn downward into the asthenosphere. However, earthquakes do not appear to result from motion between the plate and the more ductile asthenosphere. Rather, the shocks are caused by stresses within the plate. Near destructive plate margins the probable orientation of these stresses found by analyzing P- and S-waves is shown in Figure 5-17. Most earthquakes occurred where the direction of maximum stress was parallel to the inclined seismic zone, that is, parallel to the downward moving plate. Directly beneath the trenches, shallow focus earthquakes were caused by tensile stress, but at greater depths most shocks were related to compressive stress. This shows that as the plate is bent downward its upper zone is in tension, but along the deeper inclined part the plate is in a state of compression.

You can see that the principal features of seismicity can be explained by the same plate tectonic hypothesis that explains the magnetic field and topographic features we discussed in earleir sections. The existence of the brittle lithosphere and the ductile asthenosphere is confirmed by studies of seismic waves. Furthermore, seismological evidence provides us with a clear indication of subduction of the lithosphere along destructive plate borders. It indicates that, at present, the movements along oceanic ridges are in accordance with the process of transform faulting.

The Age of Ocean Basin Sediment

In most parts of the ocean a layer of unconsolidated sediment covers the basaltic bedrock of the crust. This sediment consists

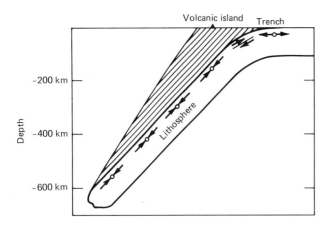

Figure 5-17 Directions of maximum stress in the downward moving plate along a destructive margin determined from earthquake focal mechanism studies. Diverging arrows indicate tension and converging arrows represent compression. Parallel opposing arrows indicate shear stresses. (Redrawn from B. Isacks, J. Oliver, and L. R. Sykes, *Journal of Geophysical Research*, page 5874, Sept. 15, 1968.)

mostly of clay particles transported from continents by rivers and wind and ocean currents, and shell fragments from a myriad of drifting marine organisms. This debris settles to the sea floor, accumulating at rates ranging from less than 1 mm/1000 yr to as much as 5 cm/1000 yr. Imbedded in the sediment are microscopic guide fossils that indicate the time of deposition.

Suppose that these particles have accumulated more or less continuously on the ocean floor. Then according to the hypothesis of sea floor spreading this layer of sediment should exhibit two features. (1) The age of the sediment resting directly on the bedrock surface should increase with distance from a ridge axis. (2) The total thickness of the layer should increase outward from a ridge axis. Because crustal rock is supposed to form beneath a ridge axis, this should be the location where sediment will first settle onto its surface. With the passing of time an increment of crust moves outward from the ridge axis, and the continuous accumulation of sediment adds to the total thickness resting upon this increment. We can see if these two features are found in the ocean floor sediment by examining cores recently obtained by drilling in the ocean basins.

Dredging and coring devices, and drilling equipment operated from ships have been developed to retrieve samples from the ocean floor. Seismic reflection and refraction measurements can be used to find the thickness of the sedimentary layer. To learn more about the nature of ocean basin sediments the **Deep Sea Drilling Project** was initiated in 1964 under United States sponsorship. This project is directed by the **Joint Oceanographic Institutions for Deep Earth Sampling** (JOIDES). The large drilling ship **Glomar Challenger** (Figure 5-18) was commissioned for the purposes of this project.

Deep sea drilling from the **Glomar Challenger** can be undertaken where the ocean floor is as much as 8000 m below sea level. Sections of drilling pipe are threaded together

Figure 5-18 The deep sea drilling ship *Glomar Challenger*. (Photograph courtesy of Glomar Marine, Inc.)

and lowered from the derrick through an opening in the ship until the drill bit at the bottom of the column reaches the ocean floor. Then the entire column is rotated, and bores under its own weight into the sea bottom. As it penetrates, a cylindrical core of sediment and perhaps bedrock is forced up into the hollow pipe above the drilling bit. When the core reaches a certain length, it is raised up through the column of pipe to the ship. Several lengths of core can be recovered before the bit becomes too dull to penetrate farther into the sea bottom. While drilling is underway the position of the ship is maintained by propellers located in the bow, stern, and along the sides. Small beacons that send out electromagnetic signals are lowered to the sea bottom, and the ship position is monitored continuously from these fixed beacons. As the ship begins to drift, its change in position is immediately detected, and the proper propeller is activated to correct the change.

What have we learned from the Deep Sea Drilling Project? Let us look at the results obtained on the cruise of the **Glomar Challenger** in the central Atlantic region during December 1968 and January 1969. Ten drilling sites were occupied at positions indicated in Figure 5-19. Nine of these sites are situated on a profile crossing the Mid-Atlantic ridge.

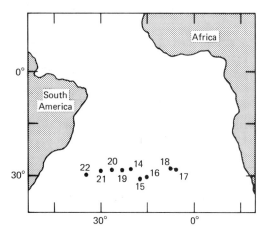

Figure 5-19 Drilling sites occupied by the *Glomar Challenger* during the cruise of December 1968 and January 1969. (Initial Reports of the Deep Sea Drillings Project, Vol. III, Introduction, Figure 1, 1970.)

The geologic cross section in Figure 5-20 illustrates the ages and thicknesses of the layers of unconsolidated sediment found in the drill cores from these nine sites.

You can see a sequence of deposits ranging in age from upper Cretaceous to Pleistocene. The total thickness of sediment increases with distance from the axis of the Mid-Atlantic ridge. The oldest sediment of upper Cretaceous age is found only at Site 21. It is at the bottom of the sequence and rests on basaltic bedrock. At Site 20 this Cretaceous layer is missing. Here you see that younger Paleocene sediment is found directly above the basaltic crustal rock. The oldest sediment at Site 19 is of Eocene age, and rests directly upon the igneous bedrock. At sites progressively nearer to the Mid-Atlantic ridge the basal sediments are successively younger. One would expect this kind of sequence if the idea of sea floor spreading is correct. By assuming that newly formed basaltic crust congeals beneath the ridge axis, and spreads laterally away from it, we would expect older sedimentary deposits to have been transported to greater distances from the ridge. In Figure 5-20 you

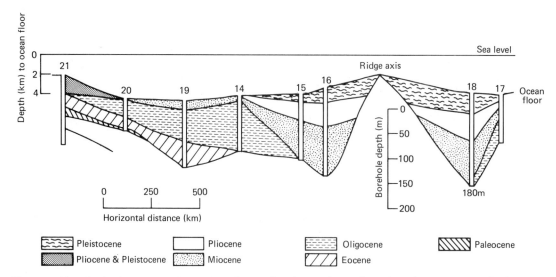

Figure 5-20 Geologic cross section across the south Atlantic Ocean based on the sequence of submarine sediments determined at drilling sites occupied by the *Glomar Challenger*. (Modified from Initial Reports of the Deep Sea Drilling Project, Vol. III, Summary and Conclusions, Figure 10, 1972.)

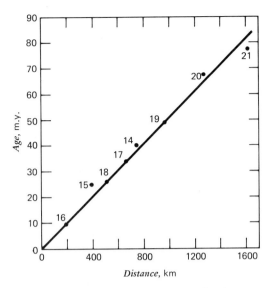

Figure 5-21 Graph showing relationship between the age of oldest fossils found at drilling sites and distance of site from ridge crest. (Numbers correspond to drilling sites described in Figures 5-19 and 5-20. From Initial Reports of the Deep Sea Drilling Project, Vol. III, 1970.)

can see clearly that the oldest sediment of upper Cretaceous age is found only at the site most distant from the ridge.

Let us continue with this analysis. Consider the ages of the oldest fossils that are found in the boreholes. The graph in Figure 5-21 shows the relationship between maximum age and distance from the ridge axis. Data from all boreholes group closely about the straight line. The reciprocal of the slope of the line can be interpreted as the velocity of sea floor spreading. A rate of 2 cm/yr is indicated. This value is in good agreement with values determined independently from analyses of topographic features and paleomagnetism. Results similar to those of the central Atlantic have been found in other oceanic regions. Therefore, you can see that the distribution of ocean floor sediments is still another feature of geology that can be explained by the plate tectonic hypothesis.

Hypothesis or Theory

We have now examined the principal features of topography, paleomagnetism and marine magnetic anomalies, seismicity, and finally the deposits of ocean floor sediment. A global tectonic process has been proposed that involves exchange of material between an outer brittle lithosphere which is divided into discrete plates, and a deeper and more ductile asthenosphere. The patterns revealed by these diverse geological features can all be explained by this process. Perhaps another global tectonic hypothesis will be proposed to explain these different features. But no one has yet developed an idea that has the broad explanatory power of the plate tectonic hypothesis. Perhaps there is now enough favorable evidence for us to call it the plate tectonic theory. But an important question remains unanswered. What makes it work?

THERMAL CONVECTION

The movement of lithospheric plates in the manner implied by the plate tectonic theory indicates a continuous transfer of mass. Some kind of thermal convection probably contributes to this mass transfer, but the nature of the process and the shapes of convection cells remain unknown at the present time. Principles of **rheology**, the science concerned with deformation and flow of materials, provide some understanding of conditions that are necessary for convection to occur in the earth.

The crust and mantle of the earth are almost entirely solid. When rock in these zones is confined under high pressure and temperature, it becomes ductile and can be compelled to flow. Because of the ductile nature of the mantle under the lithosphere, it is convenient to represent it as a very viscous fluid for pusposes of analyzing its rheology.

The **viscosity** of a fluid is a measure of its capacity to flow when stress is applied. When

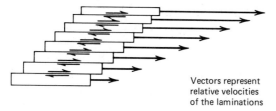

Figure 5-22 Conceptual representation of viscosity in terms of frictional resistance along surfaces between very thin fluid lamina.

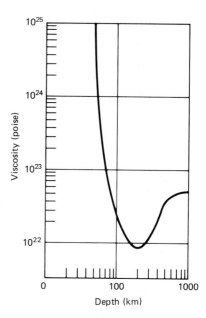

Figure 5-23 A speculative estimate of the variation of apparent viscosity with depth in the earth.

part of a fluid mass is slowly flowing relative to another part of the mass, the motion can be represented by displacements along surfaces of adjacent thin layers of fluid. This kind of motion, called laminar flow, is illustrated in Figure 5-22. The friction that resists the movement of one thin layer over another is the viscosity. The basic unit of viscosity in centimeter-gram-second units (cgs units) is the **poise**. If two parallel plates each of 1 cm² area are separated by a layer of fluid 1 cm thick, and if a force of 1 dyne applied along the surface of one plate causes it to move at a velocity of 1 cm/sec relative to the other plate, then the viscosity of the fluid is 1 poise, and has units of dyne-second/cm².

Different kinds of information are used to estimate values of viscosity in the mantle. The earth's surface sinks slightly when loads are placed on it. This happens when a reservoir fills after a dam is constructed. When a load is removed, the surface rises. We can measure the rise when reservoirs are drained. We can see similar effects where large glaciers have melted. It is evident that the adjustment of the earth to loads on its surface involves flow of ductile rock beneath the lithosphere. Scientists have analyzed the rates of rising or sinking to discover that the ductile asthenosphere has an apparent viscosity between 10^{21} poise and 10^{22} poise. The effects of loads related to filling of water reservoirs and melting of glaciers are described more fully in later sections of this chapter. Laboratory tests on rock specimens provide additional information about apparent viscosity in the lithosphere. The results of these different kinds of studies have been combined to estimate apparent viscosity in the mantle. The curve in Figure 5-23 shows one estimate.

Viscous flow in the ductile part of the mantle can occur if the distribution of mass becomes unstable. A stable arrangement of mass exists when its potential energy is minimized. The potential energy (v) associated with an element of mass (m) in the earth is given by the expression: $v = mgh$, where g is the acceleration of gravity, and h is the elevation of the center of mass above an arbitrary base level. For the case of a column of fluid such as in Figure 5-24A, the mass is calculated by multiplying volume and density, and h can be measured to the center of the column. For the case of two fluids of different density as in Figure 5-24B the potential energy is the sum of the products:

$$v = (\rho_1 V_1)gh_1 + (\rho_2 V_2)gh_2 \qquad (5-1)$$

Global Tectonic Processes 167

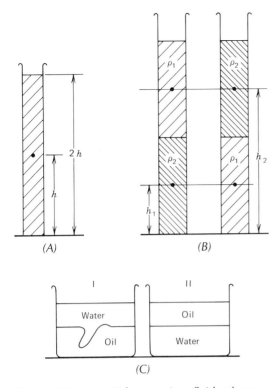

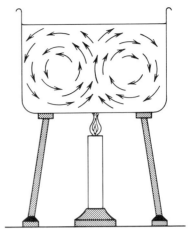

Figure 5-25 Fluid convection associated with a downward increasing thermal gradient occurs if fluid density decreases with increase in temperature. The warmer low density fluid rises, then cools, and settles, setting up a cyclic motion.

Figure 5-24 Potential energy in a fluid column (A) is the product of the mass and the height of the center of mass. Column (B) contains two fluids of equal volume, but different densities. If $\rho_1 > \rho_2$, the left column has the highest potential energy because it has a larger mass at a higher level. (C) In a pan with a high density fluid layer (water) above a low density fluid layer (oil) a slight perturbation will cause the system to invert so that the water sinks beneath and the oil rises to the top so that potential energy is reduced.

where V_1 and V_2 are volumes of fluids that have densities $\rho_1 > \rho_2$. Obviously the total potential energy of column I is greater than the potential energy of column II. If, instead of columns, we have pans as in Figure 5-24C, a similar difference in potential energy would exist. However, pan II would be stable and pan I would be in **unstable equilibrium.** A slight perturbation of pan I would set the mass into motion so that the two fluids would flow into opposite positions. This would establish stable equilibrium at a lower value of potential energy.

Instability may arise in a system as a result of thermal conditions. Consider a homogeneous fluid where density varies inversely with temperature. By heating the base of a pan of such fluid (Figure 5-25) the density of material near the base becomes lower than the density of fluid near the top. The possibility of reducing the potential energy by mass movement now exists. A slight perturbation of the system would lead to inversion of the upper and lower zones of fluid. But because heating at the base continues, the recently transported cold fluid becomes heated as it descends, and the originally warm fluid cools. The system is once again unstable. The continuing situation of basal heating and surface cooling causes continued instability and mass movement. A system of convection cells is the result.

The most probable source of heat in the earth is energy released during decay of radioactive isotopes. Most theories of thermal convection assume radioactive heat sources. Because we know little about the distribution of radioactive isotopes we cannot expect to determine precisely the mass transfer in the mantle associated with thermal effects. We do have measurements of heat flow out of the earth, however, and we know some aspects of mass transfer at least for the lithosphere. These observations provide important constraints on possible ideas of convection.

The analysis of the convection problem has been extended to account for the temperature dependence of viscosity. Depending upon the distribution of heat sources, convection cells of various asymmetrical forms are theoretically possible. Variable flow rates can be expected under the proper conditions. These theoretical studies indicate that the entire convection process needed to explain the plate tectonic motions might be contained within the asthenosphere, perhaps involving the mesosphere in a secondary way. A theoretically possible convection pattern is seen in Figure 5-26.

Laboratory experiments involving materials of temperature-dependent viscosity have revealed complicated patterns of convection. An example is seen in Figure 5-27. Such experiments reveal that flattened and elongate convection cells can be expected under certain thermal conditions.

The questions about the nature of thermal convection in the mantle will not be answered easily. Too much remains unknown about the composition and thermal conditions in this region of the earth. However, theoretical arguments based upon limited observational data and apparently reasonable assumptions indicate that the conditions of instability are within the realm of possibility. Furthermore, the apparent viscosity of the mantle is low enough to expect significant viscous flow under these conditions of instability. Therefore, it is not unreasonable to expect that some kind of thermal convective process causes movement of the plates of the

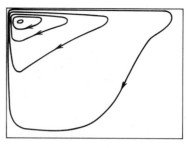

Figure 5-26 Convection cells of distorted form may result in fluids characterized by density and viscosity which are dependent upon temperature and pressure. (From K. E. Torrance and D. L. Turcotte, *Journal of Geophysical Research,* page 1159, Feb. 10, 1971.)

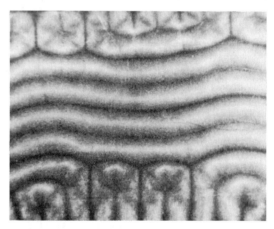

Figure 5-27 Top view of convection of silicon oil caused by thermal gradient in a 15 cm × 20 cm container. (From E. F. C. Sommerscales and D. Dropkin, *International Journal of Heat and Mass Transfer,* Vol. 9, facing page 1190, Pergamon Press, 1966.)

lithosphere, even though the details of the process are yet to be discovered.

ISOSTASY

The lithosphere, like a ship taking on cargo, sinks deeper into the asthenosphere when a load is placed on its surface; and it rises when the load is removed. We have learned about this effect by studying regions that were

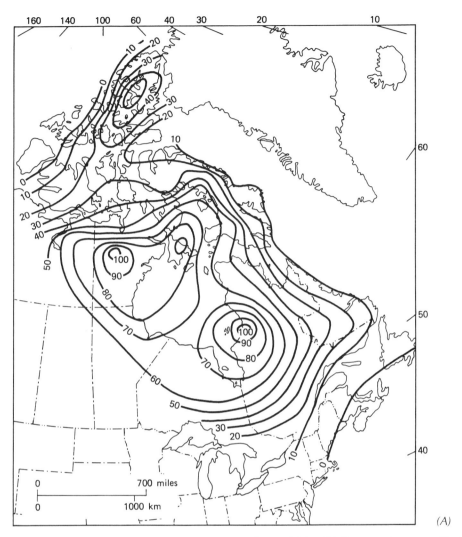

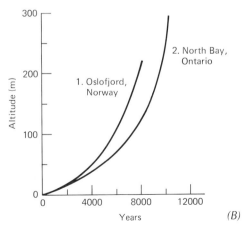

Figure 5-28 (A) Contours in meters illustrate estimated isostatic rise of the land surface during the past 6000 years in northeastern North America. (B) Isostatic rise of the land surface as a function of time (in thousands of years) at locations in Norway and Ontario, Canada. (From R. F. Flint, *Glacial and Pleistocene Geology*, pp. 345 and 363, John Wiley & Sons, Inc., copyright © 1971.)

formerly covered by large masses of ice. Vertical movements in two regions are of particular interest. In the Fennoscandian region of northern Europe (Figure 5-28) geologic evidence indicates continuous uplift of the land surface during the past 9000 years. Similarly, in the Great Lakes region of North America the land has been slowly rising for 12,000 years or more. We know from the distribution of sediments transported by glaciers that vast ice sheets intermittently spread over large areas of northern North America and Eurasia, causing the land to subside beneath the superposed load. The ice load disappeared by melting about 12,000 years ago from the Great Lakes region, and 9000 years ago from Fennoscandia. Since then, beach deposits and erosional features originally formed at sea level have been raised to heights well above the present Fennoscandian coast. Water levels of lakes in both regions have changed. Former shorelines are indicated by beach deposits situated at elevations considerably above present lake levels. Spirit level surveys of these former shorelines reveal that they are inclined relative to one another. Because they represent earlier level surfaces, their present inclinations indicate differential vertical uplift. Carbon-14 analysis of wood fragments embedded in these beach deposits indicate the times during which different shorelines existed. These kinds of information have been used to reconstruct the patterns of uplift illustrated in Figure 5-28. Rates of uplift are observed to be diminishing with time.

We can account for the postglacial uplift of the Fennoscandian and Great Lakes regions by a process involving viscous flow of ductile rock beneath the lithosphere. When a load is placed on the surface of an elastic layer floating on a viscous fluid, the layer sinks deeper into the fluid. If the load is then removed, the layer again rises to the level where it floated prior to loading. This is illustrated in Figure 5-29. The time required for the layer to rise depends upon the area of the depression and the viscosity of the fluid that must flow into the zone beneath it. The depth of the depression h_t at time t after the load has been removed is:

$$h_t = h_0 e^{-qt} \qquad (5\text{-}2)$$

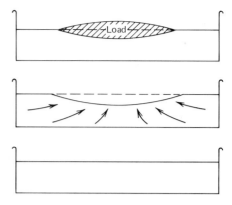

Figure 5-29 Isostatic adjustment of a viscous fluid following removal of a surface load is accomplished by viscous flow along trajectories shown by arrows, which tends to continue until a level surface is established.

where h_0 is the original maximum depth, and q is related to the fluid viscosity and density and the size and shape of the depression. This equation indicates that uplift should continue at rates which decrease with time according to a form similar to the curves seen in Figure 5-28. Information from observed uplift curves and the dimensions of the Fennoscandian depression have been used in Eq. (5-2) to obtain the estimate of 10^{22} poise for the viscosity in the zone beneath the lithosphere.

The process that we have been discussing is an isostatic adjustment process. **Isostasy** is the process by which surface topography and subsurface mass are adjusted to achieve uniform pressure on a level surface somewhere in the mantle. This surface is called the **level of isostatic compensation.** It exists when all columns of overlying rock have the same mass. Whenever there is this balance of mass and topography, the condition of isostatic equilibrium is realized. Look again at our discussion of Figure 2-18 which illustrates the concepts of J. H. Pratt and G. B. Airy. These two scientists showed idealized ways that topography and underlying density can vary

Global Tectonic Processes 171

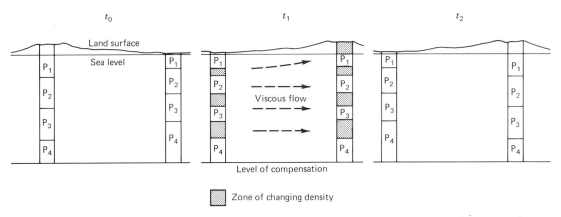

Figure 5-30 Isostatic adjustment of subsurface density in response to changing topography can result from viscous flow in various layers, each having a different viscosity. Viscous flow redistributes materials of different density so that in the shaded zones a change in density has been the result of replacing material of one density with that of another density. Topography and mass are adjusted to a condition of isostatic equilibrium at times t_0 and t_2. Isostatic adjustment is occurring at time t_1.

above a level of compensation. You can see from Figure 5-29 that when a load such as an ice sheet spreads over a continent, then the surface of this load becomes the earth's surface. Therefore, the earth's surface is changed. Now, whenever the earth's surface changes, isostatic stresses arise that act to restore a level surface of compensation. These stresses cause viscous flow in ductile layers such as illustrated in Figure 5-30. Here two hypothetical columns are compared. The topography and density at time t_0 are adjusted so that the masses of the two columns are equal. The longer one is seen to have proportionally more low density material. Between times t_0 and t_1 the topography is altered so that the two columns are no longer of equal mass. The resulting isostatic stresses then cause viscous flow in different layers so that density is changed in some zones. The redistribution of mass related to these density changes is complete by time t_2, when the condition of isostatic equilibrium again is achieved.

Evidence of uplift and subsidence of the continental land surface is abundant. Ancient sedimentary rocks formed in shallow seas are now exposed at great heights on the interior plateaus and mountainous terrain of the continents. In some places vertical movement of the land surface arises from arching and buckling of crustal rock masses in response to compressive stresses along destructive plate margins. Elsewhere, far from the borders of lithospheric plates, uplift and subsidence have not been explained by the more obvious plate tectonic processes. Vertical movements occurring at the present time are revealed by changes in benchmark elevations determined from repeated spirit level surveys. Rates at which regions of the United States are rising or sinking are indicated in Figure 2-15. Whenever vertical movement of the land surface occurs, isostasy is one of the processes contributing to this movement or retarding it. To this extent it is a global tectonic process.

The process of isostasy may be very complicated in regions that have recently undergone severe tectonic deformation. Perhaps there are several ductile zones where viscous flow occurs. A method has been devised for calculating directly from topographic data and Bouguer gravity anomalies the subsurface density changes associated with isostatic adjustment. Although the mathematical methods are beyond the scope of this discussion, the results are noteworthy. A study of the isostatic adjustment of the Great Basin region of the United States has been completed. The Great Basin extends from Arizona northward through Utah and Nevada and into Idaho. During the Cenozoic era this was a region of intensive tectonic activity involving volcanism, faulting, and mountain building. This region stands in marked contrast to the stable continental interior of central North America. No evidence of significant tectonic deformation during Cenozoic time is found in the stable interior. Therefore we might expect that the process of isostatic adjustment has been active in the Great Basin and inactive in the stable interior during the Cenozoic era. By using the stable interior as a reference region where isostatic density changes have not occurred it was possible to calculate by comparison the probable subsurface density changes in the unstable and isostatically active Great Basin. The density changes associated with isostatic adjustment are seen in Figure 5-31.

Seismic refraction measurements have been made in the stable interior and in the Great Basin. The subsurface variation of P- and S-seismic wave velocities has been determined from these measurements. Again using the stable interior as a reference, velocity differences between the two regions have been calculated. These results are also shown in Figure 5-31.

The comparison of velocity differences and subsurface density change related to isostasy reveals that significant density change has occurred in the same zones where important velocity differences are observed. The most noteworthy correlation is found for the depth range of 375 to 450 km. These comparisons may be interpreted to indicate that the isostatic adjustment process has caused changes both in density and in seismic wave velocity.

This could mean that the subsurface materials are being physically altered by the process of isostasy. Such changes might be related to alteration of the crystalline structure of the rock, as well as viscous flow processes. The results of this study point out that materials at least as deep as 450 km are involved in the process of isostasy, and that the process is a complicated one. The lithosphere and the asthenosphere are both included in the adjustment process, which may be accomplished by a combination of redistribution of mass by viscous flow, and density changes associated with physical alteration of the subsurface materials.

Our understanding of the process of isostasy is only rudimentary. Only in recent years have data and analytical techniques been available for attempting an examination of some of the more complicated aspects of the process. Studies have revealed that redistribution of mass by viscous flow is possible, and that reasonable values of apparent viscosity can be obtained.

THE GLOBAL TECTONIC SCHEME

A clearer understanding of diverse geological phenomena has come with the development of the plate tectonic theory during the decade of 1960 to 1970. The global patterns of seismicity, magnetism, topography, and the distribution of submarine sediments are related in terms of mass transfer between the lithosphere and the asthenosphere. This system has given a unification to the science of geology. The practical aspects associated with knowledge of the dynamics of the plate tectonic system are yet to unfold. We now have a better understanding of why earthquakes occur in seismic belts rather than more randomly. We know that exploration for coal beds of Pennsylvanian age in the ocean basins would be impractical because the oceanic crust has been recycled into the asthenos-

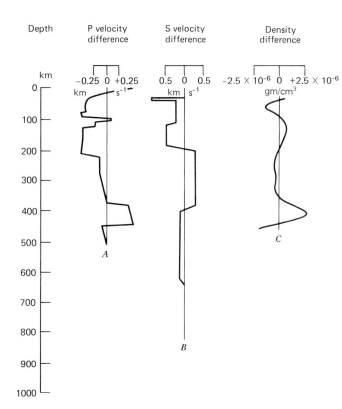

Figure 5-31 Possible subsurface density changes related to Cenozoic isostatic adjustment of topographic changes that were produced during the tectonic development of the Great Basin region of western United States. These calculated density changes are compared with measured differences in seismic P- and S-wave velocities between the tectonically active Great Basin and the tectonically stable region of central United States. The comparisons support the idea that isostatic adjustment processes extend at least 500 km down into the mantle, and may be related to changes in elastic properties that influence seismic wave velocities. (After L. M. Dorman and B. T. R. Lewis, *Journal of Geophysical Research*, page 3076, June 10, 1972.)

phere since that time. It may become reasonable to develop new concepts of geography, recognizing that plate boundaries may represent as significant a subdivision of the earth's surface for some problems as the subdivision of continents and oceans for more conventional considerations.

Our comprehension of the global tectonic machine is yet incomplete. Although we have established some general ideas about the way it works, the understanding of isostatic and convection processes is rudimentary. Questions about these processes pose challenges for the future.

STUDY EXERCISES

1. Scientists working with Professor E. C. Bullard used Euler's theorem in their analysis of the separation of South America and Africa. Explain how the results of this analysis indicate that North America is embedded in a different lithospheric plate than the plate in which South America is embedded.

2. Along the equator the Atlantic Ocean is growing wider at the approximate rate of 4 cm/yr. Use this rate and your knowledge of the plate tectonic theory to calculate the distance away from the axial trough of the Mid-Atlantic ridge to bedrock that acquired its thermoremanent magnetism at the beginning of the Pli-

ocene epoch. How many times has the main field reversed itself since that rock was formed?

3. Describe two independent kinds of evidence which indicate that sea floor spreading rates are not the same near all oceanic ridge systems.

4. Discuss some predictions of future events and features that could be based on the plate tectonic theory.

5. Use Eq. (5-2) and data from Figure 5-28 to estimate how much time will be required for the additional uplift of the land surface.

SELECTED READINGS

Jacobs, J. A., R. D. Russell, and J. T. Wilson, *Physics and Geology,* 2nd ed., McGraw-Hill Book Co., New York, 1974.

Takeuchi, H., S. Uyeda, and H. Kanamori, *Debate about the Earth,* rev. ed., Freeman, Cooper and Co., San Francisco, 1970.

Turcotte, D. L., and E. R. Oxburgh, "Continental Drift," *Physics Today,* vol. 22, pp. 30-39, 1969.

Wilson, J. T. (compiler), *Continents Adrift and Continents Aground,* readings from Scientific American, W. H. Freeman and Co., San Francisco, 1976.

Wyllie, Peter J., *The Way the Earth Works,* John Wiley and Sons, Inc., New York, 1976.

SECTION THREE
THE ROCKS OF THE EARTH

6

MINERALOGY

Physical Properties of Minerals
Crystals
Optical Properties of Minerals
X-ray Analysis of Minerals
Atomic Structure of Minerals
Some Important Mineral Groups
Silicate Minerals
Carbonate Minerals
Oxide Minerals
Sulfide Minerals
Gems
Pressure, Temperature, and Polymorphism

In the fall of 1871, two prospectors asked a San Francisco banker to place a small bag of stones in the vault for safekeeping. He at first refused. Then, a glimpse of the contents changed his mind, and the bag containing raw diamonds, rubies, and sapphires was placed securely in the vault. Word of this deposit began to spread, and in the following weeks the prospectors were persuaded to enter into an agreement with some California financiers to develop a diamond mine at a still secret location in the western United States. As part of the agreement, a well-known consulting geologist was to make an evaluation of the claim. An expedition was organized, and upon reaching the site the consultant was amazed to discover diamonds, rubies, emeralds, and sapphires lying about the surface. In his excitement he began staking claims to protect the interests of the newly formed company, not pausing to ponder the peculiar occurrence of these precious minerals with one another on a ledge of coarse grained sandstone. Upon his return to San Francisco he submitted an enthusiastic report. Final arrangements were made to assemble capital of $10 million to exploit the diamondiferous claims.

In the meantime word of the discovery reached the party of government geologists under the direction of Clarence King. They were mapping the region of the supposed precious mineral discovery. Naturally, they were curious about its location, and somewhat suspicious about the variety of different gems. Piecing together odd bits of information and fragments of conversation in San Francisco, they concluded from their knowledge of the area that the discovery was located in northwestern Colorado on the flanks of a particular mountain. They reached the area rather easily, and found the precious stones lying about the surface. It was evident almost immediately from their knowledge of minerals that this was a highly incongruous setting. Minerals never before found in close association were grouped together. Gems lay only upon the surface, and no more were found by digging. When their findings were revealed, the hoped-for diamond boom became the "great diamond hoax." Detectives later discovered that the raw gem stones had been purchased in England. These were "salted" in the area to stimulate a business venture that could be capitalized for a much larger sum than the original investment in uncut gems. The tale of the great diamond hoax was recounted in some detail by Professor Henry Faul.*

The great diamond hoax was a scheme to defraud unwary investors. It failed because some facts about the occurrence of minerals were overlooked. These facts are part of the science of mineralogy. This science is concerned with the properties of minerals. We define a **mineral** to be a homogeneous substance that occurs in nature with a distinct chemical composition, and a distinct geometrical arrangement of the atoms making up this chemical composition. You can use these properties to distinguish more than 2000 different minerals. Among these are gems that possess rare qualities of beauty and durability.

The geometrical and chemical properties of a mineral reflect the environment where it was formed. **Diamond,** the hardest mineral in nature, consists of carbon atoms in a particular geometrical grouping. This arrangement of carbon atoms forms in nature only at the elevated pressure and temperature that exist in the earth's mantle. In the lower pressure and temperature environment of the crust carbon atoms become joined in an entirely different geometrical grouping to form the mineral **graphite.** It is one of the softest minerals, and is used in pencils. If you are prospecting for gems, you might find both the **topaz** and **garnet** in the same natural setting. Even though these minerals have different chemical compositions, they can form in the same pressure and temperature environment. But you would not expect to find ruby and emerald in the same natural setting, because the environments where they can form are quite different.

In this chapter you will learn about the important properties of minerals, and how these properties are related to the atomic structure of the minerals. The fact that many different minerals have some similar properties gives us ways to classify them. Many useful and valuable minerals are difficult to recognize in their natural settings. You need to know about their distinctive properties and how they are classified to identify them. We will also look into the effects of pressure and temperature on minerals. This information gives us clues about where to look for different minerals. Or, if you happen to find them, you can figure out something about the former events and conditions that are part of the earth's history. In later chapters you will learn about the assemblages of minerals that make up the rocks of the earth.

PHYSICAL PROPERTIES OF MINERALS

Mineral collectors use some easily recognized physical properties to identify their specimens. The density of a mineral, its resistance to abrasion, and the way that it can be split apart are a few of the properties that can be tested without elaborate apparatus. We can begin with descriptions of the six physical properties most commonly used for mineral identification.

A. **Luster.** This property is the appearance of a mineral surface, which can be described as **metallic** or **nonmetallic**. Minerals that appear nonmetallic can be further described by such terms as dull, greasy, pearly, silky, vitreous, or resinous. Certainly these descriptive terms

---

*Henry Faul, "Century-old diamond hoax re-examined," *Geotimes*, Oct. 1972.

are subjective. Even with little experience, however, one can make reasonably good judgments about these different varieties of luster. Some examples are seen in Figure 6-1 A.

B. **Hardness.** The resistance of a mineral to abrasion is called hardness. It is a relative property that is tested by finding which one of a series of standard minerals can be used to scratch the surface of an unknown specimen. These standard minerals are assigned the following hardness values in a series called the **Mohs Hardness Scale**: Talc (1), Gypsum (2), Calcite (3), Fluorite (4), Apatite (5), Feldspar (6), Quartz (7), Topez (8), Corundum (9), Diamond (10). A set of stiles seen in Figure 6-1 B can be used to test hardness. You can also test the Mohs hardness of an unknown specimen with these other readily available standards: fingernail (2-2.5), copper coin (2.5-3), knife blade (4.5-5.5), window glass (5.5), steel file (6-7).

C. **Color.** One must be cautious when using color for identifying a mineral. Some minerals occur in a wide range of color owing to the presence of very small amounts of different impurities. However, many minerals possess colors within a restricted range, so that this property can aid in identifying an unknown specimen.

D. **Streak.** The color of a fine powder of a mineral is called its streak. You can test the streak by rubbing a specimen on a peice of unglazed porcelain (Figure 6-1C). The streak found from many specimens of the same mineral is usually more consistently of one color even though the individual specimens display different colors because of impurities.

E. **Density.** You will recall from Chapter 2 that this property is found by dividing the mass of a specimen by its volume.

F. **Cleavage.** Minerals possessing cleavage can be split along parallel planes (Figure 6-1D). Some minerals have more than one set of cleavage surfaces. Others that do not possess cleavage break along uneven surfaces.

In Appendix I you will find descriptions of 50 important minerals that are based on these six physical properties. The mineral descriptions are grouped so that you can analyze an unknown specimen by testing or observing the six distinguishing properties in the order that we have listed and discussed them.

Minerals possess other important physical properties that are not as easy to test as the ones already discussed. The luster, color, and streak of a mineral are all manifestations of its optical properties. Another optical manifestation is **luminescence,** which is the capacity of a mineral to emit visible light when exposed to ultraviolet light that is invisible to the human eye. In a dark room with an ultraviolet light source a luminescent mineral will glow in the dark. To learn more about the optical properties of a mineral you must study the ways that it reflects and refracts different colored light beams. This important subject is discussed in more detail in a separate section of this chapter.

We can also study the magnetic and electrical properties of minerals. Magnetic **susceptibility** and **remanent magnetic moment** describe the capacity of a mineral to acquire and retain magnetism. Recall from Chapter 4 that these properties are especially important in the ferromagnetic minerals **magnetite, hematite, ilmenite,** and **pyrrhotite.** A few minerals can transmit electrical current. The **electrical conductivity** of these minerals can be measured experimentally. But most minerals are nonconductors. You can measure the **dielectric constant** of a nonconductor, which is an indicator of its capacity to retain an electrical charge. The dielectric constants of ore minerals such as **galena,** the principal ore of lead, and **sphalerite,** the most impor-

Figure 6-1 Some easily recognized physical properties can be used to identify minerals. *(A)* Luster describes the appearance of the surface of a specimen. The two examples display shiny metallic luster, and dull *earthy* luster. *(B)* Streak is the color of the powder produced by rubbing a mineral specimen on a rough white porcelain surface. *(C)* Cleavage is the property of a mineral to split along parallel surfaces. This property is evident in the illustration of a biotite specimen. *(D)* Hardness, the resistance of a mineral to abrasion, can be tested by attempting to scratch a specimen with standard hardness minerals mounted in a set of stiles. The standard minerals are listed in the text.

tant ore of zinc, are much different than the values for the rock where these ores are likely to be embedded. Electromagnetic instruments that are sensitive to such dielectric constant differences have been designed to prospect for buried concentrations of these ore minerals. There are still other interesting electrical properties. One of these is **piezoelectricity,** the capacity of a mineral to acquire an electrical charge when it is subjected to mechanical stress. If you deform a quartz crystal electrically by compressing or distending it, the crystal becomes electrically charged. Still other minerals possess the property of **pyroelectricity,** so that they become electrically charged by abrupt changes in temperature. This happens when you quickly heat a **tourmaline** crystal. Because of their particular electrical and magnetic properties certain minerals are especially useful in electronics componenets.

All minerals possess mechanical properties that can be described in different ways. For example, the elasticity can be expressed by the bulk modulus and the shear modulus that your learned about in Chapter 3. Very few minerals have unique properties of radioactivity, but some were mentioned in Chapter 1.

These many different physical properties are all related to the atomic structure of a mineral. To learn about atomic structure you have to find out what atoms are present and how they are joined together. Our knowledge of the composition and atomic structure of minerals has come from careful study of the geometrical and optical properites of crystals, x-ray analysis, and experiments designed to test their response to changes in pressure and temperature.

CRYSTALS

Most minerals exist as small grains that have irregular shapes. These grains are the constituents of rocks. Suppose that each grain had been able to develop separately without being forced to adjust its shape to fit with adjacent grains. Then it would have formed a crystal with distinct external geometrical properties (Figure 6-2). Conditions favoring the growth of such crystals are uncommon in nature. But occasionally these unusual conditions do occur, so that well-formed cyrstals of most minerals have developed from time to time. You can see these crystals displayed in museums.

Figure 6-2 A cluster of quartz crystals.

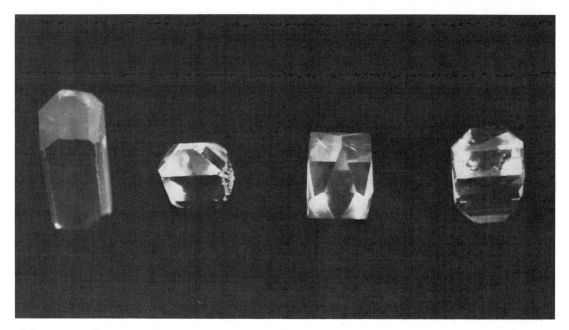

Figure 6-3 Differences in shape are exhibited by four individual quartz crystals.

Individual crystals of a mineral can occur in different shapes. Look at the quartz crystals illustrated in Figure 6-3. How can you look for geometrical similarities in crystals like these? You can begin by measuring the angles between specific surfaces called **crystal faces.** Even through the relative sizes of these faces may differ from one crystal to another, the angles between them remain the same. This feature is the basis for the **law of constancy of interfacial angles,** which was first proposed by Nicolaus Steno (1638–1687). This law states that **when measured at the same temperature, similar angles on crystals of the same substance remain constant regardless of the size or shape of the crystal.** The interfacial angle is a diagnostic physical property of a mineral. Sometimes you can recognize the intersection of two crystal faces on a grain exposed along a broken edge of a rock. Then you can estimate the angle to help in identifying the mineral. Accurate measurements of interfacial angles on a well-formed crystal are made with a **reflecting goniometer** (Figure 6-

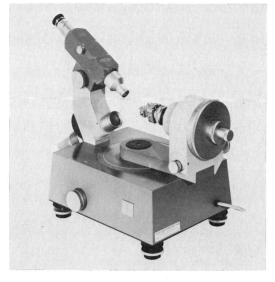

Figure 6-4 A reflecting goniometer is used to measure angles between the faces of a crystal which is mounted on a spindle so that it can be rotated into different positions. In certain positions a beam of light is reflected with maximum intensity from crystal faces into the viewing telescope. (Courtesy of Robert Huber Co., Rimsting, West Germany.)

4). A crystal mounted on a spindle can be rotated into different positions so that a beam of light is reflected from a crystal face into a viewing telescope.

How can you use angles between crystal faces to describe the geometry of a crystal in a systematic way? This is done by displaying the information on graphs that reveal symmetrical patterns. The geometrical classification of minerals is based upon crystal symmetry that becomes evident after interfacial angle measurements are plotted upon a **stereographic** grid. This can be done in the following way. Imagine the crystal to be located at the center of a sphere (Figure 6-5 A). Next, choose an axis of rotation that passes through the crystal and intersects the sphere at c and c', and let this be a rotation axis for the sphere. Now, you can draw latitude and longitude lines on the sphere relative to this axis. Then, lines from the center can be drawn outward perpendicular to all of the crystal faces. Each line is called a crystal **face pole**. It intersects the surface of the sphere at a particular latitude and longitude (Figure 6-5 B). If we look down on the sphere from an outside point on the axis, we see the grid of latitude and longitude as concentric circles and radiating lines. Intersections of the crystal face poles on the sphere appear as points (Figure 6-5 C). This diagram is a stereographic plot of the crystal face poles relative to the axis chosen.

Next let us rotate the stereographic plot. You can see that for Figure 6-5 C a rotation of 90° would bring the pattern of points into a position that appears identical to the initial position. Rotation again by 90° brings the pattern into still another identical position. An axis about which a pattern appears identically at four equal rotational intervals in 360° is called a four-fold axis of symmetry. Suppose you analyzed another sterographic plot in this way, and discovered that the points assumed identical positions for each rotation of 60°. This would indicate a sixfold axis of symmetry.

You can choose an axis of rotation in any position, and prepare a stereographic plot relative to that axis. By testing different axes you will find that for all crystals only onefold (no repeat position), twofold, threefold, fourfold, and sixfold axes occur. Some crystals will have no axes of greater than 1-fold symmetry. Others may have as many as four different axes, each with at least three fold symmetry.

The constancy of interfacial angles and the symmetry properties of axes of rotation led mineralogists to the idea of finding some elemental geometric shape that could be the fundamental building block of a crystal. If properly described, this shape repeated throughout the space occupied by a crystal could be used to construct a model having the same external geometrical properties as the crystal (Figure 6-5 D). Viewed in another way, it should be possible to describe a uniform distribution of points in the space occupied by a crystal. These points constitute a **space lattice,** and lines connecting specific combinations of lattice points would outline an elemental geometric shape. This elemental shape must be chosen to fill space perfectly with no gaps or overlaps. Mineralogists learned from geometrical considerations that a space lattice possessing the symmetry of a crystal can be represented by one of only six forms that we call **primitive** lattices. The six primitive lattices are illustrated in Figure 6-6.

The elemental shape chosen as the building block for a crystal is known at the **unit cell.** Lines along the edges of the unit cell indicate **crystallographic axes** of that mineral. The relative lengths of the axes, and the angles between them are important characteristics of the mineral. You can find the symmetry properties of a crystal by rotating it about its crystallographic axes.

The requirement that unit cells fill space perfectly restricts the variety of possible geometric shapes of crystals. From consideration of these factors mineralogists have established a geometrical classification of minerals that is based upon unit cell shapes. Accordingly, all

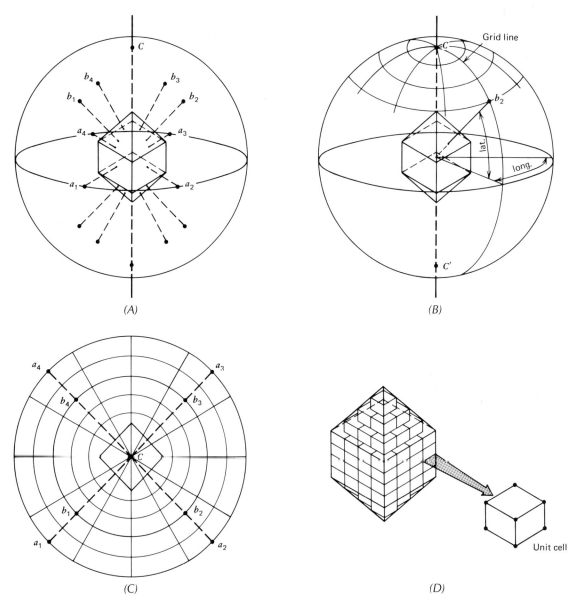

Figure 6-5 Symmetry of a crystal can be determined from stereographic plots of lines, called face poles, that are drawn perpendicular to crystal faces. *(A)* The face poles of a crystal centered in a sphere intersect the surface of the sphere at points shown on the diagram. *(B)* Positions of these face pole intersection points can be described by grid lines drawn on the sphere. *(C)* A stereographic diagram represents a view of these points and grid lines from some particular perspective. The points and grid lines have been projected onto the plane of the equator of the sphere in this example of a stereographic diagram. *(D)* Patterns seen on stereographic diagrams are used to determine the shape of the unit cell, which is the fundamental geometrical building block of a crystal. Unit cells can be fitted together to reproduce the shape of that crystal.

Mineralogy

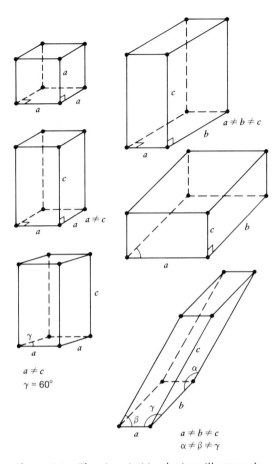

Figure 6-6 The six primitive lattices illustrate the basic shapes of the unit cells of the six crystal systems: cubic *(top left)*, tetragonal *(middle left)*, hexagonal-trigonal *(bottom left)*, orthorhombic *(top right)*, monoclinic *(middle right)*, and trictinic *(bottom right)*.

Table 6-1
Crystal Systems

1. Cubic (Isometric)
    Four threefold axes
    Five classes
2. tetragonal
    One fourfold axis
    Seven classes
3. Hexagonal
    A. Hexagonal division
        One sixfold axis
        Seven classes
    B. trigonal division
        One threefold axis
        Five classes
4. Orthorhombic
    Three twofold axes
    Three classes
5. Monoclinic
    One twofold axis
    Three classes
6. Triclinic
    No axes of symmetry
    Two classes

crystals can be classified in one of six systems depending upon the relative lengths of crystallographic axes, angles between them, and their symmetry properties. These six crystal systems are described in Table 6-1. Because of additional subtle aspects of symmetry and form, it is possible to subdivide each crystal system into from two to seven classes. A total of 32 classes is geometrically possible.

OPTICAL PROPERTIES OF MINERALS

Minerals possess distinctive optical properties. Because of these properties some mineral specimens are particularly attractive and have value as gems. Other minerals are used in the manufacture of optical instruments. You can identify and classify minerals according to their optical properties. To do this you first need some basic information about the nature of light.

The optical properites of a mineral crystal are revealed by its capacity to refract and reflect light beams. You can represent a light beam conceptually by a cluster of sinusoidally oscillating rays (Figure 6-7). A particular ray upon entering your eye stimulates a sense of color that is related to its wavelength (λ) according to data in Table 6-2. Your eye is sensitive only to rays in the visible range that include colors from violet to red. A beam of white light contains rays of all wavelengths in

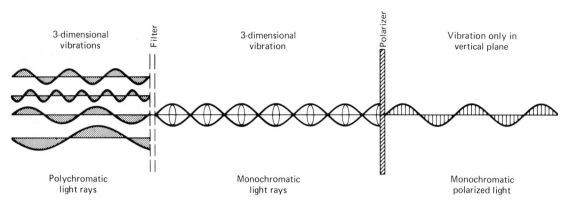

Figure 6-7 A polychromatic light beam can be represented by waves with different wavelengths that vibrate in different nonparallel planes. Such a beam can be directed on a filter that transmits only waves of a particular wavelength which produce a monochromatic beam. The monochromatic beam can be directed on a polarizer that transmits only waves vibrating in one set of parallel planes, which then produce a beam of polarized monochromatic light.

Table 6-2
Electromagnetic Waves

Wave	Color	Wavelength
Infrared	(Invisible)	7×10^{-5} to 3×10^{-2} cm
Visible	Red	7.2×10^{-5} cm
	Yellow	5.8×10^{-5} cm
	Green	5.0×10^{-5} cm
	Blue	4.5×10^{-5} cm
	Violet	4.0×10^{-5} cm
Ultraviolet	(Invisible)	1×10^{-6} to 3.5×10^{-5} cm
X rays	(Invisible)	1×10^{-9} to 1×10^{-7} cm

the visible spectrum. A **monochromatic** beam is made up of rays all having the same wavelength. **Polarized** light consists of rays oscillating in parallel planes. Monochromatic polarized light can be obtained by passing a white light beam through filtering and polarizing prisms (Figure 6-7). This kind of light beam is especially useful for studying optical properties of minerals.

A light ray passing from one medium to another separates into refracted and reflected rays (Figure 6-8A). The angles of these rays are predicted from Snell's Law:

$$\sin i / \sin r = n_r / n_i \qquad (6-1)$$

where i is the angle of the incident ray, r is the angle of a refracted or reflected ray, and n_i and n_r are the respective indices of refraction in the media through which the rays are directed. The index of refraction (n) is the ratio of the velocity of light in a vacuum (c) and the velocity in the medium along the direction of the ray (c_m); $n = c/c_m$. For all wavelengths the index of refraction equals 1 in a vacuum and, for practical purposes, in air. In a mineral the indices of refraction are different for different wavelengths. Therefore, you can see rays of different wavelengths from a beam of white light become separated (Figure 6-8B) when they are refracted through a mineral crystal. This effect can produce a "rainbow" of color. It is called **optical dispersion.** Differences in indices of refraction for different colors are larger in diamond than in most other substances, so that this mineral is especially effective in dispersing a beam of white light into a spectrum of colors.

All minerals in the cubic system have **singly refractive** optical properties. This means that a monochromatic light beam incident on a cubic crystal separates into a single refracted beam and a single reflected beam

Mineralogy 187

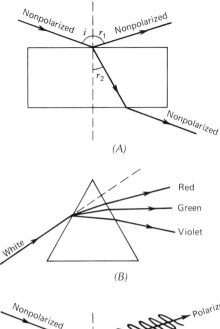

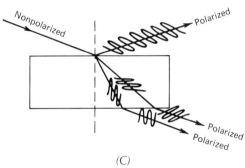

Figure 6-9 Calcite is a doubly refractive mineral. An object viewed through a calcite crystal appears as a double image.

Figure 6-8 Some important features of the refraction and reflection of light are illustrated in these diagrams. *(A)* A nonpolarized monochromatic incident beam separates into one nonpolarized refracted beam and one nonpolarized reflected beam at the surface of a singly refractive crystal. *(B)* A beam of white (polychromatic) light separates into refracted beams of the different constituent colors at the surface of a crystal. This effect is called optical dispersion. *(C)* A nonpolarized monochromatic incident beam separates into two polarized refracted beams and one polarized reflected beam at the surface of a doubly refractive crystal. The reflected beam and one refracted beam are polarized in the same plane, and the other refracted beam is polarized in a plane perpendicular to the plane of the other two beams.

(Figure 6-8A). Minerals in all other crystallographic systems are **doubly refractive**. A monochromatic beam incident on the surface of a doubly refractive crystal separates into two refracted beams and one reflected beam (Figure 6-8C). Furthermore, one refracted beam tends to polarize parallel, and the other perpendicular to the plane containing all of the beams. The reflected beam polarizes perpendicular to this plane. A doubly refractive transparent mineral such as **calcite** has the capacity to project a double image (Figure 6-9).

You can find indices of refraction of transparent minerals by measuring the angles of refracted beams produced by polarized monochromatic incident beams directed at known angles. You can also compare the brightness or intensity of the repolarized refracted beams with the intensity of the incident beam to discern other optical properties. Changes in polarization and intensity of reflected light beams can be used to establish the optical properties of opaque minerals. Mineralogists have learned from these measurements that the velocity of monochromatic light is the same in all directions through cubic minerals. The tetragonal and hexagonal minerals have one crystallographic axis that is a different length than the other axes, and is perpendicular to a plane containing the other axes. The velocity of light in the direction of this perpendicular axis is different from the velocity in the plane of the other axes. In orthorhombic, monoclinic, and triclinic minerals the velocity of monochromatic light is different in the direction of each of the crystallographic axes.

X-RAY ANALYSIS OF MINERALS

The external physical properties and geometrical shape of a mineral are manifestations of an internal structure. Optical measurements reveal some internal features, but they do not provide the fundamental information needed to describe the structure of the mineral. One may get this information by using x rays to look at the interior of a crystal. X-ray patterns produced in different experiments can be interpreted to discover what atoms exist in a crystal, and the geometrical arrangement of these atoms.

The discovery of x rays was reported by the physicist Roentgen in 1895. He realized that a beam of rays of unknown nature could penetrate shields of various materials to darken a photographic plate. During the following decades x rays were recognized as a form of electromagnetic energy similar to light, but having much shorter wavelengths (Table 6-2). Because of the very short wavelengths, x rays can penetrate much more deeply into a substance than visible light, which is reflected or absorbed very near the surface.

X rays can be generated with devices such as the **hot cathode tube** (Figure 6-10). It consists of two electrodes across which a high voltage is maintained. By heating a tungsten wire in the cathode electrons are emitted. These electrons strike the anode and dislodge other electrons. X-rays are emitted in the process of dislodging electrons from the anode, and the electrons are then conducted away from the anode target.

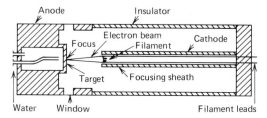

Figure 6-10 Schematic diagram of a hot cathode X-ray tube. Electrons emitted from a hot cathode wire are directed at the anode. Upon striking the anode these electrons dislodge still more electrons in a process that causes X rays to be released. (After N. F. M. Henry, H. Lipson, and W. A. Wooster, *The Interpretation of X-Ray Diffraction Photographs,* page 27, Macmillan and Co., 1960.)

The use of x-ray energy to investigate crystalline substances was pioneered by the physicist Laue early in the twentieth century. He suggested that if atoms occupy points of the space lattice of a crystal, then x rays might be deflected in orderly patterns as they penetrate through the crystal which acts as a three-dimensional grating. The systematic scattering of x rays in a beam directed on such a grating is called **diffraction.** Although some rays pass directly through the crystal, others will interfere with atoms giving rise to deflected rays. Laue's idea was confirmed by an experiment. A beam of x rays directed upon a plate cut from a crystal is separated into a pattern of diffracted beams that can be recorded photographically (Figure 6-11). The points on the photograph indicate diffracted beams of various intensity. The intensity depends upon the angles of different lattice planes along which atoms are geometrically arranged. You can see some but not all of the symmetry properties of a crystal in a Laue photograph. Because of the symmetrical arrangement of the experiment, the resulting

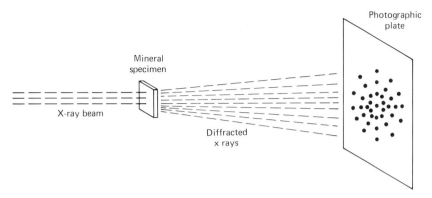

Figure 6-11 Schematic diagram of the Laue experiment for detecting x-ray diffraction through a crystal. The crystal acts as a grating that diffracts the X rays in a manner that can be recorded as a symmetrical pattern of points on a photographic plate.

pattern will tend to show more symmetry than actually exists in the crystal.

An important development in x-ray analysis was devised by W. H. Bragg and W. L. Bragg (father and son) to provide an actual measure of the distances separating individual lattice planes. They pointed out that x rays penetrating a substance could be thought of as reflecting from a succession of parallel lattice planes (Figure 6-12A). Rays reflected from successively deeper planes may join to form a single composite reflected beam (Figure 6-12B). Suppose that you use monochromatic x rays of known wavelength λ. Then, for a particular spacing (d) between lattice planes and for a particular angle of incidence (θ) the reflected rays will join so that the oscillations of all are in phase. Each oscillatory or vibrating ray can be represented by a sinusoidally curved line (Figure 6-12B). All rays are **in phase** when their peaks and troughs are in identical positions. We call this a conditon of **constructive interference.** It produces an especially bright or intense composite x-ray beam. If the oscillations of successive reflections are not in phase (Figure 6-12C), then the resulting beam is less intense, usually being almost undetectably weak. The condition for ideal constructive interference is expressed by the Bragg equation:

$$\lambda = 2d \sin \theta \qquad (6\text{-}2)$$

Laboratory apparatus has been designed so that a crystal on a spindle can be rotated in a beam of monochromatic x rays for which λ is known. When the crystal is rotated into a particular position, one can detect a reflected x-ray beam of maximum intensity. This indicates a condition of constructive interference, so the angle θ can be measured with the crystal in this position. Knowing both λ and θ experimentally, d can be calculated. By rotating the crystal into other positions the spacings for other sets of lattice planes can be measured. The procedure is shown schematically in Figure 6-13. This kind of x-ray analysis has proven to be of fundamental importance in the study of the structure of minerals.

Various instruments have been developed for routine x-ray analysis of mineral specimens. The methods are based upon one or another modification of the Bragg equation. In a modern automated instrument called a single crystal x-ray diffractometer, a crystal mounted on a spindle can be automatically

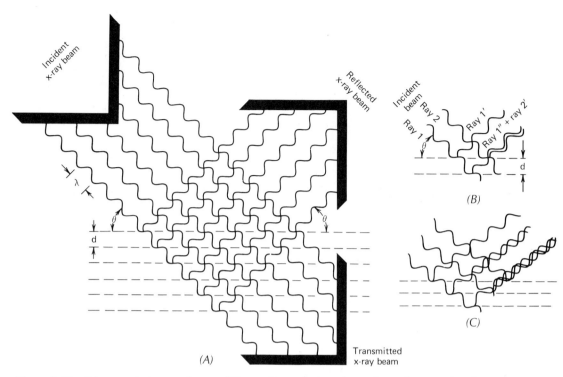

Figure 6-12 Diagram A shows a beam of X rays directed upon the surface of a crystal and penetrating into the crystal. As the beam impinges upon a particular lattice plane most of the energy penetrates it, but a small portion is reflected. Generally the reflected energy is too weak to detect except for a particular combination of the angle of incidence ϑ of the beam, X-ray wavelength λ, and lattice plane spacing d. For this critical combination the reflected X rays are in phase, that is, the peaks and troughs indicating vibration are aligned as seen in diagram B. For this condition of constructive interference a detectable reflected X-ray beam results. Diagram C illustrates destructive interference of X rays.

set in a variety of positions in an x-ray beam. X-ray intensity is recorded for each position, and the data are simultaneously recorded on charts and punched on computer cards for subsequent calculation of lattice plane spacings and angles between different sets of planes. This device may operate continuously for two or three days to complete a structural analysis of a crystal. A more commonly used instrument is the x-ray powder diffractometer (Figure 6-14). A mineral specimen is ground into a fine powder that is inserted into a cy-

Figure 6-13 *(opposite)* (A) Schematic diagram showing the components in a single crystal diffractometer. A specimen is mounted on a goniometer spindle that can be rotated. In a particular orientation a beam of X rays focused on the specimen is diffracted from a set of parallel lattice planes in the specimen. (B) A detector is moved through an arc over the specimen to measure the position and intensity of the diffracted X rays, and the results are (C) recorded on a chart that displays X-ray intensity for different detector positions. (Reprinted with permission from F. D. Bloss, *Crystallography and Crystal Chemistry*, page 490, Holt, Rinehart and Winston, Inc., 1971.)

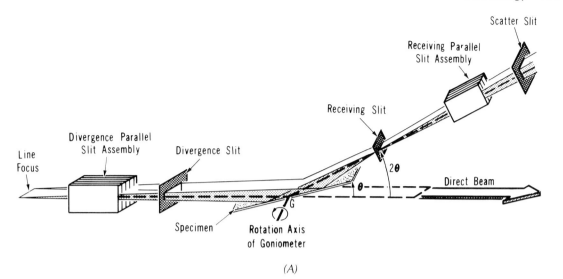

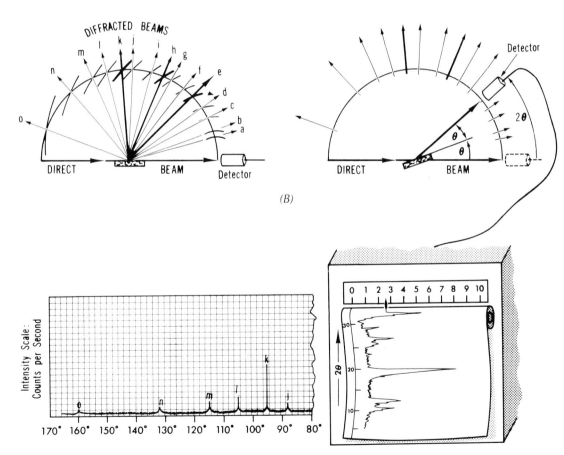

192 The Rocks of the Earth

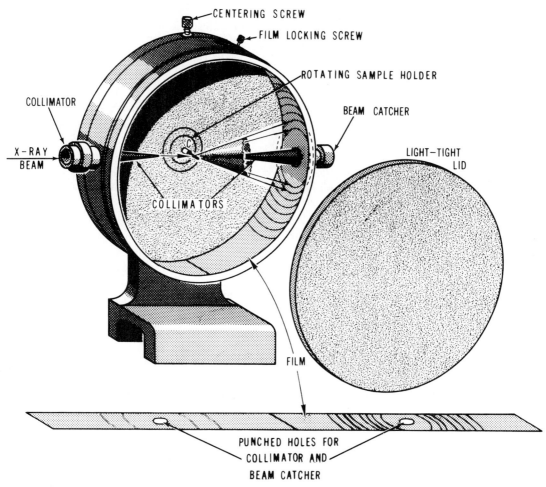

Figure 6-14 Schematic diagram of an X-ray powder diffractometer. An X-ray beam directed on a powder sample becomes separated into diffracted beams that are recorded as lines on a photographic film strip mounted in the cylindrical chamber. (Reprinted with permission from F. D. Bloss, *Crystallography and Crystal Chemistry,* page 473, Holt, Rinehart and Winston, Inc., 1971.)

lindrical tube. A single incident monochromatic x-ray beam is directed on this tube. Each grain of powder acts as a single crystal. Because they are randomly oriented, there will certainly be some grains in each of the many positions that will produce intense reflected x-ray beams from all possible sets of lattice planes. A photographic recording reveals lines with characteristic spacing. Each mineral is characterized by its own distinctive pattern, which can be viewed as a unique x-ray fingerprint. The x-ray powder diffractometer is widely used for identification of unknown mineral specimens.

To describe the internal structure of a mineral one must know the chemical elements present as well as the spacing of the lattice planes along which the atoms are distributed. Traditional tests may be used to identify chemical elements. These tests involve dissolving a specimen and mixing the solution with reagents that will produce different precipitates, depending on the elements that exist in the specimen. For many problems in mineralogy it is necessary to study tiny samples for which these traditional methods of chemical analysis are not practical. Precise chemical analysis of small particles of material can be accomplished with a device called an **electron microprobe.** It is designed on the basis of another kind of x-ray emission.

We know from experimental evidence that, if a substance is bombarded by electrons, it emits x-rays. We have also learned that each chemical element in the substance emits x-rays of a particular wavelength. Therefore, we can determine what elements are present by detecting different x-ray wavelengths.

An electron microprobe consists of an electron source and an x-ray spectrometer. Electrons can be generated in a cathode ray tube that is something like the x-ray tube seen in Figure 6-10. Because electrons have negative electric charges, they can be focused into a beam less than one micron (1 micron = 10^{-6} meter) in diameter by means of electromagnetic fields. When directed upon a specimen, such a beam causes x-ray emissions from a portion of the specimen of less than a few cubic microns in volume. The resulting emissions are recorded by the x-ray spectrometer. Then the patterns on the spectrometer record are compared with patterns obtained from standard samples of known chemical composition. These comparisons reveal what elements are present and their relative quantities. If a mineral specimen is cut into a series of parallel thin sections, one can get a detailed chemical analysis by directing the electron beam point by point over the surface of each thin section.

X-rays are indeed useful to the mineralogist. The x-ray emissions produced in an electron microprobe tell us what atoms exist in a mineral, and X-ray diffractometer measurements tell us how these atoms are arranged. From this type of information the mineralogist is able to give a detailed description of the internal crystalline structure of a mineral.

ATOMIC STRUCTURE OF MINERALS

Most atoms need companionship. Rather than drifting about as separate particles, they seek alliances. In different natural settings they join together in various ways to produce the substances we call minerals. Why do most atoms dwell in the communal environments of chemical compounds? Why do they form such geometrically precise neighborhoods? Perhaps we can begin to answer these questions by first looking into the nature of an atom.

The smallest particle that can possess the distinctive properties of a chemical element is called an **atom.** Important ideas about the structure of atoms were developed by the physicists Ernest Rutherford and Niels Bohr early in the twentieth century. According to these ideas an atom possesses a cluster of densely packed particles called a nucleus. Two kinds of particles, protons and neutrons, make up the nucleus. They have approximately the same size and mass, but protons maintain positive electrical charges in contrast to neutrons which carry no charge. Other negatively charged particles called electrons follow orbits around the nucleus. The mass of an electron is about 1/1837 of the mass of a proton or neutron. This model of an atom is illustrated in Figure 6-15.

An individual atom not associated with others will tend to have the same number of electrons and protons so that the net electrical charge on the atom is zero. The diameter of the nucleus is of the order of 10^{-13} cm, and the electrons orbit in a space of the order of 10^{-8} cm in diameter. In this space the electrons occupy specific regions more or less defined by concentric shells (Figure 6-15). Knowing these dimensions we can figure that a sand grain 1 mm in diameter contains approximately 4×10^{19} atoms.

As physicists continued to study the nature of atoms, electrons were found to be somewhat enigmatic. They possess characteristics of both matter and energy. You can estimate

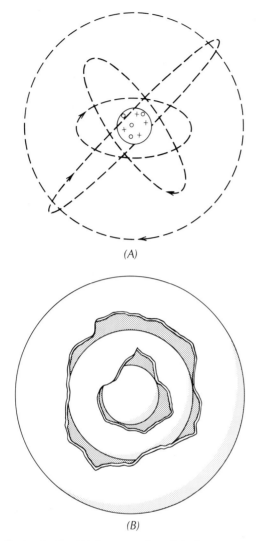

Figure 6-15 (A) Conceptual model of an atom with a nucleus of neutrons and protons orbited by electrons. (B) An individual electron orbit is restricted to a particular region surrounding the nucleus. Spherical shells describe the regions containing the different electron orbits.

the mass of an electron and also assign it a wavelength. The energy of an electron is related to the particular shell of the orbit. Energy of an electron increases with distance from the nucleus. The less massive atoms with relatively few electrons are smaller in size because the electrons are located in the inner low-energy shell regions. More massive atoms have electrons distributed in outer as well as inner shells. The maximum number of electrons (N_{max}) which can exist in a particular shell is determined from the formula:

$$N_{max} = 2n^2 \qquad (6\text{-}3)$$

where n is the shell number: $n = 1$ for the innermost shell, and assumes the values 2, 3, 4, and so on, for successive shells. Thus we can have 2 electrons in the inner shell, 8 in the second shell, 18 in the third shell, 32 in the fourth shell, and so on. Each of the shells can be further subdivided from energy considerations into subshells. The four classes of subshells are designated s, p, d, and f, and the electrons in a particular shell are restricted to one or another of these subshells. The outermost shell is an exception to Eq. (6-3). The number of electrons maintained in this shell does not exceed eight.

Because all electrons have negative electrical charges, we would not expect them to follow completely random orbits in a subshell. If two or more electrons are located in the same shell, their motions will be restricted because like charges repel each other. Their orbits will tend to be arranged to minimize the forces of repulsion. From knowledge of the nature of these forces and other energy considerations it should be possible to predict the different orbital patterns of electrons. Methods of statistical analysis have proven very important in understanding the dynamics of electron orbits. On the basis of statistical mechanics and solutions to an equation known as the **Schroedinger wave equation** physicists can predict the probability of finding an electron in a certain region. The probability, shown graphically in Figure 6-16, is divided into **radial** and **angular** parts. We can use the radial graphs (Figure 6-16A) to find the relative probability of locating an electron anywhere at a specified distance from the nucleus. You can see that for an atom with electrons occupying only two shells ($n = 1$ and

Mineralogy 195

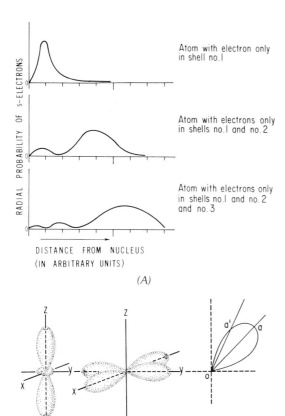

Figure 6-16 Radial and angular probability functions obtained from solutions of the Schroedinger wave equation for an atom. Diagram A presents the probability of encountering an electron anywhere in space at a particular distance from the nucleus. Diagram B indicates the probability of finding an electron anywhere along a line extending outward from the nucleus in a particular direction. Further explanation is given in the text. (After F. D. Bloss, *Crystallography and Crystal Chemistry,* pp. 166–168, Holt, Rinehart and Winston, Inc., 1971.)

$n = 2$), the probabilities of finding electrons somewhere at distances of 1 and 4 units from the nucleus are much greater than the probability of finding an electron somewhere 2 units away from the nucleus.

The angular term is illustrated by so-called **electron orbital** diagrams (Figure 6-16 B). You can interpret these diagrams in the following manner. A line is drawn outward from the center to the point where it intersects an **orbital lobe** as, for example, the line oa' in Figure 6-16 B. The length of the line oa' represents the relative probability of finding an electron anywhere along a line in that direction. The line oa indicates a direction along which there is maximum probability of finding an electron. The ratio of oa'/oa reveals the relative decrease in probability of finding an electron somewhere in the direction of oa' compared with the direction of oa. By combining the radial and the angular terms we can estimate the probability of finding an electron at a particular point in the space surrounding the nucleus. It is important to remember that these orbital lobes form symmetrical geometric patterns. This will help you to understand why atoms group together in regular patterns to form crystals.

You need to know two more things about atoms to understand why most of them seek the companionship of other atoms. Every atom wants to have the right number of electrons in each of its shells, and it wants to be electrically neutral. The right number of electrons is given by Eq. (6-3) except for the outermost shell. An atom wants to have eight electrons in this outermost shell. The atom must have the same number of electrons and protons, however, if it is to achieve its desire for electrical neutrality. The atom never feels completely satisfied unless it can have eight outer shell electrons and also be electrically neutral. Helium, neon, argon, krypton, xenon, and radon are the only elements in nature that have electrically neutral atoms with exactly eight electrons in their outermost shells. These substances, known as the **noble gases,** do not readily combine to form chemical compounds. For all other substances atoms attain greater contentment by forming groups where they can share electrons with one another. In this way they can better satisfy the needs for eight outer shell electrons

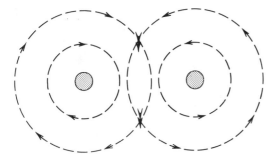

Figure 6-17 Simplified conceptual model showing electron sharing between two oxygen atoms.

and electrical neutrality. These are important conditions for stability in a chemical compound.

The conditions for stability for most elements are realized by a process of sharing or borrowing electrons. When this is done, chemical **bonds** are formed that tie one atom to another. Consider first the situation of oxygen, an atom having 8 protons, 8 neutrons, and 8 electrons. Two electrons occupy the inner shell, leaving six in the next shell, which is the outermost shell for oxygen. This second shell could hold eight electrons. Thus, a deficiency of two indicates an unstable condition. Therefore, in nature free oxygen occurs in the form of molecules each consisting of two atoms bonded together (Figure 6-17). Two electrons are shared to the extent that they orbit in outer shells of first one and then the other atom. So at least part of the time each oxygen atom has eight electrons in its outer shell. Therefore, the condition of maintaining eight outer-shell electrons is better accomplished by the bonding of two atoms than if they remained separate.

Next examine the combination of sodium and chlorine in the compound NaCl, the composition of common table salt. Considered individually (Figure 6-18) sodium has only one electron in its outer shell, and chlorine lacks one electron to fill its outer shell. When brought together the outer electron from the sodium atom shifts into the outer

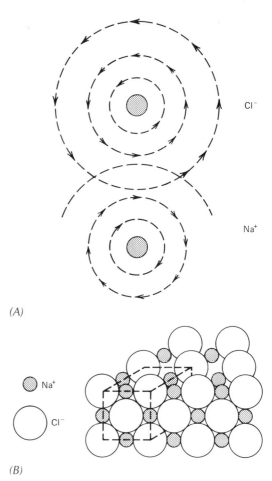

Figure 6-18 Bonding of chlorine and sodium. Diagram A shows individual Cl and Na atoms. Diagram B illustrates the crystalline structure of NaCl resulting from electron sharing and subsequent ionization of Na and Cl. An ion is a particle bearing an electrical charge because it has either yielded electrons or gained them such that electron shells are filled, but electrical neutrality is lost.

chlorine shell, thus filling it. Also sodium is left with two full shells. By losing an electron sodium attains a positive electrical charge, and by gaining an electron chlorine becomes negatively charged. These charged atoms are referred to as **ions.** Electrical neutrality is maintained only if the two ions remain close together. Thus they become bonded to better establish the conditions of full shells and electrical neutrality. This is most ideally achieved in an assemblage containing a large number of sodium and chlorine atoms by an ordered arrangement in a cubic lattice as shown in Figure 6-18.

Because the sharing of electrons is a fundamental part of the chemical bond, let us consider how it is best achieved. Some obvious considerations are (1) the availability of electrons to be shared, and (2) the possibility for an electron to follow an orbital path suitable to two or more atoms. It is clear from Figure 6-16 that the availability of electrons is restricted to certain geometrically arranged regions about the nucleus. Thus, chemical bonding is achieved when atoms are arranged in positions indicated by overlapping electron orbital lobes (Figure 6-19). Because of the geometrical arrangement of orbitals, electrons are available for sharing only if the atoms are distributed in a geometrical pattern relative to each other. Therefore, we would expect a three-dimensional array of bonded atoms to possess definite geometrical properties.

The strength of a chemical bond depends upon several factors, including the proximity of atoms to each other and the kinds of orbitals that overlap. The energy characteristics of an electron must be compatible with the atoms involved in the bond. Electrons in some orbital positions are incompatible with those of certain other orbitals. Because of this factor some orbital arrangements would lead to mutually repelling forces. Therefore, the geometry of a cluster of bonded atoms is further restricted to compatible orbital arrangements that allow constructive interference of the electromagnetic wave properties of the electrons. The strength of the resulting bond is related to the mutual attraction of the positively charged nucleus of one atom and negatively charged cloud of electrons from another. In addition, as the atoms approach each other, their similarly charged nucleii tend to be mutually repelled. These contrasting forces of at-

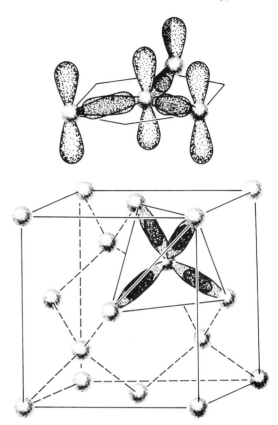

Figure 6-19 Bonding is achieved by arranging atoms in positions where electrons are available for sharing, that is, in arrangements where orbital lobes overlap. (Modified from F. D. Bloss, *Crystallography and Crystal Chemistry*, page 198, Holt, Rinehart and Winston, Inc., 1971.)

traction and repulsion are shown graphically as a function of the distance between nucleii (Figure 6-20). The strongest bond is achieved at the separation distance where energy is minimized.

The sizes of the atoms involved in a chemical compound are important in determining the strength of a bond. In order to fit various sized atoms into a structure that provides for compatible orbital overlap, some atoms may not be able to maintain the optimum internuclear separation distance where energy is

198   The Rocks of the Earth

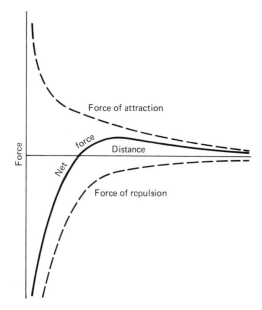

Figure 6-20 Graph showing repelling forces between two positively charged atomic nucleii, attractive forces between positively charged nucleii and negatively charged electron clouds, and the resultant net force as functions of distance between the two nucleii.

minimized. This can result in weaker bonds along some lattice planes compared with others.

Some minerals are formed from single chemical elements. **Graphite** and **diamond** are formed from the element carbon. The atoms are arranged in the patterns shown in Figure 6-21. The strength of the bonds joining the carbon atoms in a diamond is great enough to form the hardest known natural substance. In contrast the bonds in a graphite structure are much weaker, forming a soft substance useful for making pencils. These two minerals indicate that more than one geometrical arrangement of the same atoms can become bonded into a crystalline substance. This is true for some other elements and chemical compounds as well as for carbon. Minerals with different crystalline structure but the same chemical composition are

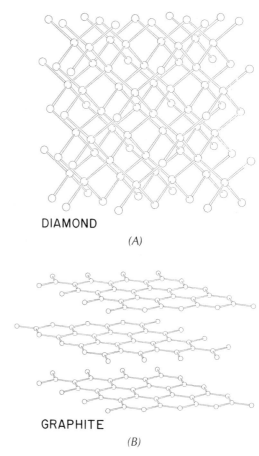

Figure 6-21 Models showing the arrangement of carbon atoms in (A) a diamond crystal and (B) a graphite crystal. Lines connecting ions in the graphite structure indicate only the strongest bonds. Other weaker bonds existing between the layers of strongly bonded ions are not shown, but they maintain the sheets in positions relative to one another that are evident in the diagram. (Courtesy of Karen Geisinger and G. V. Gibbs.)

known as **polymorphs.** Polymorphism is an important aspect of mineralogy related to the pressure and temperature conditions where a mineral is formed. This is the subject of a later section of this chapter.

Most minerals form from chemical compounds rather than single elements. The chemical bonding usually involves some atoms with relatively few outer-shell electrons which they yield to other atoms needing relatively few electrons to fill their outer shells. As a result of this transfer of electrons, some atoms become positively charged and others become negatively charged, as in the case of NaCl. We call the positively charged atoms **cations** and the negatively charged ones **anions.** In crystalline substances the grouping of cations and anions is described by rules that were proposed by the chemist Linus Pauling. Known as Pauling's Rules, they are briefly summarized.

1. Anions tend to group about a cation to form a polyhedron known as the **coordination polyhedron** of the substance.

2. In a stable structure the total electrical charge reaching an anion from surrounding cations tends to equal the charge on the anion.

3. Coordination polyhedrons in a structure tend not to share edges.

4. The number of structurally different kinds of atoms in a specific crystalline structure tends to be small.

The mineral quartz is an example to illustrate these rules. The composition SiO_2 is formed in a structure of silicon cations of $+4$ charge and oxygen anions of -2 charge. In a crystalline structure they form a coordination polyhedron known as the SiO_4 **tetrahedron** (Figure 6-22). You can see a single Si cation reaches to four oxygen anions, and its electrical charge of $+4$ is divided equally to direct a charge of $+1/2$ toward each anion. From its location in the crystalline structure each anion is seen to receive a charge of $+1/2$ from each of four cations totaling to a charge of $+2$, which is sufficient to neutralize its charge of -2. Thus the neutrality of electrical charge is maintained in the crystalline structure. You can see another polyhedral arrangement of cations and anions, and balance of

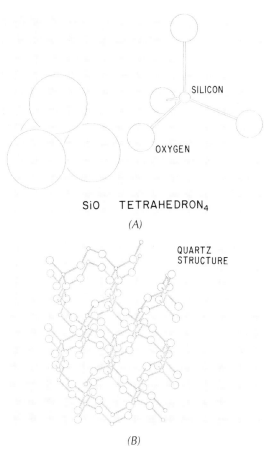

Figure 6-22 The SiO_4 tetrahedron is an important constituent in the crystal structure of silicate minerals. (A) The geometrical arrangement of ions in a single tetrahedron is shown correctly scaled in one diagram, and expanded in the other diagram to expose the silicon ion that is hidden by the larger oxygen ions in the correctly scaled diagram. In both diagrams the relative sizes of the ions are correctly scaled. (B) This diagram illustrates the arrangement of SiO_4 tetrahedra in the crystal structure of quartz. (Courtesy of Karen Geisinger and G. V. Gibbs.)

electrical charge in the geometrical arrangement of atoms in the halite (NaCl) structure (Figure 6-18B). The SiO_4 tetrahedron is an important structural unit in the silicate minerals. It is one example of a coordination polyhedron that can be viewed as a basic structural unit in a crystal. The coordination polyhedron is not the same as the unit cell which is the basic geometrical unit of the crystal.

In summary we know that chemical bonds result from the tendency of atoms to maintain neutral electrical charge and to maintain eight electrons in the outer shell. They achieve this by sharing of electrons. We know from statistical considerations (Figure 6-16) that the regions where electrons are most likely to be available are symmetrically arranged. Therefore, the atoms of common minerals tend to group in distinct geometrical arrangements to establish the most effective sharing of electrons. The strengths of the resulting bonds depend upon the kinds of orbitals that overlap, the relative sizes of the atoms in the structure, and the electrical charge differences between atoms.

We can now understand more about some of the external physical properties of minerals. The hardness and cleavage are related to the bond strengths between atoms. When the surface of a mineral is scratched, atoms are dislodged. One substance can scratch another only if its atoms are more strongly bonded. Cleavage results from weaker bonds along certain lattice planes that exist because of the geometrical distribution of atoms of different sizes. The **density** of a mineral is related to the packing and mass of the atoms. The **luster** of a mineral is related to the way that light interferes with outer shell electrons causing different degrees of constructive interference for reflecting rays of various wavelengths. Electromagnetic properties depend on the extent to which electrons move through large regions of a structure and are shared among a great many atoms. Such electron mobility is greatest in substances that have orbital shells approximately half-filled. Then, more electron sharing is required to satisfy the tendency to maintain full shells. Metals such as iron and copper are in this category. Elasticity parameters are related to the bond strengths in the structure. Color is less easily understood; it involves the capacity of a structure to absorb light of some wavelengths and allow other wavelengths to be transmitted through the substance. We know of no simple relationship between color and the crystal geometry and bond strengths of a mineral.

SOME IMPORTANT MINERAL GROUPS

We know that more than 2000 kinds of minerals occur in nature. Chemical compositions of these minerals include nearly all of the 103 elements in the Periodic Table (Appendix II). The amount of information about minerals is staggering. What basic knowledge of minerals is necessary to make use of this information? Suppose you want to find out why iron or aluminum can be economically extracted from one mineral and not another. Perhaps you need some facts about how minerals respond to different geologic processes. What general knowledge of minerals would help you to organize your thoughts about problems such as these? You will find it useful to know that relatively few minerals consisting of a limited number of elements form the bulk of earth materials. These minerals and elements are of greatest importance in understanding geologic processes.

The most abundant chemical elements that we find in the earth's crust are listed in Table 6-3. We find other chemical elements in minute trace amounts or in sparsely distributed concentrations. You can see that carbon, one of the most important elements in living organisms, is not one of the more abundant elements in the earth.

The basic rock forming minerals, as well as others of economic importance, consist mainly of the ten elements in Table 6-3. Because many of these minerals have common features of composition and crystal structure, we can classify them in groups possessing similar properties. The most important groups

Table 6-3
Common Elements in the Earth's Crust[a]

Element	Relative Abundance (Weight Percent)
Oxygen	46.60
Silicon	27.72
Aluminum	8.13
Iron	5.00
Calcium	3.63
Sodium	2.83
Potassium	2.59
Magnesium	2.09
Titanium	0.44
Hydrogen	0.14
All other elements	0.83
	100.00

[a]From Brian Mason, 1958, *Principles of Geochemistry*, 3rd ed., John Wiley & Sons, New York, p. 48.

are the silicates, carbonates, oxides, and the sulfide minerals. Knowledge of these groups will give you some basic ideas about the structure of minerals.

Silicate Minerals

The principal constituents of igneous and metamorphic rocks are the silicate minerals. When weathered or reworked, they are important in sedimentary rocks. Approximately 95 percent of the earth's crust consists of silicates. The important groups include olivines, pyroxenes, amphiboles, micas, feldspars, and the silica minerals.

Common to the crystal structure of all silicate minerals is the SiO_4 tetrahedron (Figure 6-22). Because silicon atoms in this unit are surrounded by four oxygen atoms, we say that silicon exists in **tetrahedral coordination** with oxygen. The SiO_4 tetrahedra may be joined one to another by sharing oxygen atoms that we call **bridging** oxygens, or they may form bonds with other elements most commonly Al, Fe, Mg, Ca, Na, and K. The strongest bonds in a crystal are the tetrahedral bonds involving bridging oxygens between tetrahedra.

You can distinguish the important silicate mineral groups by crystal structure features. The differences are seen in the ways that the SiO_4 tetrahedra are bonded to one another. Depending upon the mineral group, the tetrahedra may be distributed individually or they may be joined in pairs, rings, chains, sheets, or in complicated three-dimensional frameworks. Examples of these structures are illustrated in Figure 6-23.

The SiO_4 tetrahedra in **olivine** crystals are not bonded to one another. Oxygen atoms at the four corners of each tetrahedron form bonds with iron or magnesium atoms. The olivine minerals are classified as **orthosilicates** because of this kind of SiO_4 tetrahedra grouping. A representative crystal structure is seen in Figure 6-23. An olivine mineral can have any composition within a continuous range from Fe_2SiO_4 (**fayalite**) to Mg_2SiO_4 (**forsterite**) depending upon the proportions of Fe and Mg. Because the Fe and Mg atoms are approximately the same size, one is as easily accommodated in the crystal structure as the other. This means that olivine minerals make up a **continuous series** of compositions ranging between the extremes of fayalite and forsterite. The Fe and Mg atoms are distributed so that there are no planes in the crystal where relatively weak bonds are formed. For this reason the olivine minerals do not have a well-developed cleavage. The **garnet** group and the mineral **zircon** are also orthosilicates which for similar reasons do not show good cleavage. Zircon is not an abundant mineral, but it is useful for obtaining isotope age measurements (Chapter 1) because of small amounts of uranium, thorium, and their radioactive decay products that are found in the crystal structure.

The melting temperatures for olivine minerals at atmospheric pressure range from 1205° C for fayalite to 1890° C for forsterite. Because of these high values, the olivines are

202 The Rocks of the Earth

	OLIVINE	TOURMALINE
ORTHOSILICATE AND RING SILICATE STRUCTURES	Mg / O or Fe / Si	O Si
SINGLE CHAIN SILICATE STRUCTURES	CHAIN OF SiO_4 TETRAHEDRA	DIOPSIDE STRUCTURE O Mg Si or Ca
DOUBLE CHAIN SILICATE STRUCTURES	CHAIN OF SiO_4 TETRAHEDRA	TREMOLITE STRUCTURE Ca OH O Mg Si
SHEET SILICATE STRUCTURES	SHEET OF SiO_4 TETRAHEDRA	BIOTITE STRUCTURE K OH O Mg/Fe Si
FRAMEWORK SILICATE STRUCTURES	FRAMEWORK OF SiO_4 TETRAHEDRA	ORTHOCLASE STRUCTURE K O Si

Figure 6-23 *(opposite)* Crystal structures of common silicate minerals consist of SiO_4 tetrahedra and other ions. The diagrams illustrate how the SiO_4 tetrahedra are joined to one another, and how these tetrahedra are arranged with other ions in the structure of a mineral. Relative sizes of ions are correctly scaled, but spacing of the ions is expanded to show the structure more clearly. Only the bonds between tetrahedral ions are shown by lines. Other weaker bonds that maintain the arrangement of these tetrahedra with other ions in the structure are not shown. (Courtesy of Karen Geisinger and G. V. Gibbs.)

useful as refractory materials in industrial metallurgical and ceramics processes.

Another important group of silicate minerals is the **pyroxene** group. Some pyroxene minerals are listed in Table 6-4. In this group the SiO_4 tetrahedra are bonded together in chains. The pyroxene crystal structure shown in Figure 6-23 illustrates how **single chains** are formed with bridging oxygens between adjacent tetrahedra. Two oxygen atoms from each tetrahedron are shared in the chain, and the two remaining oxygens bond with other elements, primarily Fe, Mg, and Ca. Because the tetrahedral bonds are stronger than the bonds with these other elements, pyroxene crystals have two sets of cleavage planes intersecting at approximately 90° and oriented parallel to the tetrahedral chains.

The **amphibole** minerals make up another distinctive silicate mineral group. Amphiboles are characterized by **double tetrahedral chains.** You can see a typical double chain structure in Figure 6-23. Tetrahedra along the center of the chain share three bridging oxygens with other tetrahedra in the chain. Bordering tetrahedra share two bridging oxygens in the chain. The central and bordering tetrahedra have one and two unshared oxygens, respectively, which form bonds with such elements as Ca, Mg, Fe, Na, Al, and the $(OH)^-$ ion. Because of the strong tetrahedral bonds, two sets of cleavage planes parallel to the double chains and intersecting at approximately 60° are found in the amphiboles. Some important amphiboles are listed in Table 6-4.

Mineralogists have studied the chemical decomposition of amphiboles at elevated temperature and pressure. Consider the amphibole **tremolite** that decomposes into quartz, H_2O, and the pyroxenes **enstatite** and **diopside** at a pressure of approximately 500 bars (1 bar = 10^6 dynes/cm^2; standard atmospheric pressure = 1.0133 bars) and a temperature of 800° C.

$$Ca_2Mg_5(Si_4O_{11})_2(OH)_2 \rightleftarrows 3MgSiO_3$$
$$\text{tremolite} \qquad \text{enstatite}$$
$$+ \; 2CaMg(SiO_3)_2 \; + \; SiO_2 \; + \; H_2O$$
$$\text{diopside} \qquad \text{quartz}$$

Similarly, the other amphiboles yield H_2O upon decomposition. Because the amphiboles possess the constituents to produce H_2O when they decompose, we call them hydrous **silicates.** You can usually recognize a hydrous silicate by the presence of the $(OH)^-$ ion. Note that the $(OH)^-$ ion does not appear in the composition of pyroxenes. The pyroxenes do not produce H_2O when they decompose, so we call them **anhydrous** silicates. This difference suggests that amphiboles have crystallized in hydrous environments, whereas pyroxenes indicate anhydrous conditions. We know that H_2O can exist in an **aqueous fluid** at the elevated conditions of temperature and pressure deep within the earth's crust and upper mantle.

Certain nonaluminous amphiboles such as tremolite and actinolite can be processed to get fibers of asbestos. Amphibole asbestos consists of long fibers. Short-fibered asbestos is found in another more hydrated mineral called **serpentine.** Although long- and short-fibered asbestos have about the same heat resistant properties, the latter is more suitable for spinning and weaving into cloth. Conse-

Table 6-4
Some Silicate Minerals

	Mineral	Composition	Crystal System	Economic Use
Ortho-silicates	Olivine	$(Mg,Fe)_2SiO_4$	Orthorhombic	Refractory in glass
	Garnet	$(Ca,Mg,Fe)_3(Al,Fe)_2(SiO_4)_3$	Cubic	Abrasives
	Zircon	$ZrSiO_4$	Tetragonal	Refractory, source of ZrO_2 and zirconium useful in ferroalloys
Single Chain Silicates	Augite	$Ca(Mg,Fe,Al)(Al,Si)_2O_6$	Monoclinic	
	Diopside	$CaMg(SiO_3)_2$	Monoclinic	
	Enstatite	$(Mg,Fe)_2(SiO_3)_2$	Orthorhombic	
Double Chain Silicates	Hornblende	$NaCa_2(Mg,Fe,Al)_5(Si,Al)_8O_{22}(OH)_2$	Monoclinic	
	Tremolite	$Ca_2Mg_5(OH)_2(Si_4O_{11})_2$	Monoclinic	Source of long fiber asbestos
	Actinolite	$Ca_2(Mg,Fe)_5(OH)_2(Si_4O_{11})_2$	Monoclinic	
Sheet Silicates	Muscovite	$KAl_2(AlSi_3O_{10})(OH)_2$	Monoclinic	Electrical insulators, filler in paper, paint, and lubricants
	Biotite	$K(Mg,Fe)_3(OH)_2AlSi_3O_{10}$	Monoclinic	
	Phlogopite	$KMg_3(OH)_2AlSi_3O_{10}$	Monoclinic	Electrical insulator
	Chlorite	$Mg_5Al(OH)_8AlSi_3O_{10}$	Monoclinic	
	Kaolinite (a clay mineral)	$Al_2(OH)_4Si_2O_5$	Monoclinic	Important source of ceramic materials for porcelain china, tiles, refractories, paper filler
Framework Silicates	Orthoclase	$KAlSi_3O_8$	Monoclinic	Source of ceramic materials
	Microcline	$KAlSi_3O_8$	Triclinic	Ceramic materials, powder and paint filler, scouring soap
	Albite	$NaAlSi_3O_8$	Triclinic	
	Feldspars of intermediate composition			
	Anorthite	$CaAl_2Si_2O_8$	Triclinic	Have some use in ceramics
	Quartz	SiO_2	Hexagonal	Glass, road paving sand, optical products, electrical transducers, ornaments
Silica	Tridymite	SiO_2	Hexagonal and Orthorhombic forms	
	Cristobalite	SiO_2	Tetragonal	
	Coesite	SiO_2		
	Stishovite	SiO_2		

quently, we do not process amphibole asbestos commercially for industrial use. But it has potential value if proper processing techniques are developed.

The **mica** minerals form another silicate mineral group. These minerals are characterized by sheets of SiO_4 tetrahedra. Each tetrahedron shares three bridging oxygens with other tetrahedra in the sheet. Adjacent tetrahedra are oppositely arranged so that the single unshared oxygens are distributed equally on both sides of the sheet. The crystal structure of the micas (Figure 6-23) consists of alternating tetrahedral sheets and layers of cations bonded to the unshared oxygens. The important cations include Al, K, Fe, and Mg. We have learned that Al can also be arranged in tetrahedral coordination with oxygen. In most micas about 25 percent of the tetrahedral positions is occupied by Al rather than Si. The bonds between these units are strong, although not as strong as the corresponding bonds between SiO_4 tetrahedra. Because the tetrahedral sheets extend continuously through a crystal, the micas have well-developed cleavage along only one set of parallel planes. The bonds between cation layers and tetrahedral sheets are weak, so that these minerals split easily along cleavage surfaces. The most abundant micas are listed in Table 6-4. Because the $(OH)^-$ ion appears in the chemical formula for each, we identify them as hydrous silicates. The most common of the micas is the mineral **muscovite.** It is used as an insulator in electrical apparatus.

Feldspars are the most abundant minerals in the crust of the earth. The crystal structures of these minerals include a variety of complicated forms where the SiO_4 tetrahedra and aluminum in tetrahedral coordination are bonded together in three-dimensional frameworks (Figure 6-23). Therefore, the feldspars are classified as **framework silicates.** Distributed throughout the tetrahedral framework are several elements the most abundant being Al, Na, Ca, and K. Because of the three-dimensional distribution of tetrahedral bonds, the feldspars form in cohesive crystals that are not easily fractured. However, the bond density, that is, the number of bonds crossing a unit of area is minimal along some planar surfaces so that the crystals possess cleavage.

The feldspar minerals in the **plagioclase** group are alumino-silicates that also contain sodium and calcium. They can have compositions within a continuous series between $NaAlSi_3O_8$ **(albite)** and $CaAl_2Si_2O_8$ **(anorthite)** depending upon the proportions of Na and Ca. The potassium feldspars are in a different group that has the composition $KAlSi_3O_8$. This group has two polymorphs, the monoclinic form **orthoclase** and the triclinic form **microcline.** Important clays such as kaolinite are derived from the decomposition of feldspars. The potassium feldspars and some derivative clays are of considerable value in the ceramics industry.

The chemical compositions of feldspars, lacking the $(OH)^-$ ion indicate anhydrous conditions of formation. In contrast, you can see that the micas are hydrous silicates. They can decompose at elevated pressure and temperature to yield H_2O. The decomposition of muscovite at a pressure of approximately 1500 bars and a temperature in excess of 700° C will produce orthoclase:

$$KAl_2(OH)_2AlSi_3O_{10} \rightleftarrows KAlSi_3O_8 + Al_2O_3 + H_2O$$
$$\text{muscovite} \quad \text{orthoclase} \quad \text{corundum}$$

If this decomposition takes place in an environment rich in quartz, it can react with the corundum to form the compound Al_2SiO_5. This is the composition of three important polymorphs: kyanite, andalusite, and sillimanite, which we will discuss in the next section of this chapter.

The silica minerals including **quartz, tridymite, cristobalite, coesite,** and **stishovite** are polymorphs of SiO_2. With the exception of **stishovite** the crystal structures consist of frameworks of SiO_4 tetrahedra. Therefore, these minerals are usually classified as silicates rather than oxides. Tetrahedral bonds extend throughout the crystals, all oxygens forming bridges. Therefore, these structures have uniformly strong bonds and display no obvious cleavage. In the extremely rare mineral **stishovite,** silicon is in octahedral coordination with oxygen. The density of a silica mineral is related to the packing of tetrahedra. Quartz with a relatively open structure has a density of 2.65 gm/cm^3 compared with the more closely packed structure of coesite which has a density of 2.91 gm/cm³. This higher density polymorph has been found in meteorite craters. Coesite crystals were produced where the extreme impact pressure of the meteorite altered the crystal structure of pre-existing quartz grains.

Some natural forms of silica have no regular crystalline structure. Known as **amorphous** substances, they consist of SiO_4 tetrahedra which are not arranged in any regular framework. Those mineralogists preferring a strict definition of mineral that implies definite crystalline structure have proposed the term **mineraloid** to include amorphous substances. Chert, flint, agate, onyx, jasper, chalcedony, and opal are some varieties of amorphous silica that are colored in characteristic ways by impurities.

The silica minerals, primarily quartz, find many uses. The most important use is in the manufacture of glass, an amorphous form of SiO_2. In crystalline form quartz is used in electromechanical devices where its piezoelectric properties are important. Optical prisms are designed to utilize its properties of double refraction and light dispersion.

Carbonate Minerals

The element carbon is not abundant in the earth. Nevertheless, it is found in an important group of sedimentary rocks. Carbon occurs in a structural unit that we call the **carbonate ion** ($CO_3^=$). This unit combines with other elements, mainly calcium, magnesium, and to a lesser degree Fe, Zn, Pb, Ba, and Sr, to form the important carbonate minerals.

The mineral **calcite** ($CaCO_3$) is the primary constituent of limestone. In the crystal structure of calcite (Figure 6-24) calcium and carbonate ions are arranged to form a crystal classified in the hexagonal system. Its property of double refraction is illustrated in Figure 6-9. Another polymorph of $CaCO_3$ is the mineral **aragonite,** also formed from Ca and CO_3 ions. In this mineral atoms are arranged

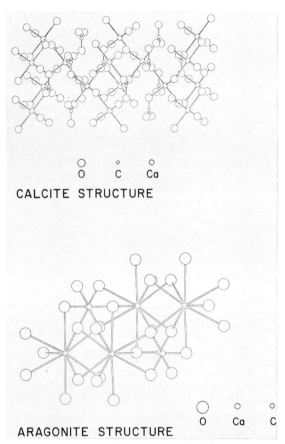

Figure 6-24 Crystal structures of calcium carbonate: calcite and aragonite. Relative sizes of individual ions are represented in the diagrams. (Courtesy of Karen Geisinger and G. V. Gibbs.)

(Figure 6-24) to form a crystal of the orthorhombic system. Aragonite is less stable and not as abundant as calcite. We have learned from experiments that it would not form in normal conditions that exist near the surface of the earth unless certain biological processes are taking place.

The mineral dolomite ($CaMg(CO_3)_2$) consists of Ca and Mg ions distributed in a regular three-dimensional structure. Some variation in the relative proportions of these ele-

Table 6-5
Some Carbonate Minerals

Mineral	Composition	Crystal System	Economic Use
Calcite	$CaCO_3$	Hexagonal	Concrete, source of lime, marble, optical instruments, and many other uses
Aragonite	$CaCO_3$	Orthorhombic	
Dolomite	$CaMg(CO_3)_2$	Hexagonal	Building stone, road paving materials, source of metallic magnesium, refractory
Magnesite	$MgCO_3$	Hexagonal	Substance used in refractory bricks, hot pipe coverings
Smithsonite	$ZnCO_3$	Hexagonal	Ore of zinc
Siderite	$FeCO_3$	Hexagonal	Minor iron ore
Malachite	$Cu(OH)_2CO_3$	Monoclinic	Ore of copper
Azurite	$Cu_3(OH)_2(CO_3)_2$	Monoclinic	Minor ore of copper
Colemanite	$Ca_2B_6O_{11} \cdot 5H_2O$	Monoclinic	
Kernite	$Na_2B_4O_7 \cdot 4H_2O$	Monoclinic	Important sources of borax
Cerusite	$PbCO_3$	Orthorhombic	Often associated with ore of lead

ments is possible, but the range of compositions is restricted because of differences in the sizes of the Ca and Mg atoms. The proportion produced depends upon the temperature of the environment as well as the amounts of Ca and Mg that are available.

Other carbonate minerals are all relatively rare. You can get an idea of their variety and composition from Table 6-5.

Oxide Minerals

The geologically important oxide minerals include compounds of Fe, Ti, Al, Mg, Cu, and Cr. These minerals are important as ores. Although they make up a relatively insignificant part of the total volume of crustal rocks, they are important because small amounts disseminated through a rock impart significant color and magnetic properties. We classify the important oxides into various groups including the **hematite group** and the **spinel group.** The properties and uses of some typical oxide minerals are outlined in Table 6-6.

The crystal structures of hematite and magnetite are shown in Figure 6-25. These minerals represent the two most important oxide groups. You were introduced to these minerals and ilmenite in Chapter 4 where we discussed their magnetic properties. The magnetism of a crystal is caused partly by the spinning of electrons. In most structures the spin directions are random, and the individual electron magnetic moments interfere one with another so that no net magnetic field is developed in the larger crystal. Because of the particular arrangement of electron orbitals in certain ion oxides, a measurable magnetic field is produced by the interaction of electron magnetic moments. This field is developed in ideally ferromagnetic minerals re-

Table 6-6
Some Oxide Minerals

Mineral	Composition	Crystal System	Economic Uses
Hematite	Fe_2O_3	Hexagonal	Important ore of iron
Corundum	Al_2O_3	Hexagonal	Abrasives
Ilmenite	$FeTiO_3$	Hexagonal	Source of TiO_2 which is used in paint, paper, rubber, ceramics, textiles, etc.
Spinel	$MgAl_2O_4$	Cubic	Refractory materials
Magnetite	Fe_3O_4	Cubic	Important iron ore
Franklinite	$(Zn,Mn)Fe_2O_4$	Cubic	Ore of zinc
Chromite	$(Mg,Fe)Cr_2O_4$	Cubic	Refractory, chrome, brick, and ferroalloys
Rutile	TiO_2	Tetragonal	Important ore of titanium
Cassiterite	SnO_2	Tetragonal	Most important ore of tin
Uraninite	UO_2	Cubic	Pitchblende is an important source of uranium
Bauxite	$Al_2O_3 \cdot nH_2O$		Chief source of aluminum

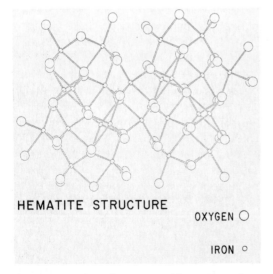
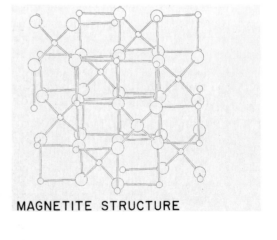

Figure 6-25 Crystal structure of hematite and magnetite. Relative sizes of individual atoms are shown in proper scale in the diagram. (Courtesy of Karen Geisinger and G. V. Gibbs.)

gardless of the existence of an external magnetic field. The magnetic field associated with the crystal structure can be increased by an external field that causes orbitals to become warped so that electron spin moments are more closely aligned.

Other common iron oxides are produced near the earth's surface by hydration of hematite and magnetite. These minerals display colors ranging from light yellow to dark purple and including red and orange hues. They commonly occur in the cementing materials of sedimentary rocks and give color to the rock.

Sulfide Minerals

These minerals include the important ores of copper, lead, zinc, silver, nickel, and mercury. Important sulfide minerals are listed in Table 6-7. Several different sulfide minerals are commonly found in close association; among these are several forms of iron sulfide.

Pyrite (FeS_2) possesses a cubic structure. Because of its color, early prospectors mistook it for gold, and thus it was called fool's gold. One can easily distinguish it from gold by its hardness (6 to 6½) and its black streak. The mineral pyrrhotite ($Fe_{1-x}S$) commonly has small quantities of Cu, Ni, or Co that occupy sites in the crystal structure in place of Fe. The amounts of these impurities in different varieties of pyrrhotite are indicated by the value of x in the subscript, which is usually between 0 and 0.2. The impurities affect the crystal structure so that although several varieties are hexagonal, some possess other structures such as monoclinic. Pyrrhotite has distinct crystal magnetism that is strong enough to cause magnetic anomalies over deposits where it is disseminated.

Some other minerals of general interest include halite (rock salt), the economically important sulfates **gypsum** and **anhydrite,** and the barium sulfate, **barite.** Some properties of these and a few other minerals of economic importance are described in Table 6-8.

Table 6-7
Some Sulfide Minerals

Mineral	Composition	Crystal System	Economic Uses
Argentite	Ag_2S	Cubic	Important ore of silver
Chalcocite	Cu_2S	Orthorhombic	Important ore of copper
Bornite	Cu_5FeS_4	Cubic	Important ore of copper
Galena	PbS	Cubic	Chief ore of metallic lead
Sphalerite	(Zn,Fe)S	Cubic	Chief ore of zinc
Chalcopyrite	$CuFeS_2$	Tetragonal	Important ore of copper
Pyrrhotite	$Fe_{1-x}S$	Hexagonal and others depending on impurities	Often has nickel as a replacement for iron making it an important nickel ore
Cinnabar	HgS	Trigonal	Chief source of metallic mercury
Stibnite	Sb_2S_3	Orthorhombic	Chief ore of antimony
Pyrite	FeS_2	Cubic	Source of sulfur dioxide used in manufacture of sulfuric acid
Molybdenite	MoS_2	Hexagonal	Chief ore of molybdenum

Table 6-8
Micellaneous Minerals of Importance

Mineral	Composition	Crystal System	Economic Uses
Halite	NaCl	Cubic	Important source of salt
Gypsum	$CaSO_4 \cdot 2H_2O$	Monoclinic	Plaster of paris, fertilizer
Anhydrite	$CaSO_4$	Orthorhombic	
Barite	$BaSO_4$	Orthorhombic	Additive in oil well drilling fluid, ready mix paint filler, component in glass, insecticides, and wallpaper.
Epsomite	$MgSO_4 \cdot 7H_2O$	Orthorhombic	Source of epsom salts
Ice	H_2O	Hexagonal	Numerous well-known uses
Apatite	$Ca_5F(PO_4)_3$	Hexagonal	Source of phosphates
Fluorite	CaF_2	Cubic	Fluxing material in steel manufacture, and metal extraction processes

Gems

Certain rare minerals possess beauty of such quality that a specimen may command a high price. A gem has the characteristics of beauty, durability, rarity, and to a certain extent, fashion. Because the durability factor is important to the preservation of a gem, the most precious gems are minerals of great hardness. Some gems are noted especially for their brilliance and "fire." To a certain extent this depends upon the optical dispersion properties, that is, the variation of index of refraction for light rays of different color. Diamond possesses high values of both index of refraction and dispersion. When properly cut, it is more effective than other transparent substances in separating white light into a rainbow spectrum of colored rays. This accounts for its brilliance and fire. Some gems have a particular richness of color. Emerald occurs in exquisite hues of green. It is a variety of the ring silicate **beryl** in which SiO_4 tetrahedra are bonded together in rings of six members. Soft hues of deep and rich red are found in ruby, a variety of **corundum** (Al_2O_3). Other gems display unique luster. When properly cut, sapphire possesses a luster known as asterism where within the crystal a star-shaped image appears in reflected light. Some of the precious gems are listed in Table 6-9.

Gems are prepared by shaping the natural stone to enhance certain qualities. The gem can be fashioned into one of two basic forms by grinding, polishing, cutting, and splitting along cleavage planes. The gem form known as the **cabochon cut** is produced by grinding and polishing curved surfaces. Gems have been prepared in this way since ancient times to enhance luster and color. More recently, the style called the **brilliant cut** has been developed. This is produced by cutting planar facets on the surface of a stone in a pattern

Table 6-9
Important Gemstones

Gem	Mineral Source
Diamond	Raw diamond
Ruby	Corundum
Sapphire	Corundum
Topaz	Corundum
Amethyst	Corundum
Emerald	Beryl
Precious opal	Opal
Agate	Quartz
Onyx	Quartz
Jade	Jadeite
Moonstone	Feldspar

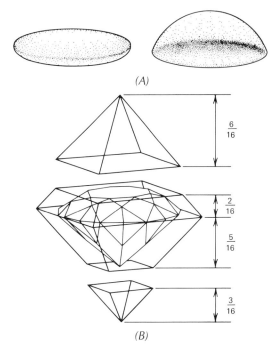

Figure 6-26 Examples of (A) cabochon forms and (B) brilliant cuts of gems. (After E. H. Kraus, W. F. Hunt, and L. S. Ramsdell, *Mineralogy*, 4th edition, page 413, McGraw-Hill Book Co., 1951.)

that emphasizes the display of color resulting from refraction. Stone can be cut with a saw blade that has imbedded diamond fragments. Some examples of cabochon and brilliant cuts are shown in Figure 6-26.

Well-developed crystals and unusual specimens of many minerals possess qualities of beauty. Most are not durable enough to be considered gems. Because of their form and color, however, they are of interest to collectors. Sometimes well-developed mineral crystals can be found as growths upon surfaces of cavities in a rock. Some cavities are completely enclosed in a rock specimen and are discovered by breaking the specimen apart (Figure 6-27). This kind of specimen is called a **geode**.

PRESSURE, TEMPERATURE, AND POLYMORPHISM

At this point you know about the physical properties, chemical composition, and crystal structure of the important mineral groups. Now you can learn about the environments where they form. We know that mineral polymorphs have distinctively different crystal structures, but the same chemical composition. Under what conditions will one or an-

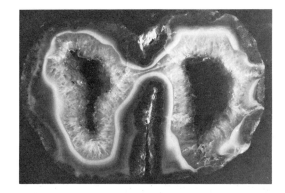

Figure 6-27 Geode containing quartz crystals.

other of these polymorphs form? You can answer this question by testing compounds in different environments of pressure and temperature. If you know what conditions are needed for a particular polymorph to form, then you can learn where these conditions existed during former periods of earth history by recognizing minerals in rock specimens.

A cylindrical furnace, a pressure pump, equipment for recording temperature and pressure, and a metal cylinder can be used to contain the specimens (Figure 6-28) to test the effects of pressure and temperature on a mineral. The cylinder containing the specimen is inserted in the furnace, where the temperature can be controlled. The furnace illustrated in Figure 6-28 can produce temperatures higher than 900° C. Then air pressure is used to pump fluid into the cylinder to control pressure on the specimen.

The apparatus shown in Figure 6-28 can produce pressures of up to 5000 bars. These are the pressure and temperature ranges for the earth's crust and upper mantle. An experiment is begun by preparing a powder from a particular mineral or compound, or a mixture of different substances. The powder is sealed in a gold vial which is inserted in the cylinder. The cylinder is then placed in the furnace and attached to the pressure pump. Then the temperature and pressure in the cylinder are raised to specified levels, and the vial containing the specimen powder is allowed to remain in that particular environment for a period of time. Afterward the vial is removed and the sample powder is analyzed by x ray and electron microprobe techniques to determine if there are any changes in grains of the powder that resulted during the experiment. In certain substances reactions may occur within a few hours. But sometimes the experiment must continue for months or even years before one can get good results.

Mineralogists have done many experiments with kyanite, andalusite, and sillimanite, the three polymorphs of Al_2SiO_5. The results illustrate the effects of pressure and temperature on this compound. Kyanite has a triclinic crystalline structure. Andalusite and sillimanite are both classified in the orthorhombic

(A)

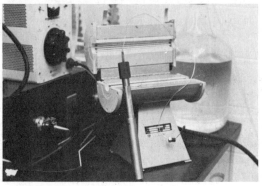
(B)

(C)

Figure 6-28 Laboratory apparatus for study of minerals at various conditions of temperature and pressure. Pressure can be increased to approximately 5 kilobars, and temperature to approximately 900° C with apparatus in (A and B). Equipment in (C) can be used to reach temperatures of 2000° C and pressures of 50 kilobars.

system, but otherwise possess distinctively different crystal structures. Powder specimens consisting of precisely measured proportions of these minerals have been tested in different conditions of temperature and pressure. When a specimen is examined after residing in a particular pressure-temperature environment, the mineral proportions are found to have changed. For example, a powder originally consisting of andalusite and sillimanite in equal amounts, after residing a few weeks at 4000 bars and 750° C, will have changed so that a larger proportion of sillimanite is detected. The result tells us that this environment favors the growth of sillimanite at the expense of some andalusite. By performing similar experiments at many different conditions the range of pressure and temperature favoring growth of one or another of these polymorphs has been established. The results are displayed on a graph (Figure 6-29) that we call a **polymorphic phase diagram.**

A particular polymorph of Al_2SiO_5 can be formed only at pressure-temperature conditions within the range indicated on the phase diagram. Once formed, the polymorph may continue to exist at pressure and temperature conditions that are outside of this range. The phase diagram indicates that high temperature is required for silliamanite to form. We know, however, that this mineral can exist almost indefinitely at low pressure and temperature. We find it in outcrops, which indicates that the rock containing sillimaanite was formerly subjected to much higher temperature and perhaps pressure. The region outlined on the phase diagram for a particular mineral indicates conditions where that mineral is a **stable** polymorph. At conditions outside this region the mineral is a **metastable** polymorph.

You know that atoms in crystalline substances are arranged in minimum energy configurations. This means that the atoms are grouped in the particular arrangement requiring the least amount of energy to maintain the bonds between them. Both the potential energy associated with electrical charges and distances of electrons from nucleii, and the kinetic energy related to thermal motion and orbiting electrons must be considered. Minimum energy configurations of the atoms making up a chemical compound are not the same at different conditions of pressure and temperature. Polymorphs of that compound represent the preferred arrangements of atoms for conditions within the separate stable regions that are seen on a phase diagram. See Figure 6-29. At 800° C and 3000 bars a minimum energy configuration exists when the atoms of Al_2SiO_5 are arranged in the crystal structure sillimanite. But at 500 ° C and 1000 bars the atoms would have to be grouped in the andalusite crystal structure for a minimum energy configuration to exist.

A metastable polymorph does not possess the minimum energy configuration for the environment where it exists. By adding enough energy to the compound to break apart the bonds and rearrange the atoms the metastable polymorph can be converted into the stable polymorph. This happened to some of the mineral grains used to obtain the experimen-

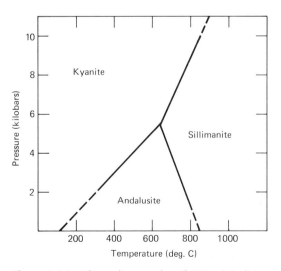

Figure 6-29 Phase diagram for Al_2SiO_5. (Modified from S. W. Richardson, M. C. Gilbert, and P. M. Bell, *American Journal of Science,* page 266, Figure 2, March 1969.)

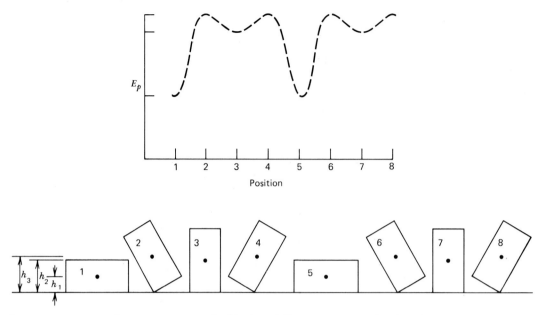

Figure 6-30 Potential energy variation of a block in different positions related to changes in the height of the center of mass of the block.

tal results displayed in Figure 6-29. However, the amount of energy required for this conversion may be much more than the energy required to simply maintain the metastable form. To clarify this concept let us represent different atomic configurations by the different positions of the block illustrated in Figure 6-30. The potential energy of the block is $E_p = gh$, where g is the acceleration of gravity and h is the height of the center of mass. When the block is in positions 1 and 5, its potential energy is minimal. These are positions of **stable equilibrium.** Potential energy is maximum at positions 2, 4, 6, and 8. These are positions of **unstable equilibrium** where the block is so delicately balanced that the slightest disturbance will cause it to shift to another position. **Metastable equilibrium** is realized in positions 3 and 7 where the block is at an intermediate potential energy level. To change the block from a metastable position into a stable position it must be moved over an **energy barrier.** This means that its potential energy must be raised to a level higher than it possesses in the metastable position. If this additional energy is not available, then the block will remain in the metastable position indefinitely even though it is not the minimum energy position.

Energy barriers exist between the metastable and stable crystalline structures represented by mineral polymorphs. A metastable structure can be converted into a stable structure only by raising the internal energy of the substance above the level of the barrier. Otherwise it will continue indefinitely to maintain the metastable structure. The energy ordinarily available near the earth's surface is insufficient to overcome the energy barrier between sillimanite or andalusite and kyanite. Therefore, sillimanite and andalusite exist as metastable polymorphs in bedrock outcrops.

Mineralogists have studied the polymorphs of many chemical compounds. The phase diagram in Figure 6-31 indicates the pressure

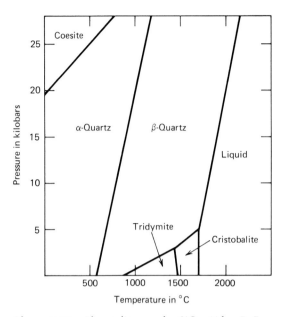

Figure 6-31 Phase diagram for SiO_2. (After F. R. Boyd and J. L. England, *Journal of Geophysical Research,* page 752, Figure 1, Feb. 1960.)

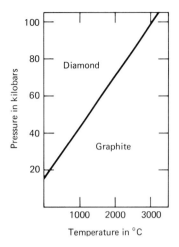

Figure 6-32 Phase diagram for crystalline forms of carbon. (After F. P. Bundy and others, *Journal of Chemical Physics,* page 384, Figure 1, Vol. 35, No. 2, 1961.)

and temperature conditions where different polymorphs of silica will form. You see that coesite can form only in a high pressure range. A specimen of coesite can be examined at room temperature and pressure in a laboratory, however, without reverting to the polymorph that we call alpha (α) quartz. Similarly, tridymite can exist metastably in a low P, T environment. We know from experiments that coesite will form only at high pressure, but once formed it can continue to exist as a metastable polymorph in P, T environments below the range indicated by the phase diagram. Similarly, tridymite and cristobalite will form only at relatively high temperatures. But after they have formed they can exist at lower temperatures. Another polymorph that we call beta (β) quartz, on the other hand, reverts easily to alpha quartz when temperature decreases. The energy barrier between these two silica polymorphs is so low that metastable beta quartz is never found in rocks near the earth's surface. Fortunately for geologic purposes many minerals can exist metastably at low P and T. The minerals whose stable regions are in high temperature or pressure ranges must have formed in environments that were quite different from the conditions at the earth's surface, where they are found in rock outcrops. Their existence in a rock, then, gives an indication of the condition in which the rock formed. Tridymite and cristobalite are sometimes found in volcanic rocks. Their presence tells us that the temperature of the material from which the rock formed must have been above 1000° C. Coesite has been found in meteorite craters, which indicates that the meteorite impact generated such intense pressure that some silica recrystallized to produce this high-pressure polymorph.

The polymorphs of carbon are diamond and graphite. They have been tested intensively because of the value of diamond as a gem and for industrial uses. The stable ranges of these minerals are shown on the phase diagram in Figure 6-32. The laboratory synthesis of diamonds achieved in recent decades requires special apparatus capable of generating much higher P, T conditions than is possible with the equipment shown in Figure 6-28. High pressure can be generated by heating carbon in a container to high temperature, then quenching the container to cause rapid thermal contraction and high pressure on the carbon within. Diamond can be synthesized by subjecting carbon to a temperature of approximately 3000° C and momentary pressure of 10^5 bars. Synthetic diamonds are useful for making abrasives and drilling bits, but they have little value as gems. Natural diamonds are found in association with volcanic rock that originated in the mantle of the earth. From our knowledge of pressure in the earth (see Figure 3-18), we know that the conditions necessary to form diamond do not exist in the crust, but are found at depths extending into the upper mantle.

You know that a solid, a liquid, and a gas are different physical states of a chemical substance that can be produced by changing pressure and temperature. Similarly, polymorphs represent different solid states of this substance. Ordinarily the energy barriers between the solid, liquid, and gaseous states are very low. Therefore the substance cannot be maintained in a metastable condition in one or another of these states because it will convert rapidly to the stable state. We know that much higher energy barriers are common between the different solid states represented by mineral polymorphs. Because of this, many minerals can exist in a metastable condition to preserve a record of former pressure and temperature that was much different than their present environment.

STUDY EXERCISES

1. If diamond red light has an index of refraction of 2.407 and green light has an index of refraction of 2.427. Which color of light travels at the highest velocity in diamond?
2. Explain why atoms tend to become geometrically arranged in mineral crystals.
3. Explain why the cleavage properties of hornblende and orthoclase are different. Describe the crystal structure exhibited by these minerals, and how these structures influence cleavage.
4. Name the polymorphs of Al_2SiO_5 that are stable at a pressure of 1 bar and a temperature of: (a) 100° C, (b) 500° C, (c) 900° C.
5. Describe the pressure and temperature conditions required to produce diamond. When you purchase a beautiful and expensive diamond, what assurance do you have that it will not undergo a polymorphic phase change into graphite?

SELECTED READINGS

Berry, L. G., and B. Mason, *Elements of Mineralogy*, W. H. Freeman and Co., San Francisco, 1968.

Bloss, F. Donald, *Crystallography and Crystal Chemistry*, Holt, Rinehart and Winston, Inc., New York, 1971.

Ernst, W. G., *Earth Materials*, Prentice-Hall, Inc., Englewood Cliffs, N.J., 1969.

Zim, Herbert S., and Paul R. Shaffer, *Rocks and Minerals*, Golden Press, New York, 1957.

7

INTERNAL PROCESSES AND IGNEOUS ROCKS

Describing and Classifying Igneous Rocks
Central Eruptions and Associated Volcanoes
Fissure Eruptions and Extrusive Sheets
Physical and Compositional Features of Volcanic Materials
Volcanic Activity and the Distribution of Extrusive Rocks
Plutons
Melting and Crystallization
Composition and Temperature in the Earth
The Origin of Magma
Ores of Igneous Origin
Geothermal Power

Early in the spring of 1980 people living in southern Washington became aware of strange happenings under nearby Mount St. Helens. They began to feel the earth tremors that occurred more and more frequently. Then on March 27 they watched steam and glowing ash burst from fissures that opened on the mountain. Geologists knew from the kinds of rock exposed here that Mount St. Helens, like most of the peaks in the Cascade Range, was once an active volcano. Was the old volcano awakening, or was this brief display of fireworks its last gasp? The answer came without warning on Sunday morning, May 18. Mount St. Helens exploded in the most dramatic eruption ever witnessed in the continental United States. During the next few minutes 77 people perished as the mountainside was devastated by fiery ash, lethal gases, and avalanches of mud shaken loose from the trembling ground. In the following weeks emanations of ash drifted over the surrounding countryside leaving a blanket of debris that threatened the health and economy of residents living more than 100 miles from Mount St. Helens.

During Tertiary and early Quaternary time volcanism was widespread in this region. Volcanic eruptions are one of nature's most awesome phenomena. Molten lava, gases, and solid fragments spew with explosive violence from vents and fissures in the earth. By studying these eruptions and the rock that they produce we can learn about major geologic processes that occur deep in the crust and upper mantle. Perhaps we can learn from these studies how to interpret the eruption from Mount St. Helens.

The volcanic outpourings spread over the surrounding countryside in a searing blanket. When it congeals, newly formed rock is added to the crust. Volcanic lavas give direct evidence that molten rock-forming material is produced in the earth. We call this molten material **magma.** We know that **igneous** rocks are formed when magma cools and crystallizes. During volcanic eruptions magma is extruded onto the earth's surface. When it congeals, the rock that is produced is called **extrusive** igneous rock. Because the magma crystallized rapidly after an eruption, most of the mineral grains in extrusive igneous rock are of microscopic size. Geologists have found an abundance of ancient as well as recently produced extrusive igneous rock. So we know that volcanism has occurred during most of geologic time.

The earth produces some magma that crystallizes before it reaches the surface. How do we know this? Geologists have never watched a mass of magma congeal underground, but they have looked at rock that was probably produced when this happened. Most of the minerals in this kind of rock are interlocking grains that can be seen without magnification. Large mineral grains such as these would crystallize from a magma that cooled slowly. One would expect it to cool this way in an insulated setting deep underground. The rock that is produced is called

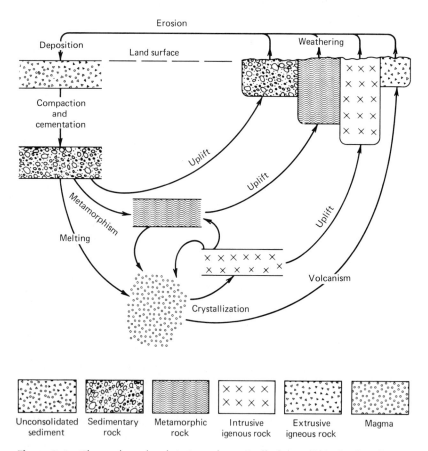

Figure 7-1 The rock cycle, showing schematically how all kinds of rocks provide the source materials for other rocks. Temperature and pressure required to fuse igneous rocks and synthesize metamorphic rocks exist in the subsurface realm of the crust and upper mantle. Sedimentary rocks are formed primarily in a near surface environment.

intrusive igneous rock, or **plutonic** rock after Pluto, the mythical god of the subterranean realm. We now find outcrops of ancient intrusive rock. This rock has been raised up by tectonic events and exposed at the earth's surface following erosion of the rock that formerly covered it. You can learn about pressure and temperature and the chemical environment in Pluto's realm from the minerals that are assembled in intrusive igneous rocks.

Through an intricate sequence of processes the earth's rocks are continually reworked (Figure 7-1). Newly formed rock is produced from more ancient rock. James Hutton called this sequence the "great geological cycle." We know that sedimentary rocks originate from deposits of particles that were produced by chemical decomposition and mechanical disintegration of older igneous, metamorphic, and sedimentary accumulations. Similarly, we have learned that metamorphic rocks can form when any kind of source rock is subjected to elevated pressure or temperature in the crust. When lithosphere is consumed into the asthenosphere in the plate tectonic cycle, ancient rocks are partially melted to produce the magma that will crystallize into still newer igneous rock. In this chapter you will learn about the part of the "great geological cycle" which involves igneous rocks and the processes that produce them. Metamorphic and sedimentary rocks and related chemical and mechanical processes are discussed in other chapters.

We will begin our study by learning about features of igneous rocks, and how these rocks are classified. This will help us to understand the important intrinsic features of these rocks and the variety of compositions that exist. Then we will discuss volcanic processes and the extrusive igneous rocks that they produce. Turning next to the intrusive igneous rocks, we will discuss where they are found and what they tell us about geologic processes deep in the crust and upper mantle. Then we will learn about the laboratory experiments that tell us how igneous rocks respond to changes in temperature and pressure. Finally, we will discuss some of the ideas that geologists have about the composition of the earth, the evolution of magma, and how some important ores are produced.

DESCRIBING AND CLASSIFYING IGNEOUS ROCKS

When you tell someone that you have a specimen of intrusive or extrusive igneous rock, this implies that you already know something about the origin of the specimen. Suppose that you are not sure about its origin. How can you describe the specimen in a way that conveys only information about which you are certain? Geologists distinguish different kinds of igneous rock by describing two intrinsic features. These features are **texture** and **composition.** The texture refers to the size, shape, and arrangement of the constituent mineral grains. Composition can be described by relative abundance of different minerals or chemical compounds related to these minerals. Texture and composition are features that are used for comparing rock specimens without referring to their origins. The illustrations in Figure 7-2 show textures and compositions of some common igneous rocks.

Most igneous rock consists of interlocking mineral grains. If these interlocking grains can be seen without magnification, then the rock has a **coarse-grained texture.** The coarse-grained texture of a granite specimen is illustrated in Figure 7-2A. Individual mineral crystals in a rock that has **fine-grained texture** cannot be recognized without magnification. Look at the basalt specimen illustrated in Figure 7-2B. Some magnification is needed to see the tiny mineral crystals that make up its fine-grained texture. No mineral grains even under high magnification can be detected in a specimen that has a **glassy texture.** Obsidian (Figure 7-2G) has this texture.

Rocks have a **porphyritic** texture if they

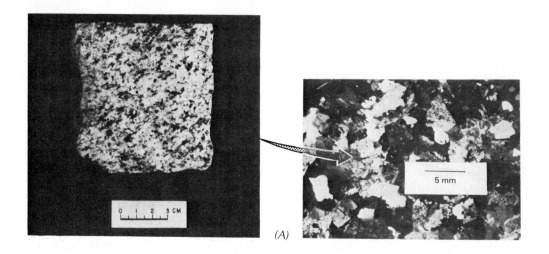

(A)

(B)

(C)

(D)

(E)

(F)

(G)

Figure 7-2 Textures of common igneous rocks. (A) Granite with equigranular coarse-grained texture. Note that the interlocking mineral grains are all more or less of the same size. (B) Basalt with a fine-grained texture consisting of a few relatively large mineral grains called *phenocrysts* interspersed through an equigranular mass of much smaller interlocking mineral grains that is called the *groundmass*. (C) Porphyritic granite with a texture consisting of a coarse-grained equigranular groundmass with some noticeably larger phenocrysts. (D) Porphyritic basalt with a texture of large phenocrysts that can be seen without magnification in a fine-grained groundmass. (E) Vesicular basalt with a texture of more or less spherical openings in a fine-grained groundmass. (F) Amygdaloidal basalt consisting of a fine-grained groundmass with vesicles filled by other minerals. (G) Obsidian which is volcanic glass so fine grained that individual minerals cannot be discerned with a standard petrographic microscope.

contain conspicuously larger crystals which are distributed throughout the mass of smaller grains. Illustrations of porphyritic granite and porphyritic basalt are shown in Figures 7-2 C and 7-2 D. Specimens with **vesicular** texture contain more or less spherical bubble cavities that are not filled. The texture is called **amygdaloidal** if mineral growths fill these cavities. These textural varieties are also pictured in Figures 7-2 E and 7-2 F.

After you have described the texture of an igneous rock you must then find its composition. You can do this by identifying the constituent minerals and measuring their proportions. Some of the most common rock-forming minerals can be identified easily in coarse-textured rock just by looking at a hand specimen. But to be sure that all of the important minerals are recognized, a geologist examines thin sections cut from the specimen. A thin section is prepared by mounting a chip of rock on a glass slide and grinding it paper-thin so that it is translucent. Then the slide can be examined with a petrographic microscope, which is a microscope equipped with attachments for viewing a thin section under polarized light of different colors. A geologist familiar with features of minerals that are revealed by magnification can identify the minerals present in a thin section.

A statistical method known as a **point count** can be used to obtain an accurate measure of mineral abundance. A plane surface is cut on a specimen and the minerals found at uniform intervals over the surface are identified. The intervals should be smaller than the mineral grains and the surface should be large enough to be representative of the rock. By tabulating the frequency of occurrence of different minerals over the surface, one can calculate the volume proportion in the rock to a precision of better than 1 percent. This kind of determination of mineral abundance is called a **modal analysis** of the rock. Point counts can be done on polished surfaces of coarse-grained specimens or on thin sections from rocks of fine-grained texture.

We can also describe the composition of a rock from quantitative chemical analysis. Sometimes it is not practical to attempt a point count, especially for very fine-grained or glassy textured rocks. But we can detect the presence of constituent chemical elements even though the mineral grains are too small to identify. We can use either conventional methods of chemical analysis or x-ray diffraction analysis of a powder prepared from the specimen to discover what elements exist. A convention introduced in the latter part of the nineteenth century and still widely followed involves expressing the rock composition in terms of **oxide components.** This procedure can be understood by considering how the chemical composition of constituent minerals is expressed in terms of oxide compounds. Orthoclase, for example, can presumably be separated into three oxides according to the chemical equation:

$$2KAlSi_3O_8 \rightleftarrows K_2O + Al_2O_3 + 6SiO_2$$
orthoclase   potash   aluminum   silica
                       oxide

Similarly, the composition of the plagioclase feldspar **anorthite** can be equated with three oxides:

$$CaAl_2Si_2O_8 \rightleftarrows CaO + Al_2O_3 + 2SiO_2$$
anorthite   lime

We found in Chapter 6 that the amphibole **tremolite** can be decomposed into the pyroxenes **enstatite** and **diopside, silica,** and H_2O. The compositions of these pyroxenes can in turn be equated to oxide components:

$$MgSiO_3 \rightleftarrows MgO + SiO_2$$
enstatite   magnesia

$$CaMg(SiO_3)_2 \rightleftarrows CaO + MgO + 2SiO_2$$
diopside

Similar expressions can be given for other silicates. Chemical tests for these and other oxides can be performed in standard laboratories. Powdered rock is decomposed by heating and dissolution in acid. Then reagents are added to cause precipitation of the diagnostic oxides or compounds related to them. These precipitates are weighed, and the proportions of the representative oxides are calculated to an accuracy of better than one-tenth of one percent. The compositions of a large number of igneous rock specimens have been obtained in this manner.

In recent decades x-ray methods of chemical analysis have played an increasingly important role in the study of rock composition and have replaced some older conventional chemical methods. However, it is still common to refer to composition in terms of oxide components even though x-ray methods test for specific elements rather than compounds.

The minerals in a rock are the actual chemical compounds existing in that rock. These minerals are distinct units that are physically different from one another. In chemistry distinctive substances such as these are called **phases.** Accordingly, the different kinds of minerals in a rock are the phases present in that rock. The chemist also can describe a material in terms of the **components.** Oxides described above are examples of the components of a rock. But it is important to understand that when you express the composition in terms of oxide components this **does not imply their existence as separate phases in the rock.** For example, a rock with a composition of 70 percent SiO_2 and 15 percent Al_2O_3 may contain 25 percent quartz grains and no corundum, the mineral forms of these compounds. The fact that Si, Al, and oxygen also exist in other silicates accounts for the high percentage of these oxides. They really do not exist as distinct compounds in the rock, but presumably one could obtain them from its decomposition in laboratory experiments. Because these oxides are not specific grains in the rock, it would perhaps be better to describe the composition simply by stating the proportions of individual elements. We persist in using oxide components more because so many rocks have already been analyzed by common chemical tests than because of any geological significance. For most geological purposes when we compare a rock of 60 percent SiO_2 with another of 50 percent SiO_2, we are interested in the fact that one has a higher proportion of the element Si. Similarly, we interpret the stated abundance of, say, MgO as an indication of the amount of the element Mg in the rock.

The geologist would like to be able to reconstruct the mineral composition of a rock from a description of its oxide components. From your knowledge of polymorphism you can appreciate the difficulty in doing this. But you can calculate what minerals could form from oxide components that are indicated from chemical analysis of the rock. When you make these calculations, you are doing a **normative** analysis of the rock to find its probable mineralogy. Compared with a modal analysis this tells you what minerals would normally occur rather than what minerals actually exist in the rock. Although the results are subject to uncertainty, they can be useful for comparing compositions of fine-grained or glassy textured rocks with compositions of coarse-grained rocks that have been determined by point counts.

Information about igneous rock composition can be displayed graphically on a **triangular composition diagram.** The relative proportions of three different groups of minerals present in a rock can be plotted on such a diagram. The preparation and interpretation of a triangular composition diagram are explained in Figure 7-3. This kind of diagram is very useful in presenting information about composition that is used in the classification of igneous rocks.

The common igneous rocks are classified according to the proportions of different minerals contained in them. Most geologists use a classification system that was accepted by the International Union of Geological Societies (IUGS) in 1972. In this system three groups of minerals are used to distinguish the most abundant igneous rocks. These mineral groups are (1) quartz, (2) plagioclase, and (3) alkali feldspars which are the potassium bearing feldspars such as orthoclase and microcline. Depending upon the relative proportions of these minerals, different kinds of coarse-grained igneous rocks can be distinguished using the triangular composition diagram in Figure 7-4. How is this classification diagram interpreted? Consider the combined volume of quartz, plagioclase, and alkali feldspar in a rock. We could determine this volume from a point count. Let us call it the QPA volume. If between 20 percent and 60 percent of this volume is quartz, then the rock is called either **granite** or **granodiorite.** We can tell from Figure 7-4 that the larger proportion of the remaining QPA volume in a granite is alkali feldspar. Compare this with granodiorite that contains a larger proportion of plagioclase. Now consider the six groups of coarse-grained igneous rocks in which quartz amounts to less than 20 percent of the QPA volume. The balance of the QPA volume in the **syenite** group is probably alkali feldspar. Rocks in the **monzonite** group have more or less equal proportions of alkali feldspar and plagioclase.

Next, look at the small region close to the plagioclase corner in Figure 7-4. It is labeled **diorite** and **gabbro.** In both of these rocks the

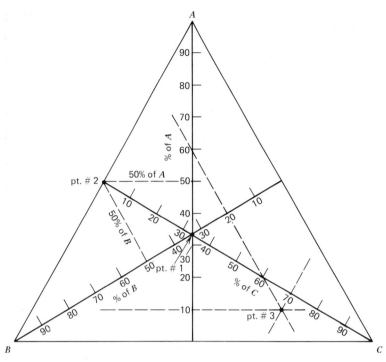

Figure 7-3 Preparation of a triangular composition diagram. The constituents indicated at the corners must add to 100 percent. Point No. 1 represents a material containing $33\frac{1}{3}$ percent each of constituents A, B, and C. Point No. 2 represents a composition made up of 50 percent each of A and B, and No C. Point No. 3 represents the composition of 10 percent A, 20 percent B, and 70 percent C. The points are plotted at the intersection of three lines. First, a line perpendicular to BC is drawn from A as illustrated. Another dotted line perpendicular to this is drawn to intersect at the 10 percent point. Any composition containing 10 percent of A will plot somewhere on this dotted line. By preparing similar dotted lines for 20 percent B and 70 percent C we find from their intersection the composition indicated at point No. 3.

QPA volume contains more than 85 percent plagioclase and less than 5 percent quartz. Another mineral group in addition to those making up the QPA volume is needed to distinguish diorite from gabbro. It includes biotite, amphiboles, pyroxenes, and olivines, all of which are rich in iron and magnesium. This group is called the **ferromagnesian** group, or the **mafic** group. The word **mafic** is an acronym based on the first letters of the chemical symbols Fe and Mg. In diorite the QPA volume is larger than the volume of mafic minerals, in contrast to gabbro which contains a larger proportion of mafic minerals.

Next, we can distinguish the **monzodiorite-monzogabbro** group that includes rocks with compositions in the range between the monzonite group and the diorite-gabbro group. Monzodiorite is distinguished from monzogabbro by having a larger QPA volume compared with its volume of mafic minerals.

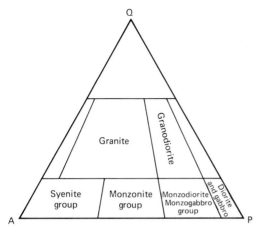

Figure 7-4 Classification diagram for coarse-grained igneous rocks based upon the International Union of Geological Societies (IUGS) classification system adopted in 1972. The triangular composition diagram displays the relative proportions of quartz (Q), alkali feldspar (A), and plagioclase (P) found in common coarse-grained igneous rocks. (Plutonic Rocks, IUGS Subcommission on the Systematics of Igneous Rocks, A. L. Streckeisen, chairman, *Geotimes*, pp. 26–30, October 1973.)

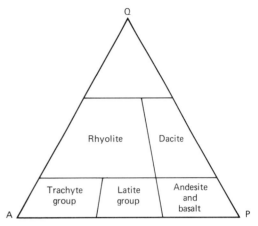

Figure 7-5 Classification diagram for fine-grained igneous rocks based upon the International Union of Geological Societies (IUGS) classification system adopted in 1972. The triangular composition diagram displays the relative proportions of quartz (Q), alkali feldspar (A), and plagioclase (P) found in common fine-grained igneous rocks.

Finally, we come to the group of coarse-grained igneous rocks that are called **ultramafic** rocks. More than 90 percentage of the total volume of an ultramafic rock consists of mafic minerals. This group of rocks is not indicated on the diagram in Figure 7-4 because the quartz, plagioclase, and alkali feldspar minerals do not occur in significant proportions. Other relatively rare igneous rocks occur with compositions that plot in the unlabeled regions of the diagram in Figure 7-4. We will not be concerned with these more unusual varieties.

Fine-grained igneous rocks can be classified in a similar manner according to the IUGS system. The triangular composition diagram in Figure 7-5 indicates the abundant kinds of fine-grained rocks with mineral compositions that correspond to those of coarse-grained rocks. By comparing Figure 7-5 with Figure 7-4 we can see that **rhyolite,** for example, is the quartz-rich, fine-grained equivalent of granite. **Andesite** and **basalt** are, respectively, the fine-grained equivalents of diorite and gabbro. There are no abundant fine-grained equivalents of ultramafic composition.

The difficulty of doing a modal analysis of very fine-grained rocks was pointed out earlier. Therefore, it is sometimes most convenient to compare the chemical compositions of fine-grained and coarse-grained rock. The average compositions of common igneous rocks are stated according to the proportions of oxide components in Table 7-1.

Now that we have described the features of texture and composition of igneous rocks we can ask what these features tell us about the way the earth works. What can these rocks tell us about conditions in the crust and upper mantle? We will now look into the nature of these igneous processes.

Table 7-1
Average Compositions of Common Igneous Rocks (Weight Percent)

	Rock Type			
	Granite Rhyolite	Diorite Andesite	Syenite Trachyte	Gabbro Basalt
SiO_2	70.8	57.6	62.5	49.0
Al_2O_3	14.6	16.9	17.6	18.2
Fe_2O_3	1.6	3.2	2.1	3.2
TiO_2	0.4	0.9	0.6	1.0
FeO	1.8	4.5	2.7	6.0
MgO	0.9	4.2	0.9	7.6
CaO	2.0	6.8	2.3	11.2
Na_2O	3.5	3.4	5.9	2.6
K_2O	4.2	2.2	5.2	0.9

From R. A. Daly, *Igneous Rocks and the Depths of the Earth*, McGraw-Hill Book Co. (1933), Table 1.

CENTRAL ERUPTIONS AND ASSOCIATED VOLCANOES

On the afternoon of February 20, 1943, a farmer, Demetrio Toral, finished plowing a furrow across a field near the Mexican village of Parícutin. Turning to continue, he watched in awe as a fracture opened along the furrows just ploughed. The crack continued to widen, and its walls began to slump. Smoke and sparks billowed out. Later that night the violence of the eruption increased, and by morning ash and glowing fragments had piled up in a cone over 40 m high. During the next few days basaltic lavas flowed out from newly forming vents near the base of the cone. The height of the cone reached 200 meters during the first week of eruption, and the lavas spread slowly outward from the base. The villages of Parícutin and San Juan Parangaricutiro, a few kilometers distant, were abandoned during the following months. They became buried under ash falls and advancing lavas. A year later only the church tower of San Juan (Figure 7-6) could be seen above the basalt beds. Eruptive activity continued intermittently until 1953. By this time a plateau of interbedded basalt layers and ash deposits had reached a thickness of over 200 meters, and the cone rose an additional 150 m above the plateau. The area finally covered by the Parícutin eruptions is shown in Figure 7-7 A. The history of growth of this volcano is shown in cross section in Figure 7-7 B.

The birth of Parícutin was preceded by minor earthquake activity in this region which is located about 300 km west of Mexico City. Commencing a few weeks prior to the initial eruption, the tremors increased in frequency and intensity. More than 300 separate shocks were recorded on the day prior to appearance of the first fracture. The seismic data are not accurate enough for seismologists to estimate the upward progress of magma. Now they can do this elsewhere with modern seismographs installed on Hawaiian volcanoes.

The volcanic activity of Parícutin is an example of a sequence of **central eruptions.** The subterranean materials come up from a central conduit to vents at the earth's surface. Here they spread radially outward to form the familiar volcanic mountains of conical shape. The shape of the volcano that is built up after many eruptions from the conduit depends on the viscosity of the lavas and the proportions of lava and solid ash fragments. Basaltic lavas tend to be more fluid than the more silicious andesitic and rhyolitic lavas. As a result, ba-

Internal Processes and Igneous Rocks 227

(A)

(B)

Figure 7-6 (A) Birth of Mt. Parícutin in February 1944. (B) The church of San Juan Parangaricutiro in June 1944 is all that extends above the ash and lava beds following 15 months of eruptive activity from Parícutin. (W. F. Foshag, U.S. Geological Survey.)

saltic lavas flow more freely and spread to farther distances from the center of eruption. After a prolonged period of basaltic eruptions a dome- or shield-shaped volcano develops. **Shield volcanoes** have gentle slopes that are generally inclined less than 10°. Among the most striking examples are the Hawaiian volcanoes and those in Iceland. The lavas from several vents in the Hawaiian group have coalesced to form a volcanic pile rising almost 10,000 m from the original ocean basin floor (4000 m above sea level). Shield volcanoes on the island of Hawaii are pictured in Figure 7-8.

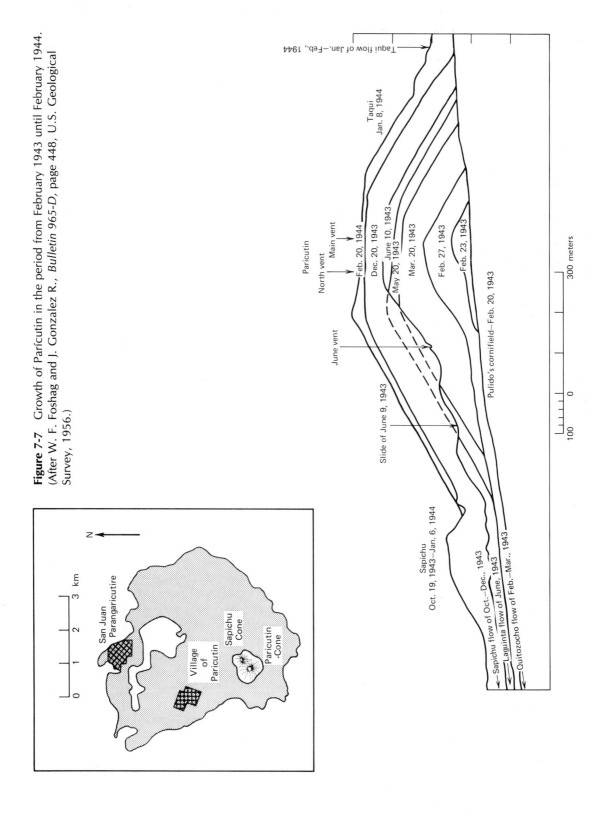

Figure 7-7 Growth of Paricutin in the period from February 1943 until February 1944. (After W. F. Foshag and J. Gonzalez R., *Bulletin 965-D*, page 448, U.S. Geological Survey, 1956.)

Figure 7-8 The shield volcano Mauna Loa and the slope of a second shield volcano Mauna Kea *(upper right)* on the island of Hawaii. (G. A. MacDonald, U.S. Geological Survey.)

Viscous lavas of high SiO_2 content tend to flow more slowly away from the erupting vent. These lavas solidify before flowing very far. The volcanoes built from these viscous lavas may eventually become high and steep-sided mountains. Ordinarily the eruptions of lava are accompanied by ejection of ash and larger solid fragments that we call **pyroclastic debris.** This debris can pile up in **cinder cones** such as the Parícutin cone (Figure 7-6). **Composite** or **strato-volcanoes** are built up from mixtures of lava flows and pyroclastic deposits. Some of these strato-volcanoes reach elevations of several thousand meters (Figure 7-9). The strato-volcano, Mount Saint Helens (Figure 7-9) began a series of violent eruptions during the spring of 1980 after a long interval of quiescence. Observe the slightly concave conical shape that is typical of many strato-volcanoes.

We can only stand outside and watch while volcanoes are built up from erupting lavas and pyroclastic debris. How can we learn about the insides of volcanoes? Most volcanoes have several vents near the top and along the sides as well. During some eruptions lava flows from several vents. By analyzing the similarities of lavas produced from the different vents we can surmise that a branching system of interconnected conduits must exist inside the volcano. Ideas about the inside of Mr. Vesuvius (Figure 7-10) have come from this kind of information. This strato-volcano in southern Italy was built up within the shell of a still older volcano called Mount Somma. During eruptions of Mt. Kilauea on Hawaii, lava flows from several interconnected vents. This tells us that lava probably moves through a system of fractures (Figure 7-11) that extend through this shield volcano.

Internal Processes and Igneous Rocks 231

Figure 7-9 *(opposite)* Mount Saint Helens erupting on May 18, 1980. Skamania County, Washington. (U.S. Geological Survey.)

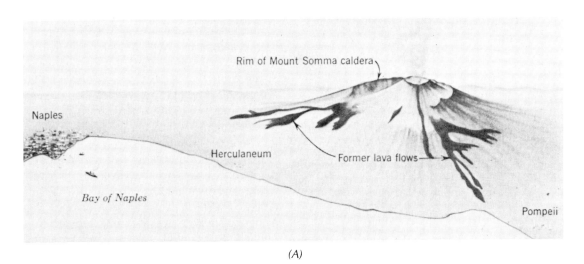

(A)

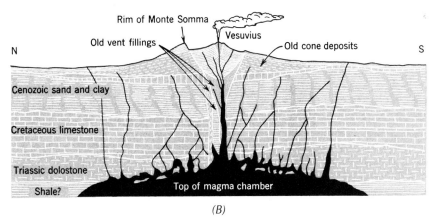

(B)

Figure 7-10 *(A)* Landscape showing the location of Mt. Vesuvius within the remnants of a more ancient volcano, Mt. Somma. (Reprinted from C. R. Longwell, R. F. Flint, and J. E. Sanders, *Physical Geology*, page 472, John Wiley & Sons, Inc., 1969.) *(B)* Cross section illustrating the kinds of fractures, lava conduits, and magma zone that might exist beneath Mt. Vesuvius. (After A. R. Rittmann, *Volcanoes and Their Activity*, page 128, John Wiley & Sons, Inc., 1962.)

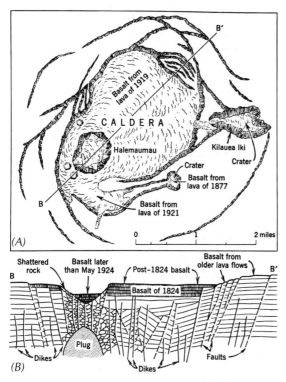

Figure 7-11 (A) Sketch showing the principal surface features of the shield volcano Halemaumau on the island of Hawaii, and (B) cross section illustrating the kinds of geologic features believed to exist beneath the volcano. (After H. T. Sterns and G. A. MacDonald, *Bulletin 7*, Hawaii Division of Hydrography, 1946.)

Geologists have discovered ancient volcanic structures that have been laid open by erosion long after eruptive activity ceased. We can see from these structures how interconnected tubular conduits and fractures extended outward through a former volcano from a central conduit that continued downward presumably to a zone where magma was produced. After the volcano ceased to be active lava solidified in this central conduit to produce a feature that we call a **volcanic pipe.** Some ancient volcanic pipes have been exposed by erosion. Where they are more resistant to erosion than the surrounding rock, these pipes now stand as striking topographic eminences such as the Devil's Tower in Wyoming, and Chimney Butte in Arizona (Figure 7-12). These remnants of long-extinct volcanoes tell us that the central conduits were more or less cylindrical and vertical. Probably the most carefully studied ones are the volcanic pipes in South Africa, which consist of a rare type of diamond bearing ultramafic rock called **kimberlite.** These diamond pipes have been mined to depths of more than one kilometer. They are nearly cylindrical and vertical. The presence of diamonds indicates that the conduits must have reached into the high pressure and high temperature environment of the upper mantle.

We know that seismicity is associated with eruptions of the Hawaiian volcanoes Kilauea and Mauna Loa. The number of small earthquakes begins to increase several weeks before an eruption. Initial earthquake foci occur at depths of 50 to 60 km, well within the upper mantle. Later foci are detected at progressively shallower depths. Presumably this indicates rise of magma in the conduit. This evidence supports the idea of a tubular conduit connecting the volcano and the zone where magma is produced. We have also learned from the patterns of seismicity that separate magma zones and conduits exist for each of these neighboring shield volcanoes. The studies of seismicity associated with Hawaiian volcanoes are probably the most complete and precise of this kind of investigation. However, we know that increased earthquake activity prior to eruption occurs in many volcanoes.

Attempts have been made to detect magma zones by seismic exploration methods. Seismologists have reported the apparent absence of some S-wave phases for waves passing beneath a volcano on the Soviet island of Kamchatka. They interpret this to indicate a zone through which S-waves will not propagate. But this kind of evidence is difficult to assess. The precise delineation of a magma zone remains to be accomplished.

How similar are the eruptions of one vol-

Internal Processes and Igneous Rocks 233

cano compared with another? Are there some typical kinds of eruptions? The events associated with particular eruptions vary considerably, but geologists can recognize some characteristic sequences. These are usually described by referring to the history of some particular group of volcanoes, although they may apply to several other groups. Those sequences associated with central eruptions are discussed below.

Hawaiian Type

Large pitlike craters contain lakes of lava. The surface is sometimes crusted over and sometimes molten. The lava is continually stirred by convection. Gases are more or less constantly bubbling from the lava. Occasionally the release of gas is violent enough to lift fiery plumes and fountains of lava. Most of the lava in these temporary fountains falls back into the crater. Earthquake activity suggests the coming of a major eruption. In the days immediately preceding the event tiltmeters indicate upward doming of the volcano. The discharge of lava wells up in the lake, filling the crater and overflowing down the flanks of the mountain. Relatively little ash and other pyroclastic debris are erupted. During major eruptions the escaping gases may hurl incandescent lava spray high above the crater. In 1924 the level of the Halemaumau lava lake dropped quickly by over 100 m. Water draining into the crater was engulfed in the lava and rapidly converted to steam which then escaped with a tremendous explosion. This fractured and enlarged the crater, dislodging considerable masses of rock. Illustrations of Halemaumau in eruption are shown in Figure 7-13.

Mediterranean Types

The eruptions of four Italian and Silician strato-volcanoes have been recorded since

(A)

(B)

Figure 7-12 (A) Chimney Butte in northeastern Arizona (photo by H. E. Gregory, U.S. Geological Survey), and (B) Devil's Tower in eastern Wyoming (photo by L. C. Huff, U.S. Geological Survey) are volcanic pipes exposed by erosion.

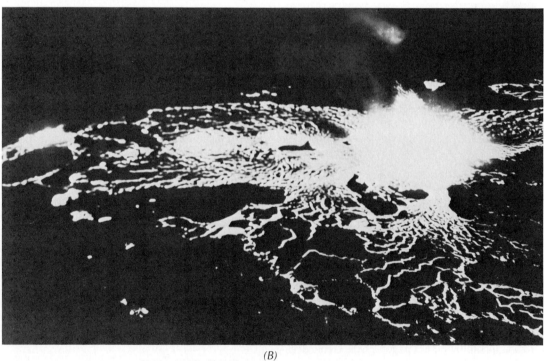

Figure 7-13 Scenes from Halemaumau on the island of Hawaii. *(A)* Large crater containing lava that has temporarily formed a solid crust. *(B)* Lava breaks the thin crust and erupts in fountains of spray during the night of July 6, 1952. (G. A. MacDonald, U.S. Geological Survey.)

Roman times. The most regular eruptive sequence, the **Strombolian** type, is associated with the island volcano Stromboli. Fairly regular and almost rhythmic explosive episodes release gases, clots of lava, and incandescent spray. Because the viscous lava tends to contain gases until a threshold of pressure is reached, the eruptions are explosive. Minor explosions may occur four or five times each hour. During more infrequent periods of increased intensity, lava may well up and overflow from the crater. Stromboli has been active since history began to be recorded. The **Vulcanian** type of eruption, exemplified by the island volcano Vulcano about 50 km southwest of Stromboli, is less regular and more violent. Thick, viscous lavas crust over soon after eruption, and this crust seals the gases that rise from the depth. Some of the lava congeals in the conduit. When gas pressure reaches a critical threshold, the volcano clears its throat with a violent explosion. Ash, clots of lava, and gas are hurled high into the air. A considerable amount of pyroclastic debris is ejected in this kind of eruption. Less intensive explosions may follow as the volcano clears out congealed material, finally waning to a quiescent period. The **Vesuvian** type is still more infrequent and violent than the Vulcanian type of eruption. Mt. Vesuvius, inactive for hundreds of years, finally releases the pent-up gases and lava in paroxysmal explosions that clear the conduit to considerable depth. Ash and pyroclastic fragments are hurled high above the mountain and quickly settle as lethal blankets over the countryside, obliterating life and settlements. So fast was Pompeii inundated in 79 A.D. that the inhabitants were buried while attempting to flee the city. After this event Mt. Vesuvius was quiescent for 1600 years before another violent eruption occurred. Perhaps the most violent kind of eruption is the **Plinian** type where the explosion tears masses of rock from the walls of the vent partially destroying the volcanic mountain itself. Illustrations of these various types of eruptions are shown in Figure 7-14.

Pelian Type

Mt. Pelée is a strato-volcano situated on the West Indian island of Martinique. In 1902 a particularly deadly kind of eruption occurred. A fiery cloud of incandescent ash, partially molten lava, and gas sprayed from the crater and hurtled down the slopes and over the surrounding countryside at speeds of 200 km/hr. We call this kind of fiery cloud a **nuée ardente** (Figure 7-15). The city of St. Pierre was destroyed (Figure 7-16), and over 28,000 inhabitants perished in a matter of moments during the 1902 nuée ardente eruption. Prior to this Mt. Pelée had been quiescent for several decades. Lava had solidified in the conduit forming a plug. Magma highly charged with gas developed tremendous pressure that forced out this plug, then spilled from the crater. Gases expanding rapidly upon release of pressure blew the magma into a seething spray which formed into nuées ardentes. This kind of eruption recurred from Mt. Pelée between 1929 and 1933. The development of a notch in the crater directed these later nuées ardentes toward uninhabited regions. These fiery clouds have erupted from other volcanoes as well.

In the course of eruptions the crater of a volcano is altered. Rock is torn loose by explosions and slumps from precipitous walls. We call a crater that has become greatly enlarged a **caldera.** The caldera of an old quiescent volcano may become the container of newly developing cones if volcanic activity resumes. You can see that this is true of Vesuvius, which grew in the caldera of Mt. Somma (Figure 7-10A). Calderas develop primarily from explosion and collapse processes. In 1883 the island of Krakatoa in the Indonesian island group was obliterated by four stupendous explosions. With unprecedented volcanic violence Krakatoa exploded, and more than 20 square kilometers of land were replaced by a pit in the sea floor almost 7 km across and 300 m deep.

Much larger calderas, some being several

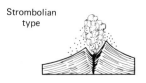
Strombolian type

Vesuvian type

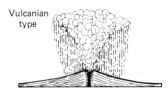
Vulcanian type

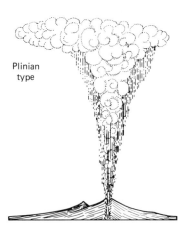
Plinian type

Figure 7-14 Volcanic eruptions of the Mediterranean types. Strombolian-type eruptions occur frequently and more or less regularly with discharge of lava and explosive release of gas. Vulcanian eruptions are less regular and more explosive. Viscous lavas tend to seal conduits until sufficient gas pressure builds up to cause violent explosions. The Vesuvian eruptions are even more infrequent, but considerably more violent when they occur. Large accumulations of gas build up over periods of many hundreds of years. When suddenly released, these gases and associated pyroclastic debris spread over the surrounding countryside to considerable distances from the volcano. Plinian eruptions are the most violent. Explosions not only release pent up gases and lava, but also break apart large masses of rock from the sides of the crater. (After Arthur Holmes, *Principles of Physical Geology,* page 305, Ronald Press Co., 1965.)

Figure 7-15 Eruption of a fiery spray of gases, molten droplets, and pyroclastic debris called a *nuee ardente* from Mt. Pele on the island of Martinique in 1902. (Alfred La Croix.)

tens of kilometers in diameter, have been discovered in older volcanic terranes. Judging from the geologic features that are exposed on the rims and within these structures, most geologists believe that these huge calderas were produced by collapse processes rather than explosions. What could cause this kind of collapse? Perhaps it occurs when magma drains from an underlying zone. This idea was suggested to explain the large caldera in Oregon that is now filled by Crater Lake (Figure 7-17). We do not have a good explaination of what caused the magma to drain from beneath this region.

Internal Processes and Igneous Rocks 237

Figure 7-16 Ruins of St. Pierre on the island of Martinique shortly after its destruction in 1902 by a nuee ardente. (Brown Brothers.)

Figure 7-17 Crater lake in Oregon now fills a nearly circular caldera which is about 8 km wide. Note the island formed from a cinder cone. (H. Miller Cowling, Washington Air National Guard.)

FISSURE ERUPTIONS AND EXTRUSIVE SHEETS

You can find extensive deposits of extrusive igneous rock on vast volcanic plateaus in western North America, the Deccan region of India, rift valleys in eastern Africa, and other continental areas (Figure 7-18). The most conspicuous volcanic plateaus developed during Cretaceous, Tertiary, and Quaternary time, although similar deposits are known from earlier periods. These plateaus are landscapes of subdued relief. Here there is little evidence of any large volcanic structures produced by central eruptions. In some places rivers have eroded deep gorges in these pla-

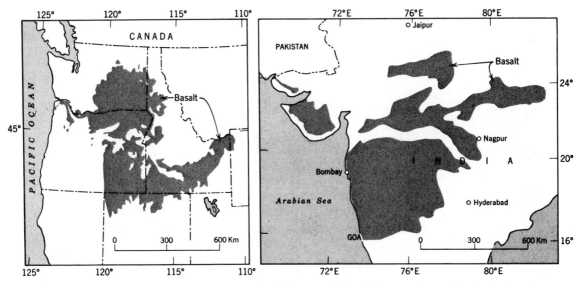

Figure 7-18 Volcanic plateaus consisting mostly of horizontal basalt flows and associated ash deposits cover broad areas of northwestern United States (after A. C. Waters, *Special Paper 62*, facing page 703, Geological Society of America, 1955) and a broad region of western India called the Deccan Plateau (after H. D. Sankalia, 1964).

teaus. On the sides of these gorges you can see massive sheets of basalt interbedded with pyroclastic deposits (Figure 7-19). Individual sheets are seldom more than 50 m thick, but some of them can be followed for several tens of kilometers. These deposits cover areas larger than 500,000 km^2 both in North America and India. The total thickness of volcanic deposits in these regions reaches a few kilometers.

You can also find vast sheets of very fine-grained or glassy porphyritic extrusive rock that is rich in SiO_2. These sheets cover areas of several thousand square kilometers. The nature of the eruptions that produced these extrusive sheets is somewhat a mystery. Geologists believe that highly fluid lavas erupted from long fissures rather than central vents, and spread rapidly over the countryside before congealing.

No one has observed fissure eruptions as large as the ones believed to have produced the vast Cretaceous and Cenozoic volcanic plateaus. The largest eruption of this kind that people have described occurred in the Laki district of Iceland in 1783. The Laki fissure extends for approximately 30 km. During an eruption that continued intermittently for about three months, basaltic lavas flooded a 565 km^2 area bordering the fissure. The flows accumulated to an average thickness of approximately 20 m, so that about 12 km^3 of lava were erupted. The associated pyroclastic deposits were an order of magnitude less in volume. The Laki eruption was quiet in comparison with central eruptions. It lacked the explosive furor of the paroxysmal discharge of pyroclastics and lavas highly charged with gas.

In all likelihood fissure eruptions play the dominant role in the origin and spreading of oceanic crust along submarine ridge systems that we discussed in Chapter 5. The Laki fissure and other Icelandic fissure zones are situated along a stretch of the Mid-Atlantic ridge that reaches above sea level. Central eruptions also occur along the ridge. Evidence of this was the birth of the volcano Surtsey,

Figure 7-19 Horizontal layers of basalt exposed along a gorge on the Columbia River Plateau in Franklin County, Washington. (F. O. Jones, U.S. Geological Survey.)

which broke the Atlantic surface on November 14, 1963. During the following 18 months the volcanic island grew to an area of more than 2 km². Even though individual volcanoes in the Icelandic group erupt intermittently, we know that volcanic activity has been continuous in the group during modern times. Other islands rising from the southern

Mid-Atlantic ridge have intermittent central eruptions. The vents probably lie along crustal fissures that open sporadically along certain portions of their lengths. So far we do not know how to compare the volume of oceanic extrusive rock coming from fissure eruptions with that produced from central vents. The Icelandic fissures and central cones show us that both kinds of eruptions occur.

We think that the extensive sheets of silicic volcanics were produced by another kind of fissure eruption. We know that lavas of high SiO_2 content are too viscous to spread over large areas before congealing. Because geologists first thought that these layers were rhyolites or andesites, it was difficult to imagine how they were erupted. Composition was determined from chemical analysis rather than modal analysis of minerals. Then the normative mineralogy was calculated, and these rocks were judged to be rhyolite or andesite. Later, a careful examination of texture (Figure 7-20) under high magnification revealed a glassy groundmass containing tiny fragments of pyroclastic debris. This texture is different than the fine-grained matrix of interlocking mineral crystals that would be expected in rhyolite or andesite. Similar glassy textures are found in the deposits that congeal from nuée ardentes. Accumulations of pyroclastic fragments contained in a glassy groundmass are called **welded tuff,** and the vast sheets of this rock are called **ignimbrites.** Although the chemical composition of ignimbrite is similar to that of rhyolite, it is a rock with a distinctively different texture. Suppose that silicic lavas became highly charged with dissolved gases. Then, when pressure was released upon eruption, one would expect the material to expand into a cloud of gas, solid bits, and partially molten droplets. In this highly fluid state the siliceous spray could spread rapidly over the countryside to form a sheet of ignimbrite. A nuée ardente of stupendous proportions, many times larger than any seen in modern times, would be required to

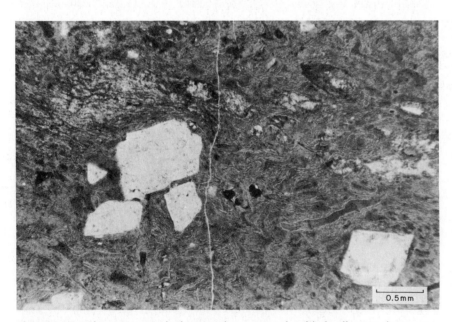

Figure 7-20 Photomicrograph showing the texture of welded tuff. Note the angular fragments imbedded in a glassy groundmass. (F. S. Simons, U.S. Geological Survey.)

explain the Cenozoic ignimbrites. All the same, no other mechanism based upon observed phenomena has been proposed. Think about the physical and psychological effects if such eruptions, which were widespread during recent history of the earth, were to occur now in a populated region. An event of this sort, although unwitnessed, probably occurred in the Alaskan Valley of Ten Thousand Smokes in 1912. Here we find welded tuff covering an area of more than 150 km^2. Nuée ardentes apparently erupted in rapid succession from several fissures and spread over the valley leaving deposits over 200 m thick in places.

PHYSICAL AND COMPOSITIONAL FEATURES OF VOLCANIC MATERIALS

Molten lavas, volcanic gases, and solid fragments are products of eruption. We know that some solid particles congeal at the time of eruption. Other fragments appear to have crystallized deep in the conduit at some time before eruption. Still other solid fragments are broken from the sides of the conduit during upwelling of the magma. The nature of these fragments that we find later in solidified beds of lava tells us about the composition of the earth's interior, and the conditions of temperature and pressure at various places between the magma zone and the volcano. We know that substances that exist as gases at atmospheric pressure originate as part of a solid lattice, or they are dissolved in fluids when they are confined under high pressure. But these constituents of the earth's interior boil off as volcanic gases during an eruption and are not incorporated into the extrusive rocks. Therefore, we have to sample the volcanic gases as well as the lavas to learn about igneous processes. Geologists have to measure lava temperature to discover the natural conditions in which certain compositions can exist as fluids. Then they can relate physical features preserved in solidified lava beds to the viscosity and rate of cooling of the lava.

It is not necessarily difficult or dangerous to collect samples from lava lakes. Crusts form over these pools of lava during times of quiescence or subdued activity. It takes more nerve and daring to get samples of lavas produced during violent eruptions from volcanoes where we do not find quasi-permanent lakes in the craters. Geologists have discovered that temperatures of basaltic lavas upon eruption are in the range of between 1100° and 1200° C. They have learned from laboratory experiments that lava at 1400° C has a viscosity of about 300 poises. This increases to 30,000 poises when the lava cools to 1150° C. Water at 68° C and atmospheric pressure has a viscosity of about 10^{-2} poise. Although much stiffer than water, these viscous lavas flow easily during eruption. They become too viscous, and cease to flow as the temperature approaches a value of about 1050° C. The more siliceous lavas are discharged at somewhat lower temperatures, generally in the range of 900° C to 1000° C. These lavas, from which andesite and rhyolite are derived, are more viscous than basalt, and they do not spread very far from the central vent.

The low viscosity lavas flow as thin glowing liquid sheets, twisting and swirling to form the ropy features seen in Figure 7-21. These are called **pahoehoe** lavas, a descriptive word taken from the Hawaiian language. More viscous lavas, that contain jagged clinkers and irregular blocks, grind over the surface in a reluctant interrupted pattern of flow. We call these **aa** (ah ah) lavas according to another Hawaiian word. Aa flows produce the irregular jagged appearance seen in Figure 7-22. These two kinds of lava are related to viscosity. It is not unusual for a flow to begin as pahoehoe lava near the vent or fissure, and become more aa farther away. This occurs because viscosity increases as the lava cools, and blocks of material congeal. Blobs of lava, cast aloft by explosive eruption and

Figure 7-21 Syrupy and twisted ropy features are the distinguishing features of pahoehoe lava. (G. A. MacDonald, U.S. Geological Survey.)

Figure 7-22 *Aa* lava has an irregular jagged and blocky appearance. (G. A. MacDonald, U.S. Geological Survey.)

Figure 7-23 Volcanic bombs formed from blobs and clots of lava that were hurled into the air by volcanic explosions, then fell, partly solidified, to the ground. (H. T. Sterns, U.S. Geological Survey.)

solidifying before reaching the surface, are called volcanic **bombs** (Figure 7-23). Lavas erupted under water form pillowlike features because a thin glassy skin quickly forms over a molten lobe. New lobes develop when the skin ruptures, and lava spills out from the openings. This produces a sequence of pillows (Figure 7-24). These newly erupted **pillow lavas** can be compared with similar features found in ancient rocks (Figure 7-25). Geologists use features like these for recognizing submarine eruptions of former times. During a 1976 expedition called Project Famous, small submarines descended into the axial trough of the Mid-Atlantic ridge. Pillow lavas were found in abundance, so we know that the axial trough is a zone where submarine eruptions are frequent.

Some fluids dissolved in the magma vaporize upon release of pressure. The bubbles of escaping gases form vesicles in the solidifying rock. **Scoria** is basalt that has an abundance of vesicles. A more siliceous rock called **pumice** forms from a lava that was blown into a froth by the vaporization of volatile constituents. Pumice is so full of vesicles that it will float upon water. The surface of a pumice block is rough from glassy edges of broken vesicles. Sailors used it to scour wooden decks of ships.

244   The Rocks of the Earth

Figure 7-24 Basalt pillows produced from lava recently erupted in the axial trough of the Galapagos Ridge in the eastern Pacific Ocean. The fracture illustrated in the center of this remarkable photograph marks the line along which sea-floor spreading is commencing. Rock in the foreground lies on the Cocos Plate, which is moving northward, and rock in the background is part of the southward moving Nazca Plate. Locations of these plates is shown in Figure 5-5. The photograph was taken by Dr. Alexander Malahoff from the small submarine Alvin. (Courtesy of Alexander Malahoff, NOAA)

We know that lava viscosity is related to the content of SiO_2. To understand this you should recall from Chapter 6 that the strongest chemical bonds in silicate minerals exist between SiO_4 tetrahedra. In fluids thermal forces which cause individual ions to move randomly are stronger than bonding forces so that crystalline structures can not form. The bonding forces still exist, however, so that momentary bonds are continually formed and then broken. This produces a more viscous fluid because energy ordinarily used to maintain thermal motion of the ions must be expended to break the continually reforming tetrahedral bonds.

When a sheet of lava congeals, systems of vertical cracks called **joints** may form from thermal contraction during cooling. These joint systems are better developed when cooling is slow and even. You can see the polygonal joint pattern on the upper surface, and rock columns separated by joints along the edge of the flow illustrated in Figure 7-26.

Geologists collect volcanic gases from lava

Figure 7-25 Lava pillows now exposed in northern Michigan were formed during a submarine eruption in Precambrian time. (N. K. Hubar, U.S. Geological Survey.)

lakes (Figure 7-27) and quiescent vents. They can also get samples of gas trapped in vesicles by crushing a rock specimen in a vacuum chamber. The composition of a typical gas sample from a Hawaiian volcano is given in Table 7-2. H_2O is the most abundant constituent in this and, generally speaking, in most volcanic gases. The ratio of H_2O/CO_2 ranges between 3 and 10 for typical gas samples. If we examine a wide selection of gases, we find that the H_2O content is between 70 and 95 percent. It is difficult to determine how much of the gas comes from the original magma and how much comes from ground water and rain water drained into lava lakes. We can learn more about this from experiments which reveal the maximum quantity of H_2O that can be dissolved in different magma compositions at different P and T conditions. Probably less than 5 percent of H_2O in volcanic gases is derived from the original magma. Some gases such as the lethal sulfur dioxide SO_2 may be produced by oxidation of sulfur within the lava lake, rather than originating in the original magma zone. In addition to the volatiles listed in Table 7-2 we know that many metals can react with chlorine to form volatile metallic chlorides. Copper, iron, mercury, zinc, and several other metals in chloride compounds form crusts on the sides of volcanic vents.

A volcano is built up from numerous eruptions that occur intermittently during a long period of time. Do all of these eruptions produce the same kind of lava? Or can you find

Table 7-2
Typical Composition of Hawaiian Volcanic Gases

Component	Percent
H_2O	79.3
CO_2	11.6
SO_2	6.5
N_2	1.3
H_2	0.6
CO	0.4
S_2	0.2
Cl_2	0.05
Ar	0.04

U.S. Geological Survey

Figure 7-26 A polygonal joint system formed in a cooling lava flow separates the columns of rock seen in this mass of basalt exposed near Hidalgo, Mexico. (C. Fries, U.S. Geological Survey.)

Internal Processes and Igneous Rocks 247

Figure 7-27 Scientists installing instruments for monitoring gas emission from a vent on the west rim of Sherman Crater in Whatcom County, Washington. (*Earthquake Information Bulletin*, U.S. Geological Survey, July–August 1975.)

evidence of systematic changes in lava composition during the growth of the volcano? You can find out by measuring the oxide components in the extrusive rocks exposed on the sides of the volcano, and by measuring the ages of these rocks. It is easier to measure oxide components rather than the modal mineralogy in these rocks. Most of them have either very fine-grained or glassy textures, thus mineral point counts are impractical.

Geologists sampled four separate lava flows that erupted from Paricútin during March 1943, November 1943, November 1944, and July 1945. The proportions of oxide components in these flows are compared in Figure 7-28. This graph shows that the relative amounts of these components vary by only a few percent in the different flows. Similar data were obtained for the 1959 eruptive period of the Hawaiian volcano Kilauea Iki. There were 17 separate eruptions during November and December. With the exception of MgO the proportions of oxide components remain constant within ±3 percent. The variability of MgO reflects the settling of olivine crystals in the lava lake rather than any real change in the lava composition. These examples reveal the usual pattern in volcanism. Lavas erupted during a short interval of time all have about the same composition.

As a volcano builds up over a period of tens—or hundreds of thousands—of years, we find a history of lava composition preserved in the sequence of flows. It is not unusual to find on a particular volcano a variety of compositions that ranges from basalt where

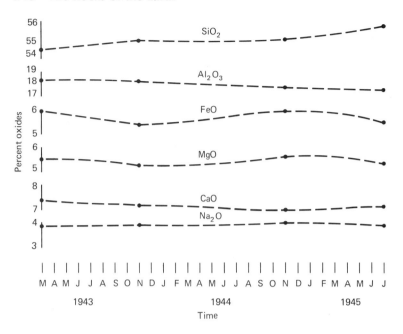

Figure 7-28 Chemical compositions of lavas erupted from Mt. Paracútin in March 1943, November 1943, November 1944, and July 1945 display only small differences in the proportions of six abundant oxide components. (Data from W. F. Foshag and J. Gonzalez R., *Bulletin 965-D*, page 361, U.S. Geological Survey, 1956.)

SiO_2 is less than 50 percent to rhyloite containing more than 70 percent SiO_2. Although the range in SiO_2 for any particular eruptive episode is within a few percent, a much wider range is encountered during a long history of activity.

Interesting relationships are observed between the proportions of SiO_2 and other oxides. We find that as the proportion of SiO_2 changes from one specimen to another, the proportions of other oxides also change in a regular way. This can be seen clearly in the analyses of specimens from a volcanic region in southern Nevada (Figure 7-29). A specimen containing 70 percent SiO_2 can also be expected to have, for example, about 16 percent Al_2O_3. Another sample with 77 percent SiO_2 will have approximately 13 percent Al_2O_3. Similar data from the Asama Volcano in Japan (Figure 7-30) cover a wider range of SiO_2 content. These examples from Nevada and Japan both show that the proportions of oxide components in lavas vary in a systematic way, and that the composition of lavas erupted from a particular volcano can change significantly during its eruptive history.

VOLCANIC ACTIVITY AND THE DISTRIBUTION OF EXTRUSIVE ROCKS

We know that more than 800 volcanoes are now active, which means that they may continue to erupt intermittently. Almost all of these active volcanoes are situated near borders of tectonic plates. The island arcs and continental margins adjacent to submarine trenches and the oceanic ridges are the settings where volcanism is most evident. The notable exception is the Hawaiian group where we observe intraplate volcanic activity.

Geologists are interested in finding associations between volcanic and tectonic processes. What things might indicate such relationships? Perhaps there might be compositional differences in the volcanic rocks erupted in different tectonic settings. We can look at four kinds of tectonic regions that include: (1) regions above the descending lithospheric plate near submarine trenches; (2) oceanic ridge regions; (3) ocean basin intraplate regions; and (4) continental intraplate regions.

Internal Processes and Igneous Rocks 249

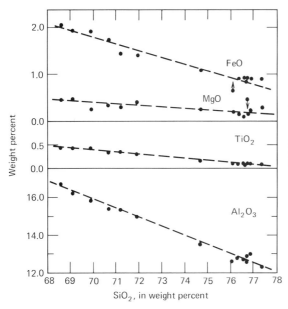

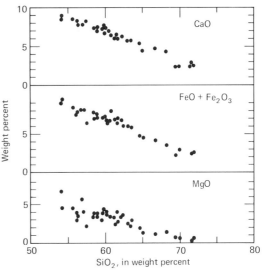

Figure 7-29 Rock specimens were collected from a layer of volcanic rock in southern Nevada that was formed from ash falls and lava flows produced during several episodes of eruption. Chemical analyses indicated significant differences in the compositions of these specimens, but a clear relationship between the proportions of various oxide components and the proportion of SiO_2 is evident. This suggests a systematic change in the composition of material from a succession of episodes of eruption. (After P. W. Lipman, R. L. Christiansen, and J. T. O'Conner, *Professional Paper 524-F*, page F31, U.S. Geological Survey, 1966.)

Figure 7-30 Relative amounts of these oxide components in rock specimens from Asama Volcano in Japan vary proportionally with the amounts of SiO_2 in the specimens. These relationships indicate systematic changes in the compositions of materials from a succession of eruptions during a long period of time. (After Shigeo Aramaki, *Tokyo University Faculty of Science Journal*, Sec. 2, Vol. 14, Part 2, page 413, 1963.)

Basalt is the most abundant kind of extrusive rock in all of these regions. Smaller proportions of andesite and rhyolite also occur. But these latter kinds of rock are much more abundant in the regions bordering submarine trenches than in the other tectonic regions. Basalts erupted near trenches have particularly low values for the ratio $K_2O/(SiO_2 + Na_2O)$. This compositional feature of basalt and the relatively high abundances of andesite and rhyolite distinguish the assemblages of extrusive rock in island arcs and continental margins bordering submarine trenches from the extrusive rock assemblages elsewhere.

The only clearly evident intraplate volcanic activity that is known at the present time is occurring in the Hawaiian group. Compositional differences between Hawaiian basalts and oceanic ridge basalts are difficult to recognize. Also, these oceanic rocks have compositions similar to the older continental intraplate basalts that are found on the volcanic plateaus of western North America and India, and in the rift valleys of Africa. The proportions of the principal silicate minerals are similar for basalts from all of these tectonic settings. There are some indications that subtle differences in the content of radioactive isotopes and other trace elements may eventually become the basis for distinguishing these rocks from one another. Otherwise the compositions of extrusive rocks from these latter three tectonic settings are remarkably similar.

The numerous seamounts on the ocean basin floor are extinct volcanoes. Some may have been produced by intraplate volcanism. But we believe that most were originally built up along oceanic ridges. Then they have moved to their present locations by sea floor spreading.

Continental intraplate volcanism is not observed at the present time. Some geologists propose that older rocks making up the continental volcanic plateaus were erupted from fissures that opened when a plate began to break apart. In some places this process continued so that new plates came into existence and a new ocean opened. This happened with the breakup of Pangea (Figure 5-1). But in other places preliminary rifting ceased before complete separation occurred. If this proposition is correct, then the continental intraplate assemblages of extrusive rock that make up the vast volcanic plateaus may have been produced by tectonic processes similar to the spreading now occurring along oceanic ridges. So most geologists believe that far more volcanic deposits have been produced by plate margin processes than by intraplate processes.

PLUTONS

Peculiar things seem to be happening in the Rio Grande rift area of southern New Mexico. Geologists have detected unusual patterns of vertical uplift of the land surface. The temperatures measured in boreholes indicate that an extraordinary amount of heat is coming out the crust, and many small earthquakes can be detected. What might cause these things to happen? According to one idea, magma is accumulating here in a zone deep in the crust. Plans were made to test this idea by making seismic reflection measurements during the summer of 1977. Special high-energy vibrators rather than explosives were used to generate the seismic waves. Unusually strong reflections echoed from a depth of about 21 km. The best explanation for a strong echo such as this is that a zone filled with the magma lies buried in the crust. Some calculations suggest that the magma temperature is higher than 1000° C. This is much hotter than temperatures one would normally expect to find 21 km down in the crust. Therefore, we can suppose that the magma is cooling. What will happen as it continues to cool? We believe that coarse-grained igneous rock will be produced.

Geologists have never watched coarse-grained igneous rock crystallize from a magma. But they know that the magma has to cool slowly to produce mineral grains large enough to be seen without magnification. Lavas extruded on to the earth's surface cool too quickly for this to happen. Therefore, we believe that coarse-textured igneous rock comes from magma trapped deep underground. Here it is insulated by the surrounding rock, and it cools slowly for periods of tens or hundreds of thousands of years. Only much later, when the cover of insulating rock is eroded away, do we see the coarse-grained igneous rock that crystallized from the hidden magma. We call this rock **intrusive** or **plutonic** igneous rock. Geologists believe that plutonic masses make up most of the continental crust. Individual bodies, called **plutons** or **intrusions,** can be either tabular or highly irregular in shape. They vary in composition from granitic to ultramafic. The ages of plutons we now see exposed on the land surface range from Precambrian to Tertiary. No doubt there are buried Quaternary plutons that have not been discovered. Perhaps one is now crystallizing beneath the Rio Grande rift area in New Mexico.

Even though we can not watch Pluto manufacture an igneous intrusion, we can figure out how he does it by looking at the ones now exposed at the surface. What kinds of

Internal Processes and Igneous Rocks 251

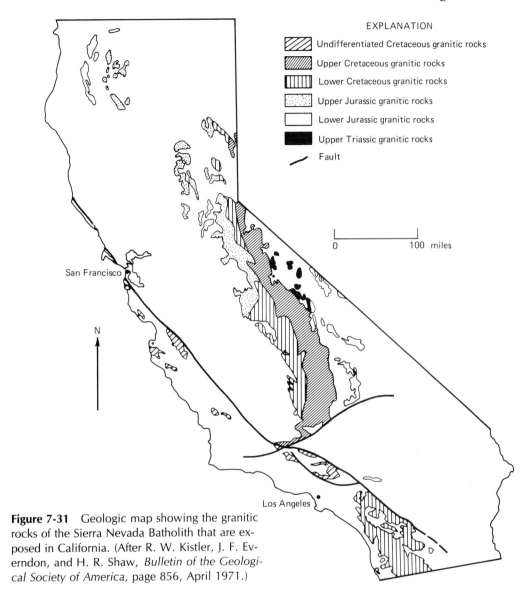

Figure 7-31 Geologic map showing the granitic rocks of the Sierra Nevada Batholith that are exposed in California. (After R. W. Kistler, J. F. Everndon, and H. R. Shaw, *Bulletin of the Geological Society of America,* page 856, April 1971.)

plutons do we see? Some are large masses called **batholiths.** One of these is named the Sierra Nevada Batholith. Exposures of intrusive rock ranging between granite and diorite in composition can be found almost everywhere in a NNW trending region of central California (Figure 7-31). We know that these rocks were emplaced during five episodes of igneous activity commencing in late Triassic time and continuing through the Cretaceous period. During each episode magmas having a range of compositions were injected into a complex of older rocks that are called the **host** rocks. These Mesozoic magmas crystallized to produce the plutons that collectively make up the Sierra Nevada Batholith. A generalized cross section (Figure 7-32) based upon seismic refraction measurements and

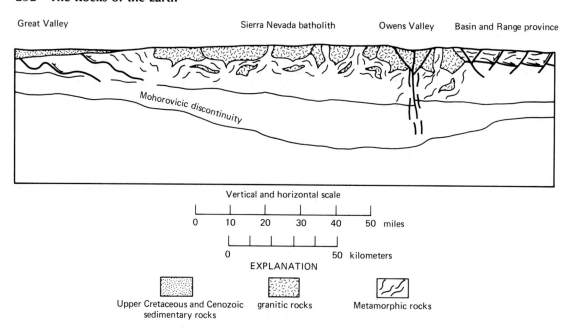

Figure 7-32 Simplified geologic cross section of the Sierra Nevada Batholith along latitude 37 degrees N. (After W. Hamilton and W. B. Myers, *Professional Paper 554-C,* page C5, U.S. Geological Survey, 1967.)

Bouguer gravity anomalies shows that the Sierra Nevada Batholith is restricted in thickness to the upper part of the crust.

Let us look at another batholith. In Figure 7-33 the surface geology in the area near the San Francisco Mountains in southwestern Utah can be seen. Local exposures of Tertiary intrusive rock, mainly granite, have significantly higher rock magnetism than the surrounding sedimentary and volcanic rocks. Analyzing aeromagnetic anomalies over these intrusive rocks has revealed that they are knobs on the irregular top of a huge pluton mostly hidden by a thin layer of sedimentary rock. The simulated magnetic field that can be calculated for the model of this pluton (Figure 7-34) compares well with the observed aeromagnetic field. This tells us that the surface exposures of intrusive rock are cupolas extending upward from a much larger batholithic mass. This batholith appears to have spread out in the upper part of the crust.

These examples are but two of the numerous plutonic masses that are called batholiths. They show some general features of large intrusions. Batholiths extend as more or less continuous masses beneath areas larger than 100 km^2. Geophysical measurements on those that are exposed at the surface indicate that these plutons are confined to the upper part of the crust. They have much greater lateral extent than thickness. They are made up from several smaller intrusions that were emplaced at different times during related episodes of igneous activity. Most, if not all of the rock in a batholith is between granite and diorite in composition.

Smaller bodies of intrusive rock are important in some localities. Plutonic masses that are thought to extend several kilometers down into the crust, but are laterally restricted to areas smaller than 100 km^2, are called **stocks.** Other plutons of restricted thickness that extend laterally along a flat

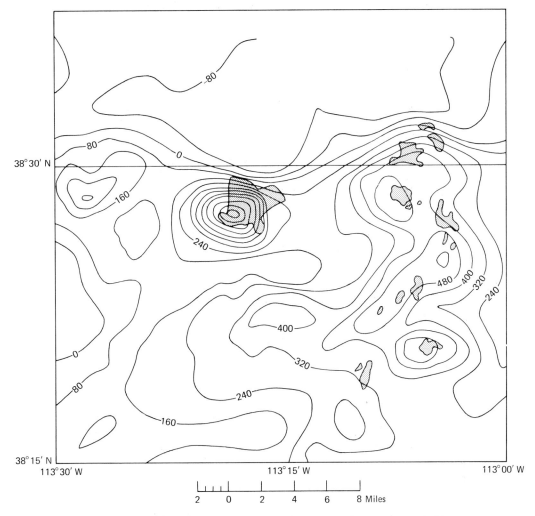

Figure 7-33 Map showing the distribution of granite and the anomalous magnetic field variations contoured at 80 gamma intervals near the San Francisco Mountains in southwestern Utah. (Modified from J. W. Schmoker, *Bulletin 98*, page 38, Utah Geological and Mineralogical Survey, November 1972.)

base from the magma source are called **laccoliths.** Stocks and laccoliths that were produced during the same episodes of igneous activity are found in the Henry Mountains of central Utah. Outcrop patterns can be used to draw diagrams (Figure 7-35) to illustrate the system of plutons that probably lies hidden underground. Many stocklike plutonic masses have been recognized on the continents. These range from granitic to gabbroic in composition. The Bingham stock near Salt Lake City, Utah, is the site of the world's largest open pit copper mine. The rock is mainly a porphyritic intrusive containing an average of less than 0.2 percent copper. This copper exists in sulfide ores that are disseminated

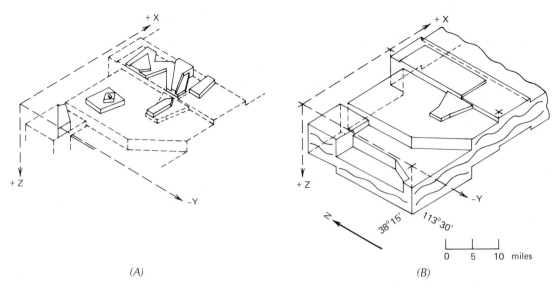

Figure 7-34 Model showing how granite with a magnetic susceptibility of 0.00345 (cgs units) would have to be distributed to produce a simulated magnetic field with the same general features that the contours in Figure 7-33 exhibit. (A) Upper and (B) lower parts of this model show the approximate shape of a batholith believed to exist beneath the San Francisco Mountains in southwestern Utah. (After J. W. Schmoker, *Bulletin 98,* page 8, Utah Geological and Mineralogical Survey, November 1972.)

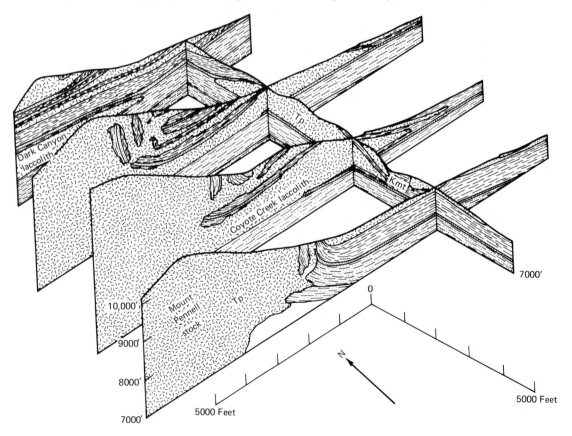

Internal Processes and Igneous Rocks 255

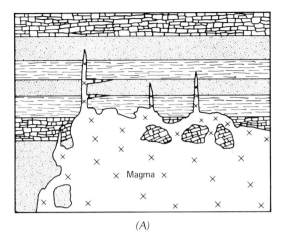

(A)

(B)

(C)

Figure 7-36 *(A)* Magma is believed to move upward in the earth's crust by a process called *magmatic stoping*. There is evidence that this process involves forceful injection of magma in a way that fractures the crustal rock and breaks off and engulfs masses of it. There is also evidence that the magma melts some of the crustal rock as it moves upward. *(B)* One photograph shows a small piece of dark-colored rock that became engulfed in a granitic magma. This piece of rock is called a xenolith. *(C)* The other photograph shows how the magma moved into small fractures in the dark-colored crustal rock.

through small fissures in the main mass of the stock.

Geologists think that movement of magma upward through the lithosphere involves fracturing and melting of the surrounding host rock. In this way the magma eats its way into the crust. We call this process **magmatic stoping** (Figure 7-36). One can look at some outcrops and see the effects of magmatic stoping. One can see how fluid was injected into fractures, and how fragments of host rock called **xenoliths** were split off and engulfed in the magma. In some places we find extrusive rocks associated with batholiths and stocks. This indicates that the intrusive magmatic stoping eventually culminated in volcanism. In many places, however, we find only plutonic rock and no related extrusive rocks. Perhaps they have been removed by erosion, or perhaps the magma crystallized entirely before it had stoped upward to the surface.

At the present time small earthquakes can be recorded in the Rio Grande rift area of New Mexico. Are these related to fracturing

Figure 7-35 *(opposite)* Geologic cross sections illustrating part of a granitic stock and associated laccoliths in the Henry Mountains of central Utah. (Modified from C. B. Hunt, P. Averitt, and R. L. Miller, *Professional Paper 228*, page 122, U.S. Geological Survey, 1953.)

that is being caused by magmatic stoping? Will the magma eventually stope all the way up to the surface, or will it cool deep in the crust to produce a batholith? We will not be here to witness the end of this current episode of igneous intrusion.

Batholiths, stocks, and laccoliths are plutons that have irregular shapes. We also know about sheets of plutonic rock that are called **tabular** intrusions. Some of these are nearly horizontal sheets of coarse-grained igneous rock called **sills.** Where they intrude sedimentary host rock, these sills are parallel to the layering. Sills of dark gabbro contrast sharply with light-colored sandstone (Figure 7-37) throughout large parts of the Transantarctic Mountains. Some of these sills are more than 100 m thick. Geologists have now examined enough sills to know that their compositions are not restricted to any particular type of plutonic rock. Compositions ranging all the way from granitic to ultramafic have been discovered.

We know about other tabular intrusions that are more steeply inclined sheets of plutonic rock. These intrusions are called **dikes.** Where they intrude sedimentary host rocks, dikes cut across the layering. In many places we find dikes that reach outward from a large irregular pluton. They fill fractures that cut into the bordering host rock. Look at the dikes illustrated in Figure 7-38. They radiate from a large pluton that is now exposed in the Spanish Peaks in Colorado.

Dikes can also fill fractures within a larger pluton or cracks that begin inside the pluton and reach outward into the host rock. Interesting dikes that consist of extremely coarse-grained rock called **pegmatite** are found in and near some batholiths and stocks. Apparently some pegmatite dikes intruded fractures in the host rock that were produced by magmatic stoping. Others fill fractures that opened in the pluton during final stages of crystallization, perhaps because of thermal contraction. Pegmatite can be recognized by the very large crystals of quartz, feldspar, and

Figure 7-37 A dark-colored gabbro sill has intruded the light-colored sandstone exposed in the Royal Society Mountains in Antarctica. (W. B. Hamilton, U.S. Geological Survey.)

muscovite. Important sulfide ores have accumulated in some pegmatite dikes.

The Great Dike of Rhodesia is a huge tabular pluton that geologists can trace for more than 500 km; it is up to 10 km wide. Elsewhere we know that large dikes extending several tens of kilometers were intruded during Paleozic time on the eastern piedmont of North America. These large dikes appear to be emplaced in great fissures perhaps similar to those from which large quantities of volcanic rock were erupted.

Dikes can be made of all different kinds of plutonic rock. Some are almost pure quartz, whereas others may consist entirely of ultramafic minerals. But most dikes contain intermediate compositions. In some places significant compositional changes can be found within the same dike. This indicates that the dike might have been produced by successive episodes of igneous activity.

Geologists have been intrigued by an interesting kind of plutonic mass that is called a **layered intrusive.** When they are seen from a distance, some igneous masses display a distinct layering or banding. One might think that they are sedimentary rocks until they are

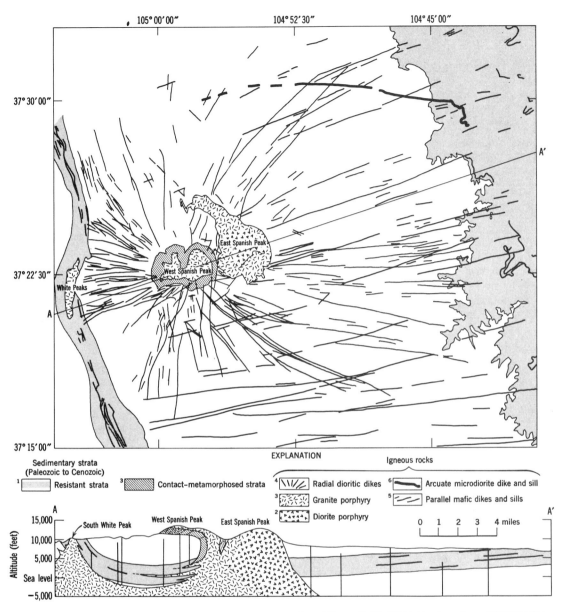

Figure 7-38 Geologic map and cross section show that the Spanish Peaks in south-central Colorado form the upper surface of a granitic batholith. Fractures caused by the intrusion radiate outward from the Spanish Peaks. (After R. B. Johnson, *Bulletin of the Geological Society of America*, page 584, April 1961.) The photograph shows nearly vertical diorite dikes formed from the magma that filled the fractures that extend outward from the Spanish Peaks. (G. W. Stoss, U.S. Geological Survey.)

examined more closely. Then a definite igneous texture can be seen. Look at the layered mountains illustrated in Figure 7-39 and the rock specimen pictured in Figure 7-49. The layering is associated with variation in mineral assemblages. The most prominent layered intrusives that are known have formed from magmas rich in mafic materials. Layering in granitic masses is rare compared with those having gabbroic to ultramafic composition. How can we explain this?

We think that layered intrusives form by a process of magma solidification where minerals crystallizing at different times settle to the bottom of the zone of liquid and become concentrated in layers. From our knowledge of melting temperatures for minerals we know that some crystallize sooner than others from a cooling magma. We also know that magma viscosity increases with SiO_2 content. Therefore, crystals could not readily settle out of viscous granitic magmas. Rather, they would tend to remain mixed with the viscous liquid. But the low SiO_2 magmas from which gabbros and ultramafic masses are derived have relatively low viscosity. Early formed crystals could more easily sink down through these liquids, and become concentrated in mafic-rich layers such as seen in layered intrusives.

Now that we have examined some of the important features of both intrusive and extrusive igneous rock masses, we can think about some perplexing problems. We know that most batholiths and stocks are between granite and diorite in composition. We find considerably smaller amounts of gabbro in sills, dikes, and some small stocks. Ultramafic rocks are rare, and we find them in dikes and sills and layered intrusions. So we can see that near the surface of the earth there is an abundance of SiO_2-rich intrusive rocks. In contrast, the most abundant kind of extrusive rock is basalt. It comes from magma produced in the mantle. Why is the composition of the most abundant intrusive rock in the upper crust so different from the composition of the most abundant extrusive rock? Is it be-

Figure 7-39 Different kinds of ultramafic rock occur in the parallel zones that are seen in these exposures of a very large, layered intrusive in the Dufek Massif, a mountain range in the western hemisphere sector of Antarctica. (A. B. Ford, U.S. Geological Survey.)

cause early forming minerals in a slowly cooling subterranean magma have time to settle, so that magma composition changes in a way that can not happen in a rapidly congealing volcanic lava? Perhaps the major differences in crust and mantle composition are the result of processes similar to those that produce layered intrusives. Geologists have proposed these ideas to explain the different kinds of igneous rocks they find near the surface, and the igneous processes that they can observe directly. To understand them more clearly we have to learn about laboratory experiments that show what happens to different igneous rocks in a wide range of temperature and pressure conditions.

MELTING AND CRYSTALLIZATION

We know that an igneous rock consists of several kinds of minerals, and that these minerals have different chemical compositions and crystal structures. What happens when a rock is heated until it begins to melt? We find that there is no specific temperature at which it changes from a completely solid mass to a completely molten mass. Instead, the rock melts bit by bit as it is heated through a range of temperatures. At the lowest temperature in this range one of the constituent minerals begins to melt, but others remain solid. As the temperature is raised, more and more minerals melt and the mixture becomes a slush. When this slush is heated to the highest temperature in its melting range, the last crystals melt and the mixture becomes entirely molten.

Now suppose that we cool a completely molten magma. We will find that it has a temperature range of crystallization. When the magma cools to the highest temperature in this range, a few mineral grains crystallize, but the rest of the mass remains molten. Continued cooling produces a slush with more and more crystals. After it cools below the lowest temperature in the crystallization range, the final droplets of fluid congeal and the mass becomes completely solid.

These processes of melting and crystallization seem simple enough, but there are complications. The temperature at which a particular mineral melts depends upon the other substances in the mixture. We have to learn about these complications to understand how different kinds of igneous rocks are produced.

An igneous rock can be thought of as a multicomponent chemical system. Each mineral is a particular **phase.** A phase is a substance that actually exists in the system. We know that these mineral phases could be decomposed into several oxides. The oxides are the components of the system; they do not actually exist as separate substances. But we could produce them if we decomposed the mineral phases. Now, suppose that we raise the temperature of this multicomponent system. Some of the minerals might melt; others might convert to polymorphs. Therefore, by raising the temperature we can change the phases in the system. But we would still get the same oxide components if we decomposed the new phases.

We can learn a lot about how igneous rocks are produced by doing some experiments at atmospheric pressure. These experiments will help us to understand **fusion** processes and crystallization processes. Fusion means the same thing as melting, and a fusion process is simply a process of melting. Apparatus (Figure 6-28) that was discussed in Chapter 6 can be used in these experiments.

Suppose that we want to find out about fusion processes. We can insert a mixture of mineral grains in a container and heat it to a temperature at which some grains melt. The mixture must remain at this temperature for several hours or even days because it takes that long for slower chemical reactions to take place. These are reactions between the fluid and the remaining minerals that can produce new mineral phases and changes in the composition of the fluid. After they are finished the system has reached equilibrium at the temperature of the experiment.

Now we have to find out what minerals remain and the composition of the fluid, so we quickly quench the slush in the container. This causes the fluid to congeal into an amorphous glass in which the solid crystals are embedded. We can use X-ray methods to find the glass composition and to identify the minerals.

The experiment can be repeated by heating

the mixture to successively higher temperatures. At each higher temperature chemical reactions that change the fluid composition and mineral phases are allowed to continue until equilibrium is reached. This is the procedure that is followed to learn about the process called **equilibrium fusion.** This is the kind of melting process that occurs if we do not add or remove anything from the mixture. After each temperature step we will find new phases, but the oxide components stay the same.

Fractional fusion is a different melting process than equilibrium fusion. To learn about this process we begin by heating a mixture of mineral grains until some of them melt. Then we remove the liquid and heat the remaining mineral grains to a higher temperature at which more of them melt. The liquid is again removed before heating the remaining minerals to a still higher temperature. This procedure is repeated at successively higher temperatures until all of the minerals have melted. In this experiment we are continually removing the fluid fraction and retaining only the solid fraction while temperature rises. So the oxide components as well as the phases of the system are continually changing during the process of fractional fusion. Therefore, the mineral phases and fluid compositions produced by this process are different from those produced at the same temperatures during equilibrium fusion.

Now let us consider the basic crystallization processes. **Equilibrium crystallization** is the reverse of equilibrium fusion. We begin by cooling a completely molten mixture. As it continues to cool, a temperature is reached at which crystals begin to form. We hold the temperature at this level long enough to establish equilibrium. Then we quench a sample of the mixture to find out the composition of the fluid and the kinds of minerals that crystallized at the temperature of the experiment. We then repeat the experiment at lower temperatures to see how the liquid and mineral phases change while the mixture cools. Nothing is added to or removed from the mixture. Therefore, the oxides in the multicomponent system stay the same during equilibrium crystallization.

A molten mixture can also solidify by the process called **fractional crystallization.** This happens when we remove the minerals that crystallize at a particular temperature before cooling the remaining fluid fraction. Similarly, at a series of lower temperatures we remove the crystals and retain only the fluid fraction. In this way we are continually changing the oxides in the multicomponent system while it cools. For this reason the mineral phases and fluid compositions produced during fractional crystallization are not the same as those that form at similar temperatures during equilibrium crystallization.

Now we have to think about the different mixtures of minerals that might be melting or crystallizing in the earth. We know that heating and cooling cause phases to change in different ways that depend on the composition of a multicomponent system. We will consider two important kinds of systems. The first system includes constituents for producing mineral phases in a **continuous series** of compositions. The second kind we call the **binary eutectic** system. It includes two minerals that are not part of a continuous series. But the presence of one alters the melting temperature of the other mineral.

We know that the olivine minerals make up a continuous series of compositions. Recall from Chapter 6 that the proportions of Fe^{+2} and Mg can vary over a continuous range from **forsterite** (Mg_2SiO_4) to **fayalite** (Fe_2SiO_4). At atmospheric pressure the melting temperature is 1890° C for pure forsterite and 1205° C for pure fayalite. It seems reasonable to presume that for olivine of intermediate composition the melting point would fall between these two extremes. This turns out to be true, but in a somewhat more complicated way than might be expected. Experiments with olivine reveal that melting persists through a range of temperatures in which both crystals and fluid coexist. But the compositions of the crystals and the fluid differ markedly. We can use a **temperature-composition diagram** (Figure 7-40) to show how olivine responds to temperature. This diagram was prepared from experiments like the ones described earlier. A point plotted above the liquidus line represents the composition of a fluid at a certain temperature. Similarly, a point below the solidus line indicates the composition of a solid at another temperature. In the region above the liquid line or below the solidus line the composition of a sample is not altered by change in tem-

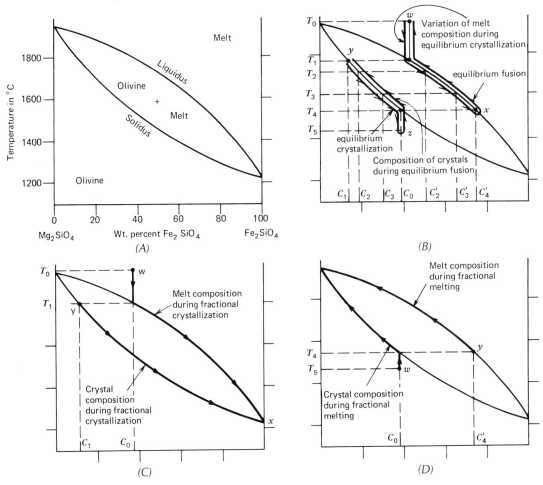

Figure 7-40 (A) Temperature-composition diagram for the Mg_2SiO_4-Fe_2SiO_4 continuous series. (B) Arrows indicate the progressive changes in fluid composition and composition of crystals during the process of *equilibrium crystallization* of a fluid with an initial composition of C_0 starting at a temperature of T_0. (C) Arrows indicate the progressive changes in fluid composition of crystals during the process of *fractional crystallization* of a fluid with an initial composition of C_0 cooling from an initial temperature of T_0. (D) Arrows indicating progressive changes in the composition of crystals and fluid composition during the process of *fractional fusion* of a mass of crystals with an initial composition of C_0 which is heated from an initial temperature of T_5. (Mg_2SiO_4-Fe_2SiO_4 diagram from N. L. Bowen and J. F. Schairer, *American Journal of Science*, page 163, February 1935.)

perature. This is not true for the region between these lines where melt and crystals coexist. Look at some examples to see what happens in this intermediate region. Consider first the equilibrium crystallization of olivine from a fluid that is 60 percent Fe_2SiO_4 and 40 percent Mg_2SiO_4 raised to 1900° C. This starting point is shown at (C_0, T_0) in Figure 7-40A. Let us now lower the temperature slowly. When it reaches T_1, crystals begin to form. We can learn the composition of these first crystals by extending a horizontal line corresponding to temperature T_1 so that it intersects the solidus curve. At this intersection we find the composition C_1 for the crystals formed at T_1. Only a few crystals of C_1 form, and they coexist in equilibrium with fluid that has the composition C_0 at T_1. As the temperature falls to T_2 we find that more crystals continue to form. But now all of the crystals have the composition C_2 (Figure 7-40A). The crystals that had formerly precipitated with composition C_1 dissolved, then reprecipitate with composition C_2. During this process of equilibrium crystallization the crystals and the fluid are free to interact. Therefore, at T_2 we have more crystals than at T_1, and they are all of composition C_2. Because these crystals are enriched in Mg_2SiO_4 compared with the fluid, it is obvious that they have removed a higher proportion of Mg than Fe from the fluid. Consequently, the fluid becomes proportionally enriched in Fe_2SiO_4. We can find the fluid composition C'_2 by projecting the horizontal line through T_2 to the point where it intersects the liquidus curve. Continued decrease in temperature produces more changes in both crystal and fluid composition providing that the two phases are free to interact. We always find at a particular temperature, say T_3, that all crystals have the same composition C_3. Because they deplete Mg in higher proportion than Fe, the fluid composition is altered to C'_3. Finally, at T_4 the remaining fluid crystallizes, and the crystals now have the composition C_0. This is the same as the original melt. What has happened is that at successively lower temperatures crystals form from a melt successively richer in Fe_2SiO_4. Consequently, the later forming crystals become progressively richer in Fe.

We can reverse the process by beginning at point (C_0, T_5) and raising the temperature. Now we are looking at equilibrium fusion of olivine. At T_4 a small amount of melt forms at composition C'_4. As temperature rises the volume of melt increases and it is progressively enriched in Mg. The remaining crystals become enriched in Mg also. At T_1 the final crystals, now of composition C_1, melt, and the fluid assumes the composition C_0. The progressive changes in solid and liquid phases during decrease and then increase in temperature from T_0 to T_5 and back to T_0 are shown by arrows in Figure 7-40A. These reactions are for the condition where fluid and crystals are continually free to interact.

Let us next consider the effect of removing crystals as they form from a cooling melt. This is **fractional** crystallization of olivine, and it would occur if the crystals settle to the bottom of the container. We can see in Figure 7-40B how the phases change. At T_1, a layer of crystals with composition C_1 forms on the bottom and is prevented from reacting with the remaining fluid. As cooling continues the fluid is progressively enriched in Fe and the accumulating layer becomes richer in Fe at successively higher levels. Finally at T = 1205°, the remaining fluid crystallizes to form an upper surface of **fayalite**. Because the crystals settle to the bottom as soon as they form, they are prevented from interacting with the remaining fluid during fractional crystallization.

The reverse of fractional crystallization is fractional fusion. In Figure 7-40C we can see what happens to olivine during this process. Start with the olivine composition C_0 at T_5. Fusion commences at T_4, and the initial fluid of composition C'_4 is removed from interaction with the remaining crystals. Because the fluid is enriched in Fe, we find that as melting progresses the remaining solid and the fluid subsequently derived from it are continually enriched in Mg until at 1890° C the final crystal to melt consists of forsterite.

The examples of equilibrium crystallization and equilibrium fusion seen in Figure 7-40A, and the fractional processes illustrated in Figures 7-40B and 7-40C are shown for olivine minerals. Similar processes can occur in other continuous series of minerals. Recall from Chapter 6 that plagioclase feldspars form a continuous range of compositions between **albite** ($NaAlSi_3O_8$) with a melting temperature of 1120° C and anorthite ($CaAl_2Si_2O_8$) with a melting temperature of 1553° C. The temperature-composition diagram for this series looks like the olivine diagram.

We must test minerals that are not part of a continuous series to learn about a **binary eutectic** system. Consider a mixture of two such minerals, fayalite and albite. We know that at atmospheric pressure pure fayalite melts at 1205° C and pure albite melts at 1120° C. Experiments performed upon mixtures of these minerals show that the presence of one decreases the melting temperature of the other. But during initial stages of crystallization only one or the other of these minerals will crystallize from a melt. We can see in Figure 7-41 how a mixture of fayalite and albite responds to changing temperature. Some examples illustrate how to interpret this diagram. Begin with a mixture of composition D_0 raised to temperature T_0 where it becomes completely molten. As the mixture cools to T_1, a small amount of melt crystallizes into fayalite. This changes the fluid composition, enriching it in $NaAlSi_3O_8$. It also decreases the proportion of Fe_2SiO_4 in the melt. This lowers the temperature at which more fayalite will crystallize. Therefore, more crystals can be produced only if the temperature is lowered. As the melt continues to cool to T_2 more fayalite forms, and the fluid composition shifts along the curve to D_2. The intersection point of a horizontal line through T_2

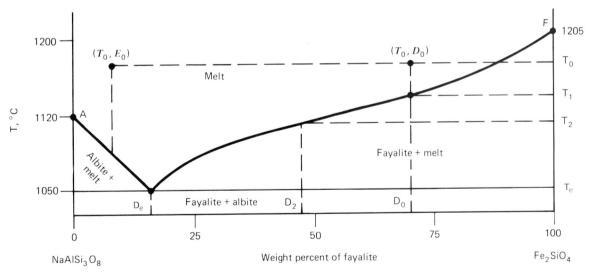

Figure 7-41 Temperature-composition diagram for the discontinuous binary system of $NaAlSi_3O_8$-Fe_2SiO_4. At temperatures above T_e only one or the other of these two constituents will crystallize. At temperatures below T_e any remaining fluid crystallizes to form both albite and fayalite in the proportions of D_e. (After N. L. Bowen and J. F. Shairer, *Proceedings of the National Academy of Sciences of the United States*, page 349, Vol. 22, No. 6, 1936.)

and the liquidus curve indicates the fluid composition at T_2. Precipitation of fayalite progresses while the temperature decreases to T_e, and the corresponding fluid composition reaches D_e. The point (D_e, T_e) is called the **eutectic point** of the system. At temperatures below T_e the remaining fluid crystallizes to form both fayalite and albite crystals in the proportion of D_e. It is important to understand that at temperatures between T_1 and T_e only fayalite precipitates. Below T_e albite and fayalite exist as separate crystals.

Cooling of the initial composition of E_0 in the temperature range between T_2 and T_e would cause precipitation of albite alone. This leads to fluid enrichment in Fe_2SiO_4. Both phases would precipitate in the proportions of D_e at temperatures below T_e.

Regardless of whether we start with D_0 or E_0 the fluid composition migrates to D_e in the final stage of crystallization. The composition of the crystals precipitated at temperatures above T_e will be either one or the other of two minerals depending upon the location of the initial composition on one side or the other of the composition at the eutectic point. Once formed, the crystals do not react with the melt. Therefore, composition is not affected by isolating the crystals from the fluid. If the crystals are allowed to settle, we can see that two distinct layers will form. The bottom one will consist of only one mineral and the top layer will be a mixture of both in the proportions indicated by D_e. Therefore, the way that phases vary during crystallization of a binary eutectic system is different from the variation of phases which crystallize when a continuous series cools.

These examples of continuous series systems and binary eutectic systems illustrate two ways that compositions in the melt and the mineral phases can change with temperature. Furthermore, we can see that a mineral such as fayalite can play two roles. It reacts in continuous series with forsterite and in discontinuous series with albite. Because the melting temperature of one constituent depends on the proportions of others, the melting range for a multicomponent system changes if its composition changes.

In nature magmas are complicated multicomponent systems. There are enough constituents for several reactions to occur simultaneously. The continuous series reactions and the reactions in binary eutectic systems described in the preceding paragraphs are simplified examples of the kinds of things that must happen in a magma. Do we have any evidence of these kinds of reactions in igneous rocks? We can find some large plagioclase crystals that are calcium rich near the center and become gradationally more enriched in sodium farther from the center. These are called **zoned** crystals. We can also find nodules of olivine in which the ratio of Mg/Fe decreases gradationally outward from the center. The growth of these nodules can be explained by fractional crystallization of a continuous series of compositions. We know that an early formed mineral grain can act as a nucleus for later crystal growth. If cooling had produced purely equilibrium crystallization, then we would not expect to find the first grains that had crystallized. These early formed grains should have dissolved as the magma continued cooling. But what if they dissolved too slowly? Suppose that they did not disappear completely before later formed grains clustered about them and insulated them from further reaction with the fluid. If the later formed grains also became insulated before completely dissolving, then indications of a tendency toward equilibrium crystallization might be preserved in the nodule.

We can also find evidence of fractional crystallization in discontinuous mineral systems. Some nodules of the pyroxene **enstatite** are found to have olivine grains in the center. These can be produced by reactions that begin with the crystallization of olivine. When the magma cools below a critical temperature, we know that the olivine crystallization abruptly ceases, and enstatite crystallizes in its place. Evidence of such a reaction is preserved in the nodules that formed when enstatite crystals clustered about incompletely dissolved olivine grains.

We get fragments of information about how igneous rocks crystallize by experimenting with simple continuous series and binary eutectic systems, and by looking at zoned crystals. Can we put these fragments together to determine whether the common rock-forming minerals crystallize from a magma in an orderly way? Will this help us to understand why these minerals occur in the particular mixtures that we find in the common igneous rocks? To answer these questions the well-known petrologist N. L. Bowen (1887-1956) devised some ingenious silicate mineral experiments and carefully studied the mineral associations in zoned crystals. He used the results of this work to figure out a sequence for the crystallization of the abundant silicate minerals from a cooling magma. We now call this sequence the **Bowen Reaction Series** (Figure 7-42). It illustrates how the important mafic minerals could crystallize in a discontinuous reaction series in the same temperature range that the plagioclase minerals crystallize in a continuous reaction series. At still lower temperatures Bowen shows a discontinuous series progressing from K-rich feldspar to muscovite, and finally quartz. To obtain all of the minerals in this reaction series we would have to begin with an ideal magma that contained all of the components. As the ideal magma cools and crystals form, some must be separated from the melt by fractionation to preserve that mineral species. Other similar crystals must continue to interact with the fluid. Here they dissolve or are changed by chemical reactions so that the constituents become available for other minerals to crystallize as cooling progresses. We do not assign specific temperatures to the Bowen Reaction Series. Such numbers would apply only to a specific mixture. This reaction series is meant to be a generalized representation of the sequence of crystallization of the most abundant rock-forming minerals.

The laboratory studies of the response to temperature of multicomponent systems help us to understand some field associations of igneous rocks. We can understand how magma composition would change because of fractional melting in a zone under a volcano. This could explain the change in lava composition during a series of eruptive episodes from the volcano. Layered intrusives could be produced by minerals settling to the bottom of a magma zone during fractional crystallization. N. L. Bowen thought that different kinds of igneous rock could be produced by a process he called **magmatic differentiation.** This is a fractionation process by which the composition of a cooling magma is progressively altered as crystals continue to precipitate and separate from the fluid. To understand this process more clearly imagine a large zone where mineral crystals are precipitating from a cooling magma in the sequence of the Bowen Reaction Series. If the mafic minerals, which are the first to precipitate, settle to the base of the magma zone, they collect to form a mass of ultramafic rock. The remaining magma is then relatively deficient in Mg and Fe, so that continued precipitation will yield a mixture of plagioclase and mafic minerals. As these crystals settle from the magma, they accumulate to form a mass with the composition of gabbro. Separation of these minerals leaves the magma further depleted of Mg, Fe, and some Ca, and relatively enriched in K, Na, and Si. Therefore, further crystallization yields mixtures of minerals found in diorite, and ultimately granite. This idealized and greatly simplified magmatic differentiation sequence indicates how different kinds of rock could be produced from a cooling magma.

In nature the crystallization of a magma can be complicated in different ways. Contamination of the original mixture results from upward movement by stoping and partial melting of the overlying host rock. Precipitation of mafic constituents causes the magma to become enriched in SiO_2. This makes the

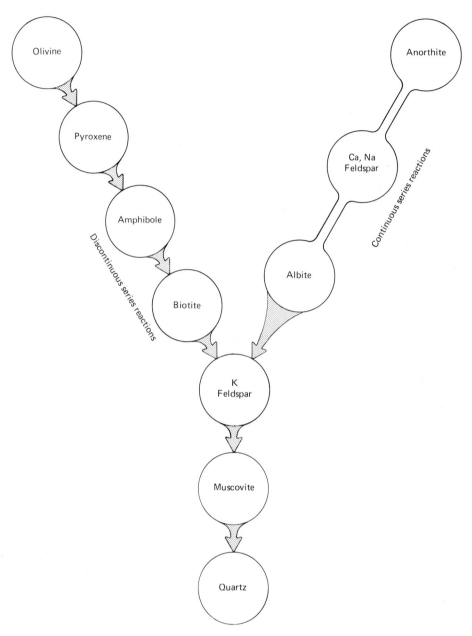

Figure 7-42 The Bowen Reaction Series indicates the sequence of crystallization of silicate minerals from a slowly cooling magma of ideal composition. First to crystallize are olivine minerals. As the magma temperature continues to drop, the pyroxenes, amphiboles, then biotite, crystallize. In this same cooling range the plagioclase minerals also crystallize. During the final stages of cooling the alkali feldspars, then muscovite, and finally quartz, crystallize. (Modified from N. L. Bowen, *The Evolution of Igneous Rocks*, page 60, Dover Publications, 1928.)

magma more viscous, so that later precipitating minerals cannot readily separate from the fluid. Furthermore, natural magmas are unlikely to possess the proportions of elements needed to produce the entire rock sequence from ultramafic to granitic compositions in a single episode of crystallization. Rather, we expect that a series of episodes involving crystallization, remelting, magma migration, and contamination probably occurs before all of the different kinds of igneous rock are produced.

COMPOSITION AND TEMPERATURE IN THE EARTH

We know less about the composition of the deep interior of the earth than we know about the makeup of the sun and other stars. How can we learn what is hidden deep inside? Lacking means of direct observation, geologists use seismic wave velocities, density and pressure estimates based upon seismic and gravimetric evidence, and assumptions about thermal conditions to piece together their ideas about the composition of the earth's interior. There is considerable controversy about the proportions of minerals or elements that exist in the mantle and core.

We believe that the outer part of the continental crust down to depths of at least 20 km consists mostly of granite and associated metamorphic rocks. It is covered by a veneer of sediments. The oceanic crust, in contrast, is primarily basalt and gabbro. Recall that the oceanic crust is generally between 5 and 10 km thick.

In the lower part of the continental crust we measure P-wave velocities of approximately 6.8 km/sec. We estimate that density is about 2.9 gm/cm^3. At pressures in the range of 7 to 10 kb both gabbro or diorite possess these velocity and density properties. Probably both of these kinds of rock exist in the lower crust, but at the present time we are not certain about their relative proportions.

We know that almost everywhere in the upper mantle P-wave velocities are in the range of 7.9 to 8.2 km/sec. Our knowledge of the density is less certain. But in general it is between 3.3 and 3.4 gm/cm^3. Rocks having these velocity and density values at pressures found in the upper mantle must consist of olivine, pyroxene, garnet, and perhaps minor amounts of amphibole. The principal ultramafic rock types having this mineralogy are **peridotite** and **eclogite.** Peridotite is an olivine-pyroxene mixture with a density of between 3.25 and 3.38 gm/cm^3. The pyroxene-garnet mixture that we call eclogite has a higher density range of 3.4 to 3.65 gm/cm^3.

We have some direct evidence of upper mantle composition. This comes from rock found in the diamond-bearing pipes described earlier in this chapter. The most abundant constituent of these pipes is a variety of mica-rich peridotite called **kimberlite.** The samples that come from diamond mines are partly decomposed so that their original composition is hard to ascertain. However, in addition to diamonds, geologists have found xenoliths of relatively fresh peridotite and eclogite in kimberlite pipes. (The proportion of peridotite is largest.) Suppose that the constituents of these pipes represent a random sampling of the upper mantle. If this is true, we could conclude that peridotite is the most abundant constituent and eclogite is an important secondary constituent.

Laboratory experiments show that depending upon the proportions of olivine and pyroxene some peridotites will yield what we call **basaltic** magma upon first partial melting. The first minerals to melt produce this magma. Suppose this first melting fraction is separated from the crystals. If it then congeals in the pressure and temperature of the upper mantle, eclogite will be produced. We believe that pods of eclogite exist in the mantle which are partial melting derivatives of peridotite.

Basalt and eclogite are different kinds of rock possessing the same chemical composition. They are similar to mineral polymorphs, and their stable and metastable pressure-temperature regions can be determined experimentally. The phase diagram in Figure 7-44 indicates that eclogite is stable at high pressure and that basalt is stable in a low pressure environment. What if the Mohorovičić Discontinuity were simply the pressure-temperature boundary between these two rock forms? If this were true, we might expect this first-order seismic discontinuity to migrate upward or downward in response to changes in pressure. But if the lower crust is diorite rather than gabbro, then this phase transformation would not take place. Diorite and eclogite are chemically different. The questions about the nature of the M-discontinuity have an important bearing upon the age of the crust. If it is a pressure-temperature boundary, then the lower crust might be quite young, forming rapidly in response to a decrease in pressure. On the other hand, if the crust and mantle are chemically different, then perhaps the crust has accumulated by magmatic differentiation. The latter process has obviously been taking place. We know of no other way for granitic plutons to accumulate in the upper crust. Without better knowledge of the composition of the lower crust, however, we cannot be sure whether some kinds of phase transformation occur between the crust and mantle.

We believe that rocks in the lower mantle consist of minerals that differ from the familiar ones that we find nearer to the surface. At the elevated pressure more than 600 km deep, Si exists in six-fold coordination with oxygen, rather than the four-fold tetrahedral coordination commonly found in the silicate minerals. Seismic waves tell us about some relatively abrupt velocity increases that are probably associated with pressure-induced phase changes in an otherwise monotonous chemistry, but this is hard to test experimentally. We can only momentarily create high pressure by shocking a substance with a small explosion. High-pressure shocking experiments have been performed on some important oxide compounds to find those that possess the density and seismic velocity properties of the lower mantle. These properties would exist in a composition having the approximate proportions of .32 MgO + .18 FeO + .50 SiO_2. The lower mantle, then, would have an iron-magnesium silicate makeup, but it does not contain minerals with the crystal structures found in the crust and upper mantle.

The pressure of the outer core is difficult to achieve even in shock wave experiments; however, some measurements have been carried out with Fe-Ni and Fe-Si mixtures. The molten outer core cannot consist entirely of Fe and Ni, but requires some lower density alloying element. Minor amounts of either Si or S mixed with Fe and Ni could satisfy the seismic velocity and density requirements. The outer core must be between 10 and 15 percent less dense than pure iron at the same pressures and temperatures.

Nature provides another unusual source of information about the earth's interior. Meteorites are useful vehicles that come from outer space at no cost to the taxpayer. When we compare the composition of the earth with meteorite compositions, we find some interesting similarities. Meteorites are either iron-rich metallic substances or stony assemblages that we call **chondrites.** The composition of chondritic meteorites is close to the composition that we believe makes up the lower mantle. Could this mean that the earth was originally put together from meteoritic debris? Perhaps its interior zones were produced later by magmatic differentiation processes. We will have to figure out new kinds of observations and experiments to test these ideas.

How can we learn about temperature in the earth? Measurements in boreholes indicate that it increases with depth in the upper crust at the approximate rate of 30°C/km.

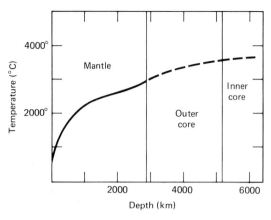

Figure 7-43 Estimated variation of temperature with depth in the mantle and core of the earth, calculated on the assumption of heat transfer by conduction. (After D. P. McKensie, *Geophysical Journal of the Royal Astronomical Society*, page 302, Vol. 14, 1967 and G. D. Garland, *Introduction to Geophysics*, page 366, W. B. Saunders, 1979.)

Deeper in the lithosphere we know that this rate must diminish. Otherwise the melting temperatures of all common silicate minerals would be reached at depths of less than 70 km. We have learned from amplitudes of seismic waves that a small portion of fluid probably coexists within a solid matrix in the transition zone between the lithosphere and the asthenosphere. This implies that the temperature is slightly higher than the solidus temperature for a peridotite-eclogite mixture. Therefore, it must lie between 1100° and 1200° C at a depth 100 km. We can explain seismic velocity discontinuities at 400 and 700 km (Figure 3-21) by crystallographic transitions between ultramafic mineral polymorphs if temperatures of about 1500° and 1900° C exist at these depths. At the core-mantle boundary the temperature must be near 3000° C to be above the melting point of the iron-nickel core, and below the melting range of the iron-magnesium silicate mantle. Similar reasoning indicates a temperature of about 4300° C at the transition between the outer and inner zones of the core. These values have been used to prepare a graph (Figure 7-43) that shows temperature conditions in the earth. We need this kind of information about temperature in the earth before we can determine what changes could initiate new episodes of melting.

THE ORIGIN OF MAGMA

We know that the lithosphere is almost entirely solid. The asthenosphere, too, is largely solid. But a small fraction of fluid is probably interspersed through the matrix of ultramafic mineral crystals. In some zones, however, substantial melting must occur from time to time to produce the magma that eventually erupts from volcanoes or crystallizes into intrusive rock masses.

What events and conditions might cause melting to begin in the upper mantle? To answer this question we need to know about the pressure-temperature conditions required for partial melting of different kinds of rock. The phase diagrams in Figure 7-44 summarize the results of laboratory experiments with granite, gabbro, and eclogite. Because pressure increases quite uniformly with depth, we can show both pressure and depth on the vertical axes of the diagrams. This way we can see immediately the depths where different pressures are found. Curves showing average temperature variation with depth, taken from Figure 7-43, are also included on these phase diagrams. We can tell from the information in Figure 7-44 that granite and basalt in the crust, and eclogite in the upper mantle, will remain solid in the normal pressure-temperature environment. Where can we find the abnormal conditions needed to start partial melting?

We know about places in the lithosphere where temperature is much different than the average shown in Figure 7-44. Beneath some oceanic ridge segments temperature reaches

270   The Rocks of the Earth

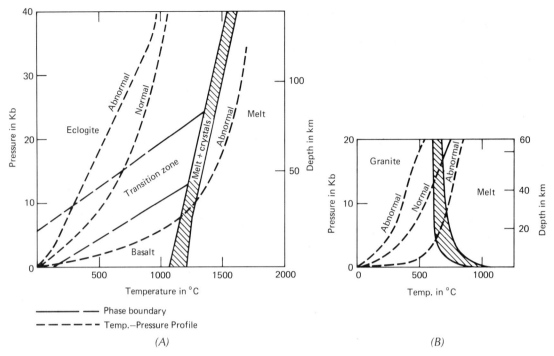

Figure 7-44 (A) Basalt-eclogite-melt phase diagram illustrates pressure and temperature conditions in which one or the other of these rocks can form. Both types of rock have the same chemical composition, but they contain different mineral phases. Superposed on the diagram is a curve showing the normal variation of temperature with depth in the crust and upper mantle, and two possible abnormal temperature depth curves. (Phase diagram after H. S. Yoder and C. E. Tilley, *Journal of Petrology*, page 498, Vol. 3, part 3, 1962.) (B) Granite-melt phase diagram with superposed normal and abnormal temperature-depth curves. (Phase diagram from W. C. Luth, R. H. Johns, and O. F. Tuttle, *Journal of Geophysical Research*, page 760, February 15, 1964.) Note that pressure is given on the left vertical axis and depth is given on the right vertical axis of each phase diagram. This can be done because pressure varies more or less uniformly with depth in the crust and upper mantle.

1200° C at depths of less than 30 km. The solidus temperature for eclogite and basalt decreases as pressure is lowered. Therefore, if a mass of eclogite beneath an oceanic ridge rises by convection into a lower pressure environment, partial melting might begin. Suppose, for example, that eclogite at 50 km where the temperature is 1250° C rises to 30 km where the temperature is 1200° C. We can see a line in Figure 7-44 A that traces this change in pressure and temperature. This is one way that basaltic magmas could be produced beneath oceanic ridges.

The phase diagrams in Figure 7-44 indicate the response to pressure and temperature of rocks devoid to H_2O. We have learned from other experiments that when small fractions of H_2O are present, the melting temperatures of these igneous rocks are decreased. The presence of H_2O might contribute to melting in plates of the lithosphere which are moving downward beneath the island arcs and continental margins bordering submarine trenches. Suppose that sea water has percolated into fractures in the ocean floor basalt, increasing the concentration of H_2O near the upper margin of the plate. Then the basaltic crustal rock would melt more readily as it is transported into the deeper environment of high pressure and temperature.

The idealized geologic cross section in Figure 7-45 summarizes some of the important features in the outer shell of the earth. Here we can see where the principal kinds of rock are situated, and where zones of melting probably exist. Most melting zones are near constructive and destructive plate margins. But there are a few zones far from plate borders where magma is produced. We have no generally accepted explanation for melting in these intraplate settings.

ORES OF IGNEOUS ORIGIN

John Augustus Sutter recalled that on January 28, 1848:

> . . . Marshall pulled out of his trousers pocket a white cotton rag which contained something rolled up in it. . . . Opening the cloth, he held it before me in his hand. . . . 'I believe this is gold,' said Marshall, 'but the people at the mill laughed at me and called me crazy.' 'I carefully examined it and said to him: 'Well, it looks like gold. Let us test it.'

You probably know what followed. First hundreds, then thousands of prospectors made the trek to California hoping to sift gold from the stream beds. Some were more observant, and realized that these bits of gold had washed down from sources in the mountains. Valuable substances such as gold that are mixed in unconsolidated stream gravels are called **placer** deposits. The more clever prospectors figured out how placer concentrations changed from place to place. They watched many others pan for gold along different stretches of a stream and noted where most of it was found. Working from patterns of placer concentration, they were sometimes able to locate rare gold-enriched zones in the hills rising beyond the streams. These gold concentrations in bedrock are called **lodes.** The gold may be laced through a quartz dike in a system of intertwining fissures, or it may occur in association with the element **tellurium,** forming the telluride ores.

The California gold rush began with a chance discovery. The gold deposits were then exploited haphazardly. This was the usual way that precious metals and ores were located and exploited during more than 5000 years of human history. But from most accounts of the search for valuable minerals we can find some attempts at recognizing patterns of occurrence. In recent times the exploration for economic deposits has become more systematic. We have records from mines where ore has already been found, and we can use a variety of geophysical and geochemical prospecting devices.

In this chapter we will consider ores of igneous origin. These include concentrations formed within igneous masses and in fractures that extend into the surrounding rock. We will discuss other ore deposits that originate from metamorphic processes and alteration of sedimentary rocks in the next two chapters.

An ore is simply a concentration of minerals from which valuable substances, most commonly metals, can be extracted by economically feasible methods. Basic metals including iron, copper, lead, zinc, nickel, molybdenum, and uranium occur as trace elements in common igneous rocks. Thus far we have not developed economical refining processes to recover these sparsely distributed elements. Consequently, we have to locate these elusive zones where high concentrations occur. Table 7-3 shows how much higher the concentrations of metallic elements are in commercial ores compared with

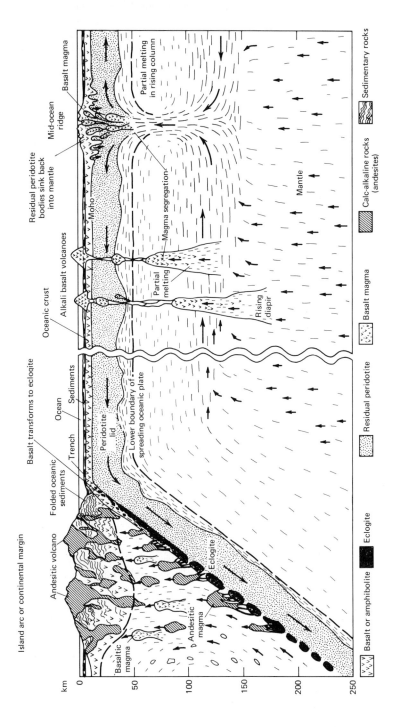

Figure 7-45 Composition of the earth's crust and upper mantle, and sources of magma are illustrated in this conceptual cross section, which is consistent with the plate tectonic theory described in Chapter 5. (Modified from A. E. Ringwood, *Geophysical Monograph No. 13*, page 12, American Geophysical Union, 1969.)

Table 7-3
Concentration of Metals in Common Ores

Metal	Weight Percent in Average Ore	Weight percent in Average Igneous Rock
Molybdenum	0.1 to 0.4	0.00015
Uranium	0.1 to 0.3	0.00027
Lead	2.5 to 10	0.00125
Zinc	4.5 to 10	0.007
Nickel	0.8 to 3.5	0.0075
Copper	0.5 to 3.0	0.0055
Iron	20 to 60	5.63

From A. H. Lang, *Prospecting in Canada*, Economic Geology Report No. 7, Geological Survey of Canada, Ottawa (1970), p. 46.

Internal Processes and Igneous Rocks

they occur comes largely from looking at features exposed in mines. Ore bodies in veins and in some layered intrusives have simple tabular shapes. A **vein** is an accumulation of ore and other minerals that fills a crack or fissure in the surrounding rock. Many ores occur in highly irregular distributions in lenses, large podlike masses, and deformed pipes. Some typical examples are shown in Figure 7-46 and Figure 7-48.

Almost all ore bodies are a mixture of valuable ore minerals and economically worthless **gangue** minerals. For example, pyrite found in gold-bearing veins is usually cast aside as a gangue mineral. The proportions of

common igneous rocks. Some important ore minerals are described in Tables 6-5, 6-6, and 6-7.

Our knowledge about the shapes of ore concentrations and the natural settings where

Figure 7-46 Shapes of sulfide ore bodies determined from mining operations in the Helena-Frisco Mine near Couer d'Alene, Idaho. (From Ransome and Calkins, *Professional Paper No. 62*, page 125, U.S. Geological Survey, 1908.)

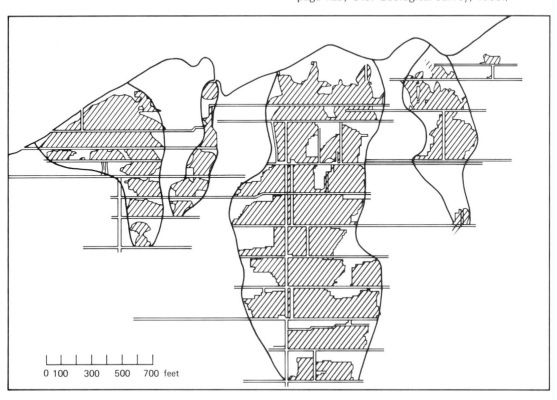

Figure 7-47 Underground mining of a large quartz vein to recover gold disseminated in fissures. (From A. H. Lang, *Economic Geology Report No. 7*, Geological Survey of Canada, page 42, 1970.)

gangue and ore minerals determine the **grade** of an ore. High-grade ore, such as found in massive sulfide bodies, contains a large proportion of ore minerals. We can see in Figure 7-47 how high-grade ore is mined underground from a vein. Lower-grade ores contain only small proportions of ore minerals. To have commercial value, they must occur in large masses readily accessible to open-pit mining. An example is the porphyry copper deposit mined in Bingham Canyon, Utah. Copper-rich minerals, constituting less than 0.2 percent of the rock mass, are disseminated in narrow fissures threaded through the main mass of the Bingham Stock. Ore mineral concentration can change from place to place in the larger structure of an ore body.

In Figure 7-48 we can see an inclined pipe that is an ore body with an irregular gold-enriched zone.

We think that igneous ore bodies are products of magmatic differentiation. We know that large layered intrusives contain tabular zones highly enriched in **chromite** ($Fe[Cr,Fe]_2O_4$) and **magnetite** (Fe_3O_4). The Stillwater layered intrusive in central Montana contains commercial chromite deposits concentrated in bands associated with olivine layers. Chromite-enriched zones occur in 13 rhythmically repeated units of peridotite. Here we can see grains of chromite, bronzite, and olivine that formed with well-developed crystal faces in separate layers (Figure 7-49). This means that each of these mineral species in its turn crystallized under conditions of free growth in the magma, then settled out. The chromite ores, therefore, appear to have been concentrated within the plutonic mass during early stages of magmatic differentiation.

We know about many pegmatite deposits that are enriched in gold and sulfide ores. These pegmatites have been produced within granitic masses. The constituents of a pegmatite are derived from the final fraction of an SiO_2-rich magma. This fluid, rich in H_2O and the elements needed for quartz, muscovite, and potassium feldspars, becomes concentrated in fissures and interstices of the almost solidified intrusion.

We can find veins enriched in sulfide minerals and quartz that extend out from intrusive bodies into the surrounding rock. They fill fractures that opened during emplacement of the igneous mass. The vein minerals crystallized from **hydrothermal** fluids. These are fluids that contain an abundance of H_2O, and are often enriched in metallic compounds. They contain the final fluid fraction of the original magma. It consists of ions that do not ordinarily fit into silicate crystal structures. As these products are "sweated out" of the intrusion in the final stages of crystallization, they become concentrated in fractures within and adjacent to the igneous mass.

We know that the sulfide ore minerals are

Internal Processes and Igneous Rocks 275

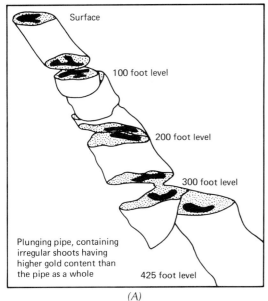

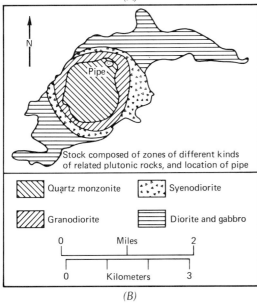

Figure 7-49 Two views of chromite enriched layers that alternate with olivine layers in the Ultramafic Zone of the Stillwater layered intrusive complex in Montana. (W. R. Jones, U.S. Geological Survey.)

7-48 (A) An inclined igneous pipe with gold enriched zones of irregular shape has been profitably mined to depths exceeding 300 ft. (B) Geologic map of the stock in which the gold-bearing pipe is situated. (From A. H. Lang, *Economic Geology Report No. 7*, page 43, Geological Survey of Canada, 1970.)

almost insoluble in water. So how do the constituents of these minerals remain dissolved in hydrothermal solutions? Recall our earlier mention of the crusts of metallic chlorides that precipitated on the walls of volcanic vents from H_2O-rich gases. Now we think that chloride compounds are important in maintaining metallic elements in hydrothermal solutions. Consider the example of galena (PbS). Because of its high melting temperature of about 1115° C and its extremely low solubility in H_2O, it is unlikely that this compound could exist in a hydrothermal solution. However, in a chlorine-rich environment the following reaction could take place:

$$\underset{\text{muscovite}}{KA1_2\,(AlSi_3O_{10}XOH)_2} + \underset{\text{quartz}}{6SiO_2} + \underset{\substack{\text{fluid} \\ \text{potassium} \\ \text{chloride}}}{2KCl} \rightarrow$$

$$\rightarrow \underset{\text{orthoclase}}{3KAlSi_3O_8} + \underset{\substack{\text{fluid} \\ \text{hydrochloric} \\ \text{acid}}}{2HCl}$$

Figure 7-50 Gold panning has remained essentially unchanged by generations of prospectors

The hydrochloric acid generated by this reaction could react with lead sulfide in the following way:

$$\underset{\text{solid}}{PbS} + \underset{\text{fluid}}{2HCl} \rightarrow \underset{\text{solute}}{PbCl_2} + \underset{\text{gas}}{H_2S}$$

The compound $PbCl_2$ is soluble in H_2O. Suppose that this compound is dissolved in hydrothermal solutions. Perhaps the lead in ores found in veins was carried in such a solution. After being transported into a fissure the dissolved lead chloride must again react with sulfur compounds to produce galena during the final stages of crystallization. Similar processes probably occur with other metallic elements, but we still have much to learn about these processes. However, we know that metallic chlorides exist in aqueous volcanic gases, and we know that these chlorides are much more soluble than sulfides in H_2O. Therefore, we think that the transport of metallic chlorides dissolved in highly mobile hydrothermal fluids is one way to produce the concentrations of metals found in the ores of many vein fillings.

The search for ores often involves a comprehensive exploration program that makes use of several geophysical and geochemical survey techniques. Because the magnetic minerals are associated with some metallic ores, aeromagnetic anomalies (Figure 4-14) are observed over concentrations of these ores. Similarly, because of their high densities, some massive sulfide deposits produce gravity anomalies. Metallic ores with electrical conductivity can be detected by electromagnetic sensing devices. Electromagnetic anomalies led to the recent discovery of a large sulfide deposit in Ontario.

Geochemical prospecting techniques have been developed to detect hidden ores. In some places gangue minerals at the surface provide evidence of an ore body. Also important are the elements released by chemical decomposition of an ore. There are places where we can detect these elements in the soil and vegetation in the vicinity of a hidden ore concentration. Sampling of soils and ground water solutions may define zones of mineral concentration.

The modern geophysical and geochemical exploration techniques cannot completely replace the romance of the gold pan. In fact, it can still be used for prospecting (Figure 7-50). If you like to hike along mountain streams, it is not much bother to carry a pan. Who knows what you might find?

GEOTHERMAL POWER

Think about the energy that the earth uses in igneous processes. Can we harness any of it for other purposes? We know that geothermal

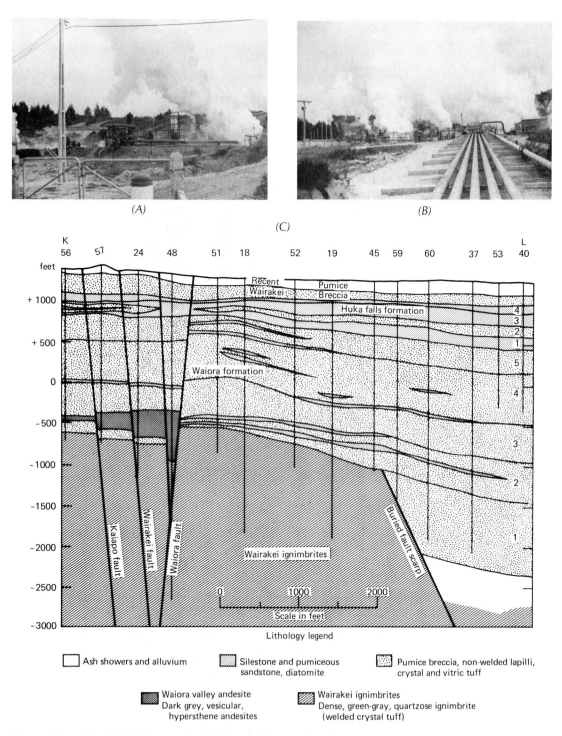

Figure 7-51 The Wairaki Hydrothermal Power Station (A) and (B), supplies electrical power to part of the North Island of New Zealand. Power is developed from steam retrieved from boreholes drilled into (C) the geologic setting seen in cross section. (Cross section reprinted from G. W. Grindley, *Bulletin n.s. 75*, page 104, New Zealand Geological Survey, 1965.)

energy is released in the earth by decay of radioactive elements, but the average amount reaching the earth's surface is quite small. About 2000 times more energy arrives here from the sun. Therefore, is it practical to think about ways of using geothermal energy? This question was answered as early as 1904 when electricity was first produced in a geothermal power plant near Larderello, Italy. Here steam is trapped under high pressure in subsurface reservoirs of fractured volcanic rock. Holes are drilled into these reservoirs to release the steam for driving turbines in the power plant.

At the present time geothermal power plants are operating in Italy and several other countries. In Iceland, geothermal steam and hot water are piped directly into homes and buildings for heating. Steam generators are used here to produce electricity. Geothermal power plants in New Zealand (Figure 7-51), Japan, the Soviet Union, and the United States all use steam from subsurface reservoirs to generate electricity. A plant at The Geysers in northern California produced close to 400 megawatts of power in 1976, which was enough to meet about one-half of the demand for the San Francisco area.

All of these power plants are situated where unusually high concentrations of heat exist near the earth's surface. These are regions with active volcanoes or hot springs nearby, or evidence of recent volcanism. Steam reservoirs are confined at depths of a few hundred to a few thousand meters. Wells drilled into these reservoirs release steam faster than it can be replenished by geothermal heating. So each well is useful only for a few years or a few decades depending on local conditions. Therefore, we must view this kind of geothermal energy as a nonrenewable resource.

Now there are very few places where we can make use of geothermal steam. Are there ways to tap the geothermal energy released in other parts of the world? We know that almost everywhere temperature increases with depth. Suppose that holes are drilled into warm rock. There we could pump water into the holes, allow it to heat, then pump it out. For this idea to be practical we have to get more energy out in the form of warm water than we put into drilling and pumping operations. At present this cannot be done in an average location where temperature increases approximately 30° C/km in the earth. However, there are places where the rate of temperature increase is considerably greater. These are places where granite plutons are exposed at the surface, or are buried at shallow depths. Granites tend to have much higher concentrations of uranium, thorium, and potassium than other common rocks. For this reason more geothermal heat is released by radioactive decay in a granite pluton than in the surrounding host rock which raises the temperature in the rock. Therefore it might be possible to recover energy by circulating water in wells drilled into granite. This is now an important subject for research and requires the cooperation of engineers and geologists interested in igneous processes.

STUDY EXERCISES

1. What specific minerals might be found in gabbro that would contribute to the measured amount of the oxide Na_2O?

2. Compare the proportions of oxides from Asama Volcano (Fig. 7-30) with proportions of oxides in average extrusive rocks (Table 7-1).

3. Would the ratio of oxide proportions K_2O/MgO be larger in batholiths or in plateau basalts?

4. In terms of the average compositions of batholiths compared with dikes and sills, which would have the highest oxide abundance ratio of SiO_2/MgO?

5. Starting with equal proportions of forsterite and fayalite what olivine composition would be found at 1400° C after (a) fractional crystallization of a melt and after (b) equilibrium fusion of a solid?

6. Prepare a graph showing the change in the oxide abundance ratio SiO_2/MgO with depth starting at the earth's surface and continuing to the core/mantle boundary.

7. Suggest a system for classifying igneous ore bodies according to shape, composition, and grade.

SELECTED READINGS

Bowen, N. L., *The Evolution of the Igneous Rocks,* Princeton University Press, Princeton, N.J., 1928 (reprinted by Dover Publications, Inc., New York, 1956).

Ernst, W. G., *Earth Materials,* Prentice-Hall, Inc., Englewood Cliffs, N.J., 1969.

Holmes, Arthur, *Principles of Physical Geology,* rev. ed., Ronald Press Co., New York, 1965.

Hyndman, Donald W., *Petrology of Igneous and Metamorphic Rocks,* McGraw-Hill Book Co., New York, 1972.

Park, C. F., and R. A. MacDiarmid, *Ore Deposits,* 3rd ed., W. H. Freeman and Co., San Francisco, 1975.

U.S. Geological Survey, *Atlas of Volcanic Phenomena,* U.S. Dept. of the Interior, Washington, D.C.,

Yoder, H. S., *Generation of Basaltic Magma,* National Academy of Sciences, Washington, D.C., 1976.

8

METAMORPHISM

Common Metamorphic Rocks
  Slate
  Phyllite
  Schist
  Gneiss
  Marble

  Metaquartzite
  Hornfels
Contact Metamorphism
Metamorphic Facies
Regional Metamorphism
Economic Deposits in Metamorphic Terranes

We know from James Hutton's idea of the great geologic cycle that practically all rocks are composed of materials taken from more ancient rocks. Look back at Figure 7-1. Magma melted from older rock crystallizes to make igneous rocks. These rocks are products of high temperatures and sometimes high pressure. In contrast, sedimentary rocks are made from broken fragments that were deposited in a low temperature and pressure environment. But what happens between these extremes? Other rock-forming processes are at work in the crust where intermediate temperature and pressure conditions exist. Here older rock is cooked and kneaded to produce new kinds of rock that have distinctive minerals and physical features. These rocks are called **metamorphic rocks.**

We can find metamorphic rocks close to some igneous plutons; they are the products of *contact metamorphism*. Hot magma raises the temperature in nearby host rock, which can cause chemical reactions, partial melting, and recrystallization that produce contact metamorphic rocks. But most metamorphic rocks are not directly associated with any particular igneous pluton. We find them exposed to regions extending over many thousands of square kilometers. These rocks are the products of **regional metamorphism.** Processes of regional metamorphism are activated by changes in temperature and stress, and circulation of chemically active fluids in large zones of the lithosphere. Sometimes there is associated igneous activity. But the regional metamorphic rocks are produced throughout the large zone, not just in places close to the igneous rock.

Recall that stress exists where force acts on an area. It is a **directed stress** if force in one direction on a particle of rock is stronger than force in another direction. A **nondirected stress** exists where force is equal in all directions. Hydrostatic pressure is nondirected stress. Insofar as temperature and nondirected stresses are the principal causes of metamorphism, we can expect some recrystallization and changes in the mineral composition of a rock. If directed stress plays an important role, however, the rock may also develop some linear or planar features as well as changes in mineralogy. These physical features and recrystallized mineral assemblages are evidence of metamorphic processes that work beneath the surface hidden from direct observation. Episodes of metamorphism can leave their imprints on a rock without obliterating all of the original features. A rock can retain evidence of more than one metamorphic event. The record of such events which is preserved in metamorphic rocks helps us to understand the earth's geologic history.

We will begin our discussion of metamorphism by describing the principal kinds of metamorphic rocks. Then we can look at some of the ways that they are distributed in the lithosphere together with some results of laboratory experiments that help us to understand how metamorphic processes work. Finally we will discuss how economic deposits are produced by these processes.

COMMON METAMORPHIC ROCKS

You can recognize the common metamorphic rocks by features that are easy to see in hand specimens or thin sections. The **foliated** metamorphic rocks contain linear or planar features. These features result from alignments of elongate or platy minerals. The **nonfoliated** metamorphic rocks do not possess these features. Of course, igneous and sedimentary rocks can also be layered or foliated, but in true metamorphic rocks the foliation is a secondary feature not necessarily related to any similar structure in the source rock from which it is derived.

Most foliated metamorphic rocks consisting of mixtures of silicate minerals belong in one of four general textural groups. In order of increasing grain size these are the **slates, phyllites, schists,** and **gneisses.** There are other groups of metamorphic rocks that have compositions dominated by a particular mineral. An example is **marble** which consists mostly of recrystallized calcite. These various groups will now be described in more detail.

Slate

Fine-grained rocks with well-developed cleavage are called slates. The mineral grains, which are too small to be seen individually without magnification, can be partially interlocking. Slates split easily along very smooth cleavage surfaces that are not necessarily parallel to any laminations which existed in the source rock. We can see from microscopic examination of slates that the cleavage surfaces are usually parallel to alignments of platy minerals such as the micas. All of these features are illustrated in Figure 8-1. Because slate can be separated into thin sheets with smooth surfaces, it has traditionally been used for chalkboards and roofing tiles.

Slates vary in color from black through hues of red, brown, blue-gray, and green. We know that slate can be produced by metamorphism of fine-grained sedimentary and igneous rocks. Perhaps the most abundant source materials are shales and clay deposits. But the term **slate** refers only to the intrinsic features of the rock after metamorphism. It does not imply anything about the source material.

Phyllite

Rocks possessing a well-developed foliation and mineral grains slightly larger than those of slate are classified as phyllites. You can see the individual mineral grains through a lens that magnifies about 5X. The foliation of a phyllite is clearly evident, but the cleavage is poorer than that of slate. Notice in Figure 8-2 the uneven surfaces that result from splitting

282 The Rocks of the Earth

Figure 8-1 *(A)* Slate is a very fine-grained metamorphic rock that can be split apart along smooth parallel cleavage surfaces. *(B)* Under high magnification the alignment of mineral grains in slate can be seen.

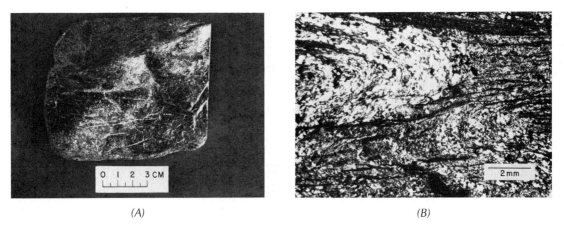

Figure 8-2 *(A)* Phyllite is a fine-grained metamorphic rock consisting of larger particles and more irregular cleavage surfaces than slate. *(B)* Under magnification the alignment of mineral grains in phyllite can be seen.

the rock along the foliation. Most phyllites have an abundance of mica, so that surfaces exposed by splitting a specimen have a silky luster or sheen.

Schist

A foliated metamorphic rock is called a schist if it consists of small mineral grains that can be seen either without magnification or with very little magnification. The individual mineral grains to a certain extent are interlocked with one another in most schists. Look at the features of schists illustrated in Figure 8-3. The foliation associated with alignment of platy and long grains is accentuated by mineral segregation. For example, thin laminations rich in mica might be found alternating with quartz-feldspar laminations.

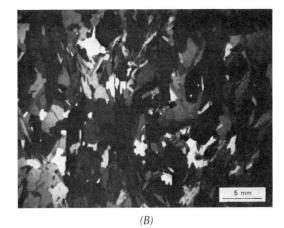

Figure 8-3 *(A)* Schist is a metamorphic rock consisting of mineral grains large enough to be seen without magnification or only low magnification. It is a foliated rock that does not possess good cleavage, but tends to break apart along somewhat irregular and subparallel surfaces. *(B)* Under magnification the subparallel alignment of flat and elongate mineral grains in a schist can be seen.

Gneiss

Coarse-textured rocks with interlocking mineral grains that are clearly segregated into laminar zones are gneisses. Most gneisses contain significant proportions of feldspar, some mafic minerals and perhaps quartz; however, the composition is not restricted to these minerals. Look at the bands in Figure 8-4 that indicate alternating zones of light- and dark-colored minerals. This is an important feature of a gneiss. The zones can be nearly planar or highly contorted.

We think that coarse-grained igneous rock is the source for most gneisses. But we know that they can be produced by metamorphism of some kinds of sedimentary rock. As with the other kinds of metamorphic rock, the term **gneiss** applies only to the features of the rock after metamorphism. It conveys no information about the source materials.

Some other kinds of metamorphic rock are sufficiently abundant for us to classify them in one or another special group.

Marble

Rocks consisting mostly of recrystallized grains of calcite ($CaCO_3$) and perhaps dolomite ($CaMg(CO_3)_2$) are classified as marbles. Limestone and dolomite are by far the most abundant source rocks. Look at the illustrations of marble in Figure 8-5. The microphotograph shows that the calcite grains are mostly interlocking. Impurities can become segregated in marble to give a banded pattern that has esthetic value in building stone. Marbles do not necessarily possess foliation, and some species are nonfoliated metamorphic rock. The important distinguishing feature is compositional rather than textural.

Metaquartzite

Rocks consisting primarily of recrystallized silica are called metaquartzites. The most abundant source material is quartz sandstone. After metamorphism the individual sand grains can become obliterated if they are completely consumed in the secondary recrystallization of silica. Amorphous silica cement is also incorporated into secondary metamorphic crystals. Depending upon the extent to which metamorphism has progressed, some proportion of quartz grains which can be seen in a specimen will be of metamorphic origin.

Impure quartzites contain other silicate minerals. The difference between an impure

Figure 8-4 Gneiss is a coarse-grained metamorphic rock typically displaying a banded appearance produced by the segregation of light and dark colored minerals into alternating parallel zones. This feature can be seen in outcrops (A) and in specimens (B). Under magnification (C) the interlocking coarse-grained texture of gneiss can be seen.

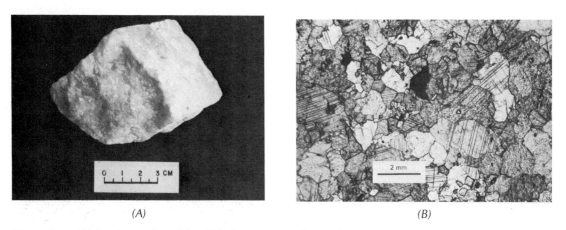

Figure 8-5 (A) Specimen of marble which is a metamorphic rock consisting mostly of interlocking grains of calcite shown in (B) the photomicrograph.

Figure 8-6 *(A)* Specimen of Metaquartzite which is a metamorphic rock consisting mostly of interlocking quartz grains shown in *(B)* the photomicrograph.

quartzite and a gneiss or schist is gradational. Look at the specimens of quartzite illustrated in Figure 8-6. Notice that the texture can be foliated or nonfoliated. Like marble, the term **quartzite** refers to composition rather than texture.

Hornfels

Fine-grained, nonfoliated contact metamorphic rock with more or less equidimensional mineral grains that show no preferred alignment is hornfels. These rocks possess some distinctive metamorphic minerals that we will discuss later in the chapter. Large crystals are sometimes found embedded in a fine-grained matrix of hornfels.

During metamorphism we can expect that minerals which were stable in the source rock will react with other minerals and fluids, or they will change by polymorphic phase transformations. New metamorphic minerals grow within the matrix of the source rock, and the surrounding grains may become deformed. Notice how the growth of a pod of metamorphic minerals seen in Figure 8-7 has deformed the laminations surrounding it. During metamorphism of the rock this pod was rotated, and it drew apart from the matrix.

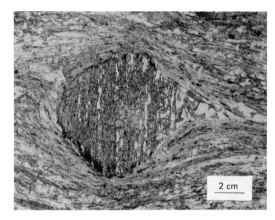

Figure 8-7 Evidence of recrystallization of minerals associated with metamorphism can be seen in this pod of minerals that displays an alignment of grains markedly different from the grains in the surrounding groundmass of rock. Growth of this pod of metamorphic minerals has deformed nearby laminations in the surrounding rock. (Photo by David Hewitt, Virginia Polytechnic Institute and State University.)

Metamorphic minerals of intermediate size have grown in the openings. These are examples of secondary recrystallization features that are found in metamorphic rocks.

Individual grains and clusters of minerals produced by secondary recrystallization give

Table 8-1
Some Important Metamorphic Minerals

Zeolite Minerals	Greenschist Minerals	Amphibolite Minerals
Kaolinite	Epidote	Sillimanite
Montmorillonite	Chlorite	Microcline
Laumonite	Tremolite	Muscovite
Prehnite	Calcite	Cordierite
Saponite	Actinolite	Biotite
Calcite	Andalusite	Almandine
Chlorite	Biotite	Anthophyllite
	Muscovite	Diopside
Blue Schist Minerals		Tremolite
Jadeite	Microcline	Grossularite
Gaulcophane	Staurolite	Calcite
		Anorthite
	Granulite Minerals	
	Kyanite	**Hornfels Minerals**
	Sillimanite	Andalusite
	Hypersthene	Sillimanite
	Orthoclase	Cordierite
	Diopside	Orthoclase
	Enstatite	Biotite
	Anorthite	Hypersthene
		Anorthite
		Grossularite
		Diopside

us clues about the origin of a metamorphic rock. Because these minerals have grown as stable phases during an episode of metamorphism, they can tell us the probable temperature and pressure conditions. Some important metamorphic minerals are listed in Table 8-1. We can use the presence of one or another of these minerals as well as texture to give a rock a descriptive name such as **staurolite schist** or **pyritic slate** or **garnet gneiss**. We can distinguish two schists by the presence of andalusite in one and sillimanite in the other. Even though these minerals comprise only minor proportions of rocks that are otherwise similar, they reveal important differences in the conditions that prevailed during metamorphism.

We learned in Chapters 6 and 7 about the stable regions of temperature and pressure for certain minerals that are important in metamorphism. We will discuss the stability of some other minerals in Table 8-1 later in this chapter. These minerals are useful for classifying rocks according to the concept of **metamorphic grade.** High-grade metamorphic rocks develop in conditions of particularly intense stress or high temperature, whereas low-grade rocks are produced in more subdued environments.

CONTACT METAMORPHISM

You can see the effects of metamorphism in rocks next to igneous intrusions. Here a contact metamorphic zone, called an **aureole,** has formed where the host rock was heated by the nearby magma. You will find metamorphic minerals in the aureole that are not found in host rocks farther from the igneous mass. We can be quite certain that the temperature was highest at the contact and diminished with distance from the igneous intrusion. From our knowledge of igneous melting processes we can estimate how high the temperature became in different parts of the aureole. The minerals found there tell us what kinds of recrystallization processes were active at these temperatures. Let us look at some specific examples of contact metamorphism that display several important features.

A mass of granodiorite exposed in central Maine has been named the Onawa Pluton. This intrusion and the surrounding contact metamorphic aureole are illustrated on the geologic map in Figure 8-8. Notice that the aureole consists of three zones that can be distinguished by differences in texture and minerals formed by secondary recrystallization. The innermost high-grade hornfels zone is characterized by sillimanite, tourmaline, and cordierite. This changes gradationally to a lower-grade hornfels zone that is more enriched in andalusite and biotite. The outermost part of the aureole consists mostly of an

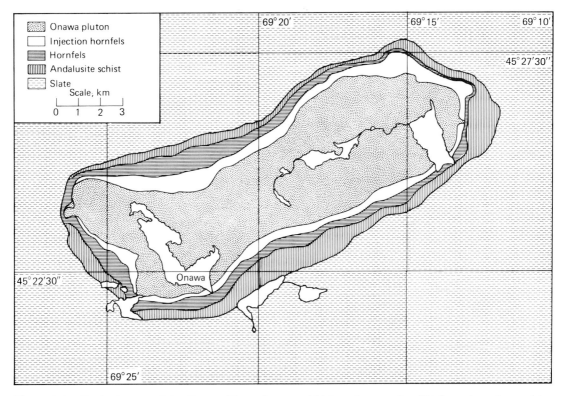

Figure 8-8 Geologic map of the Onawa Pluton in central Maine. An aureole of high-grade metamorphic rocks is superposed upon regionally metamorphosed slates as a result of the intrusion of igneous rock. (After S. S. Philbrick, *American Journal of Science,* page 2, Vol. XXXI, January 1936.)

andalusite-chlorite schist that changes gradationally into the surrounding slate.

The changes in texture and mineralogy can be tentatively related to temperature conditions within the aureole. Prior to metamorphism the rock had been of fine-grained texture. The hornfels zones indicate that relatively high temperature persisted near the contact for a sufficient length of time to allow almost complete recrystallization of this host rock. The original minerals were entirely replaced by new metamorphic minerals. Assuming that temperature decreased with distance from the magma, we can associate the assemblage of sillimanite, tourmaline, and cordierite with relatively high temperature, andalusite and biotite with intermediate temperature, and chlorite and andalusite with yet lower temperature conditions. Recall the relationship between andalusite and sillimanite (Figure 6-29). This will help to understand how the difference in composition of the two hornfels zones might be related partly to a polymorphic phase change between these two Al_2SiO_5 minerals.

The Marysville Stock is a granodioritic intrusion located in Montana. This igneous mass and its contact metamorphic aureole are illustrated on the geologic map in Figure 8-9. The granodiorite intrudes a sedimentary sequence consisting of shale and dolomite. Within the aureole the shale has been metamorphosed to hornfels enriched in cordierite and the micas. In contrast, the dolomite near-

288 The Rocks of the Earth

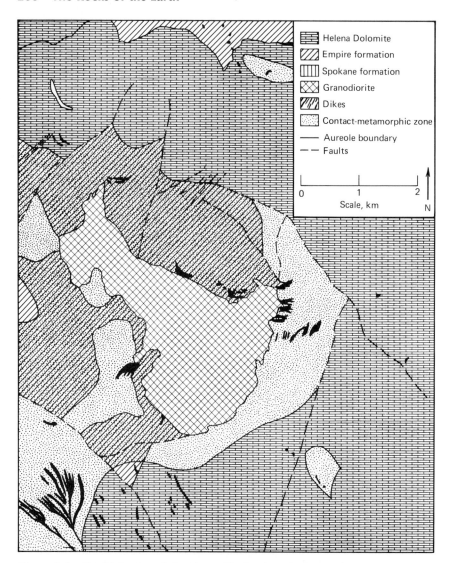

Figure 8-9 Geologic map of the Marysville Stock and adjacent metamorphic and sedimentary rocks in Montana. A contact metamorphic aureole of mixed mineralogy is seen adjacent to the pluton. Differences in mineralogy relate to initial dissimilarities in source rocks as well as diminishing temperature at farther distances from the igneous contact. (After Joseph Barrell, *Professional Paper No. 57*, U.S. Geological Survey, 1907.)

est the igneous contact has been altered to hornfels characterized by tremolite and diopside. Farther from the contact the dolomite has been metamorphosed to marble that contains growths of tremolite and diopside.

We can see by examining the aureole surrounding the Marysville Stock that fine-grained hornfels was produced from both dolomite and shale in the innermost zone. However, different metamorphic minerals are derived from these two distinctly different source rocks. Cordierite and the micas are silicates of Fe, Mg, and Al, which are relatively abundant elements in shale. Tremolite and diopside are comparatively more enriched in Ca and Mg; These elements are important constituents in dolomite. Both of these source rocks are metamorphosed in approximately the same temperature conditions. Clearly, the mineral assemblages formed during metamorphism are related to the original composition of the source rock as well as the temperature conditions. Thus, depending upon the elements available, the growth of either a tremolite-diopside assemblage or a cordierite-mica assemblage is indicative of relatively high temperature metamorphic conditions.

Look at the width of these contact aureoles (Figures 8-8 and 8-9). They extend as far as 2 km from the igneous contact. Contrast these examples with the contact metamorphism bordering the igneous sills illustrated in Figure 7-37. Here the contact zones penetrate less than one meter into the surrounding sedimentary rocks. We can relate the differences in the extent of contact metamorphism to the length of time of the metamorphic episode and the stability of the minerals occurring in rocks that border the intrusion. Thin tabular intrusive masses cool rapidly so that the elevated temperatures needed for metamorphism probably persist for only a few hundred years. Larger intrusives such as stocks take much longer to cool and crystallize. Near these intrusions temperatures between 300° and 600° C may continue several tens of thousands of years allowing time for the growth of a metamorphic aureole.

METAMORPHIC FACIES

Metamorphic minerals occur together in different combinations that tell us about pressure and temperature during metamorphism. These mineral assemblages are found in regional metamorphic rocks as well as in contact aureoles. We can use them to distinguish different groups of rocks that are called **metamorphic facies.** Rocks that are produced under similar conditions of pressure and temperature make up a particular facies. This idea originated with Professor P. E. Eskola (1883–1964), a Finnish geologist who was a pioneer in the study of metamorphic processes. The efforts of Eskola and other petrologists has led to the recognition of seven important metamorphic facies: (1) Zeolites, (2) Greenschists, (3) Blueschists, (4) Amphibolites, (5) Granulites, (6) Eclogites, and (7) Pyroxene-hornfels. The pressure-temperature environments represented by these different facies are shown in Figure 8-12.

What features of a metamorphic rock specimen tell us the facies in which it belongs? This is learned by identifying the combination of metamorphic minerals present in the specimen. Eskola defined a metamorphic facies in the following statement:*

A certain metamorphic facies comprises all rocks exhibiting a unique and characteristic correlation between chemical and mineralogical composition, in such a way that rocks of a given chemical composition have always the same mineralogical composition, and differences in chemical composition from rock to rock are reflected in systematic differences of their mineralogical composition.

Relationships between chemical composition and some of the important mineralogical compositions indicative of different metamorphic facies are summarized in Figure 8-10. The following analysis of these diagrams should give a clearer understanding of Eskola's definition of a metamorphic facies.

We can display on triangular composition diagrams (Figure 8-10) the mineral combinations that represent different metamorphic fa-

---

*T. F. W. Barth, C. W. Correns, P. Eskola. *Die Entstehung der Gesteine*. Ein Lehrbuchder Petrogenese. Springer-Verlag, 1970, p. 339.

290 The Rocks of the Earth

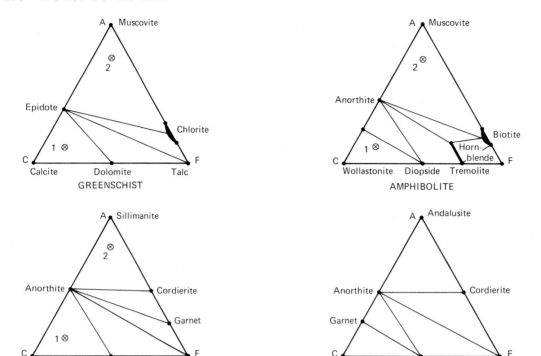

Figure 8-10 Triangular composition diagrams representing some mineral associations found in different metamorphic facies. The following symbols represent combinations of oxide components:

$$A = Al_2O_3 + Fe_2O_3, \quad C = CaO, \quad F = MgO + FeO$$

and for a particular rock $A + C + F = 100$. The diagrams indicate the minerals associated with different proportions of these oxide component combinations for different facies. (After P. Escola, The Mineral Facies of Rock, *Norsk Geologisk Tidsskrift,* page 167, Bind VI, Hefte 1-2, 1920.)

cies. These diagrams show the minerals that can be produced from the five oxide components Al_2O_3, Fe_2O_3, MgO, FeO, and CaO. These are by no means the only components involved in metamorphic processes, but they are sufficient to illustrate the important relationships between chemical composition and the minerals found in metamorphic rocks. How do we interpret these diagrams? Look first at the greenschist triangle. Point No. 1 on this diagram represents a rock with the chemical composition of 15 percent Al_2O_3 + Fe_2O_3, 15 percent MgO + FeO, and 70 per-

cent CaO. The point plots inside a smaller triangle that has calcite, epidote, and dolomite at its corners. This tells us that these three minerals are produced when the rock is metamorphosed in a greenschist facies environment. A different source rock is represented by Point No. 2. Its composition is 70 percent Al_2O_3 + Fe_2O_3, 15 percent MgO + FeO, and 15 percent CaO. Notice that this point is inside the smaller triangle with muscovite, chlorite, and epidote at its corners. These are the minerals that develop when this particular chemical composition becomes a greenschist facies rock as a result of metamorphism. We can see that the assemblage of calcite-epidote-dolomite and the assemblage of epidote-muscovite-chlorite both indicate greenschist facies rock. Whether one or the other combination forms depends upon the original chemical composition of the rock. Other compositions might plot within other smaller triangles indicating still different mineral combinations that also indicate a greenschist.

Now look at the triangle for the amphibolite facies (Figure 8-10). If the two examples represented by points No. 1 and No. 2 on the greenschist diagram are also plotted on the amphibolite diagram, they fall within the smaller triangles defined by the garnet-diopside-wollastonite assemblage and the anorthite-muscovite-biotite assemblage. The occurrence of one or another of these combinations of minerals in a specimen tells us that it is an amphibolite facies rock. Similarly, these compositions in granulite facies would contain the anorthite-diopside-wollastonite combination or the cordierite-anorthite-sillimanite combination according to points plotted on that triangle in Figure 8-10.

We can understand from these examples how rocks with the same chemical composition can represent completely different metamorphic facies. Now consider a chemical reaction that illustrates how different facies are produced by changes in pressure and temperature. First we should look more closely at the rocks surrounding the Onawa Pluton (Figure 8-8). This pluton intruded zeolite facies slate that was produced by earlier regional metamorphism. The slate contains chlorite, muscovite, and quartz. How do these minerals react to changes in pressure and temperature? We have learned the answer from laboratory experiments:

$$(MgFe)_5Al(AlSi_3O_{10})(OH)_8 + KAl_2(AlSi_3O_{12}) + 2SiO_2 \rightarrow$$
$$\text{chlorite} \quad\quad \text{muscovite} \quad\quad \text{quartz}$$
$$(Mg, Fe)Al_4Si_5O_{18} + K(Mg, Fe)_3(AlSi_3O_{10})(OH)_2$$
$$\text{cordierite} \quad\quad \text{biotite}$$
$$+ KAlSi_3O_8 + Al_2SiO_5 + H_2O.$$
$$\text{K-feldspar}$$

You can see from this formula how the cordierite, sillimanite, and andalusite in the aureole could be produced by contact metamorphism of the slate. Recall that sillimanite and andalusite are polymorphs of Al_2SiO_5.

By testing mixtures of these minerals at different pressures and temperatures we have determined the conditions that are required for the chlorite-muscovite-quartz reaction to begin. The results are presented in a phase diagram (Figure 8-11) together with the stable ranges for the Al_2SiO_5 polymorphs that we learned about in Chapter 6. What can this composite phase diagram tell us about metamorphism caused by intrusion of the Onawa Pluton? The temperature must have been raised above 500° C to produce cordierite and Al_2SiO_5 from the minerals in the slate. The outer zone of the aureole contains andalusite. Therefore, the pressure here must have been lower than 4.6 kb, which is the pressure where the two lines cross on the phase diagram. At higher pressure the Al_2SiO_5 would have crystallized as kyanite. Even at very low pressure the temperature here could not have been higher than 850°C; otherwise sillimanite would have crystallized. This polymorph is found with cordierite in the innermost zone of the aureole, so there the temperature must have been higher than 625° C. Notice on the phase diagram that this is the lowest temperature at which sillimanite can crystallize.

We can tell from the cordierite and sillimanite that hornfels facies rock makes up the innermost zone of the aureole. This changes gradationally to greenschist facies rock in the outer zone where chlorite and andalusite are found. These mineral combinations do not tell us the exact pressure and temperature that produced the different zones of the aureole. But we can use the composite phase diagram (Figure 8-11) to estimate the ranges of pressure and temperature required to produce the facies that can be identified by these minerals.

The chlorite-muscovite-quartz reaction does not occur until the temperature rises above 500° C. How can we learn about

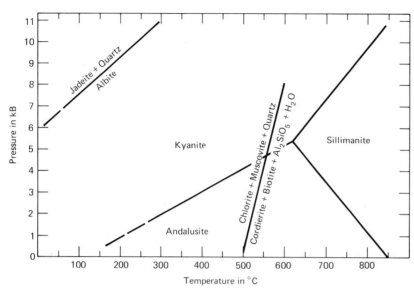

Figure 8-11 Phase diagram showing stable regions of pressure and temperature for the Al_2SiO_5 polymorphs; the reactions between muscovite, orthoclase, and quartz; and the reactions between albite, jadeite, and quartz. (Compiled from information in Kelley and others, *Report of Investigations No. 4955*, U.S. Bureau of Mines, 1953; A. Hirschberg and H. G. F. Winkler, *Contributions to Mineralogy and Petrology*, pages 31 and 36, Vol. 18, 1968; and S. W. Richardson, M. C. Gilbert, and P. M. Bell, *Americal Journal of Science*, page 266, March 1969.)

metamorphism that occurs at lower temperatures? Obviously we must know about other chemical reactions. One of these involves the decomposition of the plagioclase mineral albite:

$$NaAlSi_3O_8 \rightarrow NaAl(SiO_3)_6 + SiO_2$$
$$\text{albite} \quad \text{jadeite} \quad \text{quartz}$$

Experiments have been done to determine the pressure and temperature conditions at which this reaction occurs. Results are included on the composite phase diagram (Figure 8-11). Here we can see that jadeite and quartz will be produced in a low temperature, but in a relatively high pressure metamorphic environment.

We can tell from the number of minerals in Table 8-1 that many chemical reactions and polymorphic transformations play a role in the processes of metamorphism. Each one gives some information about pressure and temperature conditions. By combining phase diagrams for all of these reactions we can recognize the regions of pressure and temperature that correspond to the different metamorphic facies. These regions are illustrated in Figure 8-12. We see from this diagram that the zeolites represent low-grade metamorphic rocks. They formed in the mildest conditions of pressure and temperature in which metamorphism can occur. The greenschists and amphibolites are metamorphic rocks of intermediate grade. Blue schists are high-grade metamorphic rocks associated with high pressure. Hornfels facies rocks are produced at high temperature. A granulite represents a high pressure and temperature metamorphic

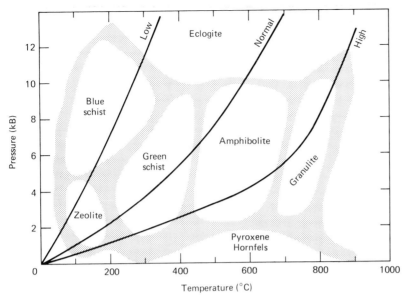

Figure 8-12 A tentative subdivision of the pressure-temperature field into regions corresponding to important metamorphic facies. Superposed curves indicate average and abnormal temperature conditions in the earth. (After F. J. Turner, *Metamorphic Petrology*, page 366, McGraw-Hill Book Co., 1968.)

environment. Depth units are also given on the vertical axis of the diagram in Figure 8-12 to indicate the zones in the earth where pressure corresponding to different metamorphic facies would exist. Also superimposed on this diagram are curves representing average and abnormal temperature conditions in the earth. Figure 8-12 summarizes the pressure and temperature conditions required to produce different metamorphic facies, and indicates depths at which these conditions may exist. Now we can look at how these facies are arranged in regional metamorphic settings.

REGIONAL METAMORPHISM

We find foliated regional metamorphic rocks on the continental platforms. They are most abundant in regions called **continental shields** and in linear **metamorphic belts**. Figure 8-13 shows where these shields and metamorphic belts are located.

Complexly deformed igneous and metamorphic rocks that evolved during Precambrian time make up the continental shields. Here we find evidence of successive episodes of regional metamorphism related to the recurring growth of ancient mountain systems. The shields have undergone no significant tectonic deformation since Precambrian time. The ancient mountain ranges have been destroyed by erosion that has now produced low-lying landscapes of subdued relief. These broad shield areas remain as the stable cores of the continents.

Belts of regionally metamorphosed rocks are hundreds of kilometers long and several tens of kilometers wide. These metamorphic belts are younger than the continental

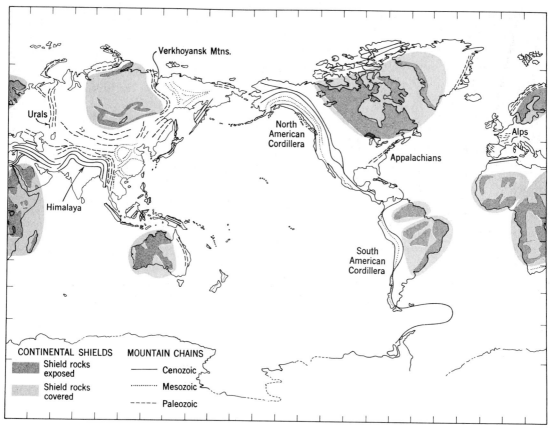

Figure 8-13 Continental shields primarily of Precambrian age rocks and mountain systems that have evolved with associated metamorphic rocks since the end of Precambrian time. (From C. R. Longwell, R. F. Flint, and J. E. Sanders, *Physical Geology,* page 546, John Wiley & Sons, Inc., 1969.)

shields. They have been produced by mountain building processes during the Paleozoic, Mesozoic, and Cenozoic eras.

The occurrence of igneous plutons is not uncommon in metamorphic belts and Precambrian shields. However, the metamorphism pervades the entire area and is not restricted to local contact aureoles. The regionally metamorphosed rock may show evidence of intense mechanical deformation. Look at the successively refolded laminations pictured in Figure 8-14. This complex mechanical deformation is a combination of plastic flow and brittle fracture caused by directed stress. We would have difficulty trying to correlate individual folds and fractures with specific events during an episode of metamorphism. We can learn more by looking at variations in mineralogy that are usually less complicated. The different mineral combinations can tell us of changes in conditions of pressure and temperature.

We can see some typical features of regional metamorphism on the eastern piedmont of North America near Roxboro, North Carolina (Figure 8-15). Here ancient sand-

Figure 8-14 Complexly deformed metamorphic rocks in St. Lawrence County, New York. (J. Gilluly, U.S. Geological Survey.)

stone beds were metamorphosed during late Precambrian time. But they did not lose all of their original characteristics. The outcrop patterns of these metasandstone layers show intense folding associated with directed stresses. Lines have been drawn on the geologic map to separate zones characterized by different metamorphic minerals. These lines are called metamorphic **isograds.** They connect points of similar metamorphic grade in the same manner that contour lines connect points of equal elevation on a topographic map. Look at the isograd crossing the northwestern part of the map. It separates a zone containing kyanite from a zone containing sillimanite. Altogether, the isograds in Figure 8-15 reveal a simple pattern of mineralogical variation. Now look at a line *(aa')* drawn perpendicular to the isograds; it points out the direction of the **metamorphic gradient.** What does this mean? A gradient indicates a transition. Here, the line begins at *a* in greenschist facies rocks and ends at *a'* in amphibolite facies rocks. The metamorphic gradient along this line is a transition in the intensity of metamorphic conditions. We shall now look at some larger areas of regional metamorphism that illustrate typical patterns and gradients. In this chapter our primary interest is with mineralogical variations. In Chapter 10 we will learn more about mechanical deformation.

The stable Precambrian core of North America is called the Canadian Shield. It extends through eastern Canada and bordering areas of the United States and Greenland (Figure 8-13). During the vast expanse of Precambrian time this area was subject to numerous episodes of metamorphism, in some places overlapping, and otherwise widely separated in time and space. Here we find some outcrops of metamorphic rock that merge gradationally with igneous plutons.

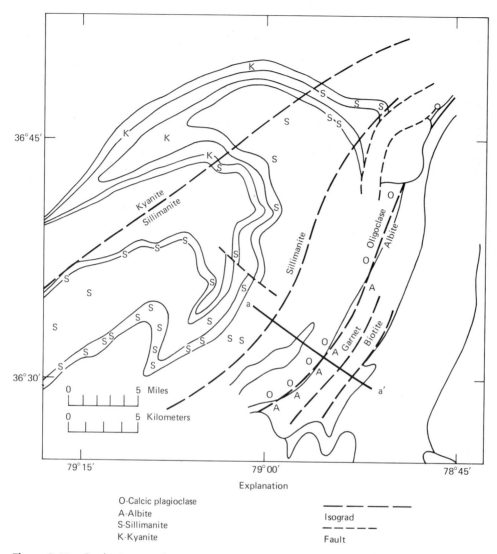

Figure 8-15 Geologic map of an area on the eastern piedmont of North America showing outcrop pattern of successively refolded metasedimentary rocks, and isograds that separate the area into zones of characteristic metamorphic minerals. (After O. T. Tobisch and L. Glover III, *Bulletin of the Geological Society of America*, page 2220, August 1971.)

Elsewhere ancient belts of metamorphic rock cut abruptly across one another. Altogether, Canadian Shield rocks occur in very complicated patterns. In regions adjacent to the exposed shield similar metamorphic and igneous rocks are found by drilling beneath a veneer of younger sediment. We say that these buried rocks make up the Precambrian "basement complex." Precambrian assemblages such as these are found only on the continent, and not in the bordering ocean basin. In fact, Precambrian rocks have not been found in ocean basins anywhere on the earth.

We will find less complicated patterns of regional metamorphism in younger metamorphic belts. An example is the series of foliated metamorphic rocks, primarily schists and slates, that outcrops along a belt extending almost 600 km through the South Island of New Zealand. Look at the important facies and subfacies of this belt that are defined by isograds in Figure 8-16. These facies range from low-grade zeolites through chlorite- and epidote-rich greenschists, and into biotite and almandine-rich amphibolites. They occur in parallel zones. Additional isograds are used to subdivide the greenschist facies into zones that indicate more subtle change in metamorphic grade. For example, in the subfacies designated **Chlorite 2** a characteristic mineral assemblage is quartz-albite-chlorite-pumpellyite-sphene. Missing from this assemblage is epidote. This mineral appears in the subfacies **Chlorite 3.** A progressive change in metamorphic grade along the line aa' in Figure 8-16 indicates a metamorphic gradient of increasing intensity toward the northwest. Radioactive isotope age determinations tell us that metamorphism began in middle-Cretaceous time, and continued through much of the Tertiary period. The rocks of this metamorphic belt are mostly of sedimentary origin.

Look again at Figure 8-12 to find out what pressures and temperatures were needed to produce the greenschists and amphibolites found in this metamorphic belt. We can see that pressures must have been higher than 2 kb, which implies that the rocks were metamorphosed at depths of more than 7 km. We have no other way to explain high pressure, but what about the corresponding temperatures? According to Figure 8-12 the temperature must be higher than 450° C to produce amphibolites. Under normal conditions this temperature is reached at some depth between 25 and 30 km, where the pressure is higher than 8 kb. Were the amphibolites produced under normal conditions near the base of the crust? Or were they produced nearer to the surface under abnormally high temperatures? Suppose we found andalusite in the amphibolites that was produced from the chlorite, muscovite, and quartz reaction (Figure 8-11). That Al_2SiO_5 polymorph would crystallize if an unusually high temperature of more than 500° C were reached at a depth of less than 15 km. On the other hand, suppose we found kyanite in the amphibolite facies. This mineral would tell us that metamorphism had occurred at greater depth under more normal temperature conditions. The mineral combinations in the New Zealand greenschists and amphibolites tell us that metamorphism took place in the upper part of the crust when temperature was unusually high.

Some particularly interesting metamorphic belts that extend along the Japanese Island Arc system have been described by Professor A. Miyashiro. There are four metamorphic belts (Figure 8-17) that can be grouped into two pairs. These are the Ryoke-Sanbagawa pair which evolved during late Mesozoic time, and the Hida-Sangun pair of late Paleozoic age. In both pairs the individual belts are in fault contact with each other. Looking first at the younger pair, we can see that the Sanbagawa belt, lying on the seaward side, is characterized by jadeite and glaucophane. These minerals indicate high pressure and low temperature metamorphism. In contrast the andalusite and sillimanite are the important minerals in the Ryoke belt; they indicate high temperature and low pressure metamorphism. A similar pattern of mineralogy is found in the older pair. The Sangun belt, located on the Pacific side, contains jadeite and glaucophane, and the adjacent Hida belt contains andalusite and sillimanite. Events that produced the Hida-Sangun pair near the end of the Paleozoic era appear to have been repeated late in the Mesozoic era when the Ryoke-Sanbagawa pair evolved. Professor Miyashiro believes that paired metamorphic

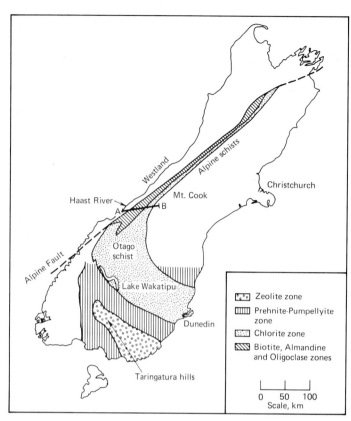
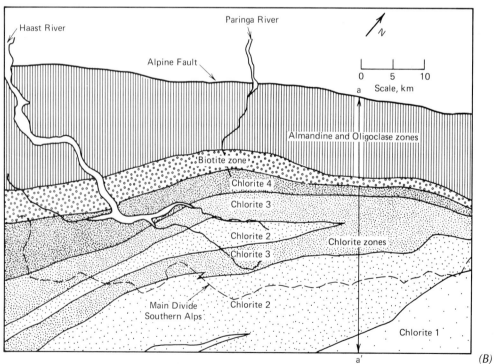

Figure 8-16 (A) Geologic map of the metamorphic belt in the South Island of New Zealand. Isograds define zones of particular facies and subfacies produced by regional metamorphism. (B) Detailed map centered on line AB shows metamorphic intensity that increases northwestward along the profile aa'. (After F. J. Turner, *Metamorphic Petrology*, page 30, McGraw-Hill Book Co., 1968; and Brian Mason, *Bulletin of the American Museum of Natural History*, Vol. 12, Article 4, facing page 218, 1962.)

Metamorphism 299

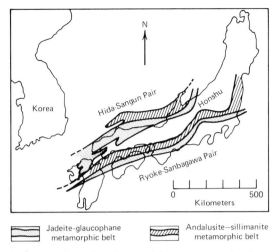

Figure 8-17 Map of part of the Japanese island arc-trench system showing the locations of paired metamorphic belts. (Modified from A. Miyashiro, *Journal of Petrology,* page 288, Vol. 2, Part 3, 1961.)

belts can be recognized in other circum-Pacific regions, but they are not so clearly defined as those in the Japanese Island Arc.

How can high pressure and low temperature metamorphic facies evolve simultaneously in fault contact with low pressure and high temperature facies? Lying to the east of these ancient paired belts is the Japan Trench. Associated with it is a zone of current volcanic activity and seismicity. Could the paired metamorphic belts have been produced by plate tectonic processes similar to those apparently active at the present time? Figure 8-18 shows how a lithospheric plate moving downward beneath a trench is in fault contact with the plate that has an island arc along its edge. Pressure on the downgoing rock increases immediately as it moves deeper. Because this rock is insulated, it remains relatively cool, and this creates the zone in Figure 8-18 where pressure is high

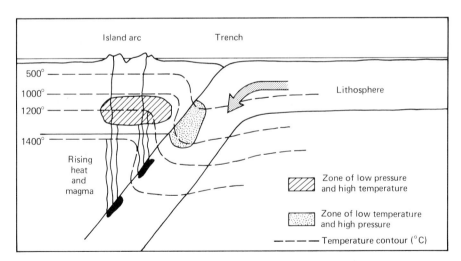

Figure 8-18 Conceptual model showing possible metamorphic zones related to plate tectonic processes. Relatively rapid subduction of lithosphere creates a zone of abnormally low temperature at high pressure. In contrast the adjacent margin of the immobile plate is heated in association with volcanism causing conditions of abnormally high temperature at low pressure. This juxtaposition of two contrasting environments could lead to the development of dissimilar facies adjacent to one another. (Modified from A. Sugimura and S. Uyeda, *Island Arcs—Japan and Its Environs,* page 167, Elsevier Scientific Publishing Co., 1973.)

but temperature is abnormally low. Here low temperature and high pressure metamorphic facies would evolve. Now look at the intensified volcanism in the plate with the island arc on its edge. Here the rising magmas create a region of abnormally high temperature where pressure is relatively low. The andalusite-sillimanite facies could develop under these conditions. In this way we might find simultaneous metamorphism in two completely different environments that are in fault contact. But according to Figure 8-18 the metamorphism occurs at considerable depth. How then could these rocks later become exposed at the surface to display the patterns we seen in the paired metamorphic belts of Japan? We do not know how nature accomplishes this task, but perhaps you may have some ideas about this after studying Chapter 10.

ECONOMIC DEPOSITS IN METAMORPHIC TERRANES

The processes of metamorphism have helped to concentrate many important economic deposits. Some of the valuable minerals in these deposits were formed by metamorphic recrystallization of chemical components initially present in the source rock. Others precipitated from chemically active fluids that migrated through the host rock during metamorphism. Ore concentrations associated with high-grade metamorphic facies have been produced by interdependent igneous and metamorphic processes.

The reaction of a host rock to chemically active fluids plays an important role in forming metamorphic ore deposits. Where do these fluids come from? Some come from nearby intruding magmas; others are produced by partial melting of the host rock. If the host rock is porous and permeable, then these fluids can percolate through it and react with a large volume of the rock. High-grade metamorphic and igneous rocks are usually not very permeable unless many fractures cut through them. But migration of chemically active fluids through impermeable rocks can take place by the process that we call **metasomatism.** This is a replacement process where some minerals are dissolved from a host rock by these fluids and almost simultaneously other minerals precipitate from the fluid to fill the space provided. In this way fluids can migrate during metasomatism by generating a temporary permeability. High fluid pressure can open cracks in the rock, which accelerates the process. We know about iron ore deposits and important disseminated sulfide deposits that have been emplaced by metasomatism.

We have to consider the solubility of a host rock to evaluate the importance of metasomatism. Of all the different igneous and sedimentary rocks, limestone and dolomite are the most frequent hosts for ores derived from contact metamorphism. These rocks are dissolved easily by most contact metamorphic fluids that come from intruding magmas. Exceptions are known, however. In Churchill County, Nevada, contact metamorphism has concentrated magnetite and vanadium ores almost entirely in the andesite layers of a host rock sequence. But this sequence also includes a variety of sandstones, shales, carbonates, and other volcanics. Two important factors have to be considered. One is the solubility of the host rock to whatever particular fluid is involved in the metasomatism. The other is the solubility of the ore in fluids which are at chemical equilibrium with the host rock.

Studies of ore concentrations in regional metamorphic rocks show that most of the important ore-forming constituents were present prior to metamorphism. The effect of regional metamorphism has been to decompose the original minerals in the source rock to produce ore minerals with the same chemical components. For example, we can decompose the carbonate mineral **siderite** ($FeCO_3$)

to produce hematite (Fe_2O_3). Metamorphic processes like this are of interest to metallurgists who are concerned with methods of extracting metals from ores. They have developed efficient methods for producing iron from hematite, but they cannot economically extract it from the abundant silicate minerals. Insofar as regional metamorphism alters the mineralogy to easily processed compounds, it plays an important part in forming an ore.

Marble is an important nonmetalliferous deposit produced by regional metamorphism. Marbles consist mostly of interlocking calcite crystals. Because of its uniform composition, marble can be easily polished to a high luster. Other rocks are more difficult to polish evenly because of variations in the hardness of the constituent minerals.

Accessory minerals and impurities add esthetic qualities to marble. The most abundant

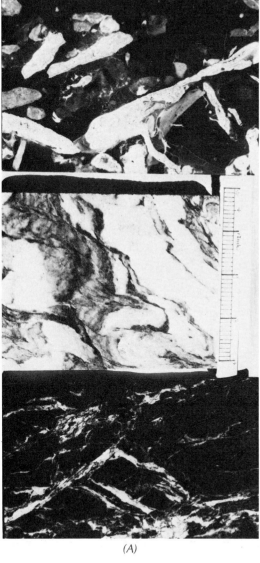

(A)

(B)

Figure 8-19 *(A)* Some varieties of ornamental marble. Patterns are the result of impurities segregated by chemical differentiation and mechanical deformation during metamorphism. Scale in actual size. *(B)* Marble quarry from which blocks are cut by rock saws into sizes useful for building construction. (T. N. Dale, U.S. Geological Survey.)

302 The Rocks of the Earth

impurities are silicates containing calcium and magnesium. These impurities existed in the original limestone or dolomite, and they too undergo metamorphic transitions when the rock is recrystallized to marble. The presence of diopside produced by metamorphism may impart a green hue to the rock, but green can also come from the growth of amphiboles. Garnets tend to cast a reddish-brown color to the marble. Directed stress in the initial stages of metamorphism can cause fracturing of the carbonate host rock or the early formed marble. Subsequent mineral growths in these fractures then impart interesting and decorative patterns that are desired in ornamental building stone. Some varieties are shown in Figure 8-19.

Other minerals of economic value include garnets, corundum, and magnetite which can be extracted from metamorphic rocks for use as abrasives. Asbestos, talc, and graphite also come chiefly from metamorphic sources. Although the demand for slate has diminished in recent decades, it is another product of regional metamorphism which has commercial value. In general, the economic features of metamorphic rocks are not considered to be as important as those more commonly associated with igneous and sedimentary rocks. Because metamorphic processes often occur interdependently with igneous or sedimentation processes, however, their effects are imprinted upon a still wider variety of economic materials.

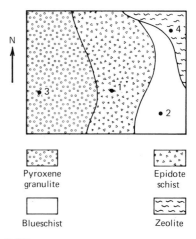

Figure 8-20

STUDY EXERCISES

1. Determine the metamorphic facies represented by the rocks in the innermost zone of the aureole surrounding the Marysville Stock in Montana. Was the rock containing cordierite produced from a source rock that possessed a small or a large proportion of K_2O? Explain!

2. To produce a gneiss containing kyanite by metamorphism of granite:
   (a) what two minerals ordinarily found in granite can react to produce the kyanite?
   (b) the environment of metamorphism should have relatively high temperature and relatively low, intermediate, or high pressure?

3. Four points are shown on the map (Figure 8-20) that illustrates the distribution of four kinds of metamorphic rock.
   (a) The rock at which point indicates the highest temperature but a relatively low pressure environment?
   (b) The rock at which point indicates the lowest pressure and temperature conditions?
   (c) The rock at which point was produced by metamorphism deepest in the earth?

4. The minerals produced in the downward moving part of a lithospheric plate near a submarine trench indicate conditions of: (a) high P and low T; (b) low P and low T; (c) low P and high T; (d) high P and high T. Explain!

SELECTED READINGS

Hyndman, Donald W., *Petrology of Igneous and Metamorphic Rocks,* McGraw-Hill Book Co., New York, 1972.

Miyashiro, A., Metamorphism and Related Magmatism in Plate Tectonics, *American Journal of Science,* vol. 272, pp. 629-656, 1972.

Turner, F. J., *Metamorphic Petrology,* McGraw-Hill Book Co., New York, 1968.

Winkler, H. G. F., *Petrogenesis of Metamorphic Rocks,* 3rd ed., Springer-Verlag, New York, 1974.

9

SEDIMENTARY ROCKS

Common Sedimentary Rocks
Sedimentary Structures
  Parallel Lamination
  Graded Bedding
  Cross Bedding
  Ripple Bedding
  Massive Bedding
  Bedding Surface Structures
  Post Depositional Sedimentary Structures
  Fossils
Origins of Sedimentary Rocks
Shale and Mudstone
Sandstone
Conglomerate and Breccia
Carbonates
Evaporites
Phosphorites
Coal
Sedimentary Facies Relationships
Sedimentary Differentiation
Sedimentary Deposits of Industrial Utility
Perspectives

There are some truly spectacular exposures of sedimentary rock. In places like the Grand Canyon we can look at layer upon layer of these rocks along towering cliffs (Figure 9-1). Scenes like this are unusual, but exposures of sedimentary rock are not. These rocks make up about 75 percent of the earth's bedrock surface. Most of them are made of clay particles, grains of sand, fragments of shells, and pebbles of various sizes. These bits of sediment are produced in the "great geologic cycle" (Figure 7-1) by mechanical disintegration and chemical decomposition of older bedrock. They are moved by flowing water, ice, and wind to places where they build up layers of sediment. As this detritus continues to accumulate, the weight on a layer increases, and the grains are more firmly compacted. Minerals that precipitate from water seeping through the layer cement the bits of sediment together. Gradually a solid layer of sedimentary rock is produced.

The economic importance of many sedimentary rocks has encouraged us to learn as much as possible about them. Almost all of our fossil fuels including coal, petroleum, and natural gas comes from these rocks. Raw materials for concrete and building stone come from sedimentary rocks, and also the most important ores of iron and aluminum. These are only a few of the useful commodities that are produced from sedimentary rocks.

We can see features in sedimentary rock layers that remind us of the rippled surface of a tidal flat, shells embedded in a beach, or

Sedimentary Rocks 305

Figure 9-1 Horizontal sedimentary rock layers exposed in the Grand Canyon in northern Arizona. (N. W. Carkhuff, U.S. Geological Survey.)

the stony bed of a stream. Features like these tell us that the rock was produced close to the earth's surface where pressure and temperature are low and water is abundant. We can actually watch many of the important steps that go into making it. Altogether, sedimentary rocks make up only about 5 percent of the earth's crust, but they are concentrated at the surface where much can be directly observed. Therefore, we have much more specific information about sedimentary rocks than we have concerning igneous and metamorphic rocks. In fact, the information about sedimentary rocks is almost overwhelming, and certainly too much to be properly introduced in a single chapter. Thus this chapter will deal mainly with the properties of sedimentary rocks and the places where these rocks are produced. Here we will only begin to discuss the creation of sediment by mechanical and chemical weathering, and the transport and deposition of this sediment by water, ice, and wind. These topics will be discussed more fully in later chapters.

COMMON SEDIMENTARY ROCKS

Nature uses both mechanical and chemical processes to manufacture sedimentary rocks. Fragments of sediment are transported and deposited by mechanical processes. The role of chemical processes is to dissolve some minerals and to cause precipitation of others from water solutions. We can usually tell from the rock texture whether its main features were produced mechanically or chemically.

Features of what is called a **clastic** texture are made by mechanical processes. In a rock possessing a clastic texture we usually can discern the original fragments of sedimentary detritus from the cementing material that binds them together. Certain features of these fragments are commonly used for describing the rock texture. Most geologists look at four features in particular. These are: (1) particle size, (2) sorting, (3) particle shape, and (4) the roundness of a particle. The size of a particle can be described according to the **Wentworth Scale,** (Table 9-1) from an estimate of its average diameter. This scale was proposed by the well-known sedimentary petrologist, C. K. Wentworth, to standardize the description of particle size. It is now widely accepted by geologists. We can see that a fragment with a 1.0 mm average diameter is called a sand grain; another with a 0.01 mm diameter is a grain of clay. In some sedimentary rocks we find sand and clay mixed together, whereas other rocks are made almost completely of either sand or clay. Sorting is the property of texture that we use to describe the extent to which fragments of different sizes are mixed. Well-sorted sediment consists of fragments that are all about the same size. Poorly sorted sediment is a heterogeneous mixture of different sizes.

The shape of a particle can be more or less equidimensional, long and slender, or flat, to mention a few possibilities. One way to

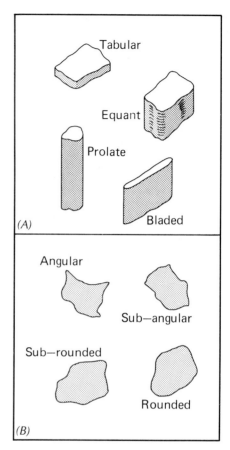

Figure 9-2 (A) The shape of a particle of sediment can be described according to the four standard geometrical forms proposed by Zingg (Schweiz. Min. Petr. Mitt., Bd. 15, pages 39–140, 1935). (B) The roundness of a particle of sediment can be described according to five gradations of irregularity.

Table 9-1
Wentworth Scale of Sedimentary Grain Size

Particle Name	Diameter Range
Cobble	more than 64 mm
Pebble	4 mm to 64 mm
Granule	2 mm to 4 mm
Sand	1/16 mm to 2 mm
Silt	1/256 mm to 1/16 mm
Clay	less than 1/256 mm

From C. K. Wentworth, A scale of grade and class terms for clastic sediments, Journal of Geology, *Vol. 30, pp. 377–392, 1922.*

describe shape objectively is to compare a particle with a set of standard shapes such as the ones outlined in Figure 9-2 A. Regardless of the shape of a particle, it can possess edges and corners. Some are sharp and angular, whereas others are smooth and rounded. We can estimate particle roundness objectively by comparison with a set of reference models such as seen in Figure 9-2 B. Always remember that shape and roundness are two different features of a particle.

Some sedimentary rocks do not have clastic textures; instead they possess a **crystalline** texture of interlocking mineral grains. Chemical processes have played the major role in producing the features of this texture. Other sedimentary rocks have transitional textures with features indicating that mechanical and chemical processes played equally important roles in putting them together. Therefore, we do not try to distinguish different kinds of sedimentary rock simply on the basis of texture. We must also consider their composition and mode of origin. All of these factors are needed to recognize five important groups. These include the **terrigenous clastic** sedimentary rocks, the **carbonate** rocks, **evaporites, coals,** and **phosphorites.**

The most abundant kinds of sedimentary rock are the terrigenous clastic rocks. They are made of fragments which come mainly from the disintegration of older silicate rocks. Therefore, the silicate minerals are the dominant constituents of these sedimentary rocks. The terrigenous clastic rocks are divided into six groups on the basis of texture. These are: (1) **claystones,** (2) **siltstones,** (3) **mudstones,** (4) **sandstones,** (5) **conglomerates,** and (6) **breccias.** Look at the features of these rocks that are illustrated in Figure 9-3. Particle size is the most important property that we use to tell one group from another. Sand-sized particles make up more than one-half of the volume of sandstone. Similarly, the dominant proportion of claystone or siltstone consists of clay- or silt-sized particles. Mudstone is a mixture of clay, silt, and sand grains, none of which accounts for more than one-half of the volume. Claystone, mudstone, and siltstone occurring in layers made up of thin (less than 1 cm) parallel laminations are commonly called **shale.** Fragments of gravel account for the dominant fraction of conglomerate and breccia. We can distinguish these coarse clastic rocks by the roundness of the gravel fragments. Conglomerates have well-rounded gravels, whereas breccias have mostly angular fragments.

Limestone and *dolomite* are the abundant kinds of carbonate sedimentary rock. Grains of calcite ($CaCO_3$) make up more than one-half of the volume of limestone. The mineral dolomite ($CaMg[CO_3]_2$) is the dominant constituent of the rock that is also called dolomite. Most carbonate rocks consist of a mixture of clay-sized particles and larger, easily discernible fragments. These larger fragments can have spherical, or disclike, or highly irregular branching, twiglike shapes. They are shells and other skeletal fragments of former organisms. Look at the typical carbonate rock textures in Figure 9-4. In one example we see the larger fragments suspended in a matrix of clay-sized grains. We call this a **mud-supported** texture. In the second example the larger grains are touching one another, and clay-sized particles fill the spaces between them. This is called a **grain-supported** texture. Both of these are clastic rock textures.

We know that the abundant ions dissolved in sea water include calcium (Ca^+) and bicarbonate (HCO_3^-). These ions react in biochemical processes to produce the calcium carbonate that makes up the shells and other skeletal parts of a myriad of marine organisms. They are broken apart by a combination of mechanical abrasion and chemical decomposition to produce particles of sediment. These particles are mechanically transported and deposited. Later chemical reactions produce dolomite within the deposits of calcium carbonate detritus. We can see that chemical processes as well as mechanical processes play an important part in making carbonate sedimentary rocks. They are transitional rocks that possess both clastic and interlocking crystalline textural features.

The other sedimentary rocks which include evaporites, coal, and phosphorite are not very abundant. But they tell us about interesting geologic processes, and they are economically important. Rock salt ($NaCl$), gypsum ($CaSO_4 \cdot 2H_2O$), and anhydrite ($CaSO_4$) are evaporites. These rocks have crystalline textures, and they appear to have formed by pre-

(A)

(B)

(C)

Figure 9-3 Features of common terrigenous clastic sedimentary rocks (A) sandstone and conglomerate, (B) siltstone and laminated mudstone (shale), (C) breccia.

cipitation from water solutions. The dissolved ions in these solutions were concentrated by evaporation. Coal is a sedimentary rock that consists mostly of carbon; the carbon has come from chemical alteration of plant remains. In beds of phosphorite the important constituent is the oxide component P_2O_5. Animal bone fragments are a source of this compound. We will learn more about the textures and compositions of coal and phosphorite in later sections where the origins of these rocks are discussed.

SEDIMENTARY STRUCTURES

Processes acting near the earth's surface leave their mark on the sedimentary rocks they help to produce. Most of these rocks occur in layers. We can distinguish the individual layers from one another by differences in composition and texture. We can also see features such as ripple marks and lamination patterns in the sedimentary beds. These marks of the environment that produced the rock are called **sedimentary structures.** We can observe similar features being engraved in modern sediments that are now accumulating in such places as river beds, deltas, and beaches. These structures can also be manu-

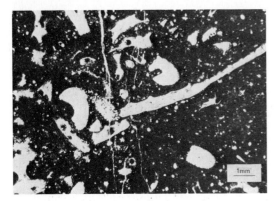 

Figure 9-4 Examples of textures common in carbonate sedimentary rocks which illustrate different proportions of clay-sized particles (carbonate mud) and larger fragments: *(A)* mud-supported textures; *(B)* grain-supported texture.

factured in laboratory experiments. By watching how sedimentary structures form we can understand more clearly how water, wind, and ice help to put together beds of sedimentary rock.

Before describing individual sedimentary structures it is useful to give a general description of the thickness of the layers where they occur. How can a geologist make a general statement about layer thickness in a sequence of sedimentary rocks without giving the thickness of each layer? Look at the layers of sandstone and siltstone in Figure 9-5. Because each of the upper layers is more than 30 cm (one foot) thick, we say that they make up a **thickly bedded** succession of strata. Contrast this with the **thin** to **medium** bedding that can be seen in the lower part of the exposure. If we could look closely at one of the thin beds, we would see that it consists of parallel **laminations,** each less than 1 cm thick. Look again at the siltstone and shale in Figure 9-3. Here the laminations are clearly evident. Parallel laminations together with thin, medium, and thick beds are included in the bedding classification in Table 9-2 that is used by many geologists. Now let us examine some of the sedimentary structures that occur in layers like these.

Parallel Lamination

Many sedimentary layers are marked by parallel laminations. These are sedimentary structures that we can recognize by variations in texture and composition. What processes could produce parallel laminations like the ones seen in Figure 9-3B? Think about the sediment carried by a muddy river. The water must be turbulent to hold this sediment in suspension. Where the river empties into the quiet water of a lake, the sediment settles to the bottom. During a season when the river is most turbulent it carries both clay and silt particles. They settle together on the lake bed where they collect to form a single lamination. The river is less turbulent at other times, and carries only clay particles. These, too, settle to the lake bed, where they form a well-sorted lamination. As the cycle continues a layer of sediment builds up. It consists of clay laminations alternating with mixed silt and clay laminations. Depending on how suddenly the turbulence in the river changes, the variations in texture between laminations may be abrupt or gradational. Look at the interbedded laminations of silt and clay in Figure 9-6. The light-colored silt is a summer deposit and the darker-colored clay is a winter

Sedimentary Rocks 311

Figure 9-5 A sequence of sedimentary rocks in southwestern Utah displays thickly bedded sandstone in the upper part, and alternating medium- and thinly bedded layers in the lower part.

Table 9-2
Scale of Sedimentary Layer Thickness

Thick bedding	Thicker than 30 cm
Medium bedding	10 cm to 30 cm thickness
Thin bedding	1 cm to 10 cm thickness
Laminated	Thinner than 1 cm

Modified from P. L. Ingram, Terminology for the thickness of stratification and parting units in sedimentary rocks, Bulletin of the Geological Society of America, v. 65, pp. 937–938, 1954.

glaciation. This is one of the ways that nature assembles beds of laminated sediment.

Graded Bedding

Some sedimentary layers display gradational changes in particle size. This sedimentary structure is called **graded bedding** (Figure 9-7A). We can find graded bedding in modern river beds and beach deposits. Look at the variation of grain size in beach sand in Figure 9-7B. It can be correlated with seasonal changes in the intensity of waves and water turbulence. Greater turbulence associated

deposit. This is called a **varved** sequence. A single **varve** consists of one pair of silt and clay layers. Varved sequences are found in ancient lake bed deposits transported by water from melting ice during former episodes of

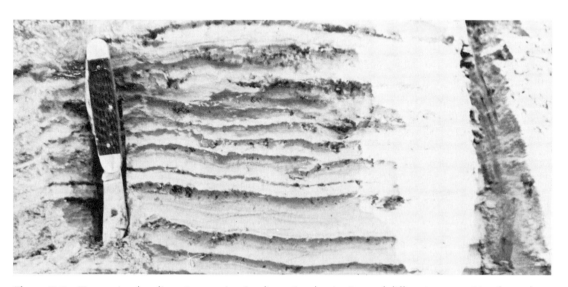

Figure 9-6 Fine-grained sediment occurring in alternating laminations of different composition formed this varve deposit in a former lake bed now exposed in Bristol County, Massachusetts. (J. H. Hartshorn, U.S. Geological Survey.)

312 The Rocks of the Earth

(A)

Figure 9-7 Graded bedding in (A) Paleozoic sandstone exposed near Pembroke, Virginia, and (B) in a modern beach along the Atlantic coast of South Carolina.

(B)

with frequent winter storms concentrates larger grains on the same part of the beach where smaller and smaller grains collect as the sea gradually becomes calmer during spring and summer months. Graded bedding can also be produced by gradual seasonal variations in water level and turbulence in a river.

Nature has another completely different and quick way of putting together a graded bed of sediment on the floor of the ocean. Recall from Chapter 2 that the submerged continental margin of the ocean has an upper shelf, a steeply inclined slope, and the lower surface that is called the continental rise. From time to time sediment breaks loose from parts of the shelf or the slope. (Sometimes it is dislodged by earthquakes.) This sediment becomes churned up in the sea water so that a heavy slurry is formed. It begins to flow down the inclined surface of the sea floor. With increasing speed this slurry spills down the continental slope. This is called a **turbidity current.** Suspended in it are fragments ranging from large cobbles and pebbles to tiny flakes of clay. When it reaches the base of the slope, it spreads out over the more gently inclined continental rise. Here the speed and turbulence of the current quickly diminish and the sediment falls to the ocean floor. Large fragments tend to settle quickly and reach the bottom first. Then smaller and smaller particles fall on the accumulating layer of sediment. When quiet is restored, a graded bed blankets the ocean floor.

Cross Bedding

Some sedimentary beds consist of thin layers or laminations that are inclined relative to the upper and lower surfaces of the bed. These structures are called **cross beds.** Some are very large and others are minute as shown in the samples in Figure 9-8. We can look at newly formed cross bedding in modern delta

Sedimentary Rocks 313

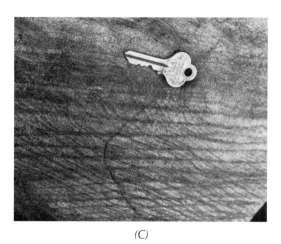

Figure 9-8 Examples of large cross beds in Paleozoic sandstone deposits in southwestern Utah (A and B), and thin cross-bedded laminations in Paleozoic sandstone exposed in western Virginia (C).

deposits, stream bed sediments, and dunes of windblown desert sand (Figure 9-9), and observe the processes that produce them. Clearly, these structures are related to the transport and deposition of particles of sediment suspended in flowing water or wind currents. Conditions for producing cross beds can be simulated in a laboratory by running water or blowing air across containers of sediment (Figure 9-10). We can see that the cross beds are inclined downward in the direction of the water current. Wind-formed cross beds tend to be thicker and more irregular than those formed in water. However, these differences can be difficult to distinguish in ancient sedimentary rocks.

Ripple Bedding

The laminations in some sedimentary layers are not simple parallel or cross-bedded structures. Instead they have a rhythmically undulating shape as you can see in Figure 9-11. These ripple beds were produced by water flowing over a bed of loose sediment. Note that they have an asymmetrical shape. These are typical of the ripples produced by cur-

(A)

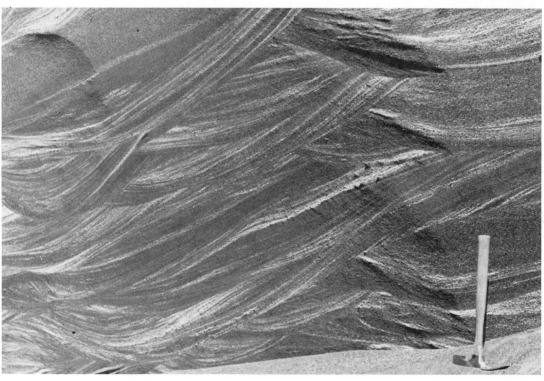

(B)

Figure 9-9 *(opposite)* (A) Wedge cross bedding in unconsolidated sand deposits, and (B) trough cross bedding in sand exposed in Elmore County, Idaho. (H. E. Malde, U.S. Geological Survey.)

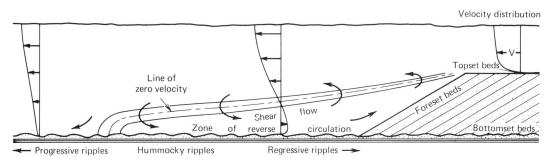

Figure 9-10 Experimental apparatus for generating cross beds consists of a long tank with a ramp on the bottom. As water containing suspended sediment flows over the ramp, turbulence and velocity decrease causing deposition. Progressive deposition eventually produces a cross-bedded layer in which individual cross beds are inclined downward in the direction of current flow. Note the ripples that form on the bottom layer of sediment. (After A. V. Jopling, *Professional Paper 424-D,* page D16, U.S. Geological Survey, 1961.)

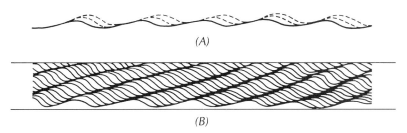

Figure 9-11 (A) Crests of ripples in a succession of laminations are shifted in the direction of the current. (B) Ripple bedding is produced in sediment deposited from flowing water.

rents of water or wind. In Figure 9-12 ripples in a Paleozoic sandstone layer are compared with those being produced at the present time by water and wind. The to-and-fro motion of waves can also produce ripples in loose sediment. These tend to have symmetrical shapes that are different from the asymmetrical shapes formed by currents.

Massive Bedding

Most sedimentary layers display some kind of structure that we can recognize by variations in texture and composition. But there are strata in which these features are absent. Such strata are said to possess **massive** bedding. They have uniform texture and composition.

Bedding Surface Structures

Bedding surfaces separate the individual layers in a sequence of sedimentary rocks. Sedimentary structures formed on bedding sur-

316 The Rocks of the Earth

(A)

(B)

(C)

Figure 9-12 Ripple marks on a bedding surface of Paleozoic sandstone exposed in western Virginia (A), and modern ripples of sand on an exposed tidal flat in eastern Scotland (B), and on desert sand dunes in central Utah (C).

faces either during deposition or at times when it was interrupted are fingerprints of the environment. Ripples are probably the most common structures that are found on bedding surfaces. But they are usually part of the internal structure of the layer (Figure 9-11) and are not restricted only to its surface. Other structures that are formed on bedding surfaces include dessication cracks that develop in drying mud, apparent raindrop impressions, and footprints. Look at the ancient and modern examples that are compared in Figure 9-13. They tell us that, similar to the present, the mud flats of former times were cracked by the baking sun, pocked by the fury of the rain, and trespassed by the animals of that age.

Postdepositional Sedimentary Structures

Following the deposition of sediment the structures are sometimes changed by postdepositional processes. Prior to lithification the layers of sediment form a quasiplastic mass that can be easily deformed by slumping. This may happen when steep embankments formed by water or wind erosion become unstable. The deformed beds and laminations seen in Figure 9-14 are evidence of slumping. We know that deposits of unconsolidated sediment can be inhabited by a variety of burrowing organisms. The activity of these creatures can alter and even obliterate patterns of stratification.

Sedimentary Rocks 317

(A)

Figure 9-13 Bedding surface structures in modern and ancient sediments. (A) Dessication cracks and rain drop impressions in modern mud flats. (B) Ancient dessication cracks and (C) raindrop impressions preserved in mudstone that outcrops in Isle Royale National Park in Michigan. (N. K. Huber, U.S. Geological Survey.)

(B)

(C)

318   The Rocks of the Earth

Figure 9-14 Deformation of sedimentary strata caused by postdepositional slumping prior to lithification is seen in rocks exposed in central Tennessee.

Fossils

The remains of ancient organisms embedded in sedimentary rocks are called fossils. They are distinctive features not found in igneous or metamorphic rocks. Fossils have been discovered in almost every kind of sedimentary rock. They are by no means present in all sedimentary sequences, but they are sufficiently abundant to indicate that at least some organisms could survive in almost all depositional environments near the surface of the earth. In ancient sediments the original organic material has almost always been dissolved out and replaced by inorganic compounds. The fossils may exist as mere impressions or as compositionally distinct components in a rock. Some carbonate beds consist almost entirely of fossil fragments. They are more sparsely distributed throughout a matrix of nonfossiliferous grains in terrigenous clastic rocks. Fossils can be of great importance in interpreting conditions of environment, but this is a subject more properly discussed in a historical geology text and will not be examined here.

The sedimentary structures may seem to be fragile features, too delicate to be preserved in the on-rushing current of a stream or in the wash of a wave. Indeed for the few that remain fixed in the ancient sediments, a vastly greater number were formed and obliterated. But the comparisons of sedimentary structures found in ancient rocks with those observed in modern environments reveal such close similarities that there can be little doubt as to their origin.

ORIGINS OF SEDIMENTARY ROCKS

We know that solid sedimentary rocks are widely distributed on the continental platforms. They are relatively rare in the ocean basins where in most places only a veneer of sediment covers igneous bedrock. Shale, sandstone, and the carbonates are the most abundant kinds of sedimentary rock. Other terrigenous clastic rocks, coal, evaporites, and phosphates are quite rare. We already know that texture and composition can be used to tell these rocks apart. Are these two properties completely independent of one another? Look at the triangular composition diagram in Figure 9-15. It shows proportions of the principal oxide components in shale, sandstone, and the carbonate rocks. Not only is the texture of sandstone different from the texture of shale, but their compositions also

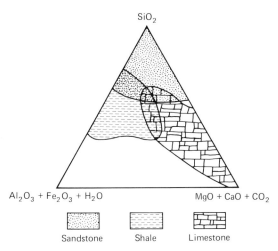

Figure 9-15 Triangular composition diagram showing ranges of chemical composition found for the abundant sedimentary rocks of silicate and carbonate composition. (After Brian Mason, *Principles of Geochemistry*, page 154, 3rd edition, John Wiley & Sons, Inc., 1966.)

tend to be different. We have seen examples showing that mechanical processes separate fragments of different sizes. Are there combinations of chemical and mechanical processes acting to concentrate sediment into deposits that have different compositions as well as different textures? Before trying to answer this question we have to look more carefully at the common sedimentary rocks to learn as much as we can about their origins.

Shale and Mudstone

The most abundant sedimentary rocks are shale and mudstone. These are fine-grained terrigenous clastic rocks. They make up approximately one-half of the bulk of sedimentary rock on the continents. We use the term **shale** to describe laminated claystones, mudstones, and siltstones. Where these rocks are found in strata without distinct laminations they are commonly called **mudstones.** At one time or another almost all exposures of fine-grained terrigenous clastic rock have been loosely referred to as shale. Where we see them interbedded with sandstone and carbonate layers, the beds of shale and mudstone ordinarily appear to be the least resistant to weathering. In Figures 9-3 and 9-5 the more durable sandstone layers jut out, whereas the shale and mudstone parts are recessed. The capacity of a shale specimen to be easily split along surfaces parallel to the stratification is called **fissility.** This property is related to the alignment of platy particles.

We can determine the composition of a shale or mudstone most easily by bulk chemical analysis or x-ray analysis. Recall that both methods tell us the oxide components that are present rather than the minerals. The results of a large number of such analyses are summarized in Figure 9-15. We would expect to find these components in rocks that consist mostly of silicate minerals. But it is difficult to measure the proportions of constituent minerals by conventional methods of modal analysis. The microscopic grains are not easily separated for detailed identification. Nevertheless, some modal determinations are available. They tell us that the micas and clay minerals are important constituents even though they may not necessarily make up the dominant proportion. Be sure that you do not confuse the clay minerals with clay-sized particles. They form a distinct group of silicate minerals which are listed in Table 9-3. Clay mineral grains may or may not be clay-sized according to the Wentworth scale. But they tend more often than not to be of very small size. The average modal composition based upon analysis of more than 400 representative specimens of shale is summarized in Table 9-4.

The colors of shale and mudstone range gradationally from black into shades of gray, blue, purple, red, brown, green and yellow. The darker colors usually indicate increased content of organic material. Black shale is rich in carbonaceous constituents. An impor-

Table 9-3
Typical Chemical Compositions of Some Important Clay Minerals
Weight Percent of Oxide Components

	Kaolinite[a]	Montmorillonite[b]	Illite[c]
SiO_2	46.90	50.20	51.95
Al_2O_3	37.40	16.19	17.81
Fe_2O_3	0.65	4.13	6.17
FeO	0	0	3.87
MgO	0.27	4.12	4.76
CaO	0.29	2.18	0.53
K_2O	0.84	0.16	4.04
Na_2O	0.44	0.17	1.06
TiO_2	0.18	0.20	0
H_2O	12.95	23.15	10.06

[a]C. S. Ross and P. F. Kerr, U.S. Geological Survey Professional Paper 165E, 1931.
[b]C. S. Ross and S. B. Hendricks, U.S. Geological Survey Professional Paper 205B, 1945.
[c]S. B. Hendricks and C. S. Ross, *American Mineralogist*, v. 26, pp. 683–708, 1941.

Table 9-4
Average Modal Mineralogy for 400 Shale and Mudstone Specimens

Mineral	Weight Percent
Quartz	30.8
Feldspar	4.5
Clay minerals	60.9
Iron oxides	less than 0.5
Carbonates	3.6
Other minerals	less than 2.0
Organic matter	1.0

D. B. Shaw and C. E. Weaver, The mineralogical composition of shales, Journal of Sedimentary Petrology, v. 35, no. 1, p. 221, Table 4, 1965.

tant carbonaceous variety is known as **oil shale.** It is discussed in more detail in a later section where properties related to its growing economic importance are examined. Green coloring is often related to the presence of chlorite and serpentine. Oxides of iron can impart a range of hues from purple, red, brown, into grades of yellow. These colors depend partly on the extent of hydration of the iron oxide, that is, the amount of water that has become incorporated into its crystal structure.

The sedimentary structures found in shale and mudstone mainly indicate parallel stratification. Bedding surfaces may contain dessication cracks and impressions from raindrop impact or footprints. All of these properties point to the origin of shale and mudstone from sediment deposition in quiet water environments. Fossils are not uncommon. The mud and clay deposits found beneath bays, lagoons, and lakes have similar structural and compositional features. Also, ponds formed from flooding rivers may be sites of fine-grained sediment deposition.

Mud and clay deposits may contain 70 to 90 percent water by volume at the time of deposition. Some of the water is maintained in the lattice of clay minerals, forming part of the crystalline structure. Also, some water adheres to the grain surfaces because of electrostatic forces. The largest portion, however, exists as free water in the pore space between grains. As deposition of sediment progresses the deeper layers become compressed under the overlying burden, and the free water is squeezed out. Compaction is the first stage of **lithification** which is the process of turning loose sediment into solid rock. The amount of free water in the pore space between grains is reduced to 5 to 10 percent of its original volume at depths of burial of approximately 1000 m. Lithification continues into a second stage of dehydration during which still more free water, as well as water adhering to grain surfaces, is removed. Chemical dehydration commences when temperature increases above 100°C. We could expect to find this temperature at a depth of about 3 km. The compaction and dehydration may continue for many thousands of years or more before the lithification is complete. During this time precipitation of substances from the free water provides some cementing materials that bind the grains more firmly together. The final result is the layer of solid rock that is called shale or mudstone.

Sandstone

We can read scientific reports about sandstone that were written more than two centuries ago. What is it about this kind of rock that continues to stimulate scientific curiosity? Probably many things. Durable sandstone cliffs make spectacular scenery (Figure 9-1). The rock is abundant; about 25 percent of the earth's total mass of sedimentary rock is sandstone. But perhaps we are more curious about the smaller features that are quite easy to see without expensive instruments. The sand grains themselves, and the sedimentary structures in the layers, resemble so closely the features found on a sand bar in a river or along a beach.

The modern geologist calls a terrigenous clastic rock sandstone if sand-sized fragments (Table 9-1) make up more than one-half of its volume. Sand fragments are large enough so that most of their important features can be seen through a hand lens or a low-power microscope. Therefore, geologists have been able to describe the textures and mineralogy of sandstones in great detail. They have discovered three important kinds of sand fragments in these rocks. They are: (1) quartz particles, (2) feldspar particles, and (3) lithic sand particles. A **lithic** sand particle is a fine-grained rock fragment that contains many tiny minerals. For example, a single lithic sand particle of andesite or basalt might contain interlocking pyroxene, amphibole, and feldspar crystals.

Sand fragments do not fit together perfectly like the interlocking grains of an igneous rock. Instead, there is open pore space between them. In sandstone part of this pore space becomes filled by the cement that binds the grains together. Calcium carbonate ($CaCO_3$) and silica (SiO_2) are the most abundant kinds of cement. Ordinarily the cement does not completely fill the space between grains. Some of this space remains open to percolating water solutions, and perhaps petroleum and natural gas. But in some sandstones part of the space between sand grains is filled by silt and clay particles. An important property of sandstone texture is the proportion of fine-grained sediment in the rock. Two groups of sandstones are distinguished by this textural property. These are the **arenites** in which less than 15 percent of the solid volume consists of silt and clay, and the **greywackes** that contain more than 15 percent silt and clay.

Different kinds of arenites and greywackes can be recognized by composition. One abundant variety is **quartz arenite.** Quartz grains make up more than 90 percent of the rock matrix. These grains tend to be equidimensional and rounded (Figure 9-2), and they are usually concentrated in well-sorted beds. Quartz arenites commonly occur as blankets of sandstone that extend over large areas. Some prominent beds are found on the Colorado Plateau in southwestern United States. Here we can see them exposed along spectacular cliffs such as those of Zion Canyon (Figure 9-16). Cross bedding and ripple marks are found in many beds of quartz arenite, and some layers contain fossils. These features tell us that quartz arenites have been made from beach deposits, river bed sediment, and wind

Figure 9-16 *(above and opposite)* Spectacular canyons such as Zion Canyon seen in these two views in southwestern Utah have been carved in thick beds of sandstone.

transported sands that accumulated on deserts. Nature has produced this kind of sandstone during all eras of geologic time.

Most arenites contain some feldspar and lithic grains as well as quartz grains. The rocks that are less than 90 percent quartz are called **feldspathic arenites** if they have a larger proportion of feldspar than lithic constituents. One interesting kind of feldspathic arenite is called **arkose**. It is a coarse-grained and poorly sorted rock that consists mostly of angular grains of quartz and feldspar. Typically the feldspars make up more than 25 percent of the rock. These mineral grains give it a light gray or pink color. Some beds of arkose possess crudely developed cross bedding. Most of these layers cover only small

areas situated near granitic terranes. We can tell from the angular particles that they have not been transported very far, otherwise the particles would be more rounded, such as the ones seen in most quartz arenites. The sediment in a typical arkose consists of weathered fragments of granite or granitic gneiss that have been washed by streams or waves into nearby topographic depressions. Here they accumulate in beds of irregular thickness and width.

Lithic arenite is another abundant type of sandstone. In this rock lithic sand grains make up a larger proportion than feldspar grains. Mixtures of light- and dark-colored grains give some lithic arenites a speckled appearance, so they are called "salt and pepper

sandstones." Lithic arenites are found in many parts of the world. They are exposed in the major mountain belts and they have been discovered by drilling in continental lowland areas. Many lithic arenite beds exist in the vast blanket of Cenozoic age sediment that covers the floor of the Gulf of Mexico and the lands bordering the Gulf Coast. Much of this sand has been carried to the Gulf by the Mississippi River. If we look at the mixtures of sand now accumulating on the beds of large rivers, we will find constituents similar to those in lithic arenites. A recent analysis of sand samples taken from different parts of the Ohio River bed showed an average mixture of 62 percent quartz, 31 percent lithic sand, 6 percent feldspar, and 1 percent other mineral grains. Many geologists believe that most lithic arenites have been produced by lithification of the deposits of large rivers.

Greywackes are troublesome sandstones. They seem to be the subject of more argument than any other kind of sandstone. Why are greywackes such difficult rocks? Most of the difficulty is with the matrix of silt and clay that exists between the sand grains. Where did this matrix come from? What does it tell us about the origin of a greywacke? Two features of the matrix make it difficult to answer these questions. (1) The particles are very small. Geologists cannot easily separate them and describe all of their important features. (2) Many of the particles are neither chemically nor mechanically durable. It is hard to tell whether these particles were mechanically deposited with the sand or whether they were produced by later chemical reactions.

Is there such a thing as a typical greywacke? There are certainly some features that are common to the larger proportion of these rocks. But there are also many exceptions. If we dare to describe a typical greywacke, we would say it has a dark gray color. Quartz would be its dominant constituent, but it would also have significant proportions of feldspar and lithic sediment. Depending on which of these latter two constituents is larger we would call the rock a feldspathic greywacke or a lithic greywacke. Angular sand fragments are abundant, and the sediment is poorly sorted. What about sedimentary structures? Many greywackes possess graded bedding. Cross bedding and ripple structures are less common, but they are certainly important features in some beds. Still other layers have no easily recognized structures.

The largest proportion of greywackes is found with other tectonically deformed rocks exposed in mountain chains. Some are interbedded with pillow lavas and other volcanic debris. But others occur in sequences of marine sediments where volcanic rocks are absent. Almost all greywacke sediment was deposited in marine environments. Certainly many beds were deposited from turbidity currents that were initiated by earthquakes and submarine volcanism. The heterogenous mixture of sediment carried by such a submarine current would contain silt and clay as well as sand grains.

What other processes might produce greywacke? Perhaps the fine-grained sediment was not part of the original deposit. Suppose instead that it was produced by chemical decomposition and mechanical disintegration of feldspar and lithic sand grains. Would it be possible to make a greywacke from a feldspathic or lithic arenite? We know that clay minerals are products of the chemical decomposition of feldspars and other minerals found in lithic sand grains. Sand particles undergoing decomposition would become weakened and then crushed into clay-sized particles by further compaction of a layer of feldspathic arenite or lithic arenite. We do not know if a process such as this has ever produced all of the clay in a greywacke. But we are certain that it has played a significant role in making some of the clay that exists in many greywackes. Under high magnification we can see that the fine-grained texture is the product of chemical as well as mechanical processes. Growths of minerals such as chlorite, sericite, and perhaps calcite developed after the sediment was deposited. Postdepositional processes such as these can alter or destroy sedimentary structures produced during deposition. Perhaps this is the reason that some greywacke beds display no easily recognizable sedimentary structures.

Lithic arenites and quartz arenites are the most abundant kinds of sandstone. Together they make up about 60 percent of the total volume. Greywackes are more abundant than feldspathic sandstones. All of these rocks are manufactured by lithification of deposits that were originally loose sediment. Before this loose sediment becomes completely lithified it undergoes various changes. Grains may be crushed or mechanically deformed by compaction. Chemical reactions involving dissolution and precipitation also change the sediment. Calcium carbonate, silica, and small amounts of iron oxides precipitate from water solutions to cement the sand grains together. Small concentrations of hematite and limonite color the sediment in hues of red, brown, and yellow. All of these processes that change the sediment are collectively called **diagenesis.** In the paragraph above we discussed the possibility that greywacke could be partly a product of the diagenesis of lithic or feldspathic sands. After studying Chapter 12 we will have a better understanding of some of the chemical processes of diagenesis.

Conglomerate and Breccia

Beds of conglomerate and breccia are formed from accumulations of gravel and boulders. Layers tend to be discontinuous and relatively thin. Altogether they make up a very small proportion of the total bulk of sedimentary rock.

Some beds of conglomerate reveal sedimentary structures such as cross bedding and imbrication. Both of these structures indicate transport and deposition of particles from flowing and turbulent water. **Imbrication** is a structure formed by inclined and overlapping pebbles and cobbles that have elongate or flattened shapes (Figure 9-17). We can find this feature in modern stream bed deposits where the flattened pebbles are inclined upward in the direction of current flow. We can observe how coarse gravel, pebbles, cobbles, and boulders accumulate in stream beds and along rugged coasts. These deposits form irregular beds that twist with the tortuous course of the stream, or collect on rocky coastal promontories.

Breccias are formed from angular fragments, a substantial proportion of which is larger than 2 mm in diameter. Stratification is poorly developed if not entirely absent from these beds. The large, angular fragments could not have been transported very far before being deposited because the sharp edges would have become rounded by abrasion. Some important deposits of breccia were pro-

Figure 9-17 Imbrication in a conglomerate is recognized from a pattern of overlapping cobbles that is inclined upward in the direction of current flow at the time of deposition. (D. A. Lindsey, U.S. Geological Survey.)

duced by landslides and rock falling from cliffs. Breccia can also form during postdepositional deformation of rock. We can find zones of crushed rock and clay along faults. The rock in these zones is called **tectonic breccia;** it is not an abundant kind of sedimentary rock.

The large detrital fragments in conglomerates are too heavy to become suspended in turbulent water for longer than a few moments. These cobbles and boulders can be rolled and bounced by the action of a strong coastal surf or by streams during periods of high water and exaggerated turbulence. Although granules and small pebbles may eventually be transported several hundred kilometers from their bedrock source, the larger cobbles and boulders are probably deposited within a few tens of kilometers from the source. Because of abrasion, these fragments become more rounded as they are transported over increasing distances.

The character of glacially transported sediment is different. Large and small fragments alike may become frozen into glacier and then transported by the slow plastic flow of the ice. Once frozen in an ice mass even large boulders can retain an angular shape while being transported. When the ice eventually melts, the sediment is deposited in an unsorted layer containing angular fragments from different bedrock source areas separated by distances of several tens or even hundreds of kilometers.

Carbonates

Limestone plays a prominent role in a delicately balanced geologic and biologic cycle. This kind of sedimentary rock consists mostly of calcite produced by marine organisms. They make the calcite from calcium (Ca^{2+}) and bicarbonate (HCO_3^-) ions dissolved in sea water. What is the source of these ions? They come from layers of limestone and dolomite that are widely distributed on the continents. In fact, these carbonate rocks comprise about 22 percent of the earth's total volume of sedimentary rock. They react chemically with water solutions seeping through them, and parts of the rock slowly dissolve. (We will learn more about these chemical reactions in Chapter 12.) The calcium and bicarbonate ions that they produce are washed from the rock into rivers, and then transported to the oceans. It is interesting to note that concentrations of these ions in sea water appear to have stayed about the same at least since early Paleozoic time. This means that marine organisms must be removing them at the same rate as rivers feed them into the ocean. Eventually the remains of these creatures collect on the ocean floor in beds of carbonate sediment. In this way nature manufactures new limestone using products from older carbonate rocks. This seems to be a one-way trip where calcium carbonate is moved from the continents into the oceans. But from time to time nature replenishes the stockpile of calcium carbonate on the continents. The beds of ancient limestone and dolomite were produced mostly in shallow seas that formerly spread over vast areas of the continents.

Naturally occurring calcite, the principal constituent of limestone, is usually not pure calcium carbonate. It contains a small proportion of magnesium (Mg^{2+}) ions that occupy lattice sites ordinarily filled by calcium. This proportion can exceed 10 percent, but it is usually much smaller. If the magnesium content is more than 4 percent, we call the mineral **high magnesium calcite.** If the magnesium occupies less than 4 percent of the lattice sites, we call it **low magnesium calcite.** Magnesium is another of the abundant ions in sea water. The extent to which it is incorporated into calcite grains depends upon the biochemical processes occurring in marine organisms. These processes are not the same in all organisms, and they produce different kinds of calcite. For example, most clams manufacture low magnesium calcite, but many starfish produce high magnesium calcite. The most abundant skeletal material in marine organisms is calcite. It is important to know that this mineral is crystallographically different from dolomite even though it may contain some magnesium impurity.

Unconsolidated carbonate sediment which is now accumulating consists almost entirely of calcite and aragonite. Dolomite is absent from these modern deposits; it becomes abundant only in the more ancient lithified beds. But we find almost no aragonite in lithified carbonate rocks. All of the common sedimentary structures including cross bedding, ripple marks, lamination features, and dessication cracks have been observed in carbonate strata. In addition other structures are found that are unique to the carbonate rocks. These structures are reminders of the varied forms found in colonies of marine organisms such as coral. They indicate that most carbonates have originated from deposition in shallow water marine environments.

In the modern marine environment we find deposits of carbonate sand and carbonate mud mixed in varying proportions. Larger, irregular skeletal frameworks resulting primarily from coral growths are imbedded in the sand and mud. The carbonate sands consist mostly of skeletal fragments, the broken shells of a myriad of marine invertebrates that lived upon the sea bottom or suspended in the water. An exception is a variety of carbonate sand grain known as an **ooid.** This kind of grain consists of concentric shells of calcium carbonate surrounding a nucleus of some other particle. These calcium carbonate coatings appear to form primarily from inorganic precipitation in shallow flowing water. Most modern ooid accumulations are forming on shallow banks washed by the ocean tide. They are the only significant carbonate deposits of inorganic origin, and account for a relatively small proportion of the total bulk of carbonate sand.

Carbonate mud consists of very small grains of calcite and aragonite that average approximately 3 microns in length. These microcrystalline grains ordinarily show no fossil markings even though they are almost entirely the products of organic decay and algal secretions. Invertebrate organisms build their shells from these secretions of calcium carbonate. The shell subsequently disintegrates with the decay of other organic matter that binds these tiny calcium carbonate particles together. The decay results from bacterial action, boring algae, and mechanical abrasion. It leaves a residue of loose carbonate mud grains. In addition to these very tiny grains there are larger detrital fragments of silt and clay produced by mechanical disintegration. They form significant proportions of some carbonate muds. There are no known important carbonate mud deposits of inorganic origin. Carbonate mud deposits are found on the continental shelves and in portions of the ocean basins. The well-known chalk deposits of Britain and western Europe are formed from such deposits.

Coral reefs are a favorite subject for undersea photographers. Most of us have seen pictures of the colorful aquatic community of plants and animals living within the shelter of a reef (Figure 9-18), which is an environment where calcium carbonate is produced and deposited. Not only is the reef itself made of calcium carbonate, but so are the skeletal parts of many invertebrate organisms living close to it. Most reef-building corals require water that is warmer than 18.5°C. Therefore we find them mostly in the warmer latitudes between 30°N and 30°S. Unlike most animals corals are not free to swim or drift in the sea. Instead they remain fixed to the ocean floor and feed upon nutrients that circulate with currents and waves. As a living coral grows, it secretes calcium carbonate which it uses to build a skeleton. This skeleton is attached to rock in shallow water, or to deeper skeletal remains of dead coral. Being an animal, the coral needs oxygen for survival, which it gets by participating in a **symbiotic** relationship with certain kinds of algae. This is a relationship in which each participant depends upon the other. Here the coral protects the algae and the algae generate oxygen needed by the coral. The oxygen is produced by photosynthesis, which requires light from the sun. For this reason the coral and algae can survive only in water less than about 20 m deep. Corals grow upward building a reef structure that reaches almost to the sea surface. The upward growth is terminated when the reef becomes exposed above the sea even temporarily at low tide. Here wave erosion and tidal currents break up portions of the reef structure and deposit the detrital fragments in and about its deeper carbonate framework. The

Figure 9-18 A modern reef community consisting of coral and algal growths on a submerged reef bordering Bikini Atoll. (K. O. Emery, U.S. Geological Survey.)

resulting carbonate deposits may then retain portions of buried reef structure as sedimentary structures. Such reef structures remain preserved as irregular mounds and lens-shaped features in ancient carbonate strata (Figure 9-19).

Algae can grow as encrustations upon the surfaces of a coral framework or upon the shells of detached organisms. Many algae also secrete calcium carbonate that strengthens a reef and cements deposits of loose detritus. Mats of algal growth are found in shallow water and on tidal flats. A modern algal mat dessicated by exposure above sea level is compared with similar growths preserved in Paleozoic rock as shown in Figure 9-20. Mound structures that geologists believe were formed from massive algal colonies are found in ancient carbonate beds.

Figure 9-19 The form of an ancient reef imbedded in other carbonate deposits can be discerned in strata exposed in northern Michigan. (J. E. Gair, U.S. Geological Survey.)

Figure 9-20 Algae may accumulate in growths that extend as organic mats over portions of tidal flats in warm seas. *(A)* Modern algal mats near Bikini Atoll have been dessicated on exposure to the atmosphere. (K. O. Emery, U.S. Geological Survey.) *(B)* Ancient algal mat growths are preserved in Paleozoic rocks in central Montana. The original organic material has been replaced. (W. H. Easton, U.S. Geological Survey.)

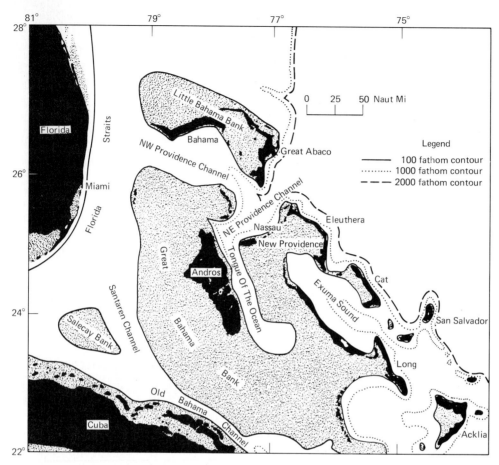

Figure 9-21 Map showing the distribution of modern carbonate deposits on the Bahama Platform in the western Atlantic Ocean. (Modified from N. D. Newell, *Special Paper No. 62,* page 306, Geological Society of America, 1955.)

The conditions most favorable to the accumulation of carbonate sediment are found in shallow water marine environments. But we find sequences of carbonate beds several hundreds and even thousands of meters thick that cover large areas. One modern setting where great thicknesses of carbonate sediment are still accumulating is the Bahama Platform. This is a large area in the western Atlantic Ocean (Figure 9-21). The water is typically shallow, and numerous islands and large reefs are found in the area. Boreholes on one of these, Andros Island, reveal carbonate deposits extending to at least 5 km below sea level. A geologic cross-section of the island is seen in Figure 9-22. How can such a large thickness of shallow-water deposits accumulate? Clearly there has been subsidence of the sea bottom as well as carbonate deposition, but the cause of subsidence is not understood. Isostatic adjustment of the increasing sedimentary load may partly explain it, but other tectonic processes must also contribute. Carbonate sedimentation has kept

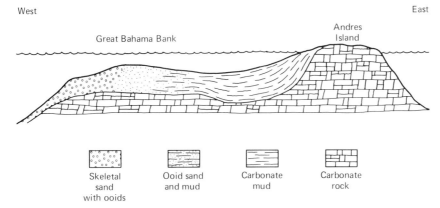

Figure 9-22 Carbonate accumulation in the vicinity of Andros Island on the Bahama Platform.

pace with the subsidence so that a shallow-water environment has been maintained over this large area.

The primary depositional features of carbonate deposits can be changed or even obliterated by diagenetic processes. The post-depositional changes are much more pronounced in carbonates than in deposits of silicate composition. Two important reasons for this are the relatively high solubility of calcium carbonate in water and the metastability of aragonite and high magnesium calcite in the near-surface pressure and temperature environment. The three important diagenetic processes include **dissolution, cementation,** and **replacement,** all of which are related to the percolation of water through the porous mass of sediment. The water solutions involved in diagenesis usually have dissolved ions in different concentrations than in the sea water that was present at the time of deposition.

Cementation results from the precipitation of carbonate minerals and sometimes silica in the pore space between grains. Precipitation is caused by physical or chemical processes that we will learn more about in Chapter 12. Changes in sea level sometimes expose beds of unconsolidated sediment. Then the sediment can become cemented partly by substances produced by evaporation of interstitial water. Changes in temperature can also cause precipitation. Through these and other processes the characteristic patterns of cementing seen in Figure 9-23 are developed.

Dissolution of carbonate rocks is caused by reactions with water undersaturated in calcium carbonate. The result is the enlargement of pore space between grains and perhaps the formation of large cavities in the rock. Some carbonate beds have been so weakened by dissolution that they could no longer support the overlying strata. When such a bed collapses, its crushed remains, together with broken fragments of overlying rock, make a new sedimentary bed that is called **collapse breccia.**

Replacement is a process of simultaneous dissolution of one mineral grain and precipitation of another in its place. Metastable aragonite can be altered to calcite in this way. A more perplexing replacement process is **dolomitization.** Beds of dolomite are produced by diagenesis, but the exact processes are not clearly understood. The replacement of calcite by dolomite very near the earth's surface appears to require seepage of water solutions with a high Mg/Ca ratio. There is evidence

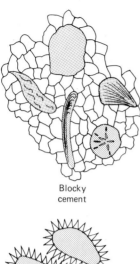
Blocky cement

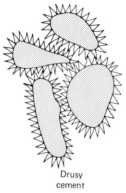
Drusy cement

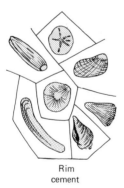
Rim cement

Figure 9-23 Characteristic cementing patterns found in carbonate sedimentary rocks.

The rise and fall of the tide causes beds of calcium carbonate sediment to be repeatedly washed by the magnesium-rich water. If the magnesium concentration is high enough, then chemical reactions are triggered that lead to replacement of calcium by magnesium in the crystal structure of the grains. But this process cannot account for all of the dolomite that is found in the earth today.

We have learned from recent experiments that temperature can be very important in dolomitization processes. At temperatures above 100°C magnesium carried in small concentrations in water solutions readily replaces calcium in grains of calcite. We already know that temperature increases with depth at the approximate rate of 30°C/km. This means that calcium carbonate sediment buried at depths of about 3 km could be dolomitized by ground water solutions that have ordinary concentrations of magnesium. So we might expect dolomite to be produced by hot water solutions seeping through buried calcium carbonate deposits.

Evaporites

A small proportion of the sedimentary rock found in the earth is formed, as the name **evaporite** implies, by precipitation from evaporating water. The most abundant of these deposits are **rock salt** or halite (NaCl), gypsum ($CaSO_4 \cdot 2H_2O$), and anhydrite ($CaSO_4$). Beds of rock salt are known in many parts of the world, some being more than 1000 m thick. Gypsum and anhydrite are also widespread but less abundant than rock salt. These evaporite deposits are common to all geologic periods since Precambrian time.

The largest evaporite deposits in North America occur in salt basins of Silurian age that extend from Michigan to New York, and Permian salt basins in west Texas and New Mexico. Halite beds averaging more than 100 m in thickness underlie approximately 30,000 km^2 of the northeastern United States.

that this kind of dolomitization is occurring on some modern tidal flats near evaporating seas. Here concentrations of magnesium are higher than we normally find in sea water.

Sedimentary Rocks 333

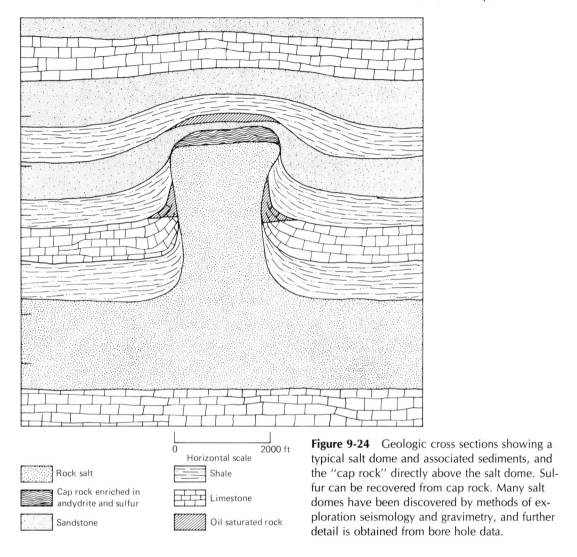

Figure 9-24 Geologic cross sections showing a typical salt dome and associated sediments, and the "cap rock" directly above the salt dome. Sulfur can be recovered from cap rock. Many salt domes have been discovered by methods of exploration seismology and gravimetry, and further detail is obtained from bore hole data.

Evaporite deposits totaling more than 100 m in thickness underlie large areas of west Texas and New Mexico. About 70 percent of these deposits consist of halite, and the remaining beds are primarily anhydrite. Other large deposits are known in Europe and Asia, and on submerged continental shelf regions.

An interesting feature of rock salt is the manner in which it can be deformed by plastic flow. Differential pressure on buried salt beds has caused plumes of salt to rise vertically, piercing the overlying sediments and forming structures that we call salt domes. Recall from Chapter 2 how we can use gravity measurements to detect these masses. Look at the example illustrated in Figure 9-24 to see the dimensions of some salt domes. In recent years a large number of salt domes have been discovered in offshore regions of the Gulf of Mexico as well as beneath the sediments of Texas, Louisiana, and Mississippi. Later in this chapter and in Chapter 10 we will discuss the sulfur deposits and the petroleum and natural gas reservoirs associated with salt domes. Because of the plasticity of halite and the ease with which recrystallization can take place, primary sedimentary structures are generally absent from these beds. Stratification is evident in some places because of impurities in the halite or from the presence of interbedded laminations of other sediments.

Gypsum is most commonly found in near-surface deposits. It has been discovered as deep as 1200 m, but such occurrences are rare. We can find it in massive beds or in veins that fill fractures and cavities in some other rock. The deeper and thicker deposits of calcium sulfate occur as anhydrite, a dense crystalline rock.

The precipitation of sodium chloride and calcium sulfate from sea water has been studied experimentally. Normal sea water, containing approximately 35 gm of dissolved ions per kilogram of H_2O, will precipitate these substances upon evaporation. We would get thicknesses of approximately 116 cm of halite and 4 cm of gypsum from the complete evaporation of a 100 m thick layer of normal sea water. Because this solution is undersaturated in these dissolved compounds, precipitation of gypsum will not commence until almost 70 percent of the water is evaporated. Only after the volume is reduced by about 90 percent due to evaporation will halite begin to crystallize.

Evaporation of sea water yields halite and gypsum in the ratio of 29:1. However, in naturally occurring evaporite deposits the proportion of gypsum or anhydrite is substantially larger. Processes contributing to the concentration of these deposits are more complicated than simply the evaporation of a basin of sea water. We find newly deposited evaporites near the shores of the Dead Sea, along the margins of shallow bays connected to the Persian Gulf, and close to the shores of some saline lakes such as the Great Salt Lake in central Utah. All of these are arid places where evaporation has produced highly concentrated brine. But we find that no significant amounts of halite or gypsum are precipitating directly from the brine. But we have discovered that precipitation of these minerals occurs in near surface layers of sediment where the pore space is only partially saturated by water. Interaction of the ground water solutions and concentrated brines from nearby shallow basins plays a role in causing the subsurface precipitation of chloride and sulfate compounds.

The origin of gypsum and anhydrite is not clearly understood. Some geologists argue that calcium sulfate was originally deposited as anhydrite. Gypsum then formed by diagenetic hydration of these deposits in a low-pressure environment. However, gypsum is the mineral that we obtain by evaporation in laboratory experiments. Under pressure and temperature conditions most common in bodies of water on the earth's surface, gypsum should be the mineral to precipitate, but no one has observed this happening in nature. Diagenetic dehydration of gypsum to form anhydrite would cause a 38 percent decrease in volume as water is removed from the crystals of gypsum. In some regions a cycle of diagenetic conversion of gypsum to anhydrite then anhydrite to gypsum appears to occur. This cycle depends upon the geothermal gradient, the dissolved ions in circulating ground waters, and tectonic uplift and subsidence. Field evidence from west Texas indicates that such changes may occur in the depth range of 500 to 2500 m.

What conditions are needed to manufacture a thick evaporite bed? It is made of constituents from a brine that was concentrated by evaporation in a shallow bay. But as water evaporates from the bay, more must flow in from the sea to bring the constituents that will add to the thickness of the deposit. But water cannot flow in too rapidly because the brine would become diluted. At the same time the bay must be subsiding to provide space for the growing deposit. So there must be a delicate balance between the evaporation rate, the influx of water from the ocean, and the rate of subsidence. Such a balance is not common in nature. Therefore, evaporites are not abundant sedimentary rocks.

Phosphorites

Rocks in which calcium phosphate is an abundant constituent are called phosphorites. These sedimentary rocks are not abundant, but some beds such as the Phosphoria formation of Permian age in the western United States are important. Phosphorite beds are primarily of marine origin. They may consist of guano deposits and submarine accumulations of vertebrate skeletal remains. Guano is made of animal excrement, mostly that of sea birds. Calcium phosphate is the primary constituent of vertebrate bone material. The sedimentary beds consist of pellets and nodules, which indicates that there has been considerable diagenesis if the original beds were bone accumulations. Bone fragments, sharks teeth, and phosphatic brachiopods are the only common fossils. These remains ordinarily do not form a significant proportion of the otherwise granular and stratified phosphorite beds. These strata are usually black or dark brown.

Coal

The volume of coal in the world is not large when it is compared with the volume of other kinds of sedimentary rock; however, it is widely distributed in continental areas. Individual layers range from paper-thin laminations to beds thicker than 150 m. The average thickness is less than 1 m. The most abundant coals were formed during the Mississippian and Pennsylvanian periods. Younger coal beds are not common, but accumulations older than Mississippian are rare. One of the most important coals in North America is the Pittsburgh Seam. We can follow it as a more or less continuous deposit beneath more than 60,000 km^2 of Pennsylvania, Ohio, and West Virginia. Coal occurs interbedded with other sedimentary rocks, most commonly shale and sandstone.

We can recognize different kinds of coal. The important varieties are **lignite, bituminous,** and **anthracite** coals. These different kinds of coal can be distinguished by physical properties and chemical composition. The constituents of coal undergo postdepositional changes that we could call diagenesis or metamorphism. The varieties of coal display different **ranks** of alteration, where the term **rank** refers to the extent of postdepositional changes. Lignite is low-rank coal usually dark brown in color. The structures of plant fragments and woody material are quite easy to see and the H$_2$O content is high. Bituminous coal consists of comparatively higher proportion of carbon and lower proportion of H$_2$O. It is a middle-rank coal. Usually banding is displayed in bituminous coal. The banding is related to alternating lamina of mineral charcoal and other substances that have a vitreous or earthy luster. Fossils, mostly plant impressions, are common in bituminous coal. Anthracite is a high-rank coal that has a large proportion of fixed carbon. It has a submetallic luster and it breaks with conchoidal fracture. The stratification of anthracite is not so obvious as in lower-rank coals. These different ranks of coal are part of a gradational sequence of compositions. The average compositions of the three ranks are summarized in Table 9-5. The primary constituents in order of abundance are carbon, oxygen, hydrogen, and nitrogen. Other elements such as sulfur are found in minor amounts. Gaseous hydrocarbons such as methane are sometimes retained in coal-bearing sedimentary sequences. These lethal and combustible gases add to the dangers that coal miners may encounter.

Table 9-5
Average Composition of Wood, Peat, and Coals

Material	Weight Percent			
	Carbon	Hydrogen	Nitrogen	Oxygen
Wood	49.65	6.23	0.92	43.20
Peat	55.44	6.28	1.72	36.56
Lignite	72.95	5.24	1.31	20.50
Bituminous coal	84.24	5.55	1.52	8.69
Anthracite coal	93.50	2.81	0.97	2.72

F. W. Clarke, The Data of Geochemistry, U.S. Geological Survey Bulletin 770, p. 773, 1924.

There is little question that coal is formed primarily from plant remains. The most important depositional environments are fresh water swamps. In these settings the plant remains settle to the bottom of ponds. Because they are shielded from the atmosphere by the water, oxidation is restricted. Decomposition takes place slowly and remains incomplete. The partially decomposed organic matter is first altered to form **peat,** an unconsolidated carbonaceous residium. Increasing pressure and temperature caused by burial lead to further **coalification** from which lignite is produced. The continued increase in these conditions as well as chemical reactions with ground water solutions are required to form bituminous coal. Anthracite appears to be the product of still higher pressure. This high rank is found only in beds that have been severely folded by tectonic processes.

Sedimentary Facies Relationships

When we examine a layer of sedimentary rock that covers a large area, we will probably see changes in composition and texture from place to place. For example, a bed consisting of pebble conglomerate in one area may change gradationally to sandstone in another area. Farther along the sandstone might grade into siltstone. We recognize these different zones as separate **sedimentary facies** within the layer. A sedimentary facies is part of an assemblage of rocks that possesses characteristics which we can use to distinguish it from other bordering facies. In a layer of limestone we might find an ooid facies, a mud-supported facies, and a grain-supported facies. A layer of shale might consist of a red shale facies and a black shale facies. We can use fossils, color, sedimentary structures, composition, texture, and other distinctive features to describe a sedimentary facies. Geologists analyze these features to determine the combination of depositional and diagenetic processes that produced an assemblage of sedimentary rocks.

We can learn about important geologic events by observing how different sedimentary facies are arranged relative to one another. One interesting pattern is the sequence of overlapping facies produced by transgression of the ocean onto a continent. We know that deposition of sand, clay, and carbonate mud along some coasts can be related to water depth and distance from the shore. This is illustrated in Figure 9-25 A. Sand is deposited close to shore where the water is shallow and turbulent. Clay particles carried into the sea

Figure 9-25 (opposite) Conceptual model relating sedimentary facies to the transgression and regression of a sea. The accumulated sediment at progressively later times is seen in the cross sections, each of which is associated with a different sea level. Labels t_0 through t_5 refer to progressively later times in the history of sedimentation.

Sedimentary Rocks 337

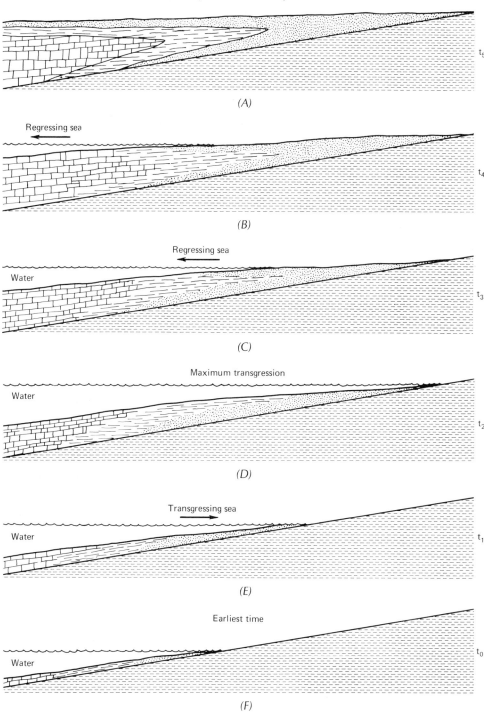

by rivers or produced by waves beating on the coast remain suspended in the turbulent water. They settle to the bottom farther offshore where the water is deeper and less turbulent. In still deeper water carbonate mud is accumulating. So the layer of sediment being deposited has a sand facies, a clay facies, and a carbonate mud facies. This is one of the typical patterns of deposition that is found in coastal areas.

A rise in sea level relative to the continent will cause the zones of sand, clay, and carbonate mud deposition to shift in a landward direction. We can see this in Figure 9-25C. Now the deep water sediments are collecting on top of older shallow water deposits. Continued rise in sea level extends these overlapping facies still farther inland. Facies that overlap in this manner make up a **transgressive** sequence of sedimentary rocks. Now suppose that sea level begins to fall. The result will be a **regressive** sedimentary sequence such as seen in the upper part of Figure 9-25E. The entire diagram shows how a wedge-shaped arrangement of sedimentary facies is produced by a sea that is first transgressing onto a continent, and then is regressing from it.

Sedimentary rocks in the northern Rocky Mountains make up the transgressive sequence illustrated in Figure 9-26. These facies range from Mississippian to Triassic in age. We can see that during a particular interval of time a layer was formed by simultaneous deposition of three or four different facies. But a particular facies such as the Phosphoria Limestone was produced by deposition of one kind of sediment in different places at different times.

Facies deposited during early Paleozoic time in Wyoming and Montana reveal transgression and regression of the sea. Look at the diagram in Figure 9-27. It shows the ages rather than the thicknesses of the sedimentary rocks along a vertical cross section. Here we see that the basal layer is mostly

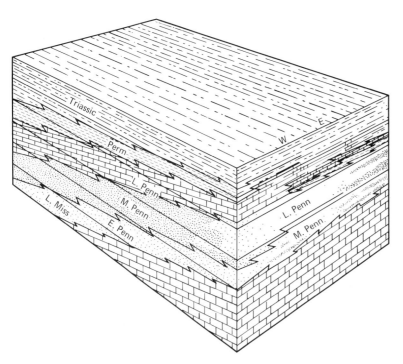

Figure 9-26 Block diagram showing sedimentary facies deposited during a late Paleozoic and early Mesozoic marine transgression and regression in the area of the present northern Rocky Mountains. (After A. C. Munyan, *Guidebook of the 8th Annual Field Conference,* page 71, Billings Geological Society, 1957.)

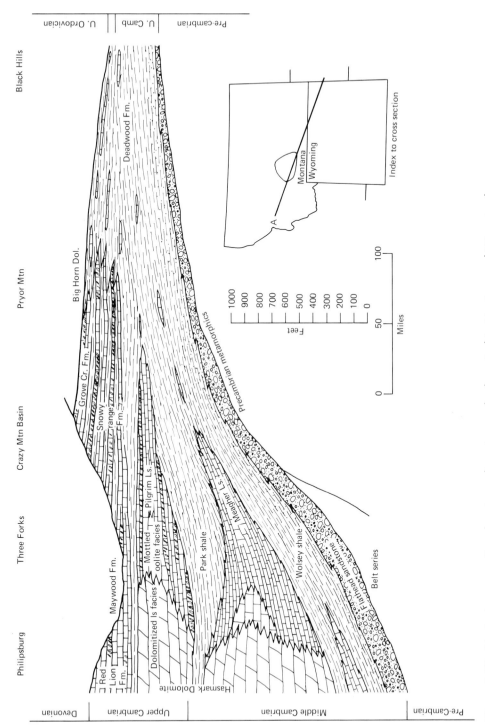

Figure 9-27 Vertical cross section showing sedimentary rocks deposited in Montana and Wyoming by a transgressing Cambrian sea. The vertical scale on the diagram reveals both time and thickness of sediment. (After A. E. Roberts, *Guidebook of the 8th Annual Field Conference*, page 48, Billings Geological Society, 1957.)

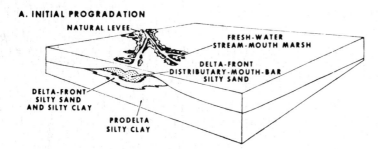

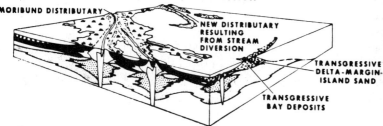
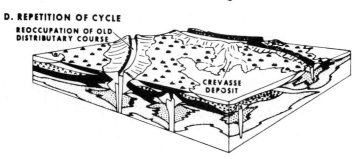

Figure 9-28 Conceptual models of the accumulation of sedimentary facies on the Mississippi delta during late Cenozoic time. (After D. E. Frazier, *Transactions of the Gulf Coast Association of Geological Societies,* page 290, Vol. XVII, 1967.)

quartz arenite with some zones of conglomerate. It is a continuous facies that formed progressively in an easterly direction during early, middle, and then late Cambrian time near the shore of a transgressing sea. Another feature seen in Figure 9-27 is the dolomite facies along the western part of the profile. Notice how it forms an interfingering gradational border with the adjacent limestone. No doubt it was forming by diagenesis of the carbonate deposits at the same time the quartz arenite facies was being deposited along another part of the profile.

Nature has put together some amazingly

complicated assemblages of sedimentary facies. Think about all of the processes of erosion, deposition, and diagenesis that are active on a delta. Here we find rivers eroding some parts and depositing the sediment in other places. There are lagoons and marshes where still different kinds of sediment are deposited. Then, on the seaward margin of the delta, ocean waves and tides redistribute the sediment in other ways. Look at the four stages in the growth of the Mississippi Delta illustrated in Figure 9-28. This interpretation is based on data from a large number of wells that reveal the numerous lenses and wedges of buried sediment. The assemblage of sedimentary facies produced on a delta can be complicated indeed. But a geologist can learn to recognize ancient delta deposits from distinctive combinations of facies.

The delta is one of several common environments of deposition where distinctive combinations of sedimentary facies are produced. We can also recognize other coastal and offshore environments such as carbonate platforms (Figure 9-21). On the continents there are arid desert regions as well as temperate and tropical areas where sediment carried by rivers, glaciers, and wind is distributed in different combinations of facies.

SEDIMENTARY DIFFERENTIATION

Typical quartz arenite is very different from typical shale, but the constituents for both kinds of sedimentary rock could come from the same source. Think about the decomposition of granite. We know that it possesses the quartz needed to make quartz arenite. Its feldspars, micas, and amphiboles have the ingredients needed to make shale. But how do these constituents of granite become separated into different kinds of sediment? This happens through processes of sedimentary differentiation. Recall from Chapter 7 how the constituents of igneous rocks can be separated by differentiation of magma. Sedimentary rocks show that another kind of differentiation takes place in the environment close to the earth's surface.

Earlier in this chapter we asked if there might be a relationship between composition and particle size in terrigenous clastic rocks. We can tell from Figure 9-15 that the composition of a typical sandstone differs from the composition of a typical shale. What factors might contribute to a relationship between composition and particle size? Think about what can happen to particles that are being carried by flowing water, wind, or ice. During transportation the particles become rounded and broken into smaller fragments by mechanical abrasion and chemical decomposition and dissolution. The ability of a mineral grain to withstand rounding and fragmentation is related to various factors including hardness, cleavage, and density. Also important is its chemical stability in the pressure and temperature conditions found near the surface of the earth. The mechanical and chemical processes tend to reduce a mineral grain to end products that are stable in the environment. This is accomplished by breaking and abrading the original grain into a size and shape that can withstand further mechanical deformation. It can also involve chemical decomposition of the grain into other minerals that are more stable. We use the term **maturity** to describe how compatible a sedimentary rock is with its environment. Immature rock is still relatively unstable. More mature rock is less susceptible to further alteration provided the environment does not change.

We can learn something about how detrital grains resist fracture and abrasion by examining the typical sizes and shapes found along beaches and stream beds at known distances from a source area. Also, we can use laboratory apparatus such as the rotating drum illustrated in Figure 9-29 to simulate some natural conditions for abrasion. The stability of fragments of different composition tested in this apparatus is described in Table 9-6. Distance of transport was computed from drum revo-

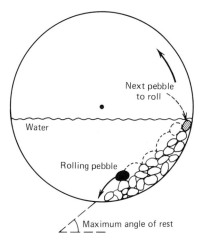

Figure 9-29 Schematic diagram of rotating drum apparatus used in experimental studies of pebble abrasion. (P. H. Kuenen, *The Journal of Geology*, page 337, July 1956.)

lutions, and loss of mass from grains was measured by periodically interrupting the experiment to weigh the fragments. Quartz is clearly the most durable material, while limestone fragments proved to be least resistant to abrasion.

Other tests of limestone fragments are summarized in Figure 9-30. These graphs indicate that rounding of originally angular limestone grains is most pronounced in the initial kilometer of transport. Particle size diminishes at a more uniform rate. Particle shape, expressed in terms of sphericity, is slow to change during transport compared with the other features. Here grains that are closer to the shape of a sphere are described by higher numerical values of sphericity on a scale ranging from 0 to 1. In nature most carbonate deposits are relatively close to the sources of calcite and aragonite, whereas quartz arenite and shale are formed from detritus that may have been transported for large distances. In the case of quartz the distance of transport can sometimes be estimated from the roundness and sphericity of the grains. Feldspar grains are usually angular fragments that have been deposited relatively close to the source area.

Quartz is the most durable of the abundant rock-forming minerals in the near-surface environment of the earth. It is chemically stable at low pressure and temperature, and relatively insoluble in most river and sea water solutions. It possesses a high value of hardness and no cleavage planes. Grains of quartz in a deposit of sediment could have been transported from both near and distant source areas. Granite is the most abundant igneous rock in the upper part of the earth's continental crust, and it is the most important source of quartz grains. These grains are commonly in the size range of granules and sand parti-

Table 9-6
Weight Loss of Sedimentary Particles in Rotating Barrel

Material	Weight Loss (milligrams)		Original Weight (mg)	Percent Loss per Day
	2nd Day	4th Day		
Quartzite	283	272	24197	0.017
Quartz	296	260	17747	0.026
Graywacke	822	797	22044	0.056
Obsidian	1932	1418	31073	0.081
Limestone	1928	1398	22157	0.120

From Ph. H. Kuenen, Experimental Abrasion of Pebbles, *Journal of Geology*, v. 64, no. 4, p. 339, Table 1, 1956.

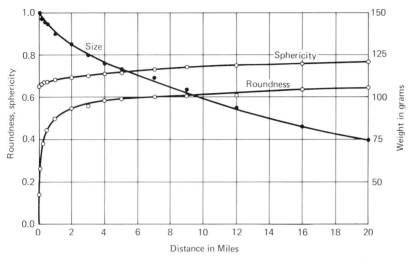

Figure 9-30 Graphical display of experimental results showing the effect of distance of transport upon grain size, roundness, and sphericity. (After W. C. Krumbein, *The Journal of Geology*, page 492, July–August 1941.)

cles in the Wentworth Scale. As granite weathers, these grains are loosened, then removed from the rock, but they are slow to diminish in size during transport. Because of their size, these quartz sand particles can settle out of water or wind that is turbulent enough to retain smaller grains in suspension. Sand-size grains accumulate in shallow coastal settings where wave turbulence is high, and in river beds. Because quartz can be transported from near and far sources it tends to concentrate in large masses in these depositional environments.

Calcite and aragonite are relatively soft minerals (hardness 3-4) with well-developed cleavage. Most fragments made of these minerals are reduced to particles smaller than sand grains before they can be transported very far. Therefore carbonate sand deposits are usually found close to the sources of sediment. When carbonate sand grains and quartz sand grains come together in an environment of turbulent water, the shifting and grinding of the loose particles reduce the carbonate grains to mud, whereas the quartz sand is retained. The mud is then held in suspension and becomes transported to less turbulent water. In this way mechanical processes tend to remove carbonate grains leaving concentrations of quartz sand. Because of their reduced size, the mud-sized carbonate grains become concentrated in deeper zones of quiet water. These zones of deposition are free of sand grains that settled elsewhere from more turbulent water. Carbonate mud deposits therefore accumulate separately from quartz sand deposits.

Feldspars originate by precipitation from magma. These minerals are metastable at low pressure and temperature in the presence of abundant water and oxygen. They have hardness values of about 6 and possess cleavage, although poorly developed. Owing to the chemical metastability of feldspars, they tend to decompose into the hydrated clay minerals in the presence of water. The clay minerals have low hardness values and well-developed cleavage. For these reasons feldspars ordinarily do not survive to be transported very far. Rather, they break down into clay mineral grains that are rapidly reduced to clay-sized particles. These tiny grains are chemically stable at low pressure and temperature and can be transported long distances before settling from quiet waters. Arkose and feldspathic greywackes, which contain a significant proportion of feldspar grains, are immature sediments with angular grains. Shale, in contrast, is a more mature sediment containing much of the weathered residue of feldspar in the form of clay minerals that are stable in the near-surface environment.

We have discussed only a few of the numerous factors that contribute to sedimentary differentiation. They illustrate how a mixture of constituents can be separated into different deposits that have distinctive features of composition and texture. The end products are rocks that are more stable in the various environments near the surface of the earth.

SEDIMENTARY DEPOSITS OF INDUSTRIAL UTILITY

The raw materials for numerous commodities and manufacturing processes come from sedimentary rocks. Some of these deposits, coal and rock salt, for example, are useful because they are found in nature and require almost no refining. Other kinds of beds such as gypsum and low magnesium limestone have more utility after some processing. In still other situations the sedimentary rock itself is of little direct value. Instead it serves as a host for ore minerals disseminated in veins or as small grains throughout its mass. Accumulations of petroleum, natural gas, and fresh water are found in sedimentary deposits filling the pore space between grains, in fractures, and in solution cavities.

In the following discussion we will consider those sedimentary deposits that are used in bulk as industrial raw materials and fuels. Among the most important are those listed in Table 9-7. The amount of bulk deposits obtained by mining or quarrying is usually measured in tons. To visualize the volume of rock involved think of a limestone with a density of 2.8 gm/cm^3. One cubic meter, then, has a mass of 2.8×10^6 grams, or 2.8 metric tons. Volume of petroleum is usually stated in barrels where 1 barrel = 42 U.S. gallons = 159 liters.

Table 9-7
Approximate Annual Production in the United States of Some Sedimentary Materials for Industrial Use Between 1960 and 1970

Rock	Annual Metric Tonnage
Limestone, dolomite, and sandstone for building materials	863×10^6
Gypsum	10.7×10^6
Rock salt	13.5×10^6
Phosphate rock	38.7×10^6
Sulfur	10×10^6
Coal	600×10^6

Data from *Commodity Data Summaries,* Bureau of Mines, U.S. Department of the Interior.

Two very important groups of raw materials not included in Table 9-7 are bauxite, the primary ore of aluminum, hematite, and the other ores of iron. These ores are found in sedimentary rock sequences as well as in associations with igneous and metamorphic rocks. The role of weathering in concentrating oxides of iron and aluminum cannot be overemphasized. Therefore, discussion of the origin and distribution of these deposits is more properly included in Chapter 12. Important clay mineral deposits, also the products of weathering, are reviewed in Chapter 12.

A substantial volume of sedimentary rock is quarried annually for use as building stone. More than 99 percent of this volume is processed as crushed stone, whereas the remainder, known as **dimension stone,** is produced in blocks of various sizes. More than 70 percent of the crushed stone is limestone and dolomite. Igneous rock, primarily granite, and sandstone are other sources of crushed stone. Thickly bedded rock that can be broken into more or less equidimensional fragments is the most desirable source. Because of its fissility, shale is not useful for this purpose. Important varieties of dimension stone are granite, limestone and dolomite, marble, sandstone, and slate. Supply and production of building stone is the largest nonfuel and nonmetallic mineral industry in North America.

Concrete is the most universal of modern building materials. The annual tonnage of concrete exceeds the collective mass of all other building materials. Concrete consists of an aggregate of crushed stone, sand, and gravel in a matrix of cement. The cement is known as **portland cement.** Different varieties are described in Table 9-8. The primary source of lime (CaO) used in portland cement is low magnesium limestone. About 15 percent of the limestone quarried annually is used to produce lime. It is obtained from decomposition of $CaCO_3$ at high temperature. Silica, aluminum oxide, and minor amounts of iron oxides are obtained from clays and sands mixed in the concrete aggregate. If unconsolidated sand and clay are not available, quartz arenite and shale may be crushed for use in the aggregate. This is one of the few industrial uses for shale. Because magnesium is undesirable in portland cement, dolomite and high magnesium limestone cannot be used.

Gypsum is a constituent used in most concrete mixtures and in the manufacture of plaster of paris. In recent decades it has become a major material in wall board. When gypsum is baked in kilns at temperatures above 160°C, about 75 percent of the H_2O is removed from its crystalline structure, and it decomposes to a powder. When mixed with water at a later time, this powder is rapidly hydrated and sets as a rigid mass that is called plaster.

Common salt (NaCl) for numerous industrial and domestic needs is produced by solar evaporation of sea water and subsurface

Table 9-8
Portland Cement Compounds

Tricalcium silicate	$(CaO)_3SiO_2$
Dicalcium silicate	$(CaO)_2SiO_2$
Tricalcium aluminate	$(CaO)_3Al_2O_3$
Tetracalcium aluminoferrite	$(CaO)_4(Al_2O_3)(Fe_2O_3)$

brines, and by mining rock salt. About 25 percent of the annual production comes from mining operations. Salt is extracted by direct excavation or by solution mining. Because of its high solubility in water, it can be retrieved by drilling holes into halite beds, circulating water in these holes to recover the salt in solution, then evaporating the water.

An interesting aspect of halite is its plasticity. Under differential pressure it will flow into vertical cylindrical salt masses that penetrate and deform overlying sediments (Figure 9-24). Concentrations of sulfur have been found in the rock directly over some salt domes that intrude interbedded limestone, anhydrite, and gypsum deposits. The largest sources of industrial sulfur in North America are found in zones called **cap rock** which lie on the tops of these salt domes. Because sulfur fuses at 114.5°C, it can be melted by pumping super-heated water into a well under pressure. Liquid sulfur is then retrieved from the well. There are other kinds of sulfur deposits that can be exploited by more conventional quarrying techniques.

Phosphorous is an element that is essential for plant growth. Phosphorous compounds together with lime, and some sulfates are important in agricultural fertilizers. Sedimentary phosphorite deposits are the most important sources of these phosphorous compounds. Deposits of guano and accumulations of the mineral apatite in igneous associations are secondary sources.

The important fossil fuels are petroleum, natural gas, and coal, all of which occur almost exclusively in sedimentary settings. They are the products of partial decomposition of organic debris into compounds of carbon, hydrogen, and oxygen. The presence of sulfur impurities is undesirable in coal because it causes less efficient combustion and contributes to furnace corrosion and air pollution. Low sulfur coals are of highest value. The major accumulations of petroleum and natural gas are found in beds of sandstone, limestone, and dolomite where they occupy the pore space. Owing to its low porosity and permeability, shale is not a good reservoir rock for these fluids. However, it plays an important role by sealing petroleum and gas in the interbedded porous zones thus preventing these fuels from percolating to the surface to be lost in the atmosphere. We will learn more about the migration and entrapment of oil and gas in Chapter 10.

How does nature manufacture petroleum and natural gas? These fluids appear to form initially in deposits of mud that are rich in organic matter. Marine microorganisms, primarily plankton, are the most important source of this organic matter. The organic compounds contain the hydrogen and carbon that must be synthesized to form the hydrocarbon compounds of petroleum and natural gas. But they contain oxygen and nitrogen as well. If the marine mud deposits have accumulated in an oxygen-deficient environment, then anaerobic bacteria can remove nitrogen and oxygen. In laboratory experiments certain of these bacteria have synthesized some hydrocarbon oil and gases from natural lipids and fatty acids. Such bacterial activity appears to play an important role in starting production of petroleum and natural gas within marine mud deposits.

Following subsidence and further sedimentation, these marine mud deposits experience increased pressure and temperature. In this environment the proto-petroleum droplets are probably decomposed and distilled into more refined hydrocarbon compounds. As the sediment is compacted, these fluids may be squeezed from the interstices into adjacent carbonate or sandstone layers that will ultimately retain a higher porosity than the mudstone. This could account for the fact that although organic constituents needed for petroleum synthesis are much more likely to accumulate in mud deposits, the petroleum and natural gas are recovered mostly from carbonate and sandstone reservoirs.

The chemical processes that contribute to the genesis and evolution of petroleum and natural gas are complex and by no means clearly understood. The same processes and chemical constituents are not found everywhere as we can tell from the existence of different kinds of gas and crude oil. Most of these fluids are mixtures of various hydrocarbon compounds. We can describe them in terms of the proportions of hydrogen and carbon. A *paraffin base* oil has a high hydrogen content, and a *naphthenic* or *asphaltic base* oil has a high carbon content. Other oils with more even proportions of these elements are intermediate base oils. Paraffin base crude oil is more enriched in the gasoline and kerosene constituents. The more viscous lubricating oils contain a larger proportion of asphaltic base crude oil.

A variety of shale known as *oil shale* is becoming recognized as an important natural resource of the future. Vast accumulations are known in western North America, for the most part in the Green River Formation of Colorado, Utah, and Wyoming. These shale beds are sufficiently rich in organic matter to yield at least one-half barrel of oil from a ton of processed rock. The extraction processes have been too expensive for this source to be competitive with petroleum obtained directly in the liquid state. As demand increases and supply dwindles, production of oil from this shale may become a more practical reality. Shale for this production is quarried by conventional means.

Many other industrial minerals are found in sedimentary settings, but are processed in smaller volumes than the materials described in the above discussion. Some of these minerals are noted in Chapter 6.

PERSPECTIVES

Sedimentary rocks are formed on or near the surface of the earth. We can observe, directly, numerous processes that contribute to the production of these stratified rocks. Because sedimentary structures found in ancient rocks are so similar to those now forming in modern depositional environments, we now realize the major role of rivers, waves, and wind in forming the present and past landscape. The thin film of sediment that accumulates about a coral reef or on a stream bed during a human lifetime may seem insignificant. The massive thicknesses of sedimentary rock focus our attention on these commonplace phenomena, emphasizing their persistence during most of the history of the earth. For many, the features displayed in sedimentary rocks have provided the incentive to examine more closely those numerous processes that occur without notice almost everywhere during every year. Some of these will be explored more fully in subsequent chapters.

For some the challenge of sedimentary rocks is found in the search for industrial materials. Another aspect, perhaps less practical, is the unfolding of earth history seen in the rock record. Sediments ranging in age from Precambrian to Quaternary time are widely distributed on the continental platforms. The locations of these deposits provide a basis for estimating the extent of marine transgressions during former times.

Estimates of the position of the equator indicate the probable drift of continents related to plate tectonic processes. We know that the present time is somewhat unique. During much of the past shallow arms of the ocean, known as *epicontinental* seas, extended over much of the area of the continents. The sedimentary remnants of these ancient seas reveal features that help to connect the present with the past. We can almost visualize a remote, yet strangely similar scene in the words of James Hall, as he described his thoughts in 1843 while examining an exposure of Silurian age in western New York.

*In standing upon the exposed surface of the quarry, one can almost fancy himself still upon the shore of some quiet bay or arm of the sea, where the waves of the receding tide have left these little ridges of sand, which on their return will be obliterated and mingled with the mass around. The shells and fragments, and the clouded sand, all lie around him with a freshness of appearance that might almost make him doubt. But his foot is upon the firm rock, and his hand cannot obliterate the faint waveline, nor remove a single shell from its place. Every thing is firm and fixed, and he is forced to recollect that millions of ages have rolled on, since the sea washed this shore, and the shells lay upon the glistening sand as he may have seen them in the haunts of his childhood. How beautiful, how simple, and how grand this exhibition; and how much does it illumine the mind as to the mode of production of these older formations which have been considered so obscure. Here was an ocean supplied with all the materials for forming rocky strata: in its deeper parts were going on the finer depositions, and on its shores were produced the sandy beaches, and the pebbly banks. All, for aught we know, was as bright and beautiful as upon our ocean shores of the present day; the tide ebbed and flowed, its waters ruffled by the gentle breeze, and nature wrought in all her various forms as at the present time, though man was not there to say, how beautiful!**

---

*James Hall, *Geology of New York, Part IV, Comprising the Survey of the Fourth Geological District*, p. 57, Albany, 1843.

The Rocks of the Earth

STUDY EXERCISES

1. Discuss why the sedimentary structures most commonly found in limestone with a mud-supported texture might be different from the sedimentary structures more commonly found in limestone with a grain-supported texture.

2. Why would a sandstone consisting of equal proportions of quartz grains and ooid grains be unlikely to form in nature?

3. Explain why sedimentary structures in greywackes are typically more poorly preserved than the sedimentary structures in quartz arenites.

4. Textures and sedimentary structures in some limestone beds indicate that these beds are produced partly by organic processes involving two different kinds of organisms. What are these organisms and why did they work together to produce the limestone?

5. Discuss:
   (a) why shale is not a useful source of dimension stone;
   (b) why dolomite is not useful for making Portland cement;
   (c) why we do not find outcrops of halite.

SELECTED READINGS

Blatt, Harvey, Gerald Middleton, and Raymond Murray, *Origin of Sedimentary Rocks,* Prentice-Hall, Inc., Englewood Cliffs, N.J., 1972.

Dunbar, C. O., and J. Rodgers, *Principles of Stratigraphy,* John Wiley & Sons, Inc., 1957.

Levorsen, A. I., *Geology of Petroleum,* W. H. Freeman and Co., San Francisco, 1967.

Pettijohn, F. J., and P. E. Potter, *Atlas and Glossary of Primary Sedimentary Structures,* Springer-Verlag, New York, 1964.

Weller, J. M., *Stratigraphic Principles and Practice,* Harper & Row, New York, 1960.

10

TECTONISM AND THE ARCHITECTURE OF THE EARTH'S CRUST

Tectonic Structures
  Inclined Surfaces
  Geologic Maps
  Folds
  Faults
  Horizontal Strata
  Joints
  Tectonic Fabrics
  Evidence of Repeated Episodes of Deformation
  Continental Cratons and Orogenic Belts
Mechanical Analysis of Tectonic Structures
  Nondirected Stress
  Directed Stress
  Stress and Strain in Rock Specimens
  Model Studies of Tectonism
Orogeny
  Orogenic Belt Form
  Orogenic Belt Sedimentation
  Tectonism and Epeirogeny
  Igneous Activity and Metamorphism
  Mountain Building Cycles
  Orogeny on Destructive Plate Boundaries
  Orogeny on Constructive Plate Margins
  Conservative Plate Margin Tectonics
Natural Resources and Tectonic Structure
  Occurrence of Petroleum and Natural Gas
  Structural Settings of Ores
  Tectonic Structures and Industrial Development
Perspectives

Mountains! Where do they come from? The loftiest mountains are made of rock that was raised from deep in the crust. This rock has been crumpled into folds and broken by fractures. Folding, fracturing, and uplift are produced by the processes of **tectonism.** These are the processes that cause mechanical deformation of the earth's lithosphere. The global tectonic processes that we learned about in Chapter 5 have the strength to move entire continents, and to split apart the floor of the ocean. Imagine how these processes could contribute to mountain building!

There are places where we can observe tectonism near the earth's surface. Recall our earlier discussions about vertical movement of the land surface (Figure 2-15), and the buildup of strain near faults (Figure 3-10). But nature has taken tens of thousands, or perhaps millions of years of sustained or intermittent tectonism to produce the complex deformation that we observe in some rock. Most of this deformation takes place deeper in the crust where it cannot be directly observed. Earthquakes and brief episodes of volcanism are but small steps in the ongoing tectonism of the lithosphere.

To learn about the tectonic deformation of

rocks we will first look at the typical features of folding and fracturing that we can piece together from outcrops and supplementary borehole and geophysical survey data. From these sources of information we can learn about the architecture of the earth's crust. Then we can try to determine how patterns of deformation are produced by testing laboratory scale models and using our knowledge of the mechanical behavior of solids.

TECTONIC STRUCTURES

Mechanical deformation of rock is indicated by folds, fractures, and distortion of the constituent particles. On a small scale we can see these features in hand specimens or even as microscopic crenulations and dislocations in minute crystals. Larger folds and fractures can be seen in outcrops or inferred from the edges of tilted strata exposed at the earth's surface. These features that make up the architecture of the crust are called **tectonic structures.** Systematic description of tectonic structures is fundamental to understanding the processes of tectonism.

The purpose of this section is to examine the principal kinds of tectonic structures and the conventional methods of recognizing, describing, and classifying them. We will be concerned with folds, faults, joints, and tectonic fabrics. A **fold** is a bend or flexure that is commonly evident from the shape of surfaces such as bedding surfaces which were originally planar. A **fault** is a fracture along which the rock on one side has been displaced relative to the rock on the other side. Recall from Chapter 3 that sudden movement along faults causes earthquakes. Fractures along which there has been almost no displacement are **joints.** Patterns of grain orientation and shape distortion arising from mechanical deformation are recognized as **tectonic fabrics.** Examples of these tectonic structures are illustrated in Figure 10-1. There are similar structures that extend deep in the crust and are too large to be seen in a single

(A)

(B)

(C)

Figure 10-1 Tectonic structure in rocks: *(A)* a fold in Paleozoic sediments exposed near Buchanan, Virginia; *(B)* a fault in Paleozoic carbonates near Hollins, Virginia; *(C)* joints transecting a massive sandstone bed in southwestern Utah; *(D) (next page)* tectonic fabric revealed by oriented mineral grains partially recrystallized in response to tectonic stress near Kaladar, Ontario, Canada. (Photograph D by D. A. Hewitt, Virginia Polytechnic Institute and State University.)

Figure 10-1 *(continued)* (D)

outcrop. Geologists find evidence of these large tectonic structures in the patterns of rock exposed at the surface. Such patterns are apparent on aerial photographs of some regions of sparse vegetation. Large folds are evident in Figure 10-2 from the edges of deformed strata. In areas of subdued relief that are covered by regolith and vegetation, these patterns are obscure. They are seen more clearly on bedrock geologic maps prepared from information gathered from whatever outcrops can be found. In the following discussion we will consider typical folds and faults, and their distinctive patterns on bedrock geologic maps.

Inclined Surfaces

Measurement of the orientation of surfaces within rock assemblages is a very important aspect of the description of tectonic structures. Inclined sedimentary bedding surfaces, for example, indicate mechanical deformation because we know that these surfaces were originally horizontal. Foliation surfaces in metamorphic rock and the surfaces of faults and joints also help us to recognize deformation patterns. The orientation of a surface is described from measurements of its strike and dip. The **strike** is the direction of the line of intersection of the inclined surface and a horizontal plane. The **dip** is the angle between these two surfaces taken in the downward direction. In Figure 10-3 we see a sedimentary bedding surface dipping into a lake. The water line on this bedding surface points in the direction of strike because the surface of the lake is a horizontal plane. We can also see in Figure 10-3 how strike and dip are measured with a compass-inclinometer.

As a result of severe deformation, sedimentary strata may become steeply inclined and perhaps overturned. The original upper surface can be identified from bedding surface features such as ripple marks and dessication cracks that we learned about in Chapter 9.

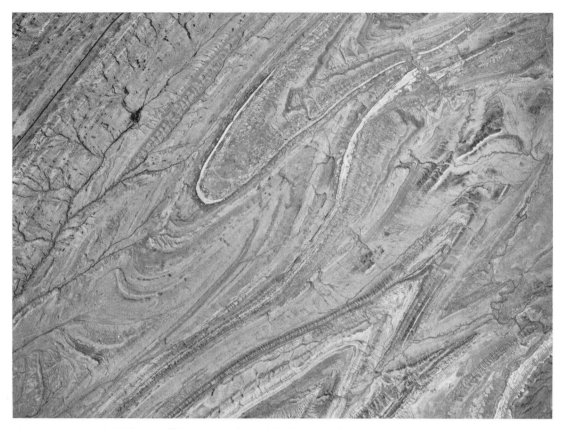

Figure 10-2 Large folds are evident from edges of beds exposed in Wyoming. (U.S. Geological Survey.)

Geologic Maps

Exposures of bedrock account for a very small proportion of the land surface of continents. In most places the bedrock is hidden under a veneer of unconsolidated regolith and vegetation. Little of the complexity of subsurface tectonic structure is evident from a casual look at the landscape. We can see this structure more clearly on a geologic map that displays the bedrock surface as it would appear in the absence of all regolith and vegetation. Obviously the preparation of such a map involves conjecture about the probable location of different bedrock units that is based upon interpolation between outcrops.

The accuracy of a bedrock geologic map depends upon the distribution of outcrops and the condition of the exposed rock. Some outcrops display relatively unweathered rock. Here lithology can be clearly recognized, orientations of bedding surfaces or foliation can be measured, and perhaps fossils or isotope samples can be obtained for age determination. In other exposures the rock may be so badly weathered that lithology is difficult to identify, and bedding surfaces or foliation are virtually impossible to discern. A geologist is forced to use fragmentary data of varying quality to prepare a bedrock geologic map.

An example of geologic map preparation is illustrated in Figure 10-4. Here we can see a

354 The Rocks of the Earth

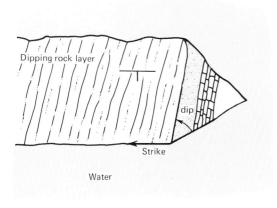

(A)

(B)

(C)

Figure 10-3 (A) The strike of an inclined surface is the azimuth of a horizontal line in this surface, and the dip indicates direction of downward inclination. (B) The strike and (C) dip of an inclined surface can be measured with a compass-inclinometer.

landscape with several topographic features and scattered outcrops. From information gathered at these outcrops the geologist can group the various exposed rocks into units of sufficient extent and thickness to be illustrated on a map. The rocks included in a particular mappable unit give it characteristics such as lithology, and perhaps mode of origin so that it can be distinguished from other nearby units. These units are usually called **formations.** One formation in an area might consist of thickly bedded sandstone, and another could be made up of thin alternating shale and limestone layers.

Having decided upon the criteria for distinguishing one formation from another, the geologist must next estimate the location of **contacts.** These are the boundaries between formations on the geologic map. The methods of stratigraphic correlation that we learned about in Chapter 1 are used to determine the relationship of one outcrop to another. In some places topographic features are helpful in correlation. For example, all outcrops located on the same ridge might represent the same formation, although the rock exposed in some of these is too badly weathered to display all of the correlation criteria evident in other exposures. Topographic maps and aerial photographs are helpful to the geologist attempting to correlate outcrop data. Then, guided by the locations of outcrops of the different formations and diagnostic topographic patterns, the geologist sketches on a map the probable formation boundaries, as seen in Figure 10-4. The patterns on such a bedrock geologic map indicate different tectonic structures. Before these patterns can be interpreted, we have to learn more about the typical tectonic structures that we will now examine.

Folds

We can easily recognize small folds such as the one in Figure 10-1 from the curved sur-

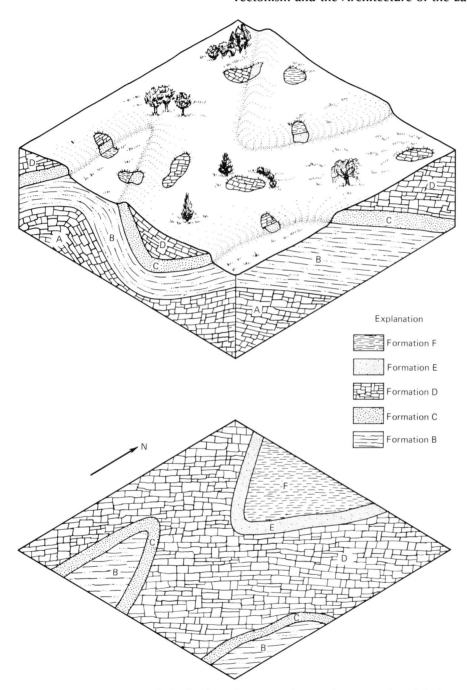

Figure 10-4 Preparation of a bedrock geologic map: basic information about lithology, strike and dip, and geologic age is obtained from scattered outcrops; rocks exposed in the area are grouped into formations according to lithological characteristics; exposures of the different formations are located on a map, and probable locations of formation contacts are drawn.

356 The Rocks of the Earth

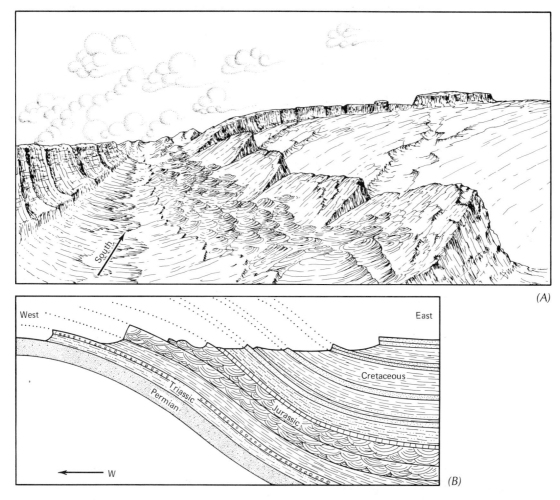

Figure 10-5 *(A)* The Waterpocket Fold is a monocline exposed in southern Utah. *(B)* A vertical cross section and schematic reconstruction of the beds in this structure illustrate the uniform dip direction characteristic of a monocline. (Redrawn from G. K. Gilbert, U.S. Geographical and Geological Survey of the Rocky Mt. Region [now U.S. Geological Survey], pages 12 and 13 in *Geology of the Henry Mountains*, 2nd edition, 1880.)

faces exposed in a single outcrop. Shapes of larger folds can be inferred from the edges of tilted strata exposed at a number of different locations. We can learn to recognize folds from patterns of curved contacts evident on geologic maps and aerial photographs. Regardless of size, folds are described and classified according to an objective geometrical scheme based upon shape and orientation.

The fold with the most simple shape is the **monocline.** This structure is a single flexure in which folded surfaces dip consistently in one direction. The Waterpocket Fold in Utah (Figure 10-5) is an example of a large monocline. A schematic reconstruction of the sedimentary beds in this structure shows how

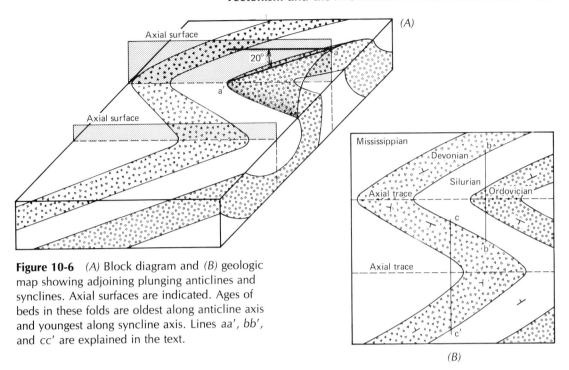

Figure 10-6 (A) Block diagram and (B) geologic map showing adjoining plunging anticlines and synclines. Axial surfaces are indicated. Ages of beds in these folds are oldest along anticline axis and youngest along syncline axis. Lines aa', bb', and cc' are explained in the text.

they existed prior to being partly destroyed by erosion. Look carefully at the ages of the beds exposed at the surface along a profile crossing this fold. Be sure to observe how successively younger beds are encountered as we cross the monocline in the direction of dip. As a result of folding and erosion, older strata, formerly buried under younger rock, have been tilted upward and exposed at the surface.

Other kinds of folds have surfaces that dip in different directions from one location to another. Look at the folded beds illustrated in Figure 10-6. Here you see two basic kinds of folds. One of these is called an **anticline**. Observe how the beds dip outward and away from the hypothetical surface that bisects the anticline. This bisecting surface is called the **axial surface** of the fold. The other basic kind of fold is called a **syncline**. It has beds that dip inward toward its axial surface. Now look at line **aa'** in Figure 10-6 that shows where a fold surface intersects an axial surface. If this line is inclined, the fold is called a **plunging** fold. This indicates that the entire structure is tilted. The direction of downward inclination is called the **plunge** direction, and the angle of plunge is measured from the horizontal. Observe that line **aa'** indicates an anticline that is plunging westward at an angle of 20 deg.

How can we rocognize anticlines and synclines from information on geologic maps? Look at the map in Figure 10-6. Note that the **axial traces** of the folds are the lines showing where the axial surfaces and the map surface intersect. From place to place there are map symbols that indicate strike and dip of the beds. Observe how these symbols indicate beds dipping inward toward the axial trace of the syncline and outward from the axial trace of the anticline. But suppose that these symbols are not given on a geologic map. What other information could be used to recognize anticlines and synclines? We can use the ages of the beds in the following way. Look at line

bb' in Figure 10-6. It crosses the anticline in a direction perpendicular to the axial trace. Note that the oldest bed is exposed at the axial trace. Younger and younger beds appear farther and farther away from the axial trace. Now look at line **cc'** that crosses the syncline. Here we see that the youngest bed is exposed at the axial trace, and older beds appear farther away. We can also use age relationships to determine the plunge direction of a fold. We already know from the block diagram in Figure 10-6 that the anticline and the syncline both plunge toward the west. Observe that younger and younger beds are exposed along the axial traces on the geologic map as we proceed in the direction of plunge. This is true for both the anticline and the syncline.

Typically, a series of plunging folds consisting of adjoining anticlines and synclines can be recognized from aerial photograph or geologic map patterns similar to those seen in Figure 10-7. Here younger sedimentary strata are located along syncline axes and older strata are exposed along anticline axes. See how the contacts form more or less V-shaped zigzag patterns. For anticlines the V's point in the down-plunge direction. Similar V-shaped contacts along synclines open outward in the down-plunge direction. Look again at Figure 10-6 where these relationships are especially clear.

Depending upon the inclination and curvature of axial surfaces, other kinds of folds can be defined. Examples of asymmetrical, overturned, recumbent, and completely inverted folds are illustrated in Figure 10-8. Look at the structure that is called a synformal anticline (Figure 10-8 D). Here the rocks were first folded into an anticline, then it was later turned upside down. Although the structure now looks like a syncline, observe that the original bedding surfaces are facing downward. A complexly deformed sequence revealing some of these exotic folds in a metamorphic terrane is shown in Figure 8-14.

Anticlinal structures in which the beds dip outward from a central point are called **structural domes.** We can recognize these struc-

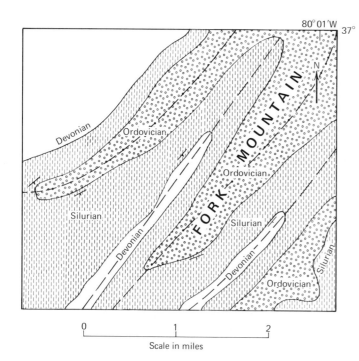

Figure 10-7 Geologic map of an area in western Virginia. The bedrock pattern indicates adjoining anticlines and synclines plunging in a southwesterly direction. (Modified from F. G. Lesure, *Geologic Map and Sections of the Clifton Forge Iron District of Virginia,* Division of Mineral Resources, Commonwealth of Virginia, 1957.)

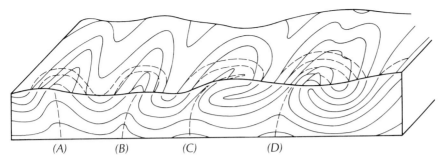

Figure 10-8 Gradations of fold geometry include (A) asymmetrical, (B) overturned, (C) recumbent, and (D) inverted folds. Note that fold (D) originated as an anticline, but now appears to be a syncline. It is called a synformal anticline.

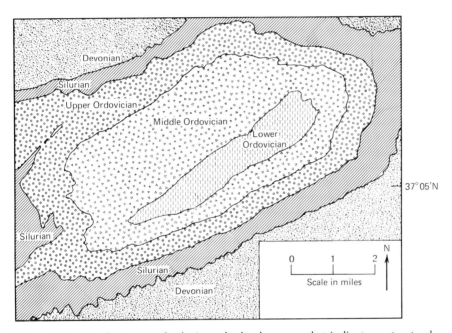

Figure 10-9 Geologic map displaying a bedrock pattern that indicates a structural dome near Burkes Garden, Virginia. (Modified from B. N. Cooper, *Bulletin 60*, Plate 1, Division of Mineral Resources, Commonwealth of Virginia, 1940.)

tures from geologic map patterns similar to those in Figure 10-9. Observe that the oldest sedimentary rocks are exposed near the center of the dome. These are bordered by younger rocks on the outer margin of the structure.

The topographic features and the formation contacts seen in Figure 10-9 reveal an oval pattern. The strata dip away from the center of a dome.

Structural basins are formed from folded

beds that dip inward toward the center of the structure. They can be recognized on geologic maps by more or less oval patterns. Younger strata in the center of a basin are bordered by progressively older sediments on the flanks of the structure. Look at the bedrock geologic map of Michigan in Figure 10-10. It reveals a broad structural basin. Known as the Michigan Basin, this large structure is easily recognized from the age relationships of the sedimentary rocks. Observe that the youngest rocks appear in the center of the basin. We could not recognize this basin by looking at aerial photographs. Here bedrock is hidden under Pleistocene glacial deposits that blanket most of the region. The geologic map was compiled from information about the lithology and age of Paleozoic sedimentary rocks obtained from widely distributed boreholes. Outcrops in this region are scarce.

It is important to realize that there is not necessarily a simple correspondence between

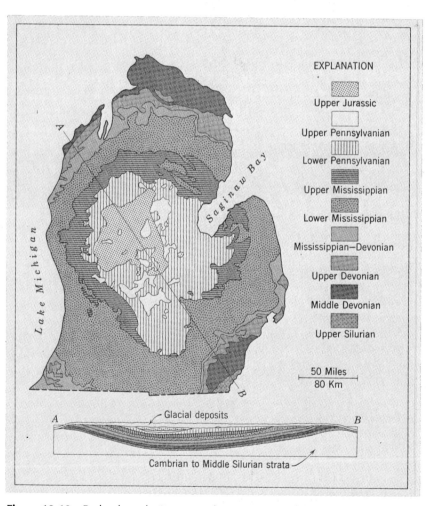

Figure 10-10 Bedrock geologic map and cross section of the Michigan Basin. (Reprinted from J. H. Zumberge and C. A. Nelson, *Elements of Geology*, 3rd edition, page 56, John Wiley and Sons, Inc., 1972.)

the form of the landscape and the form of an underlying tectonic structure. The axis of a syncline may lie along a topographic ridge in one place, and in a valley elsewhere. You can see this in Figure 10-4. Hills and ridges are usually supported by rock which is resistant to erosion. Because some rock units are more durable than others, the location of hills and ridges depends upon where these units are exposed. A fold is determined by the form of surfaces within the rock assemblage, and not necessarily by their topographic expression.

Faults

Fracture zones along which there has been relative displacement of rock masses are tectonic structures that are called faults. We can recognize a fault from abrupt dislocations of features that would otherwise be continuous. The small fault in Figure 10-1 is evident from the dislocation of sedimentary bedding surfaces. Larger faults along which massive crustal blocks have been shifted are revealed in some places by topographic discontinuities. Recall the topographic character of submarine fracture systems that were discussed in Chapter 5. The San Andreas fault in southern California is clearly marked by the abrupt truncation of landscape features seen in Figure 3-9. Most faults are not so easily recognized as these examples. They are discovered by geologists only after careful study of the structural featues and age relationships of rocks exposed in an area.

There is an objective way to describe different kinds of faults, which makes use of one important feature. This feature is the direction of displacement relative to the strike and dip of the fault surface. Look at the different faults illustrated in Figure 10-11. Here we can see that the displacement on a **strike slip** fault is horizontal and in the direction of strike. The San Andreas fault and the large submarine transform faults described in Chapter 5 are strike slip faults.

The displacement on a **dip slip** fault is along the fault surface in the direction of dip. One side that we call the upthrown side has been displaced upward relative to the other side that is called the downthrown side. The two kinds of dip slip faults illustrated in Figure 10-11 are the **normal** fault and the **reverse** fault. We can see that the surface of a normal fault dips toward the downthrown side and the surface of a reverse fault dips toward the upthrown side. The small fault in Figure 10-1 is a reverse fault. In central Utah the Wasatch Mountains rise abruptly along a normal fault that is situated along the eastern side of the Salt Lake Valley (Figure 10-12). In this part of North America the Wasatch Mountains form one of numerous mountain ranges that are separated from one another by broad valleys bordered by normal faults. The geologic cross section in Figure 10-13 illustrates how the crust in this region is broken into large blocks by steeply dipping normal faults. This is called a **block-faulted** region. The uplifted blocks are known as **horsts** and the down faulted valleys are named **graben.** These terms originated with German geologists during the nineteenth century.

Two other kinds of faults are included in Figure 10-11. The displacement is diagonal to the directions of strike and dip of the surface of an **oblique slip fault.** Rotational displacement along a fracture is characteristic of a **hinge** fault.

A special kind of reverse fault is the **thrust.** The upthrown side, called the thrust plate, has overridden the downthrown side in a manner illustrated in Figure 10-14. The fault surface may be nearly horizontal or curved in a gently dipping form. Thrust plates displaced by several kilometers or even tens of kilometers have been discovered. As a result of thrusting, sequences of younger sedimentary rock may become covered by more ancient strata. We can see in Figure 10-14 how Cambrian rock has been thrusted over younger Paleozoic sediments.

Some faults can be recognized from **scarps**

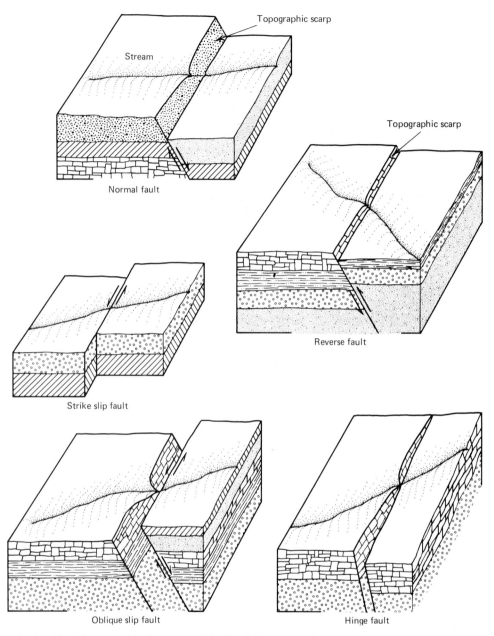

Figure 10-11 Classification of faults. *Normal fault* exhibits vertical displacement, and dips toward the downthrown side. *Reverse fault* exhibits vertical displacement, and dips toward the upthrown side. *Strike slip* fault exhibits horizontal displacement in the direction of the strike of the fault surface. An *oblique slip fault exhibits* both vertical and horizontal components of displacement. A *hinge fault* exhibits rotational displacement. A fault *scarp* is a topographic feature bordering the fault. It can develop on either the upthrown side or the downthrown side depending on which is most resistant to erosion.

Tectonism and the Architecture of the Earth's Crust   363

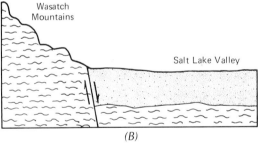

Figure 10-12 (A) The Wasatch Mountains are separated from the adjacent Salt Lake Valley by the Wasatch Fault in central Utah. (B) The geologic cross section illustrates the displacement on this normal fault. (Photograph by H. J. Bissell, U.S. Geological Survey.)

which are evident in the landscape. A fault scarp is a pronounced topographic slope, more or less linear, and located along the surface trace of the fracture. These features are labeled in Figure 10-11. In Figure 10-12 the mountains that face Salt Lake City are the eroded scarp of the Wasatch fault. As a result of faulting, relatively durable rocks are shifted into contact with rocks less resistant to erosion. The scarp is then developed by differential erosion of the land surface. The elevated side of the scarp does not necessarily correspond to the upthrown side of the fault. If the rock exposed on the downthrown side is more durable, the topographic relief could be opposite to the relative displacement on the fault. This possibility is illustrated in Figure 10-11.

Scarps are by no means evident for all faults that intersect the earth's surface. Other features including age relationships of exposed rock and offset formation contacts must be painstakingly documented to establish the existence of obscure faults that would otherwise remain hidden from the casual observer.

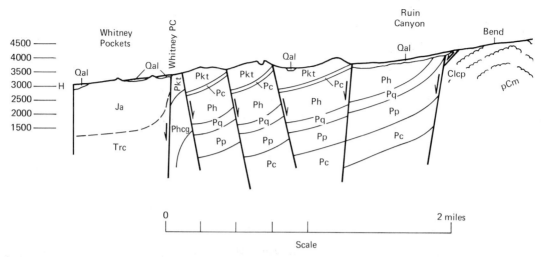

Figure 10-13 Geologic cross section illustrating block faulted structure in the Northern Virginia Mountains near the Arizona-Nevada border. Here the upper part of the crust is broken by normal faults. This structure is typical of the Basin and Range region that extends over parts of Utah, Nevada, Idaho, Arizona, and New Mexico. (After W. R. Seager, *Bulletin of the Geological Society of America*, page 1519, May, 1970.)

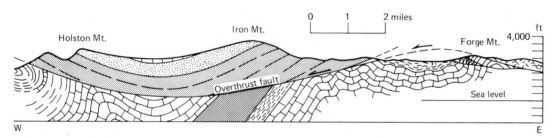

Figure 10-14 Generalized geologic cross section from Bristol, Virginia to Mountain City, Tennessee illustrates a thrust fault where the thrust plate has been displaced from east to west. (From C. Butts, G. W. Stose, and A. I. Jonas, Southern Appalachian Region, *Guidebook 3*, International Geological Congress, XVI session, 1933.)

The example in Figure 10-15 illustrates how a fault can be recognized from a geologic map, and how the directions of relative displacement can be determined. An adjoining anticline and syncline which plunge toward the south are offset along a fault that has eastward strike and southward dip. We can find the direction of horizontal displacement from the offset of the fold axes. Vertical displacement can be estimated by comparing rock ages at points where axial traces of folds intersect the fault on opposite sides. Older rock is exposed on the upthrown side, and from the information on the geologic map we see that this fault is an oblique reverse fault. The upthrown southern side has also been displaced eastward relative to the downthrown northern side. This is an example of the rea-

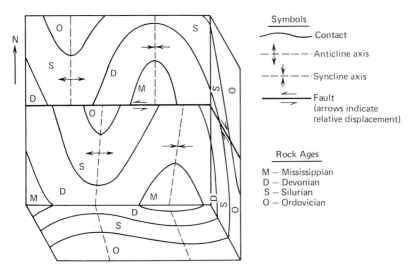

Figure 10-15 Faults can be recognized from information obtained from geologic mapping. Truncated, and offset formations are evidence of faulting. Horizontal displacement along a fault can be determined from the offset of fold axes that terminate against the fault. Vertical displacement can be estimated from differences in the ages of rocks on either side of the fault at points where a fold axis terminates against the fault. Older rock is exposed on the upthrown side.

soning used by a geologist to recognize and describe faults from data gathered on the earth's surface.

Tectonic deformation of a region can produce groups of faults oriented in systematic patterns. A pattern of fracturing located in southern California is seen in Figure 10-16. Here a fracture zone approximately 3 km wide extends for 20 km in a northwesterly direction. Within this zone a set of north-striking normal faults intersects a set of northwest-striking normal faults. Elsewhere geologists have mapped other patterns including circular and radial fault systems. In a later section we will look at scale-model experiments which point out some relationships between fault systems, folding, and the orientation of stresses that cause these features of strain.

Horizontal Strata

Many regions on the continental platforms are covered by sequences of nearly horizontal sedimentary strata. The Colorado Plateau in the southwestern United States is formed from thickly bedded sandstone formations interbedded with shales and carbonates. Nearly flatlying layers are exposed at elevations of up to 3000 m above sea level. Many of these sediments are clearly of marine origin. It is obvious that tectonic processes have lifted the beds to their present elevations. The Appalachian Plateau in the eastern United States is also formed from nearly horizontal strata. In this region these beds are found mostly at elevations that are less than 1000 m above sea level. We discussed the question of vertical

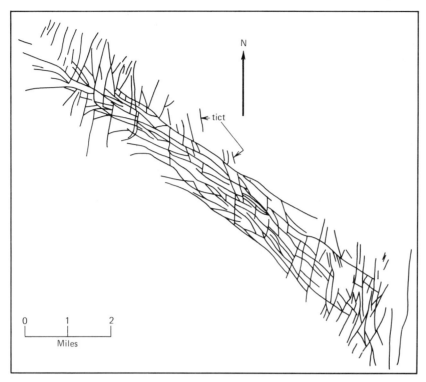

Figure 10-16 Intersecting normal faults in the Kettleman Hills of southern California. (After Ernst Cloos, *Bulletin of the Geological Society of America,* page 251, March 1955.)

uplift in Chapter 5, and you will recall that except for a few obvious examples of isostatic adjustment, the causes of uplift are not clearly understood.

We can recognize distinctive geologic map patterns for uplifted regions covered by horizontal strata. The edges of deeper layers are exposed in the topographic depressions and gorges. Because the surfaces of contact between the beds are horizontal, the formation contacts on a geologic map parallel the topographic contours. Thus we find a direct correspondence between outcrop patterns and topography. We can see this in Figure 10-17 for an area in Arizona. Here the gulleys and gorges have been cut into the horizontal strata by eroding streams. Accordingly, the outcrop patterns are bisected by the streams.

Joints

Almost all rock masses near the surface of the earth are broken by fractures along which there has been no displacement. These fractures are called joints. Many joints oriented parallel to one another at regular intervals make up a **joint set.** It is not unusual to find systems of two or more intersecting joint sets. Figure 10-18 shows how intersecting joint sets have separated the rock into large vertical columns.

Joints come in all sizes. We can find small feather joints that penetrate only a few centimeters or less in any one direction. Much larger are master joints which extend several kilometers or more. Examples of joint sets of different sizes are illustrated in Figure 10-19.

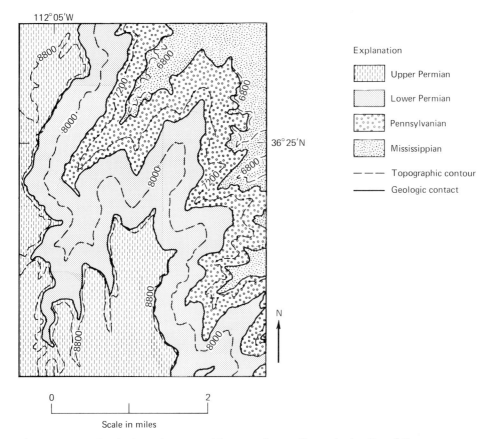

Figure 10-17 Geologic and topographic map of a small area in the Grand Canyon of Arizona. Note that contacts between formations closely parallel the topographic contours. Here the sedimentary strata are nearly horizontal. (Modified from the *Geologic Map of the Grand Canyon National Park,* Arizona, Grand Canyon Natural History Association.)

Here we can see how the spacing varies from a few centimeters or less in a set of feather joints to several meters or more in sets of larger joints. If the spacing is less than 1 cm, the set is usually called **fracture cleavage.** Intersecting joint sets that are symmetrically oriented about some other tectonic structure make up a **conjugate joint system.** The diagrams in Figure 10-20 show a conjugate joint system related to a fold.

The lineations shown on the aerial photograph in Figure 10-19 indicate regional joints extending through the Zion Canyon region of southwestern Utah. These large joints are the loci for secondary streams in the area. They have an important effect on the development of the landscape insofar as they constitute zones of weakness where the rock is more readily eroded.

Several joint sets formed at different times and by different processes may cut through a rock mass. Although some are related to particular local tectonic structures such as small folds, others are part of a larger regional sys-

368 The Rocks of the Earth

Figure 10-18 Two views of cliff faces in Zion Canyon, Utah, illustrating massive sandstone cut by intersecting joint sets.

tem. During successive episodes of tectonism older sets can become folded or offset along faults.

Tectonic Fabrics

We use the term **fabric** to describe orientations of the constituent grains in a rock. Primary depositional fabrics in sedimentary rocks such as imbrication in conglomerates (Figure 9-17) are formed during accumulation of sediment. Primary fabrics in igneous rocks come from orientation of elongate or platy mineral grains during flow of a partially molten mass. These fabrics are described as pri-

Tectonism and the Architecture of the Earth's Crust 369

(A)

(B)

Figure 10-19 Joints of differing size are illustrated by (A) regional joints in the Zion Canyon, Utah, area which extend several kilometers, and (B) feather joints in a shale bed located in western Virginia. (Aerial photograph by U.S. Geological Survey.)

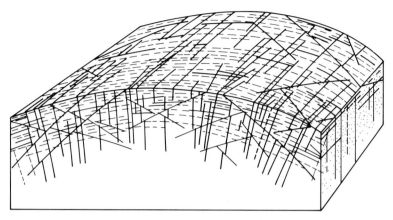

Figure 10-20 Some possible orientations of conjugate joint sets relative to a fold. (After R. Balk, *Memoir 5,* page 73, Geological Society of America, 1937.)

mary because they are developed while the rock is initially forming. Subsequent mechanical deformation may reorient the grains or deform them into shapes more compatible with the later tectonic stresses. The patterns of orientation related to such deformation are called **tectonic fabrics.**

The photograph of mineral grains in Figure 10-1 shows a tectonic fabric. Sometimes we can find a pattern like this even where larger folds and fractures are not obvious. Other features that indicate a tectonic fabric include deformed rock fragments and fossils (Figure 10-21). We know that the fossils were bilaterally symmetrical at the time of deposition. Their deformed shapes are clear evidence of subsequent mechanical deformation. The cobbles come from a conglomerate of Paleozoic age; they were fractured and deformed

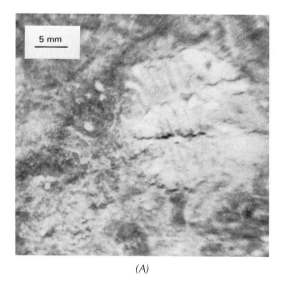

(A)

(B)

Figure 10-21 *(A)* Tectonic deformation of a trilobite fossil that was originally bilaterally symmetrical. *(B)* Cobbles in a conglomerate exposed near Fincastle, Virginia, have been flattened, fractured, and rotated as a result of postdepositional tectonic activity.

when the bed was folded, and they now make up a tectonic fabric.

Evidence of Repeated Episodes of Deformation

Sequences of deformed rock may contain tectonic structures from more than one episode of tectonism. There are places where we can observe sets of fractures formed during one tectonic event that were subsequently folded and offset along younger fractures produced during a later stage of deformation. Elsewhere **angular unconformities** such as the one pictured in Figure 10-22 separate different episodes of tectonism. Here sediments deposited during early Paleozoic time were later tilted and eroded. Another layer of sediment was then deposited over the upturned edges of the older inclined strata. The entire sequence was then further deformed. We know this because we can see that both sets of strata are now inclined. An angular unconformity is not itself a tectonic structure. It is the result of successive episodes of sedimentation, or perhaps emplacement of igneous rock, and tectonism. Sometimes unconformities can be confused with faults. Features of dissimilar orientation come in contact along both angular unconformities and faults, but faults are commonly marked by pulverized rock close to the fracture. This is not found along unconformities.

Continental Cratons and Orogenic Belts

Tectonic structure is most dramatically dis-

Figure 10-22 Angular unconformity along the Waimakariri River, New Zealand, reveals horizontal Quaternary sediment resting upon inclined beds of Teritiary age.

Figure 10-23 Generalized tectonic map of North America showing the Canadian Shield and adjacent plains and lowlands that define the cratonic region of the continent. Bordering the craton are mountainous orogenic belt regions. (Modified from Arthur Holmes, *Principles of Physical Geology*, page 1120, Ronald Press Co., 1965.)

played in the rugged relief of mountain chains that stretch in sinuous patterns over the surface of the solid earth. These ranges are the topographic expression of zones in the crust where rock is severely folded and fractured. Elsewhere on the continents the landscape for the most part is subdued, and the rocks that have accumulated since the end of Precambrian time reveal relatively minor tectonic deformation. Continental **cratons** are regions that have remained relatively stable and unaffected by tectonism since the beginning of the Paleozoic era. We can distinguish cratonic regions from **orogenic belts** where post-Precambrian tectonism has produced large-scale folding and faulting. The North American cratonic region and bordering orogenic belts are shown on the tectonic map in Figure 10-23.

An orogenic belt is produced by tectonic processes that are sufficient to cause large-scale folding and faulting. We can expect to find zones of deformed sediments, metamorphic belts, and igneous assemblages in an orogenic belt. The sequence of tectonic processes and events associated with the evolution of an orogenic belt is called an **orogeny**. In some older orogenic belts we find evidence of repeated orogenies. Each individual orogeny presumably formed a range of mountains. These ancient ranges have been reduced by erosion following cessation of tectonism, and then reformed in response to renewed tectonic activity.

The geologic cross sections in Figure 10-24 illustrate some of the tectonic structures found in orogenic belts. The folded and thrust-faulted sedimentary strata seen in Figure 10-24A are typical of the southern part of the Appalachian Orogenic Belt in North America. If we were to reconstruct these beds as they probably appeared prior to tectonism, the region would be considerably broader than the present width of this part of the orogenic belt. Clearly there has been lateral shortening of the upper zone of the crust along this profile. Large-scale folding and faulting in this region commenced in late Paleozoic time.

Now look at the typical orogenic belt structure in the Coast Ranges of California illustrated in Figure 10-24B. Here we find sediments and interbedded volcanics folded, cut by plutonic intrusions, and broken by faults. Much of the tectonic deformation in this orogenic belt on the western margin of North America occurred during Cenozoic time. The spectacular tectonic structure pictured in Figure 10-24C developed during Cenozoic time with the formation of the Alps of Switzerland. **Nappes,** which are very large recumbent folds extending several kilometers or more, are found in this region. Associated with nappes are thrusts along which displacements of several tens of kilometers have been measured. The structural features illustrated in Figure 10-24 give us some idea of the crustal deformation in orogenic belts. We can see in Figure 8-13 where the major orogenic belts in the world are located.

Cratonic regions of continents have remained relatively stable since the end of the Precambrian era. However, at one time or another portions of these regions have been subjected to mild tectonic stresses. We call these mild tectonic episodes **epeirogenic** events. Mainly they cause vertical uplift or subsidence with virtually no large-scale folding. The development of the Michigan Basin (Figure 10-10) resulted from accumulation of sediment in a cratonic depression produced by epeirogenic subsidence. The vertical subsidence amounted to more than 3 km. Other cratonic depressions illustrated in Figure 10-25 have been zones of sediment accumulation at different times since the beginning of Paleozoic time. Vast reserves of petroleum and natural gas have accumulated in these basins, as well as coal and evaporite deposits. Epeirogenic uplift raised the horizontal strata that we see on the Appalachian Plateau (Figure 10-17).

The core of a craton is the continental shield of Precambrian rock. Recall our discussion of continental shields in Chapter 8 where it was pointed out that they appear to be complexes of Precambrian orogenic belts. On the Canadian Shield rocks of similar age are distributed in approximately parallel belts. It seems reasonable to believe that Precambrian tectonism in these regions was in many ways similar to the tectonic activity in younger orogenic belts that have evolved since the beginning of Paleozoic time. This implies that the distinction between continental cratons and orogenic belts is relative. Tectonism was common to both regions but at different times. Post-Precambrian tectonic activity has occurred in both regions, but with different intensity.

The tectonic structures that we have examined in the foregoing discussion are the components of architecture in the earth's crust. In different combinations these features occur in sizes from microscopic to massive. Although most of the examples used to illustrate these structures show deformed sedimentary rocks, the structures themselves are not restricted to any particular kind of rock. Folds, faults, joints, and tectonic fabrics are found in igneous, metamorphic, and sedimentary rocks alike. The patterns of deformation revealed

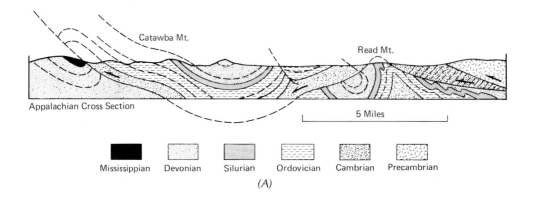
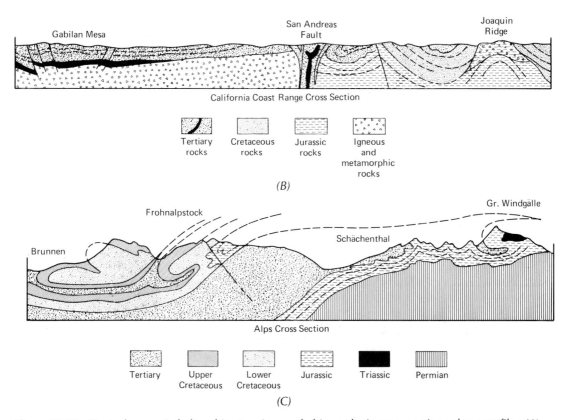

Figure 10-24 Typical orogenic belt architecture is revealed in geologic cross sections along profiles *(A)* crossing the southern Appalachian Mountains in western Virginia, *(B)* crossing the central Coast Range in California, and *(C)* crossing the Swiss Alps. (From *(A)* Charles Butts, G. W. Stose, and Anna I. Jonas, Southern Appalachian Region, *Guidebook 3*, International Geological Congress, XVI session, 1933; *(B)* N. L. Taliaferro, Geologic History and Structure of the central Coast Ranges of California, California Dept. of Natural Resources, Division of Mines, *Bulletin 118*, Plate II, Section V, 1943; *(C)* M. Lugeon, *Soc. Geol. de France Bulletin*, Ser. 4, Vol. 1, PP. 723-813, 1901)

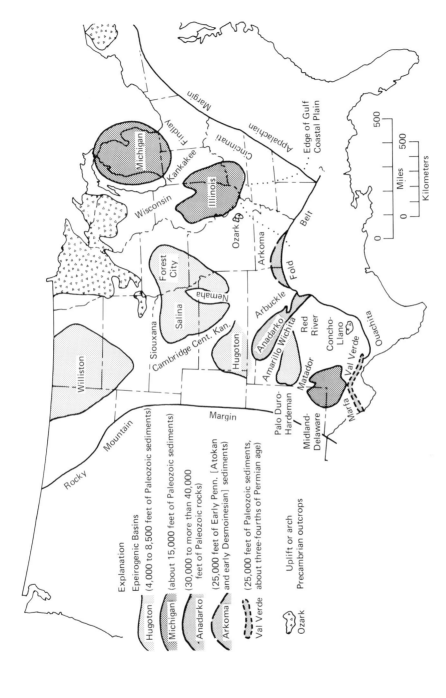

Figure 10-25 Regional eperiogenic basins and arches in the cratonic region of the United States. (After W. E. Ham and J. L. Wilson, *American Journal of Science*, pages 334–335, May 1967.)

by these tectonic structures must be identified so that we can determine how different kinds of rock are arranged in the crust. Furthermore, we learn from these patterns about the tectonic processes that were active during earlier periods of earth history.

MECHANICAL ANALYSIS OF TECTONIC STRUCTURES

Folds, fractures, and tectonic fabrics tell us about the response of rocks to **stress.** Recall that stress is described by a force acting upon a surface. The resulting deformation is defined as **strain.** Tectonic structures are evidence of strain. From a study of these features the geologist attempts to learn about the stresses that produced them. You were introduced to relationships between stress and elastic strain in Chapter 3. Then in Chapter 5 we discussed the response of viscous fluids to stress. In the following discussion we will look at some ways that stress can produce plastic strain and fractures in rocks.

We do not observe tectonic stresses directly. Instead, we infer that they existed from the deformation they produced. The strain that we can observe within a particular mass of rock resulted from external stresses acting on the outer surface of the mass, and from internal stresses that existed within the mass. External mechanical stresses on this mass are related to the way that surrounding materials pull it or press against it. Internal stresses are related to the ways that constituent particles press against one another. Our purpose here is to determine what stresses may have existed to produce patterns of strain revealed by tectonic structures.

Relationships between external stress and strain patterns can be established experimentally by observing the response of masses of simple shape to stresses of known magnitude and direction. Some examples will help us to understand the nature of external stresses and the strain they can produce.

Nondirected Stress

Let us begin by examining the stress over the surface of the small sphere illustrated in Figure 10-26. Let the sphere be submerged in a fluid of density ρ at the earth's surface where the acceleration of gravity is g. How does the fluid press against this sphere? The weight of the fluid on the small area designated A_1 is a force with the magnitude: $F_1 = mg$, where m is the mass of the vertical column of fluid above A_1. This mass is: $m = \rho h A_1$, where h is the length of the vertical fluid column and corresponds to the depth of A_1 beneath the upper surface. Pressure is: $P_1 = F_1/A_1 = \rho g h$. We can see that pressure is a stress because it is expressed as force divided by area.

We note that the force on some other similar small area elsewhere on the sphere, for example A_2, must have the same magnitude as F_1 and must be directed perpendicular to that surface. If this were not so, the sphere would be displaced to another position. Insofar as the fluid is not flowing the forces acting on the sphere from different directions must mutually cancel one another. This is not to say that the forces do not exist; they do.

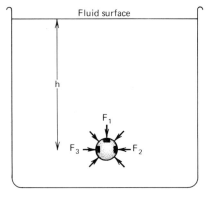

Figure 10-26 Nondirected stress on the surface of a sphere submerged in a fluid. If the sphere is taken to be infinitely small, then forces acting on the sphere from all directions are of equal magnitude. Small increments of area are shown in black.

Their net effect, however, is to maintain the center of the small sphere in the same position. In this sense the force is directed equally from all directions on the surface of the sphere. The pressure on this surface is an external stress on the sphere. It is directed perpendicular to the surface, and is of equal magnitude from all directions on the surface of the sphere. It is called a **nondirected stress.** Pressure within any fluid mass in which there is no flow is a nondirected stress.

The strain caused by nondirected stress is revealed by a change in volume. This volume change in an elastic substance is related to the nondirected stress by a constant of proportionality, the bulk modulus, which we learned about in Chapter 3.

Directed Stress

Let us next consider external stresses that are more intense in one direction than in another. We can use the experimental apparatus illustrated in Figure 10-27 to apply force over only part of the surface of a specimen.

The external stress is perpendicular to the ends of a cylindrical rock specimen in Figure 10-27 A. Stresses resulting from forces acting perpendicular to a surface are called **normal stresses.** They are **compressive** if in-directed and **tensile** if out-directed. In Figure 10-27 B the force is applied in a direction parallel to the ends of the specimen. This causes a **shear stress.** In both of these examples parts of the specimens are free of external stress.

By encasing the specimen in a container filled with fluid, as illustrated in Figure 10-27 C, external stress can be generated on the sides as well as at the ends of the specimen by increasing the fluid pressure. The entire surface of the specimen can be subjected to stress; however, the stress can be of different magnitude and direction from one part of the

Figure 10-27 Experimental apparatus for applying (A) directed normal stress; (B) directed shear stress; and (C) nondirected and directed normal stresses on cylindrical rock samples. The nondirected stress is produced by fluid pressure in the chamber containing the specimen.

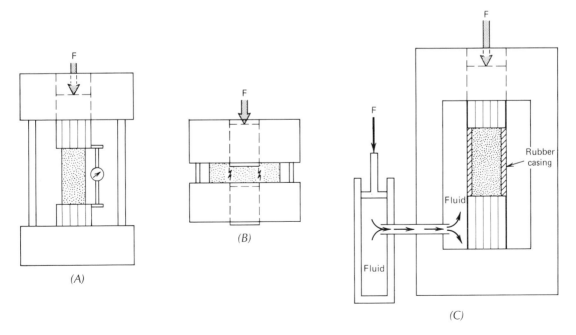

specimen to another. Under these conditions the total external stress is seen to be the sum of the nondirected fluid pressure, and some other directed stress. We refer to the nondirected component as the **confining stress.** Apparatus such as this can be used to simulate conditions within the earth. Rock masses at depth are subjected to confining stresses related to the weight of the overlying materials, as well as other normal and shear stresses which are more intense in one direction than in another. We can simulate temperature conditions in the earth by controlling fluid temperature in the apparatus.

The strain caused by confining stress is a volume change. Directed stresses cause strain that gives the specimen a distorted shape. Insofar as the strain is directly proportional to the stress, persisting only as long as the stress exists, we call it elastic strain. Plastic deformation is strain that is not directly proportional to stress. Ordinarily a plastically deformed substance retains some distortion after the stress ceases. In this sense folds and tectonic fabrics indicate plastic strain. Fractures result when materials break in response to stress.

Stress and Strain in Rock Specimens

The way that a rock responds to directed stress depends upon the associated confining stress and the temperature of the environment. Geologists have tested the response of rock specimens with apparatus similar to that pictured in Figure 10-27C. Some results are presented in Figure 10-28 for specimens of

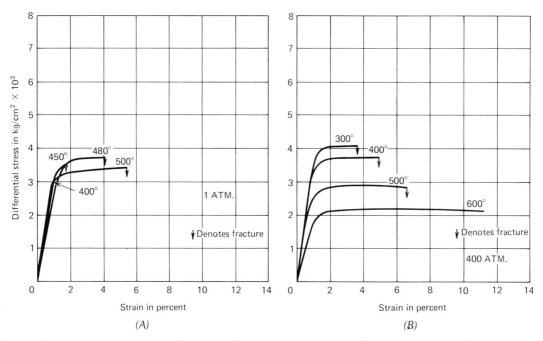

Figure 10-28 Stress-strain curves for Solenhofen Limestone specimens in compression at different temperatures and confining stresses: (A) at $P \simeq 1$ bar (temperature indicated by each curve); (B) at $P \simeq 400$ bar (temperature indicated by each curve). (After H. C. Heard, *Memoir 79* [D. Griggs and J. Handin, editors], page 201, Geological Society of America, 1960.)

Tectonism and the Architecture of the Earth's Crust

Solenhofen Limestone, a formation of particularly uniform composition and texture. Cylindrical specimens were subjected to compressive normal stress on the ends. Experiments were repeated under different conditions of temperature and confining stress. The resulting strain is expressed by the percent change in specimen length parallel to the direction of normal stress. Look at the parts of the graphs where strain is smaller than 1 percent. What kind of strain is this? Observe that the curves appear to be straight lines. This tells us that the strain is directly proportional to stress. Therefore we can see that strains smaller than 1 percent in these specimens are elastic strains.

Plastic deformation began after the specimens had undergone strains of more than 1 percent. Finally, fractures developed at strains of between 1 and 11 percent, depending on the temperature and the confining stress. But we have learned from other experiments that the time taken to build up stress is also important. If the stress increases slowly, then plastic strain can greatly exceed the limits shown in Figure 10-28.

The levels of temperature and stress used in these experiments also exist in the earth's crust. The results indicate how much elastic and plastic strain crustal rocks can endure before breaking. The graphs in Figure 10-28 tell us that rocks behave as brittle substances near the surface. But deeper in the earth these same rocks deform plastically because of the higher temperature and confining stress.

We have learned that the orientation of fractures is related to the direction of stress. Examples in Figure 10-29 illustrate fractures produced by external compressive, tensile, and shear stresses. The fracture surfaces that form under compression make angles of approximately 30 deg with the stress direction. Tensile stress produces fractures perpendicular to its direction. Fractures formed by external shear stresses make angles of more than about 60 deg or less than about 20 deg with the direction of the force causing the shear stress.

The relationships between external stress and strain patterns are summarized in Figure 10-30. The diagrams indicate that masses initially spherical in shape are deformed into ellipsoids by external stresses. This is true regardless of whether they are compressive, tensile, or shear stresses, or a combination of these. The fractures that are produced can be described with respect to the axes of the ellipsoid. We call this ellipsoid the **strain ellipsoid.** Fractures that form perpendicular to its long axis are called **tension fractures** regardless of the nature of the external stress that actually caused the deformation. Fractures oriented at about 30 degrees to the short axis and parallel to the intermediate axis are called **shear fractures** even though they may have been caused by normal external stresses

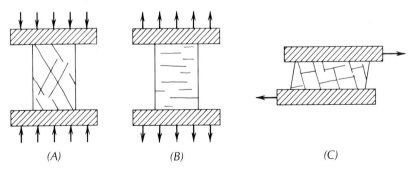

Figure 10-29 Fractures in rock specimens under (A) compression, (B) tension, and (C) shear.

380 The Rocks of the Earth

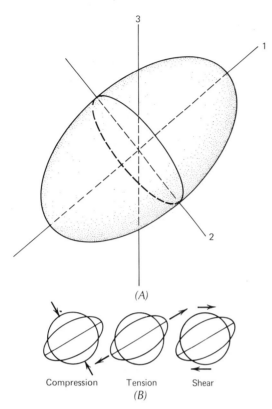

Figure 10-30 The strain and fracture of a material caused by stress can be illustrated by ellipsoids obtained from deformation of originally spherical units. (A) The strain ellipsoid is described by the 1-axis (long), the 2-axis (intermediate), and the 3-axis (short). Tension fractures, if formed, are perpendicular to the 1-axis. Shear fractures, if formed, are along surfaces parallel to the 2-axis, and at angles of less than 45 deg with the 3-axis. (B) The strain ellipsoid can be formed from purely tensional, compressional, or shear stress, or a combination of these. Fractures in a particular substance maintain a consistent orientation relative to the ellipsoid regardless of the causitive stress system.

as well as by shear stresses. The fractures, then, are classified according to their orientation in the strain ellipsoid. The same strain ellipsoid could be formed by different combinations of external compressive, tensile,

and shear stresses. Consequently, we cannot relate a particular pattern of fractures and tectonic fabric to a unique system of external stresses without some additional information about the origin of the stresses.

The tectonic fabrics illustrated in Figures 10-1 and 10-21 indicate strain that could have been caused by different combinations of external stress. If we know the original shapes of objects in the rock such as deformed fossils, it is then possible to find the orientation of the hypothetical strain ellipsoid that represents the distortion of that particular part of the rock mass. Knowing this we can infer directions of either normal or shear stresses which independently could have produced the tectonic fabric. But we cannot tell if the deformation resulted from a combination of external normal and shear stresses.

Model Studies of Tectonism

What stresses in the crust could produce the block faulted terrane (Figure 10-13) typical of the Great Basin of North America? How do zones of intersecting faults (Figure 10-16) develop? Can folds and thrust faults be produced by the same external stresses? We can find some answers to these questions by testing laboratory scale models. Look at the experiments illustrated in Figure 10-31. Moist clay and loose sand have proved to be good materials for testing the deformation caused by different stresses. Before a model is deformed reference circles are marked on the surface of the clay in different places. Later as the mass is deformed we can see how these circles become altered to ellipses that indicate the strain. Consider first the deformation produced by a condition of stress that we call **pure shear** (Figure 10-31 A). As the grating upon which the clay rests is extended, circles marked on the clay surface are deformed into ellipses of strain. The axes of these ellipses remain parallel and perpendicular to the direction of extension. A pattern of intersecting fractures is oriented at angles of approxi-

Tectonism and the Architecture of the Earth's Crust 381

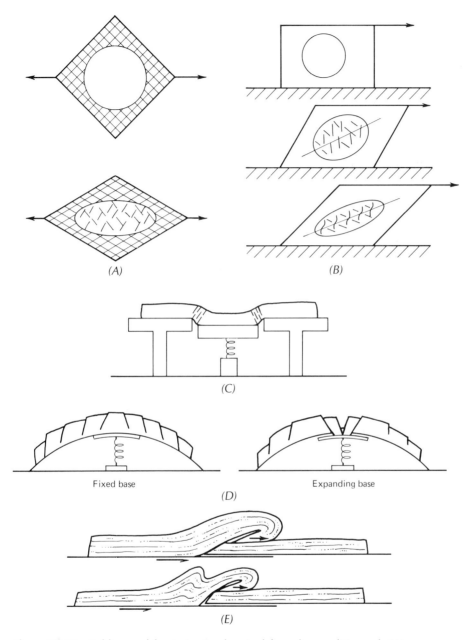

Figure 10-31 Folding and fracturing in clay models under conditions of: (A) pure shear; (B) simple shear; (C) subsidence at the base of the layer; (D) doming of the layer; (E) converging basal plates. (After E. Cloos, *Bulletin of the Geological Society of America*, March 1955.)

mately 30 deg to the short axis of the ellipse. This is the principal axis of compression. The orientation of these fractures remains unchanged as deformation progresses.

Next, look at the deformation produced by a stress condition that we call **simple shear** (Figure 10-31 B). Here we observe that the axes of the strain ellipses rotate with continuing deformation. Fractures, too, are rotated. The first ones to form later become sigmoidal or S-shaped. Eventually, zones of intersecting fractures develop that are similar to the pattern seen in Figure 10-16. A fault zone such as this is called a shear zone.

Fractures can be developed near the margins of a subsiding block in another experiment (Figure 10-31 C). Notice how they form an intersecting system that does not rotate while deformation progresses. But one fracture set in this system is dominant. Look at the displacement along these dominant fractures. We can see that they are reverse faults. Initially the fractures develop in the lower portion of the clay cake while the upper portion is simply downwarped into a synclinal fold. With continued subsidence the reverse faults penetrate the entire thickness of the clay cake.

What kind of faults are produced by vertical uplift? In Figure 10-31 D we can see two ways that uplift of a clay cake can be accomplished. In one experiment the original length of the base is preserved, and in the other experiment this length is continuously extended during uplift. In both experiments two intersecting sets of fractures develop, and one set is dominant. A slight rotation of early formed fractures results as deformation continues. We can tell from the displacements along the dominant fractures that they are normal faults. They are, therefore, distinctly different from the reverse faults (Figure 10-31 C) that develop during subsidence.

Finally we will consider an experiment designed to produce a thrust fault and nappe structure (Figure 10-31 E). The experiment begins with a clay cake resting upon two sheets. One sheet then begins to slide beneath the other, which produces compression in the clay at the junction. An anticlinal fold develops initially, and it then becomes altered to an overturned and finally a recumbent fold. Still further movement produces a thrust fault and the growth of a second fold. These features bear some resemblance to the structures in Figure 10-24. Observe that the folding and fracturing occur in the immediate vicinity of the junction of the sheets beneath the clay. The clay remains essentially undeformed farther from this junction. This indicates that the internal stresses sufficient to produce plastic deformation and fractures are limited to the zone close to the largest concentration of external stress.

At this point we have looked at the important kinds of tectonic structure, and we have examined some simple relationships between stress and strain. We turn next to speculation about how stress builds up in the lithosphere to produce tectonic structures.

OROGENY

Mountain building is an important part of the evolution of an orogenic belt. Modern ideas about how orogenic belts evolve come from the plate tectonic theory that we learned about in Chapter 5. Before we discuss these ideas we need to summarize our knowledge about the important features of orogenic belts.

Orogenic Belt Form

Orogenic belts are linear or gently curved arcuate regions. Most of them are on continental platforms near the margins of these platforms. They consist of more or less parallel zones. Each zone has distinctive lithologies and patterns of deformation. The zones are asymmetrically arranged so that the orogenic belt itself is asymmetrical in cross section.

Orogenic Belt Sedimentation

Sedimentary rocks, mostly of marine origin, are widespread in an orogenic belt. Here the total thickness of sedimentary rock is considerably greater than in regions adjacent to the orogenic belt. The oldest sediments were deposited before the beginning of intensive tectonism. These preorogenic sediments accumulated in broad elongate areas that spanned much of the region where the orogenic belt evolved later. Other sediments in the orogenic belt were deposited while intensive tectonism was in progress. These are syntectonic sediments. They are mostly impure sandstones and mudstones with sedimentary structures that indicate rapid accumulation. Debris apparently transported by turbidity currents is abundant. The sediments were deposited in troughs parallel to zones that were being uplifted and tectonically deformed. The youngest sediments in an orogenic belt were deposited after the time of intense tectonism. These post-tectonic rocks are relatively undeformed in comparison with the pre- and syntectonic sediments. The proportions of pre-, syn-, and post-tectonic sedimentary rocks vary from one orogenic belt to another.

Tectonism and Epeirogeny

Episodes of intense tectonism are short compared with the long intervals of time during which only minor epeirogenic events occurred, and preorogenic sediment accumulated. During the short episodes of tectonism extensive systems of folds and thrusts are developed. The deformed rocks indicate lateral shortening, that is, a decrease in the width of the zone of deformation. This may, but does not necessarily imply crustal shortening. Tectonic structures developed during the episode of orogeny tend to be asymmetrically arranged. This asymmetry is indicated by the directions of overturned folds and displacement on thrusts that are toward one side of the zone of deformation rather than the other.

During an episode of orogeny the intensity of tectonism varies from one zone to another in an orogenic belt. Not all zones are contemporaneously deformed. One temporarily quiescent zone may receive deposits of sediment from a bordering tectonically active zone. Later, tectonic stresses act on this zone and deform these new sedimentary deposits. Furthermore, episodes of orogeny are not necessarily contemporaneous along the length of an orogenic belt. One end of the belt can be tectonically active while the other end is quiescent.

Following orogeny is a period of regional uplift. During this time there is no large-scale tectonism. In the mountain building stage, when deformed rocks are lifted to great heights, the processes of erosion commence to carve them into a mountainous landscape.

Igneous Activity and Metamorphism

Igneous activity in orogenic belts is contemporaneous with episodes of tectonism and periods of postorogenic uplift. Lavas erupted during orogenic volcanism are derived from magmas generated in the upper mantle and from lower continental crustal zones as well. Syntectonic and post-tectonic granitic plutons are implaced in the upper continental crust. Relatively little upper crustal material is melted and assimilated into these magmas.

Regional metamorphism occurs at intermediate continental crustal depths. Suites of rocks known as **migmatites** are formed by injection of granitic fluids along partings parallel to foliation in metamorphic masses. The granitic fluids may come from partial melting within the zone of metamorphism, but the processes are by no means clearly understood. Zones of plutonic rock, migmatites, and other metamorphic rock are uplifted during the postorogenic period. Later they are exposed by erosion of the landscape.

Suites of mafic and ultramafic rocks called

ophiolites occur in some orogenic belts. They are rare but important because they indicate mass transport of considerable magnitude. The constituents of ophiolites are similar to oceanic crustal rocks and upper mantle rocks. The ophiolites found in the upper continental crust along orogenic belts may be upthrusted slivers of adjacent oceanic lithosphere.

Igneous activity and metamorphism can be particularly intense in one zone of an orogenic belt, whereas other zones are relatively quiescent.

Mountain Building Cycles

The evolution of a mountain system can be traced through different stages. Ages of sedimentary rocks involved in this cycle can be obtained from fossils. Emplacement of igneous rock and metamorphic events are dated from radioactive isotope analysis. From these sources of information preorogenic, orogenic, and postorogenic stages of orogenic belt evolution can be recognized. The first is usually much longer than the latter two, commonly lasting 100 to 200 million years or more. Stages of orogeny and postorogenic uplift take less than 10 million years. A cycle of mountain building, then, appears to begin slowly with a prolonged preorogenic stage, then terminates rather abruptly with orogeny and postorogenic uplift. This cycle must be viewed as an oversimplified generalization. Some orogenic belts display evidence of repeated mountain building and episodes of orogeny at highly irregular intervals. In other younger orogenic belts only one episode of mountain building can be discerned.

We have now examined the principal features of orogenic belts that must be explained by theories of mountain building. Now let us consider how these features might be produced by plate tectonic processes. At the present time we observe indications of tectonic activity including seismicity and volcanism mostly near tectonic plate boundaries. Under what conditions might orogenic belts evolve from plate margin processes? To answer this question recall that lithospheric plates can be bounded by destructive, constructive, or conservative margins. Movements of plates are probably related to movement of material in the underlying asthenosphere, perhaps by convection processes. Along destructive boundaries oceanic lithosphere is subducted and then assimilated into the asthenosphere. Continental subduction has yet to be demonstrated and probably does not occur extensively because of the buoyancy of the granitic continental crust. Recently geologists have suggested possible sequences of plate movements that could lead to mountain building. We will consider individually the sequences associated with destructive, constructive, and conservative borders.

Orogeny on Destructive Plate Margins

We shall examine three different ways that mountain building can occur along destructive plate boundaries. These are summarized briefly in the following statements.

1. A new destructive plate boundary forms along the margin of a continent and ocean basin in a formerly unbroken lithospheric plate. An orogenic belt evolves as oceanic lithosphere is subducted into the asthenosphere.

2. Two continents rafted upon converging plates collide. An orogenic belt evolves as the continents approach and then merge.

3. An island arc and a continent rafted upon converging plates collide. An orogenic belt evolves as these crustal units draw together and merge.

First we consider what could happen when the lithosphere breaks so that a new destruc-

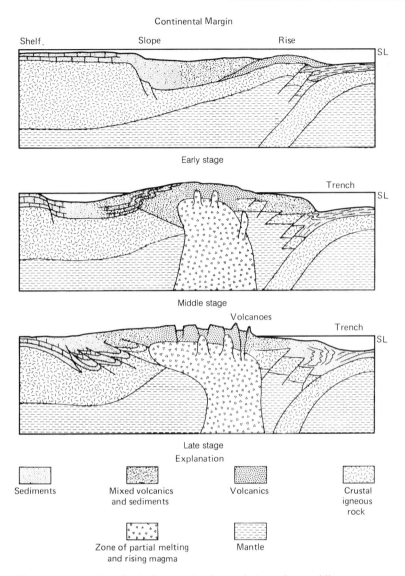

Figure 10-32 Hypothetical stages in the evolution of a cordilleran-type orogenic belt related to subduction of oceanic lithosphere along the margin of a continental plate. (After J. F. Dewey and J. M. Bird, *Journal of Geophysical Research*, page 2638, May 10, 1970.)

tive plate margin is formed along the border of a continent and an ocean basin. Stages in the evolution of an orogenic belt following such an event are illustrated in Figure 10-32. Initially the lithosphere yields to stresses by arching upward along the continental slope and breaking along the outer margin of the slope. Before this happened sediment that washed from the adjacent continent had been accumulating in preorogenic deposits on the submerged shelf and slope. Then, as arching commences masses of sediment break loose and are transported in turbidity currents and submarine mud flows. Then they are redeposited and make up beds of poorly sorted syntectonic sediment. As newly subducted oceanic lithosphere is transported to depths of 100 to 200 km, melting commences (Figure 10-32 A) and rising magmas are extruded in submarine eruptions. The migration of magma accelerates upward flow of heat that causes partial melting in the lower continental crust. Gabbroic and granodioritic magmas rise from this zone to form a mobile core of plutonic rock and migmatite along the arched zone of the continental slope. Continued arching of this zone (Figure 10-32 B) forms an orogenic **welt** or wrinkle in the lithosphere that eventually rises above sea level. Now there is a new source area from which sediment can be eroded and transported to bordering submerged zones.

Persistent horizontal compressive stress continues to be exerted on the continental plate by the converging oceanic plate, which causes asymmetrical distortion of the mobile core. The zone of deformation shifts toward the continent while overfolding and thrusting occurs in the older continental shelf sediments and in the younger syntectonic deposits eroded from the adjacent orogenic welt. Finally the entire region begins to rise (Figure 10-32 C), probably in response to isostatic stresses. Late forming SiO_2-rich magmas rise into the upper part of the orogenic welt to form granitic plutons, or erupt to form rhyolitic and andesitic volcanic features. A mountainous landscape forms from erosion, and sediment is transported to adjacent zones of lower elevation where it accumulates in post-tectonic deposits.

The sequence of events outlined above and in Figure 10-32 could explain many of the features of the Andes along the western margin of South America. Here the geologic evidence indicates that subduction of oceanic lithosphere commenced in early Mesozoic time. Growth of the Andes has continued into the Pleistocene epoch, and the subduction processes continue actively at the present time along this orogenic belt. This kind of mountain system is called a **cordilleran**-type belt. We do not know how much longer tectonic activity will continue in this orogenic belt. In its initial stage of development there were shallow submarine zones where sediment could accumulate. But in later stages these zones disappear, and the post-tectonic sediments appear to be too far inland to be deformed by horizontal stresses created by the converging plates. Perhaps the Andean mountain building cycle is nearly complete.

We consider next the evolution of an orogenic belt as continents on converging plates approach and collide. Look at the hypothetical sequence of events illustrated in Figure 10-33. One continental margin borders the destructive plate boundary. As the two continents approach there is an ever narrower ocean basin where preorogenic sediment accumulates. Sediments in this basin are compressed and tectonically deformed as closing of the sea continues. Eventually, as the two continents converge, the intervening marine sediments become highly deformed and uplifted. Remnants of oceanic crust, arched upward just prior to collision, are broken and thrusted as ophiolite slivers into the mass of deformed sediment. Subduction ceases when the continents finally collide, because the relatively low density granitic crust is too buoyant to be drawn into the asthenosphere. With the joining of continental crustal plates horizontal stresses are rapidly dissipated, and the episode of tectonism draws to a close. In the final stage of mountain building there is regional uplift related to isostatic adjustment. This is followed by erosion and deposition of post-tectonic sediment in bordering regions. This kind of continental collision model can account for important features of the Himalayan Mountain chain and the Alpine chain of southern Europe.

The third model of orogenic belt evolution, involving the approach and collision of a continent and an island arc on converging

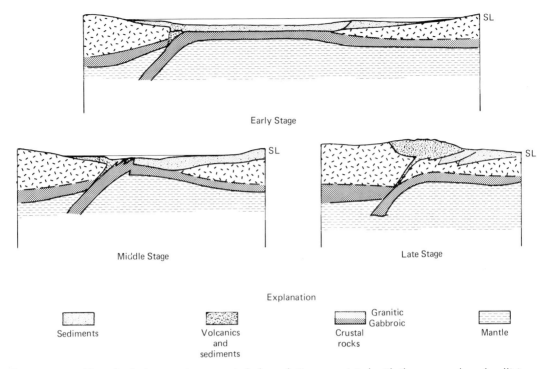

Figure 10-33 Hypothetical stages in orogenic belt evolution associated with the approach and collision of two continents rafted on converging lithospheric plates. (After J. F. Dewey and J. M. Bird, *Journal of Geophysical Research*, page 2642, May 10, 1970.

plates, is illustrated in Figure 10-34. As with the continent-continent collision model, this one has a narrowing ocean basin bordered by land masses that can supply preorogenic sediment. The preorogenic sediment accumulates on the margins of this sea in the early stages, then fills the narrow basin shortly before the island arc and the continent merge. These accumulations of sediment become highly deformed and uplifted as a result of collision. Owing to the buoyancy of both the continent and the island arc, however, subduction along their juncture ceases after collision. Persisting stresses on the lithosphere might then cause a new subduction zone to develop on the seaward margin of the island arc. Perhaps this has happened near the Japanese island arc-trench system. Do you think that this idea is supported by the arrangement of paired metamorphic belts in Japan? Look again at the discussion of these features near the end of Chapter 8.

Orogeny on Constructive Plate Margins

Submarine ridges, the submerged mountain systems of the ocean basins, lie along constructive plate margins. Magma rising from the asthenosphere crystallizes to form new lithosphere beneath these impressive topographic features. Shallow seismicity and volcanism tell us about ongoing tectonic activity along these plate borders. Is it possible for constructive plate margins to develop beneath a continental platform? Evidence from

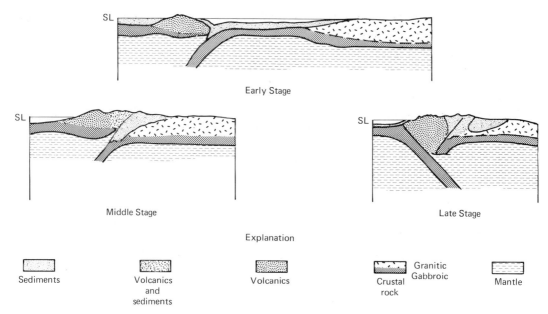

Figure 10-34 Hypothetical stages in the development of an orogenic belt related to the approach and collision of an island arc and a continent rafted on converging lithospheric plates. Note the cessation of subduction at the collision margin and commencement of subduction along the seaward margin of the island arc. (After J. F. Dewey and J. M. Bird, *Journal of Geophysical Research,* page 2641, May 10, 1970.)

different parts of the world suggests that this may be happening. The hypothetical model shown in Figure 10-35 illustrates possible stages of development. Extension of the continental crust first produces a block-faulted structure with tilted crustal blocks, grabens, and horsts. These structures contribute to topographic irregularity so that continental sediment eroded from the ridges accumulates in the valleys. Further extension of the crust exaggerates these structural features and is accompanied by injection of igneous dikes and volcanic fissure eruptions. The block-faulted mountains at this stage would be similar to those now seen in the Great Basin of the western United States. Typical structural patterns for this region are seen in Figures 10-12 and 10-13.

The continental crust eventually becomes extended and thinned to the point of being completely torn apart. Continued spreading is then accompanied by generation of new oceanic crust. Here we find the beginning of a new ocean basin, which will continue to widen with further spreading of lithosphere. Seismic, aeromagnetic, and gravimetric measurements have been used to learn about the crustal structure beneath the Red Sea, the Gulf of Aden, and bordering regions. The results have been interpreted to indicate a proto-ocean basin (Figure 10-36) that is just beginning to form. Perhaps the Atlantic Ocean had similar beginnings during the Cretaceous period.

Conservative Plate Margin Tectonics

Can mountain systems form along a conservative plate boundary? Modern seismicity

Tectonism and the Architecture of the Earth's Crust

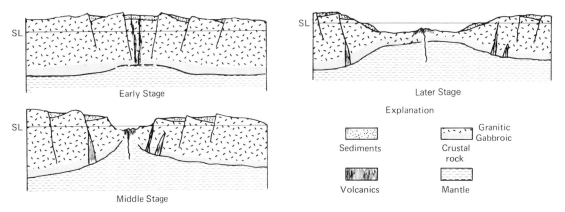

Figure 10-35 Hypothetical stages of continental rifting, associated block faulting, and eventual growth of oceanic crust related to formation of a constructive plate boundary beneath a continent, and subsequent lateral spreading of lithospheric plates away from this margin. (After J. F. Dewey and J. M. Bird, *Journal of Geophysical Research,* page 2629, May 10, 1970.)

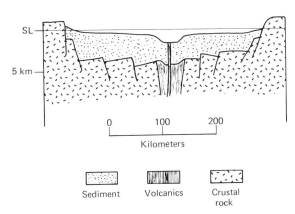

Figure 10-36 Cross section illustrating geologic structure across the Red Sea. (After R. V. Girdler, *Paper 66-14,* pages 65–75, Geological Survey of Canada, 1966.)

studies tell us that fault displacement is mostly horizontal along these boundaries. This activity is accompanied by relatively minor volcanism. Where are the compressive stresses to buckle and thrust the crustal rock, or tensional stresses to tilt crustal rocks into block-faulted ridges? Answers to this question are not readily evident. Furthermore, it is not clear if any mountain systems have, in fact, evolved along conservative plate margins.

A grand mountain system, the Transantarctic Range, extends almost 3000 km through the center of Antarctica. Peaks rising to over 4000 m are found in this range. Some typical cross sections are seen in Figure 10-37. Here we find a mountain system of asymmetrical cross section. A relatively thick continental crust (40–45 km) abuts a thin continental crust (28–30 km). Pleistocene volcanic rock is abundant. The dominant rock unit, however, is a thickly bedded sandstone more than 1500 m thick in places. It displays no significant folds, and it is broken only by normal faults. This relatively undeformed sedimentary unit has been uplifted to heights of more than 4000 m. An exotic landscape has been carved in it by glacial erosion. How was it uplifted with only minor tectonic deformation? Has it evolved along a conservative plate boundary? These questions are yet to be answered, and continue to challenge the imagination of the geologist.

In the foregoing discussion we have reviewed some modern ideas about orogenic belt evolution and mountain building that are based on the plate tectonic theory. Many questions remain. The nature of conservative plate margin tectonics has received little at-

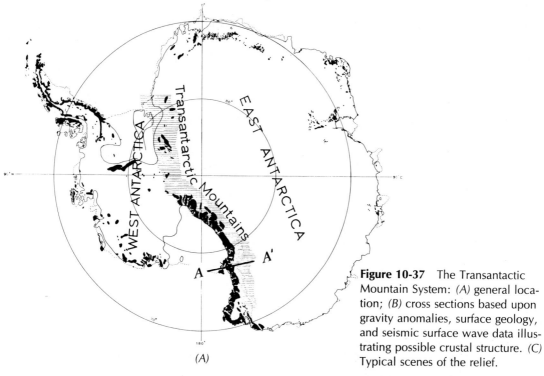

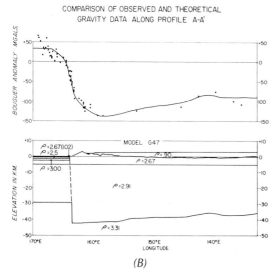

Figure 10-37 The Transantactic Mountain System: *(A)* general location; *(B)* cross sections based upon gravity anomalies, surface geology, and seismic surface wave data illustrating possible crustal structure. *(C)* Typical scenes of the relief.

tention. Mechanisms for intracratonic epeirogeny, possibly involving downward convection of the asthenosphere, are not clearly understood.

NATURAL RESOURCES AND TECTONIC STRUCTURE

Tectonic structures control the emplacement of natural resources in many regions of the earth's crust. To an important extent exploration for certain raw materials is based upon knowledge of their patterns of occurrence in different tectonic settings. This knowledge comes mainly from discoveries made in the course of drilling, mining, and quarrying operations. In previous chapters we have discussed some geochemical and depositional factors that contribute to the concentration of economically valuable commodities. The purpose of this section is to present some of the typical structural settings of petroleum and natural gas and ore deposits.

Occurrence of Petroleum and Natural Gas

Petroleum and natural gas are found almost exclusively in sedimentary settings. They are concentrated in the pore space of subsurface reservoir rocks, mostly carbonates and sandstones. Recall from Chapter 9 that these fluids probably form in marine mud deposits that are eventually lithified into beds of mudstone. Some mudstones have retained petroleum which then turns out to be very difficult to extract from the rock. Examples of such mudstones are the Green River oil shales of Colorado, Utah, and Wyoming. But ordinarily the petroleum and natural gas escape from the mother rock during compaction, and migrate into the carbonate and sandstone reservoirs. We will consider some current ideas about the migration and concentration of these fluids. The important North American reservoirs are found in the regions outlined in Figure 10-38.

The migration and concentration of petroleum and natural gas are governed by the extent to which fluid can percolate through interconnected pores in the rock and the different densities of the subsurface fluids. Pore space is occupied mostly by water containing various dissolved ions. Most petroleum has lower density than the ground water solution and is not very soluble in it. Consequently, droplets of oil tend to rise through water. When oil and water are first thoroughly mixed in a pan and then allowed to stand, the oil droplets migrate to the top, and form a separate layer. Similarly, petroleum will migrate upward through the water in a porous rock if certain conditions are found. The petroleum must be of lower density than the water, and it must be of sufficiently low viscosity to flow. The pores in the rock must be interconnected, that is, the rock must be permeable so that fluids can percolate through it.

Different kinds of petroleum are usually described by the term **gravity** rather than by density. The gravity is measured in units called degrees on the **Baume Scale.** Pure water has a gravity of 10 degrees. Most oils are of higher gravity, ranging from 10 degrees to more than 60 degrees. This implies lower density. The gravity (G) and the density (ρ) are related in the conversion formula:

$$\rho = \frac{141.5}{G + 131.5}$$

There are two basic kinds of petroleum. One is called **paraffin base crude oil** and the other is called **asphaltic base crude oil.** Paraffin base crude oil tends to have higher gravity, and therefore lower density compared with asphaltic base crude oil.

The viscosity of petroleum depends upon chemical composition and temperature. Asphaltic base oils contain a relatively higher proportion of viscous constituents than paraffin base oils. However, some of these more viscous constituents are also more volatile. That is, they change from liquid to gas at

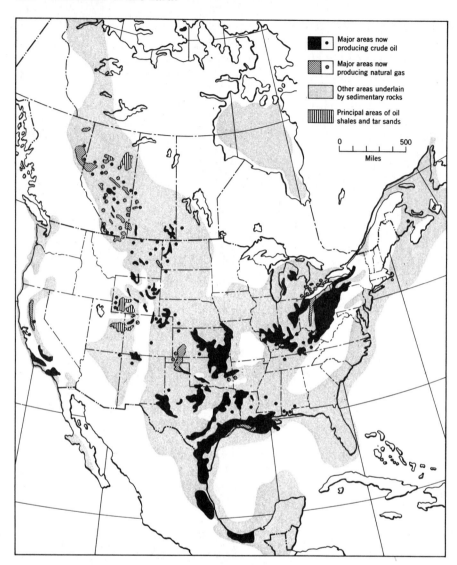

Figure 10-38 Areas of petroleum and natural gas production, oil shale and tar sand accumulations, and sedimentary rock assemblages where future hydrocarbon reserves may be discovered in North America excluding Alaska and northern Canada. (Reprinted from R. F. Flint and B. M. Skinner, *Physical Geology,* page 404, John Wiley & Sons, Inc., 1974.)

lower temperatures than other less viscous but also less volatile components. Some oils that flow easily at the higher temperatures of deep reservoirs become too viscous to flow in lower temperature settings. Other mixtures that were originally of high viscosity became more fluid when the more viscous volatile constituents "boil off" in the higher temperatures of deep reservoirs. We can understand that the extent to which oil is fluid enough to percolate through permeable rock depends upon temperature, composition, and the time available to alter the composition through loss of volatile constituents.

Tectonic structures have important effects on the migration and concentration of petroleum and natural gas. Fractures contribute to the porosity and permeability, and folds and faults influence the distribution of permeable and impervious zones. The subsurface rocks can be viewed as an assemblage of permeable conduits interspersed with impervious rock. Permeable and impervious zones alike possess some pore space that is filled by water solutions, and perhaps some petroleum and natural gas. These latter fluids tend to percolate upward through the water in the permeable zones until further migration is halted by impervious rock. Here the petroleum and gas become trapped. Some typical structures in which oil and gas can become trapped are illustrated in Figure 10-39. We can see from these examples that the hydrocarbon fluids, having higher gravity than the water, migrate along an inclined permeable layer to the highest part. This may be along the crest of an anticlinal fold, or it may be at a place where the reservoir rock is truncated by a fault over which impervious rock has been displaced so that it seals the reservoir. These structural settings in which the concentration of oil and gas is governed by folds and faults are called **structural traps.** The Elk Basin Anticline (Figure 10-40) is a structural trap located on the Wyoming-Montana border. Here an elongate faulted dome is clearly evident from the aerial photograph. The accompanying **structure contour map** shows with elevation contours the form of the top surface of the buried Frontier Formation. Oil and gas are recovered from two sandstone layers within this formation, and from three

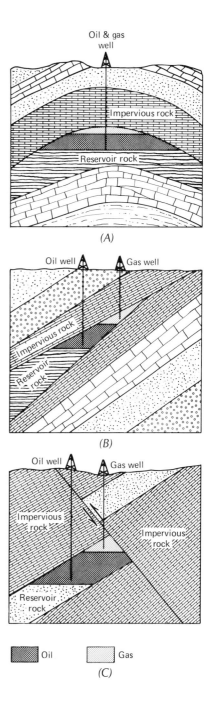

Figure 10-39 Some subsurface structural and stratigraphic settings in which petroleum and natural gas can become trapped: *(A)* anticlinal structural trap; *(B)* pinch-out of permeable reservoir rock bounded by impervious rock forming a stratigraphic trap; *(C)* structural trap related to faulting.

394 The Rocks of the Earth

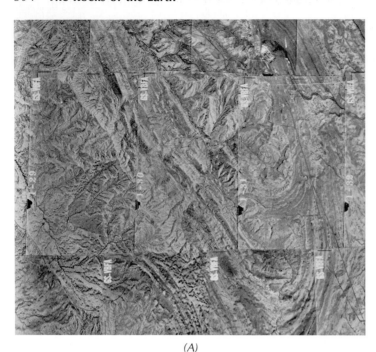

(A)

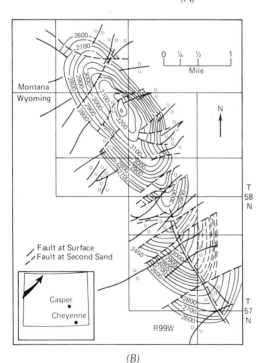

(B)

Figure 10-40 The Elk Basin Anticline on the Montana-Wyoming border is a faulted elongate structural dome where petroleum has been trapped in five sedimentary units. (A) Aerial photograph illustrates topographic expression of this structure. (B) Structure contour map illustrates the form of the upper surface of the Second Sand Unit of the Frontier Formation. (After W. S. McCabe, *Bulletin of the American Association of Petroleum Geologists,* page 64, January 1948.)

deeper sandstone reservoirs. These producing layers are separated from one another by impervious sediments that keep the oil and gas from migrating from one reservoir to another.

Salt core structures are another important kind of structural trap. Recall how beds of rock salt deform plastically and form salt domes. When this happens, overlying sediments are domed, pierced, and fractured as we can see from the example in Figure 10-41. Oil found in the Spindletop Dome in Texas is trapped in the cap rock and in upturned flanking sediments. The 1901 discovery of oil in Spindletop marked the beginning of the oil industry in the Gulf Coast region of United States. Since that time several hundred salt dome structures containing oil and gas have been located. Gravimetric surveys have been used to explore for these structures. Methods of exploration seismology are widely used in the search. These geophysical tools are essential because most Gulf Coast salt structures are not evident from topography or rock exposed at the surface.

Other settings in which oil and gas can accumulate are **variable permeability traps.** Some important ones have developed more from sedimentation processes than from tectonic deformation. Petroleum can be trapped in permeable wedges or in reef carbonates

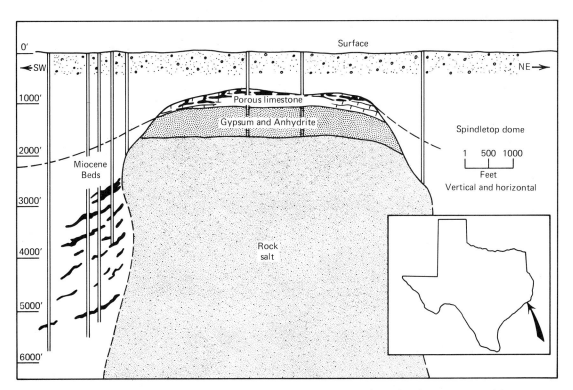

Figure 10-41 Spindletop Dome, Texas is a salt structure where petroleum has been recovered from flanking sediments and cap rock reservoirs. Production commenced in 1901 with a gusher flowing at the rate of 75,000 barrels/day. An estimated volume of more than 50 million barrels of oil accumulated in cap rock reservoirs of dolomite with high cavity porosity. An even larger volume accumulated in flanking sediments. (After K. K. Landes, *Petroleum Geology,* page 266, John Wiley & Sons, Inc., 1951, based on data from J. B. Eby and M. T. Halbouty, *Bulletin of the American Association of Petroleum Geologists,* pages 475–490, April, 1937.)

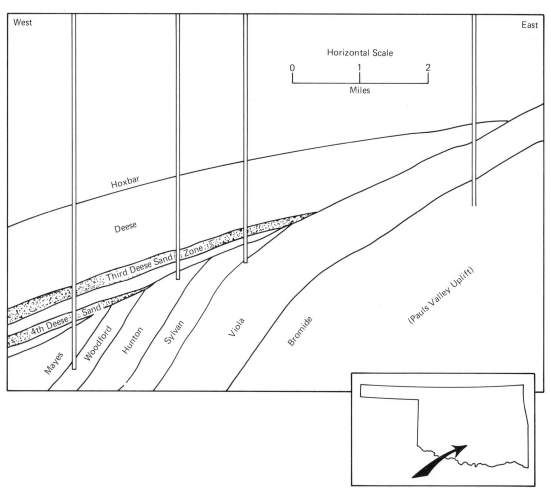

Figure 10-42 Permeable sandstone reservoirs of the Deese Formation in central Oklahoma truncate against older impervious rocks along an unconformity forming traps for petroleum accumulation. (After K. K. Landes, *Petroleum Geology,* page 314, John Wiley & Sons., 1951, and Carter Oil Co., *The Link,* May 1947.)

that are buried under impervious mud deposits. Migration of oil and gas can be impeded by sedimentary facies changes within a layer. Figure 10-42 shows a geologic section across the Pauls Valley area in Oklahoma. Observe how sandstone strata in the Deese Formation terminate against an older erosion surface of impermeable rock. These strata were deposited during transgression and regression of a Pennsylvanian sea. Oil is produced from these sandstones, but not from the deeper, more steeply inclined strata. This example suggests another form of combined structural and variable permeability trap, although it is not found here. Perhaps elsewhere oil could be trapped in steeply inclined strata sealed by overlying impervious rock at an angular unconformity. Such traps have been discovered.

The geologist plays a key role in the search for oil and natural gas. Structural and stratigraphic traps are discovered from bedrock geologic map patterns and geophysical surveys. However, many structures suitable for trapping oil turn out not to be oil-bearing structures. Many prospects have proven to be geologic successes but economic failures. In recent years some progress has been made in discovering what kind of fluid is in a reservoir rock from analysis of the amplitudes of seismic waves reflected from these rocks. Otherwise the nature of the fluid can be discovered only by drilling. Although the skeptic will say "oil is where you find it," some knowledge of tectonic structures and the distribution of sedimentary rock can greatly increase the chance of finding it.

Structural Settings of Ores

We shall consider two aspects of ore distribution and tectonic structure. The first is concerned with identifying regional settings where ore is likely to occur. The second aspect concerns how ore is actually distributed in particular tectonic structures.

Past discoveries tell us that sulfide ore bodies are more likely to be discovered near the margins of batholiths and stocks than in other geologic settings. They can occur within the pluton, in associated volcanics, or in the adjacent country rock. At the time of igneous activity or during a later episode of metamorphism ore solutions tend to migrate most readily through fracture systems and porous rocks units. However, ores are not found in all fracture zones or porous rocks near a particular pluton. There are many plutonic settings where no ore has been discovered regardless of the tectonic structures there. Furthermore, some important ore bodies have been found in sedimentary rocks so far from plutonic rock that a direct connection is not obvious. Most ore bodies are discovered from mineral clues in outcrops, from the effects of geochemical solutions on ground water and vegetation, and from geophysical surveys. Exploration is concentrated in continental shield areas and orogenic belts.

There are places where regional structural trends form a basis for mineral exploration. In central Nevada volcanic and sedimentary rocks bearing disseminated gold were discovered in zones formerly covered by a thrust plate. They were later exposed where erosion destroyed portions of this thrust plate. Look at the geologic map of the area in Figure 10-43. Geologists found from careful study of the bedrock geology that the places where the thrust plate was destroyed by erosion were crudely aligned in a northwesterly direction. They appeared to lie along a subtle arch in the thrust. Guided by the alignment of windows where gold-bearing rock had been found, an additional area was discovered further along this arch were the thrust plate had been eroded away. Here important disseminated gold deposits were found. This example illustrates how a subtle structural alignment can point out the best direction to search for more ore.

Knowledge of local tectonic structure in the vicinity of a known ore deposit can prove useful in locating other nearby deposits. Figure 10-44 shows two maps of mine workings on an underground level. First it was thought that there was only one sulfide ore vein (Figure 10-44A) which was offset along several faults. Following careful study of fault displacement directions it was later realized that the apparent extension of vein No. 8 in the lower part of the map was actually another vein. This prompted further exploration of the zones on either side of the vein, leading to the discovery of a second vein, No. 10 (Figure 10-44B). Here we see how faulting altered the original distribution of ore. Another example, seen in Figure 10-45, illustrates fold control of ore emplacement. Here ore solutions migrated through limestone beds before becoming concentrated along axes of anticlines. Clearly a knowledge of folds in this area could prove useful in exploring for further ore deposits. Numerous examples such as these reveal the relationships between tectonic structure and ore concentration.

Geophysical exploration for ore deposits involves magnetic and electromagnetic sur-

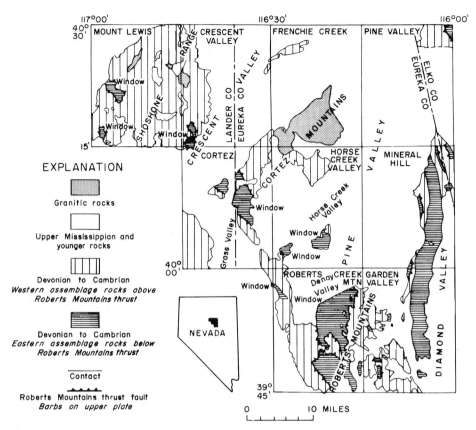

Figure 10-43 Generalized geologic map of central Nevada showing windows in the Roberts Thrust plate aligned in a northwesterly direction. Valuable accumulations of disseminated gold have been discovered in some of these windows. (Reprinted from E. S. Robinson, *Bulletin of the Geological Society of America,* page 2046, July 1970.)

veying, to a lesser extent gravity field measurements, and very limited use of seismic methods. To a greater extent than in oil exploration, these surveys seek to measure properties of the ore itself, rather than to detect the structure where it might occur.

Tectonic Structures and Industrial Development

The geologist plays a role in appraising the engineer and architect of the terrane where construction projects are to be undertaken. Among the important kinds of projects are dams, tunnels, highways, as well as buildings. All of these structures rest upon bedrock or regolith. Depending upon the stresses they can be expected to exert upon the ground, factors concerning rock strength must be determined by the geologist. Joints, faults, and to some extent foliation and bedding define zones of weaknesses. These features also contribute to porosity and permeability of a rock mass, and therefore have an important effect on subsurface water movement. This

Tectonism and the Architecture of the Earth's Crust 399

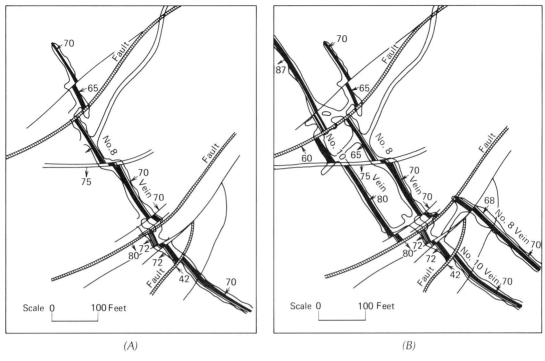

Figure 10-44 Vein offsets along faults discovered after study of relative fault displacement. (A) Early mining activity indicated only one vein offset small distances along a series of parallel faults. (B) Later mine development based upon careful study of fault displacement reveals two parallel veins. (After F. A. Linforth, in *Ore Deposits of the Western States,* Part II, page 696, American Institute of Mining and Metallurgical Engineers, 1933.)

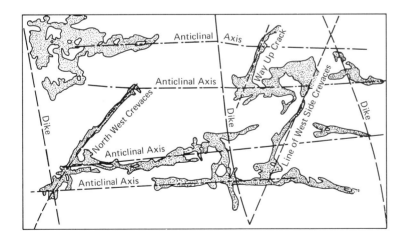

Figure 10-45 Ore bodies in a mining district near Tombstone, Arizona, are located in limestone along axes of anticlines. Apparently migration of ore solutions in the limestone was controlled by preexisting structure. (After Church, American Institute of Mining and Metallurgical Engineers.)

can be particularly important in selecting sites suitable for water storage reservoirs and in planning highway drainage facilities. These and other geological factors of engineering importance will be examined further in later chapters.

400 The Rocks of the Earth

STUDY EXERCISES

1. On the bedrock geologic map (Figure 10–46), ages of the different rock units are symbolized by: O, Ordovician; S, Silurian; D, Devonian; M, Mississippian. The arrow on the fault indicates the dip of the fault plane.

   (a) What kind of fold is exposed in the northwest part of the map?

   (b) What is the plunge direction of the fold exposed on the southeastern part of the map?

   (c) Which side of the map is on the upthrown side of the fault? Is the fault normal or reverse?

   (d) Which site; No. 1, No. 2, or No. 3, would be the best place to drill for oil believed to be trapped in Devonian rock?

2. Suppose that a drill encounters strata in the following sequence of ages as it drills deeper into the earth: Triassic; Permian; Pennsylvanian; Permian; Triassic; Permian; Cretaceous.

   (a) Does this sequence indicate the existence of a fold? If so, what kind of fold is indicated?

   (b) Does this sequence indicate the existence of a fault? If so, what kind of fault exists and where is it located?

3. What kinds of faults and folds are most likely to develop in the sedimentary strata overlying a rising salt dome?

4. Discuss why continental mountain ranges are asymmetrical in cross section, but oceanic ridge systems are more or less bilaterally symmetrical.

5. Discuss the role of isostasy in each stage of a mountain building cycle, and the extent to which isostatic processes aid or oppose other tectonic processes during these stages.

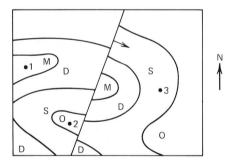

Figure 10-46

SELECTED READINGS

Billings, M. P., *Structural Geology*, 3rd ed., Prentice-Hall, Inc., Englewood Cliffs, N.J., 1972.

Dewey, J. F., and J. M. Bird, Mountain Belts and the New Global Tectonics, *Journal of Geophysical Research*, vol. 75, no. 14, pp. 2625–2647, 1970.

Landes, Kenneth K., *Petroleum Geology of the United States*, Wiley Interscience, New York, 1970.

Ramsey, John G., *Folding and Fracturing of Rocks*, McGraw-Hill Book Co., New York, 1967.

Spencer, E. W., *Introduction to the Structure of the Earth*, McGraw-Hill Book Co., New York, 1977.

SECTION FOUR
THE FACE OF THE EARTH

11

GROUND WATER

Porosity and Permeability
Ground Water Zones
Circulation in Unconfined Ground Water Systems
Circulation in Confined Ground Water Systems
Piezometric Surface Fluctuation
Water Wells
Ground Water Quality
Geysers and Hot Springs

People living in rural areas of North America get most of the water for their daily needs from wells. And wells supply most of the water used by communities with populations of under 100,000 people. What is the source of this water that is pumped from wells? Can we continue to rely on this source as our demands for fresh water increase?

Water is abundant near the earth's surface. The moisture contained in the solid earth is called **ground water.** Very close to the surface it exists as a film that coats particles of rock, and as droplets that partly fill the cracks and interstices between the particles. Farther down these small openings are completely saturated with ground water. Here is an important source of the fresh water that has numerous agricultural, industrial, and household uses. But it also plays an important role in geologic processes. Ground water is not static; it is physically and chemically active. Everywhere it is in motion, seeping through the openings between grains of rock. It is continually dissolving some bedrock constituents and precipitating other substances. Ground water is part of the earth's hydrologic cycle (Figure 11-1). It comes from rain and melting snow that seep into the ground.

Our purpose in this chapter is to discuss the zones where ground water exists and how the ground water moves through these zones. Chemical processes involving ground water, and other parts of the hydrologic cycle involving streams, glaciers, and oceans are presented in other chapters. To understand how ground water moves we must first learn about the porosity and permeability of the bedrock and regolith that contain this water, and the different zones where it is distributed. Then we need to learn about the dynamics of ground water movement. Finally, we should learn something about how to choose locations for water wells, and how to evaluate the quality of the ground water produced from them.

The Face of the Earth

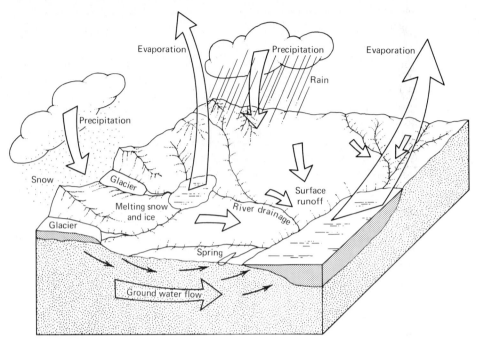

Figure 11-1 Schematic representation of the hydrologic cycle which illustrates how water moves through the ground, the atmosphere, and the oceans.

POROSITY AND PERMEABILITY

Regolith and bedrock near the earth's surface contain some open space where water can collect. This space exists between grains of sediment, along fractures, and in cavities formed by dissolving parts of the rock. We call it **pore** space. The ratio of this pore space to the total volume of a rock including both pores and solid material is the **porosity** of that rock. Figure 11-2 illustrates different kinds of pore space found in rocks that have intergranular, fracture, and cavity porosity.

Intergranular porosity in sedimentary rock and regolith depends upon grain size and arrangement, and the proportion of the volume filled by cementing materials. Intergranular porosity can be higher than 0.5 in unconsolidated sediment consisting of grains that are relatively uniform in size. We can expect much lower values in poorly sorted sedimentary rock that is made of grains of various

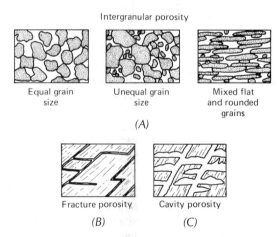

Figure 11-2 Representative examples of pore geometry in rocks with (A) intergranular, (B) fracture, and (C) cavity porosity.

sizes and contains interstitial cement. Typical values of porosity for the different kinds of terrigenous clastic sediments are summarized in Figure 11-3. The porosity is usually higher

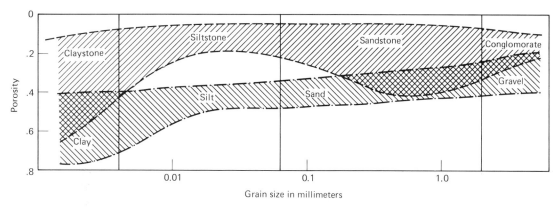

Figure 11-3 Typical ranges of porosity for common terrigenous clastic sedimentary rocks, and for accumulations of unconsolidated sediment.

in unconsolidated sediment than in lithified rock of similar grain size. Fine-grained sedimentary deposits tend to be more porous than coarse-grained deposits. Intergranular porosity can be important in deposits of volcanic ash and tuff as well as in common sedimentary rocks. It is absent in igneous and metamorphic rocks that consist of interlocking mineral grains.

Joints and faults account for **fracture porosity** in bedrock. This is the only significant kind of porosity in intrusive igneous and metamorphic rock. In some carbonate rocks we can find fracture porosity together with **cavity porosity** that exists where the rock has been partly dissolved by ground water seeping through preexisting fractures. Porosity can be higher than 0.5 in carbonate rocks. Another form of cavity porosity is found in volcanic rock where gas bubbles have been retained as the lava congealed. Such cavity porosity exceeds 0.9 in pumice, and it accounts for most of the porosity in vesicular basalt.

Water percolates through regolith and bedrock by following an intricate network of intertwining channels consisting of connected pores. The intersecting paths are so tortuous that it is impractical to attempt to predict the route followed by some particular droplet of ground water. It is more useful to examine

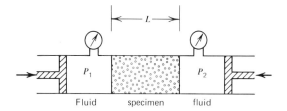

Figure 11-4 Measuring permeability in a substance can be done by forcing a fluid through a permeable specimen. Fluid pressure P_1 at one end of the specimen is maintained at a higher level than pressure P_2 at the other end of the specimen. This causes fluid to flow from the high pressure end toward the low pressure end.

the integrated movement of a mass of water that is seeping through porous rock. The **permeability** of the rock is a measure of its capacity to transmit fluid. Permeability depends upon the extent to which pores are connected and the size and irregularity of the resulting channels. Perhaps we could describe permeability in terms of the intricacies of pore geometry, but we would find such a scheme difficult and time consuming. A more practical method is to measure the rate at which we can force a fluid to flow through a porous rock under known pressure conditions. Figure 11-4 shows how this might be done. Fluid of viscosity μ flows through a specimen of length L because the pressure P_1 on one end of the specimen is greater than the pressure P_2 on the other end of the specimen. The flow of fluid through the specimen is measured in terms of the discharge $q = (V/\alpha)/t$, which is the volume (V) of fluid that passes through the end of the specimen in time t. The cross-sectional area at the end of the specimen is α. The discharge is related to the pressure difference across the specimen according to the equation:

$$q = (k/\mu)(P_1 - P_2)/L \qquad (11\text{-}1)$$

where k is the permeability and is expressed in units of area. The **darcy** (1 darcy = 0.987 × 10^{-8} cm²) is the conventional unit of permeability. This manner of defining permeability makes no reference to the internal geometry of the specimen. It is concerned only with the capacity to transmit fluid under pressure without consideration of percolation paths of individual droplets. Typical values of permeability for different kinds of regolith and bedrock are given in Table 11-1.

Compare the data in Figure 11-3 and Table 11-1. Observe that the sediments which are most porous are not the most permeable. In fact, the fine-grained clays and claystones that are most porous tend to be the least permeable of sediments. We can understand this partly from knowledge of the dimensions of the interconnected pore channels. Even though a great many channels may exist in fine-grained sediment, they are very small in diameter. Molecular forces on solid particle surfaces tend to hold a thin film of fluid. If the pore diameters are about the same as the thickness of this film of fluid, it tends to adhere to the particles rather than to migrate through interconnected pores. This effect would greatly diminish water mobility in clays. But it would be an insignificant impediment in coarse-grained sand and gravel deposits that are among the most permeable materials.

Vesicular basalt and pumice, although highly porous, may have very low permeability because the cavities formed by trapped gas bubbles are not connected to one another. If these kinds of rock are found to be permeable, it is probably because of fractures that are of secondary importance in terms of the total pore volume of the rock.

GROUND WATER ZONES

There are two ground water zones beneath the earth's surface. Nearest the surface is the **vadose zone** in which pore space is only partly filled by ground water. Beneath is the **phreatic zone** where pores are completely saturated with ground water. We call the upper surface of the phreatic zone the **water table.** These zones make up the ground water profile shown in Figure 11-5.

In the vadose zone very near the land surface a thin film of moisture adheres to the solid particles. It is called soil water. Here the pore space is otherwise almost devoid of ground water. The soil water subzone merges gradationally into an intermediate vadose subzone where pore space is partially filled. Still deeper this subzone is transitional into the **capillary fringe** which forms the base of the vadose zone.

The lower part of the capillary fringe is completely saturated with ground water even though it lies above the water table. This results from the phenomenon of capillarity, which is the capacity of liquid surfaces to become elevated or depressed when in contact

Table 11-1
Permeability in Sediment

Gravel	10^3–10^5 darcy
Sand	1–10^3 darcy
Silt	10^{-4}–1 darcy
Clay	Less than 10^{-4} darcy

Modified from S. N. Davis and R. J. M. DeWiest, Hydrogeology, John Wiley & Sons, Inc., p. 164, Table 6.1, 1964.

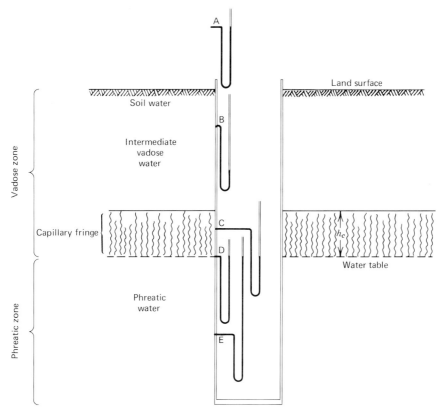

Figure 11-5 The principal ground water zones. Schematic manometer tubes indicate fluid pressure at different subsurface points. They show how high water would rise in open tubes connected to different places in the ground. (After S. N. Davis and R. J. M. DeWiest, *Hydrogeology,* page 42, John Wiley & Sons, Inc., 1966.)

with solids. The cause of capillarity is not clearly understood, but it is probably related to forces of molecular attraction. The effect is easily observed by placing a clean glass tube of small radius r in a pan of pure water of density ρ. Water can be observed to rise to a height $h = 2T/\rho gr$ above the water surface in a pan. The surface tension T of the water is approximately 0.075 gm/cm, and g is the acceleration of gravity. The capillary rise of impure water will be slightly smaller than the rise of pure water. If we think of the channels of interconnected pores as tubes that extend from the vadose zone into the phreatic zone, we can understand how pore space in the capillary fringe becomes saturated because of the capillarity effect.

The thickness of the vadose zone is not the same everywhere. In arid regions it may extend several hundred meters beneath the land surface. But in swamps and marshes the water table reaches the surface, and no vadose zone exists. The thickness of the capillary fringe is similarly variable from place to place depending upon the dimensions and geometry of interconnected pore space. It is typically a few centimeters thick in coarse sand, but it can be several meters thick in clay deposits. Knowing that the capillary rise height is inversely proportional to the radius of the conducting channel, we can understand why the capillary fringe is thicker in fine-grained sediment than in coarse-grained deposits.

If we could place pressure gauges underground, we would find that ground water is subjected to pressure that changes with depth. The pressure on the ground water in different zones is illustrated by hypothetical manometer tubes seen in Figure 11-5. This kind of manometer is a fluid pressure gauge that consists of a U-shaped tube with a diameter large enough so that capillary rise in the tube is negligibly small. One end of the manometer is open to the atmosphere. Ground water flows freely into the other end and rises to a level determined by the pressure on the water at its point of entry into the manometer. If pressure at the input end equals atmospheric pressure on the open end, water rises to a level equal to the input level. We would observe this condition at *A* in Figure 11-5. Ground water pressure is lower than atmospheric pressure where water level in the open end of the manometer is lower than the input level. We would find this condition at *B* and *C* in the vadose zone of the illustration. In a permeable phreatic zone water rises in the manometer to the level of the water table regardless of the input level, as seen at *D* and *E* in Figure 11-5. The water table, then, would be the top of the saturated zone **in the absence of capillarity.** Water level in a well penetrating a permeable phreatic zone stands at the level of the water table, even though saturated pores may be found above this level in the capillary fringe.

CIRCULATION IN UNCONFINED GROUND WATER SYSTEMS

How does water move when it is underground? We can begin to understand how this works by first looking at an idealized system where the material underground is uniformly porous and permeable. This is called an **unconfined** ground water system because the flow of water is not blocked or diverted by impervious zones. Here ground water moves by the process of **laminar flow.** This means that a water droplet follows a path of simple form. This is different from the complicated irregular route it would follow if it moved by the process of **turbulent flow.** Ground water moves slowly, typically a fraction of a cm/sec, because movement is impeded by the solid matrix of the permeable rock. Where water is free to move faster through open channels, such as in streams, it tends to swirl and revolve turbulently. The path of a water particle can be illustrated by a **streamline** or line of flow. Streamlines representing laminar and turbulent flow are compared in Figure 11-6. Actually, a particle of water must thread its way around grains of rock as it follows a path through the interconnected pores. But it follows the simplest pos-

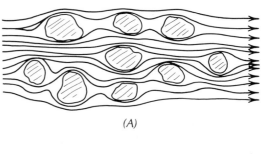

(A)

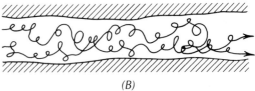

(B)

Figure 11-6 Paths indicating typical movement of a fluid particle during (A) laminar flow and (B) turbulent flow. Laminar flow of fluid through a porous medium can be represented by streamlines of simple form indicating that an individual droplet of fluid tends to follow the most direct path through a labyrinth of interconnected pores.

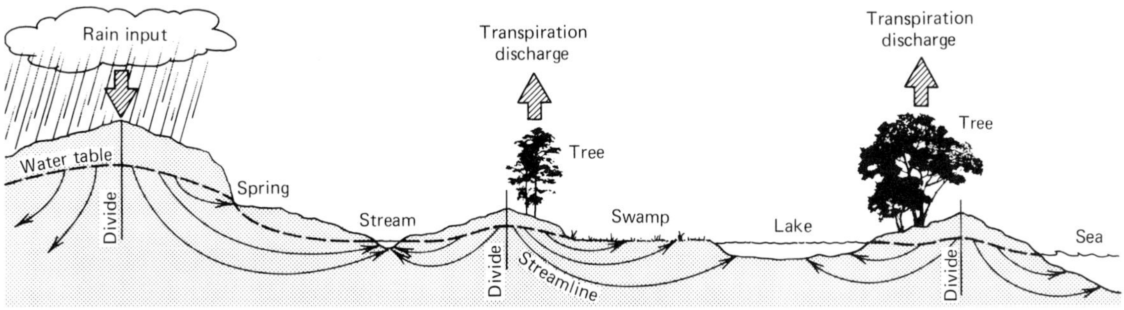

Figure 11-7 Accumulation, flow, and discharge of ground water in an unconfined system. Discharge is from springs and by seepage into streams, marshes, lakes and the sea, and by transpiration from plants. Note that the curved streamlines indicate upward movement of water in some places.

sible path when moving by the process of laminar flow.

Water enters the vadose zone from precipitation that falls on the land surface. Some of it seeps vertically downward until it reaches the water table. But part of the water is retained as soil moisture in the vadose zone. And some of it evaporates and returns to the atmosphere through open pore channels. Still other water is absorbed in plant roots and returns to the atmosphere through the leaves of plants by a process called **transpiration.** The proportion of vadose water that eventually reaches the phreatic zone depends upon climate and vegetation.

Movement of ground water in the phreatic zone is related to the shape of the water table. This surface usually bears a subdued likeness to the land surface. Under hills it is high and under valleys it is low. Ordinarily, differences in elevation of the water table are smaller than the landscape relief. It tends to be farther beneath the land surface in upland areas than in lowlands. Observe in Figure 11-7 how high and low parts of the water table are situated under high and low places on the land surface.

How does water flow in the phreatic zone? We can begin to explain this by slowly running water into one end of a tub and letting it slowly drain out of the other end (Figure 11-8A). Notice that the water level is a flat surface. Water does not pile up at the end where it is put into the tub and the level does not sink down at the end where it drains out. Because nothing gets in the way, all water particles in the tub are able to move rapidly enough from one place to another so that water level stays flat.

Now suppose that we fill the tub with loose sand (Figure 11-8B). Again, let water run in at one end and drain out from the other end. The water seeps into the sand and fills the pores between grains. But now we see water piling up at the end where it is put into the tub. And notice that the level of the water surface, which is the water table in this example, is much lower at the end where it drains out. This happens because water particles cannot move rapidly from one place to another. Instead, they have to thread their way around and between the sand grains. If we could watch individual water particles, we would see them moving slowly along directions shown by streamlines. Now suppose that we shut off the tap and plug the drain (Figure 11-8C). Slowly the water particles will move in such a way that the water level eventually flattens out.

In nature the same sort of thing happens in the phreatic zone that you see happening in Figure 11-8B. A rain shower rather than a tap puts water into the system. A stream rather than a drain takes it out (Figure 11-7). In the

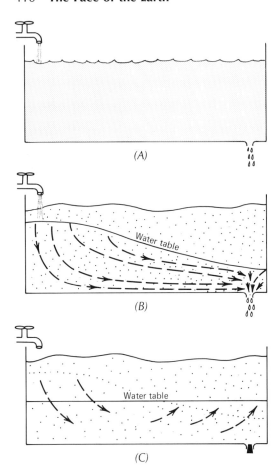

Figure 11-8 Movement of water in (A) an open tank and (B) a tank filled with porous sand. (C) By shutting off the tap and closing the drain so that no water is put in or removed from the tank, the water already in the tank moves in a way that establishes a flat water table. Dashed lines are streamlines.

phreatic zone the water particles move in a way that would eventually flatten the water table. But they move too slowly. Therefore, water piles up under the hills, and it stays low in the valleys, where it drains out of the system through streams. But all of the water particles are constantly moving. Look at a point on the water table in Figure 11-7. The droplet of water at that point slowly moves in a direction indicated by one of the streamlines. But it is replaced by another droplet that has seeped down through the vadose zone. As long as water keeps seeping down to the water table at the same rate it flows away from the water table along the streamlines, the shape of the water table will stay the same.

How can we find the directions of streamlines that represent the movement of ground water? We can determine the horizontal component of the flow direction from a contour map of water table elevation. The water movement will be in the direction of downward inclination of the water table. Consequently, streamlines must be oriented perpendicular to the contour lines on the map. Water table contours and horizontal streamlines are shown in Figure 11-9. Observe how phreatic water flows away from domes and ridges. Ridge crests on the water table are called **ground water drainage divides.** An area bounded by such ridges is a **ground water drainage basin.**

The flow of ground water along a vertical cross section can also be illustrated graphically by laminar streamlines. In Figure 11-7 streamlines are shown for a uniformly porous and permeable phreatic zone, but the location of these streamlines is more difficult to determine. A complete explanation is beyond the scope of this discussion, but the following discussion will help to explain how this is done.

The problem can be treated in terms of a factor ø that we will call the **movement potential** in the phreatic zone. This factor is related to energy conditions in the phreatic zone. At some particular point beneath the water table we can compute the value of ø if we know the pressure P, the height Z of the point, velocity v of ground water movement, permeability k, and the density ρ and viscosity μ of the ground water using the equation:

$$\text{ø} = k\frac{\rho g}{\mu}\left[\frac{P}{\rho g} + Z + \frac{v^2}{2g}\right] + C \quad (11\text{-}2)$$

Ground Water   411

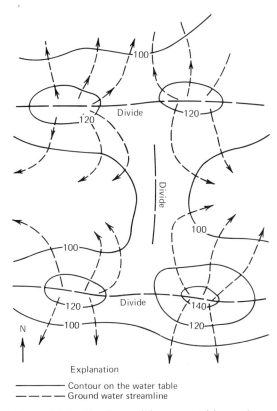

Figure 11-9 The form of the water table can be represented by topograhic contours. Streamlines drawn perpendicular to contours and directed downslope indicate the horizontal direction of ground water movement.

where g is the acceleration of gravity, and C is a constant related to the level from which height is measured. The term in the brackets is equal to the **hydraulic head** h:

$$h = \frac{P}{\rho g} + Z + \frac{v^2}{2g} \qquad (11\text{-}3)$$

which is the height of the water table. But it also has another meaning. It corresponds to the energy per unit weight of water in the phreatic zone. After evaluating ø at different points in the phreatic zone we can use contour lines to illustrate how it varies from place to place (Figure 11-10). These contours are called **equipotential** lines because they connect points where the value of the movement potential is constant. Ground water flows from high to low potential. Therefore, the streamlines showing flow direction must intersect equipotential lines at 90 degree angles as seen in Figure 11-10. A set of intersecting streamlines and equipotential lines is called a **flow net.** We will not go into the details concerning proper spacing of lines on a flow net. However, it should be noted that, if correctly prepared, the flow velocity as well as the direction can be determined from such a graph.

We can see from Figures 11-7 and 11-9 that ground water flows from water table drainage divides to points of discharge. It should be understood that the entire mass of water in the phreatic zone is flowing; there are *no stagnant regions*. The water migrates downward in some places, and elsewhere it moves horizontally or even upward. But there are places where water leaks out of the phreatic zone. It is eventually discharged at locations where the theoretical position of the water is above the surface of the earth. Such

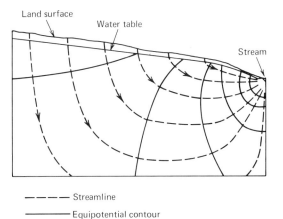

Figure 11-10 Flow of phreatic water in an open aquifer system. Intersecting streamlines and equipotential lines represent a flow net. Form of the flow net depends upon the shape of the water table and the rate of flow. Note the subdued correspondence of the landscape and water table relief.

is the case in streams or water standing in lakes, swamps, or seas. In some places such as steep hillsides and cliffs the water table is truncated at the land surface. These are locations of **seeps** and **springs** where we can see water seeping out from the ground. Typical natural locations of ground water discharge are illustrated in Figure 11-7.

The velocity of ground water flow is governed by a relationship originally published in 1856. Now known as Darcy's Law, this relationship was proposed by Henri Darcy from experiments he performed while a city engineer in Dijon, France. Experimental apparatus pictured in Figure 11-11A consists of a tube filled with a permeable material and two manometer pipes. Water flows downward in a direction that is parallel to the axis of the inclined tube. We should think of the axis as a streamline. The flow velocity v can be expressed by Darcy's Law:

$$v = -k \frac{\rho g}{\mu} \frac{\Delta h}{\Delta L} \qquad (11\text{-}4)$$

where k is permeability, ρ is fluid density, μ is viscosity, and g is acceleration of gravity; ΔL is the length of the flow path between the manometers, and Δh is the difference in heights to which water rises in the manometers. In each manometer the height (h) of the water is the hydraulic head at that location. Because pressure in the system is caused only by gravity acting on the mass of the water, the pressure difference (ΔP) between the two manometers will be equal to $\rho g \Delta h$. In this case we can see that Eq. (11-4) becomes the same as Eq. (11-1).

In an unconfined ground water system we can use Darcy's Law to estimate the velocity of ground water flow. In areas where the streamlines are approximately parallel and close to being horizontal, we can get the data needed for velocity calculations from wells aligned in the direction of flow. Look at Figure 11-11B where we can find the change in hydraulic head, that is, the change in the wa-

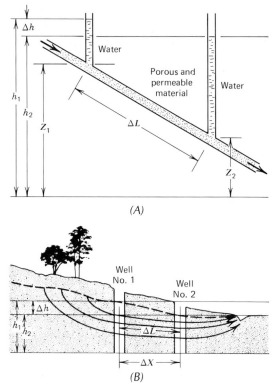

Figure 11-11 Flow of water in response to gravity (A) in a tube filled by a porous solid and (B) between test wells in an unconfined ground water system.

ter table elevation from the heights of water in two nearby wells. This change in hydraulic head, $\Delta h = h_1 - h_2$, is along the distance ΔL measured parallel to streamlines. Insofar as these streamlines are nearly horizontal, ΔL is approximately the same as the horizontal distance ΔX between the wells. For this case we see that $\Delta h/\Delta L \approx \Delta h/\Delta X$ which is the slope of the water table. We have learned from calculations based upon typical values of permeability for well-sorted, coarse-grained sand deposits that where the water table slope is less than 10 degrees the velocity of ground water flow ordinarily will be less than 0.1 cm/sec. Much lower velocities would be expected in finer-grained sediments. Normally, ground water migrates at velocities between 5×10^{-6} cm/sec (about 1.5 m/yr) and 2×10^{-3} cm/sec (about 0.6 km/yr). In exceptionally permeable sediment the velocity might exceed 0.1 cm/sec.

Keep in mind that ground water circulates everywhere in a permeable phreatic zone. The velocity of ground water movement is normally less than 2×10^{-3} cm/sec. Perhaps this movement seems to be too slow to be of any importance. But think about how far a mass of water would move during a decade. The distance would be more than 6 km.

CIRCULATION IN CONFINED GROUND WATER SYSTEMS

A uniformly porous and permeable ground water system of any significant size is highly unusual. Ordinarily, the movement of ground water is somewhat more irregular than indicated by the streamlines in Figure 11-7 because of changes in permeability from place to place. Zones that are porous enough to store a significant volume of water and permeable enough to permit measurable ground water flow are called **aquifers.** Other zones that might be porous enough to hold a significant quantity of water, but are too impermeable to permit any measurable flow, are termed **aquicludes.** The differences between aquifers and aquicludes are relative, and they are usually distinguished by practical considerations. For example, aquifers are zones capable of yielding water to a well at a rate sufficient to meet the needs for which the well was drilled. In contrast, we cannot pump enough water from a well drilled into an aquiclude for the well to be considered useful. Look again at Figure 11-3 and Table 11-1. Now suppose that a ground water system occupies alternating layers of sand and clay. Here the permeable sand layers would be aquifers and the relatively impermeable clay layers would be aquicludes.

Aquifers and aquicludes alike are saturated with water. Aquicludes are never completely

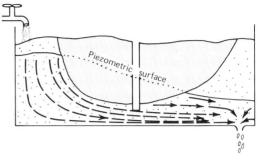

Figure 11-12 Movement of water in a tank filled with masses of permeable sand and impermeable clay. Water flows around the clay, but would rise to the height of the piezometric surface in a hole drilled through the clay into the sand.

impermeable, and ground water eventually infiltrates the interstices, albeit at an almost imperceptible rate. But the aquicludes drastically impede the movement of water. Therefore, in a system consisting of aquifers and aquicludes most ground water flow is confined to the aquifer zone. For this reason we call such a system a **confined ground water system.** In a confined ground water system water is forced to flow in directions that it would not follow in an unconfined system subjected to the same external conditions. The example in Figure 11-12 will help us to understand what this means. It is similar to the example in Figure 11-8 except that an impermeable mass of clay occupies part of the tub. Otherwise, the water is put into the system and drained out of it in the same places so that external conditions are the same. But notice that the streamlines are different. The water would like to flow in the way indicated by the streamlines in Figure 11-8B, but the clay aquiclude blocks such movement. It pushes some streamlines down into the sand aquifer so that ground water flow is confined to the aquifer.

Now suppose that a hole is drilled through the clay into the sand. Water will rise in the hole to a level corresponding to the water table in Figure 11-8B. But in another hole that

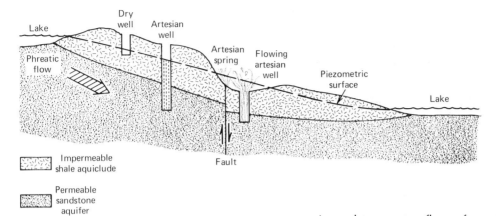

Figure 11-13 Ground water movement in a confined ground water system penetrated by observation wells.

does not reach the aquifer, there is no water. The dotted line in Figure 11-12 shows the height to which water would rise in any hole that reaches into the aquifer. The surface represented by this dotted line is called the **piezometric surface** of the confined aquifer.

The same features seen in Figure 11-12 are found in nature. In a confined ground water system (Figure 11-13) water rises in a well to a level considerably above the top of the aquifer. We can use water heights in a series of wells to find the shape of the piezometric surface. Therefore, we say that it indicates the hydraulic head of the aquifer. It is entirely possible for the piezometric surface to be above the land surface in certain places. Water will flow without pumping from wells drilled in these places (Figure 11-13). Such wells have traditionally been called **artesian wells**. According to more formal classification an **artesian aquifer** is a porous and permeable zone where ground water flow is confined by bordering aquicludes. All wells penetrating such an aquifer are formally considered to be artesian wells whether water flows from them or simply rises to a piezometric surface that is somewhere between the aquifer and the land surface. An **artesian** spring exists where water flows from the ground through fractures extending from the land surface into the aquifer or where a gully cuts into the aquifer (Figure 11-13).

In nature the phreatic zone is usually a composite system containing both confined and unconfined ground water. Look at the composite system illustrated in Figure 11-14. It consists of an upper unconfined aquifer and deeper artesian aquifers separated by aquicludes. Discontinuous impermeable lenses retard the infiltration of water through the vadose zone. Ground water that accumulates above these lenses produces localized saturated zones called **perched water bodies**. Each aquifer in a composite system has a distinct piezometric surface related to pressure conditions within it. Water will rise to the highest piezometric surface in wells that penetrate all of the aquifers. The level of water will be substantially lower in wells that reach only the upper artesian aquifer, which is under lower pressure than the deeper aquifer. The water table is the piezometric surface of an unconfined aquifer.

PIEZOMETRIC SURFACE FLUCTUATION

The position of a piezometric surface fluctuates in response to changes in the amount of water that is added to the phreatic zone and the amount of water that is withdrawn from it. We can measure these fluctuations by

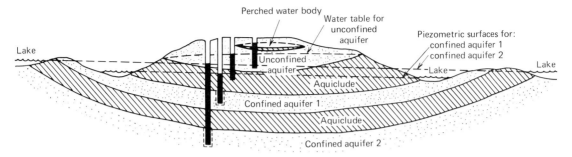

Figure 11-14 A composite ground water system with confined and unconfined aquifers including a perched water body. Note that different aquifers can have different piezometric surfaces.

recording water level in observation wells, which are simply boreholes from which water is neither removed nor injected. We can compare records from several wells to discover patterns of water level fluctuation in a ground water drainage basin.

Changes in the balance between influx and discharge of phreatic water ordinarily have the most important effect upon the position of the piezometric surface. Short-term fluctuations related to rainfall and atmospheric pressure variations can be detected. Seasonal fluctuations are commonly observed, which can be related to rates of rainfall, evaporation, and transpiration that vary during the year. In some agricultural regions a significant volume of water is pumped from wells for crop irrigation during the growing season. All of these factors contribute to the fluctuations illustrated in Figure 11-15. Here we can see that the position of the piezometric surface changes by more than 10 m in the course of a year. Agricultural development was greatly expanded in this area in the years following 1940. Observe the increase in annual fluctuation of the piezometric surface after 1940. This is related to increased pumping of water for irrigation. In most places the seasonal fluctuations are not nearly as large as those seen in Figure 11-15, but some fluctuation is observed almost everywhere.

Long-term changes in the level of a piezometric surface also result from an imbalance of influx and discharge of phreatic water. These alterations can result from natural changes in climate and from changing domestic, agricultural, and industrial needs. The long-term decrease in the piezometric surface, clearly evident in Figure 11-15, indicates that discharge exceeds influx of phreatic water. Here the ground water reservoir is being depleted. Withdrawal of water for irrigation is a major factor in this depletion. Evaporation from irrigation ponds and increased transpiration through crops also increase the withdrawal of vadose water. Here land subsidence is accompanying the long-term depletion of phreatic water, as can be seen from the changes in the elevation of a benchmark. This results from compaction of water deficient sediment.

WATER WELLS

We use fresh water for agricultural, industrial, and domestic purposes at a rate exceeding 1200 billion liters/day (more than 300 billion U.S. gallons/day) in North America. Approximately 20 percent of this volume is removed from ground water aquifers, and the remaining 80 percent is drawn from streams, lakes, and reservoirs. Most ground water used by an industrialized society is pumped from wells.

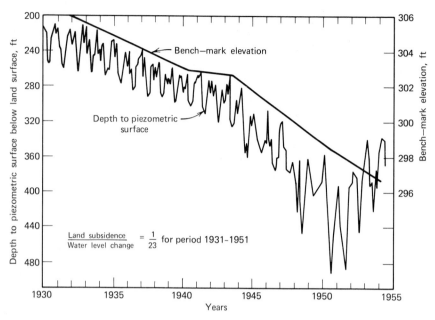

Figure 11-15 Records of water table fluctuation and subsidence of the land surface in an area near Delano, California. (After D. K. Todd, *Ground Water Hydrology*, page 151, John Wiley & Sons, Inc., 1967. Redrawn from J. F. Poland and G. H. Davis, *Transactions of the American Geophysical Union*, page 292, June 1956.)

The needs of an individual household may be satisfied by a well capable of producing less than 10 liters/minute (about 2½ U.S. gallons/minute). Larger commerical developments may require a well that yields more than 100 liters/minute (26 gallons/minute). How do we choose the locations for wells that will produce the needed ground water?

The oldest and perhaps most inexpensive method for locating wells is called "dousing" or "water witching." Equipped with a forked stick or short metal rods, the "douser" paces over the ground until a place is reached where the apparatus starts to behave in some peculiar way. The forked stick is pulled toward the ground or the rods cross at the most probable site for locating the well. No claim has been made that everyone is capable of well witching, and there is no convincing evidence that disproves the possibility that some people may possess this gift. In former times when wells were located mostly for individual household needs, some dousers were able to claim phenomenal success. But witching methods have been much less successful in the search for wells capable of high yields. Can you suggest reasons for this based upon your understanding of ground water systems?

Geologists use modern methods of stratigraphic correlation to explore for aquifers. A good way to begin is by examining data from wells that have already been drilled in areas where new wells are needed. Suppose that we can learn from such data the depths at which these different wells penetrate a particular aquifer. We can use this information to prepare contour maps illustrating the depth and thickness of the aquifer in the area of interest. In some places we can get additional data for preparing these maps from seismic reflection and refraction surveys. After we have prepared maps that illustrate the depth and shape of an aquifer, we can use them to choose sites for drilling new wells. But so far, no completely reliable methods have been devised for determining just how much water a well will yield before it is drilled. Only after the well is drilled and a pump is installed can we accurately test the rate at which water can be withdrawn.

Since ancient times shallow wells, ordinarily less than 10 m deep, have been dug with pick and shovel. This prosiac method is obviously of limited use in modern times. Today water wells are drilled by **rotary** drills, **percussion** drills, and fluid jets (Figure 11-16). Wells drilled in unconsolidated sediment usually have to be cased. This involves placing pipe in the hole to prevent the sides from collapsing. Casing pipe is then perforated where it penetrates aquifers so the water can seep into the well. Casing is not necessary for boreholes in solid bedrock unless we want to restrict flow from certain zones into the well.

When pumping commences, a water well becomes a discharge point in the hydrologic cycle. Former flow patterns in the phreatic

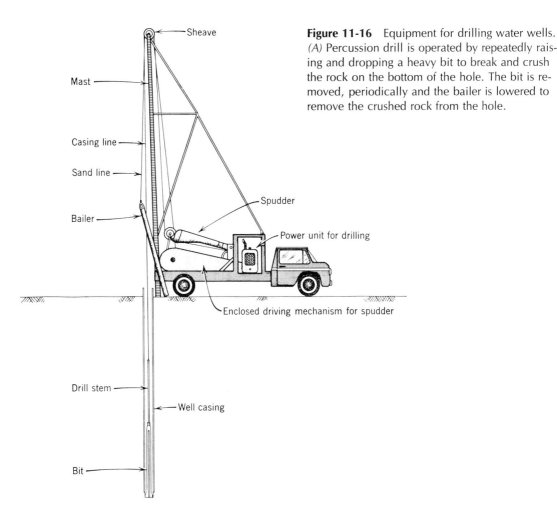

Figure 11-16 Equipment for drilling water wells. (A) Percussion drill is operated by repeatedly raising and dropping a heavy bit to break and crush the rock on the bottom of the hole. The bit is removed, periodically and the bailer is lowered to remove the crushed rock from the hole.

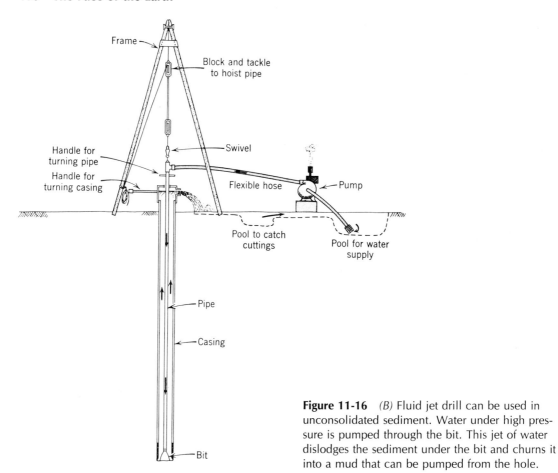

Figure 11-16 *(B)* Fluid jet drill can be used in unconsolidated sediment. Water under high pressure is pumped through the bit. This jet of water dislodges the sediment under the bit and churns it into a mud that can be pumped from the hole.

zone are altered as the ground water system adjusts to the new location of discharge. The result is a depression of the piezometric surface centered about the well. This is called a **cone of depression.** It is shaped so that the slope of the piezometric surface near the well is sufficient to cause water to flow into the well at the same rate as it is removed by pumping

Let us consider the cone of depression related to a pumping well in an unconfined and uniformly permeable aquifer. For simplicity we assume a horizontal water table and phreatic zone that was in hydrostatic equilibrium so that prior to pumping there was no flow of phreatic water in the system. Then, after we begin pumping water from the well the system will change in the manner illustrated in Figure 11-17. When pumping first begins, water level in the well immediately drops. This creates an abrupt vertical slope in the water along the sides of the well (Figure 11-17 A). According to Darcy's Law, flow of phreatic water occurs when the water table is not horizontal, that is, when $\Delta h/\Delta L \neq 0$ in Eq. (11-4). As water flows into the well along streamlines seen in Figure 11-17 B a conical depression in the water table forms around the well. Eventually this cone of depression stabilizes in a position such that the flow into

Ground Water

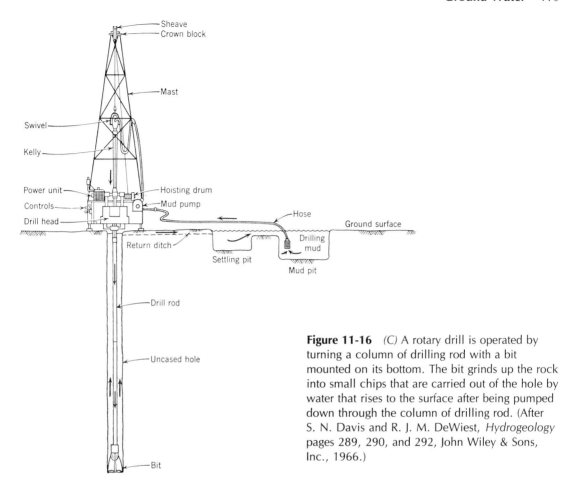

Figure 11-16 *(C)* A rotary drill is operated by turning a column of drilling rod with a bit mounted on its bottom. The bit grinds up the rock into small chips that are carried out of the hole by water that rises to the surface after being pumped down through the column of drilling rod. (After S. N. Davis and R. J. M. DeWiest, *Hydrogeology* pages 289, 290, and 292, John Wiley & Sons, Inc., 1966.)

the well is the same as the discharge by pumping (Figure 11-17C).

The development of a stable cone of depression requires that water pumped from the aquifer be replenished by water flowing into it from somewhere else. Consequently, to analyze the potential yield of a well we must consider various factors in the hydrologic cycle, as well as pumping effects. If water is withdrawn at a higher rate than it is replenished, the aquifer will eventually be depleted.

To evaluate the ground water resources of a region we must estimate the **safe yield** of wells in that region. The safe yield is the rate at which water can be withdrawn from the phreatic zone without undesirable repercussions. Estimating safe yield can be more difficult than simply balancing withdrawal and influx of water. Because wells establish new discharge points, they alter the volume of discharge into streams. If, for example, water in a river must remain deeper than the minimum depth needed for navigation of barges, then water discharge from nearby wells cannot be allowed to slow the flow of phreatic water into the river below a certain rate. Another consideration is related to the effect of a large cone of depression on other shallower wells in an area. A large industrial user may with-

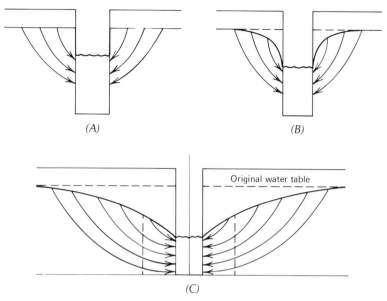

Figure 11-17 Development of a cone of depression in response to a pumping well (A) begins with the commencement of pumping, (B) widens and deepens to create a slope in the water table and corresponding flow of ground water toward the well, finally reaching (C) an equilibrium form where water table slope is sufficient to cause water to flow into the well at the same rate as it is being pumped.

draw so much ground water that the cone of depression spreads several tens of kilometers away from the well. As a result, the water table is lowered below the bottoms of wells used in individual households and farms in this area. Even though total discharge and influx are in balance, the needs of many individuals may be thwarted. Estimation of safe yield must take into consideration these kinds of undesirable effects.

GROUND WATER QUALITY

Ground water is a natural resource that has many uses. We judge the quality of this resource according to the content of dissolved constituents. Four kinds of water are described in Table 11-2 on the basis of the total concentration of dissolved material. Sodium and chloride are the most abundant ions dissolved in brackish water, salty water, and brine, but not necessarily in fresh water. Also present are many other ions that come mostly from chemical reactions between ground wa-

Table 11-2
Classification of Water

Name	Concentration of Total Dissolved Solids in Parts per Million
Fresh water	0-1000
Brackish water	1000-10,000
Salty water	10,000-100,000
Brine	more than 100,000

S. N. Davis and R. J. M. DeWiest, Hydrogeology, John Wiley & Sons, Inc., p. 118, Table 4.4, 1964.

Table 11-3
Principal Dissolved Ions in Ground Water

Anions	Cations
Chloride Cl^-	Calcium Ca^+
Bicarbonate	Magnesium Mg^-
HCO_3^{-3}	Sodium Na^+
Sulfate SO_4^{-2}	Potassium K^+
	Silicon Si^{+4}
	Iron Fe^{+2} and Fe^{+3}
	Hydrogen H^+

Modified from S. N. Davis and R.J. M. DeWiest, Hydrogeology, John Wiley & Sons., Inc., p. 97, Table 4.1, 1964.

ter and the rock through which it moves. The most abundant ions found in ground water are listed in Table 11-3. We will discuss more fully in Chapter 12 the chemical processes that produce some of these ions.

In many places calcium and magnesium are the dominant ions dissolved in ground water. We use the concentrations of these two ions to determine a property called **hardness** of water. Hardness (H) can be calculated in ppm (milligrams per liter of water) from measurements of calcium (Ca_{ppm}) and magnesium (Mg_{ppm}) concentrations in ppm, and their chemical equivalent weight ratios with $CaCO_3$, which are obtained from the atomic weights of the elements involved:

$$H = 2.497\, Ca_{ppm} + 4.115\, Mg_{ppm} \quad (11\text{-}5)$$

An estimate of hardness based upon a formula such as this conveys no information about other substances in the ground water solution.

Standards for judging ground water quality are best established by considering the purposes for which the water will be used. For agricultural purposes some crops are more tolerant of particular dissolved substances than others. Local standards of quality in one region may not be suitable for another where climate and agriculture are different. The needs of agriculture may differ from requirements for domestic use where substances corrosive to plumbing, possible reactions with detergents, and taste are important considerations. The diverse uses of ground water make it difficult to propose a simple system for judging quality.

An important factor in ground water quality near a coast is the possible encroachment of salt water in fresh water aquifers. Because of the growth of cities and industrial activity along sea coasts, salt water infiltration of fresh water supplies has become a serious problem. Ordinarily, saline water and fresh water come in contact in an aquifer beneath a coast as shown in Figure 11-18. In such an aquifer the transition from fresh to saline water is relatively abrupt, the mixing zone being less than a few meters thick. Usually this transition zone is inclined in the landward direction.

Early in the twentieth century surveys of these transition zones had been made in unconfined ground water systems bordering the North Atlantic Ocean. Because of the difference between densities of sea water and fresh

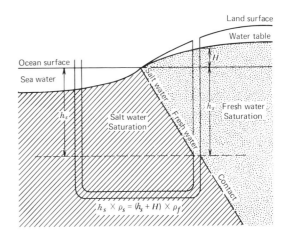

Figure 11-18 Distribution of fresh water and salt water in an unconfined coastal ground water system. The contact zone is at a depth where the weight of the overlying fresh water is equal to the weight of a column of salt water extending from sea level to that depth.

water the depth to the transition zone in coastal wells was found to be about 40 times the height of the water table above sea level. Excessive pumping from coastal aquifers can cause salt water contamination of fresh water wells. As the water table is lowered, the depth to the fresh-saline water transition changes accordingly. It may rise above the bottoms of wells that formerly penetrated only the fresh water zone.

GEYSERS AND HOT SPRINGS

Heated ground water flows from the earth through hot springs and warm springs. Most of these are located in areas that are now or have recently been volcanically active.

The flow of water from those particular hot springs called **geysers** is intermittent. When a geyser is active, there are spectacular bursts of steam and spray. Following such an eruption is a time of quiescence. Some geysers, for example the well-known Old Faithful near the Wyoming-Montana border in Yellowstone Park, follow a regular eruptive cycle. Old Faithful discharges steam and water at regular 63-minute intervals. A particular eruption such as the one shown in Figure 11-19 may last a few minutes. Many geysers do not follow such regular discharge cycles.

The discharge of water and steam from

Figure 11-19 Eruption of Old Faithful Geyser in Yellowstone National Park near the Montana-Wyoming border. (J. R. Stacy, U.S. Geological Survey.)

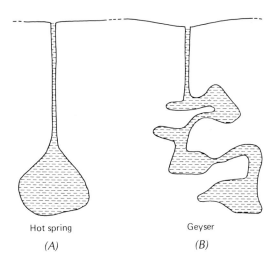

Figure 11-20 Simplified models of (A) a continually steaming hot spring and (B) a geyser subject to intermittent eruption. (After R. J. Foster, *General Geology,* 3rd edition, page 233, Charles E. Merrill Publishing Co., 1978.)

hundred meters beneath the surface. Owing to local volcanism the geothermal gradient is extreme in the area so that temperatures are higher than 100° C in rock very near the surface.

Water and steam are discharged more irregularly from spring systems such as the one seen in Figure 11-20 B. Geyser eruptions can result where temperature and pressure conditions together with the configuration of the system cause a large volume of water to be rapidly converted to steam. The boiling point of water in this system increases with depth because pressure increases. During a quiescent interval of time, water in the system is heated. Owing to variation of pressure with depth the boiling point may be reached in the upper part of the water mass first, whereas water deeper in the system remains just below its boiling point. Initial generation of steam forces some water out of the upper conduit, whereupon pressure in the entire system is suddenly lowered. As a result the boiling point drops and deeper water converts to steam so rapidly that it bursts with eruptive violence from the system. The channels then refill with water and the cycle begins anew. A sequence such as this is more probable in an irregular system of channels where early formed steam can become trapped so as to maintain heat in the system. This kind of sequence could explain the intermittent discharge that is observed for some geysers.

geysers and other hot springs is regular or intermittent depending upon the irregularity and geometry of the subsurface channels feeding the system. Steam may bubble at a uniform rate from water in a hot spring system that has the simple form seen in Figure 11-20 A. Here water in a reservoir is heated to its boiling point and yields steam through the conduit to the land surface. Such a system may extend to depths of a few tens to a few

STUDY EXERCISES

1. Compare the velocity of ground water flow through an unconfined system in a greywacke terrane with the velocity of ground water flow through an unconfined system in a quartz arenite terrane. Assume that the shape of the water table is the same in both terranes. Discuss the influence of grain size, shape, and sorting on the flow of ground water in these two terranes.

2. Consider two regions in the phreatic zone. Region I has low permeability and low porosity, and Region II has low permeability and high porosity. Under similar pressure conditions will water be

discharged more rapidly into a well drilled in Region I than a well drilled in Region II? If water is pumped at the same rate from both wells, will the cone of depression in Region I be different from the cone of depression in Region II? Explain!

3. In a ground water system consisting of interbedded aquifers and aquicludes a well is drilled to a depth of 50 m. Water rises in this well to a point 20 m below the land surface. Nearby, another well is drilled to a depth of 100 m, and water rises in this second well to a point only 5 m below the land surface. Explain why different water levels could be found in wells so close together.

4. A well situated near the Atlantic coast was drilled to a depth of 350 m below sea level at a place where the contact zone between fresh water and salt water was 400 m below sea level. If the ground water system is unconfined, how high above sea level is the water table? If pumping from the well caused the water table to fall by 1 m, would salt water contamination result?

SELECTED READINGS

Davis, S. N., and R. J. M. DeWiest, *Hydrogeology,* John Wiley & Sons, Inc., New York, 1966.

Piper, A. M. "Has the United States Enough Water?," pp. 244–270 in *Focus on Environmental Geology,* R. W. Tank (ed.), Oxford University Press, New York, 1973.

12

WEATHERING

Mechanical Weathering
   Crystal Growth Phenomena
   Release of Residual Stress
   Differential Thermal Expansion
   Collapse of Unstable Masses
   Minor Processes
Chemical Weathering
   General Weathering Zone Conditions
   Hydrolysis
   Ion Exchange Reactions
   Hydration
   Carbonation
   Oxidation and Reduction
Karst Landscapes and Caves
Soils
   Leaching and Precipitation
   Principal Soils
Perspectives

Have you ever visited an old cemetery? Some gravestones lie crumbled on the ground. Others are so badly worn that you cannot read their inscriptions. Certainly this was not expected by the people who carved the stones long ago. But rock is not as durable as one might believe. Bedrock, too, is eventually broken apart. How does this happen? Almost imperceptibly water seeps through the cracks and pores of rock near the earth's surface. Reacting with the mineral grains and cement, it dissolves the supportive constituents. A few grains become loosened and are washed away by the rain. The process continues and more grains are dissolved or dislodged. The rock may disintegrate slowly, or perhaps a mass is weakened to the point of sudden collapse. In these steady and intermittent steps the processes of decay are active close to the surface of the earth. These are **weathering** processes. They lead to mechanical disintegration and chemical decomposition of rock. The primary vehicle of weathering is the ground water that percolates through the subsurface interstices, precipitating, dissolving, and transferring the elements involved in the reactions of decomposition. From this activity soils are produced and the materials near the earth's surface are reformed to be more stable in the environment.

Regolith is the product of mechanical and chemical weathering of rock. The mechanical fragmentation may result from warming and cooling a rock made of minerals with different coefficients of thermal expansion. Freezing of water and growth of plant roots in fractures contribute to this kind of disintegration. Chemical processes include the dissolution of water-soluble constituents, hydration of water-deficient crystals, and processes of oxidation and reduction. The important ores of iron and aluminum are products of chemical weathering. Chemical processes may selectively remove some constituents to the extent that the rock mass is no longer mechanically stable. Rock usually decays from the simultaneous attack of chemical and mechanical weathering processes.

MECHANICAL WEATHERING

You can find evidence of disintegration in most exposures of bedrock. You can see where blocks of rock have fallen away from a bedrock mass that is broken by intersecting joint sets. On the rock surface are small fragments and grains that have worked loose. These are products of mechanical weathering. The principal mechanical weathering processes include (a) crystal growth phenomena, (b) release of residual stresses, (c) differential thermal expansion, and (d) collapse of unstable masses.

Crystal Growth Phenomena

Crystals growing in the pore space of a rock can break it apart in different ways. Most important in temperate and cold regions is the growth of ice crystals caused by freezing ground water. Recall that H_2O expands when it freezes. Grains of rock become loosened by the repeated freezing and thawing of ground water that results from daily and seasonal temperature fluctuations. Ordinarily the zone of freezing reaches only a few centimeters into the rock. But as the surface grains are broken apart and washed away, this weathering process gradually wears down the bedrock.

There are places in perennially cold regions where ice weathering penetrates deeper into exposed bedrock. Here wedges of ice reach 2 m or 3 m into joints and other fractures that cut the rock. Year by year these ice wedges grow wider as more water seeps into the ground during the brief summer thaw, and then freezes. The wedges remain frozen even during the summer. Eventually they grow to widths of several centimeters, or even as much as one meter. In the process large blocks of bedrock can become broken apart.

Mechanical weathering can result from other kinds of crystal growth. If sufficiently high concentrations of sodium, chloride, and other salt-forming ions exist in the ground water, salt crystals can be precipitated in the interstices of the rock. The first crystals that form act as nucleii for growth of aggregates of crystals. These aggregates can eventually break apart the matrix of rock. Ancient temples in Egypt have been severely damaged by salt weathering. Here archaeologists believe that disintegration of the building stone may have resulted from an annual cycle of alternating crystallization and dissolution of sodium chloride in the water seeping through the rock.

Recrystallization of some minerals in a rock can contribute to mechanical disintegration. One example is the hydration of anhydrite to form gypsum. This can cause volume changes of between 30 and 60 percent. Another example is the capacity of some clay minerals to absorb water and increase in volume. This is especially important in montmorillonite, which can expand as much as ten times by drawing water into its crystal structure. We can easily understand how such expansion could impart significant mechanical stress on the surrounding mass of rock.

Release of Residual Stress

When nature puts rock together deep in the crust, the constituent grains become arranged in a matrix that is more or less in equilibrium with the stresses of the subterranean environ-

Figure 12-1 Sheeting in a massive granite formation in Yosemite National Park in California. (G. K. Gilbert, U.S. Geological Survey.)

ment. These stresses are removed when the rock is later exposed at the surface. Geologists have observed fractures that are parallel to the exposed surfaces of large unstratified rock masses. These fractures have been attributed to the internal adjustment of the rock to stress release. The result is an effect that is called **sheeting.** It is illustrated in Figure 12-1. Presumably the constituent minerals, because of their different elastic properties, respond differently to the change in stress. Therefore, internal stresses develop in the matrix of interlocking grains, which might produce sheeting fractures.

Differential Thermal Expansion

We can expect rocks, like other substances, to expand when they are heated and to contract upon cooling. The dimensions of a rock mass vary with temperature according to a constant of proportionality that is called the **coefficient of thermal expansion.** The value of this coefficient differs from one material to another. Stresses related to differences in the thermal expansion or contraction of the various minerals in a rock have been suggested as a cause of mechanical weathering.

Professor David Griggs conducted an experiment to test thermal weathering of rock specimens by alternately heating them to 100° C and then cooling them to 30° C with dry air blowing from a fan. After 89,400 cycles (equivalent to 244 years) almost no disintegration could be detected. Apparently the stresses did not exceed the limits for elastic deformation of the constituent grains of the rock matrix. When the experiment was modified by using water to cool the specimens, however, weathering effects were evident after less than 1000 cycles (equivalent to about

2.5 years). The results suggested the importance of chemical processes associated with the mechanical cycle.

Collapse of Unstable Masses

Observe the ledge along a roadside (Figure 12-2 A) that remains after the underlying rock has been destroyed by erosion and weathering. Eventually rock that has been undermined in this way cracks and falls and is further broken apart in the process of collapsing (Figure 12-2 B). This too is mechanical weathering. The extent to which a ledge can project before breaking under its own weight depends upon its dimensions, tensile strength, and the degree to which it is weakened by joint sets.

A ledge can be undercut by stream erosion. Effects of chemical weathering can also cause undercutting, which sometimes occurs close to seeps. Here rock is partly dissolved by acidic ground water and the loosened grains are washed away by the water flowing from the seep. Seep undercutting and collapse are important in developing the exotic landscape of the Colorado Plateau. Common to this region are the arches in thickly bedded sandstone (Figure 12-3). They develop in the faces of cliffs that are parallel to sets of large vertical joints. These joints separate the sandstone into a series of vertical plates. Where undercutting occurs along seeps near the base of such a cliff the effect of gravity sets up stresses which cause rock to split along arcuate surfaces from the vertical plate bordered by the cliff. The examples in Figure 12-3 show different stages of arch development beginning with a small slightly arcuate undercut and progressing to the Great Arch which is more than 100 m high. These landscape features are primarily the result of mechanical weathering stimulated by prior effects of chemical weathering.

In locations where conjugate sets of vertical joints are parallel and perpendicular to canyon walls, the rock is separated into vertical columns. Undercutting leads to collapse of these columns forming recesses, and downdropped and detached blocks. Some column features in Zion Canyon are illustrated in Figure 12-4.

The debris from collapse collects at the base of cliffs to form a pile of broken rock that is called **talus**. The surface of the talus pile ordinarily slopes at an angle of less than

(A)

(B)

Figure 12-2 (A) Ledge of sandstone undercut by more rapid weathering of the underlying shale along a highway near Charleston, West Virginia. (B) In places the unsupported ledge has collapsed.

Figure 12-3 (opposite) Stages in arch development beginning with (A) seep undercutting, then continuing (B and C) by rock collapse along arcuate fractures, and eventually reaching the size (D) of the 100 m high Great Arch of Zion Canyon.

(A)

(B)

(C)

(D)

(A)

(B)

(C)

(D)

Figure 12-4 *(opposite)* Column features seen in massive sandstone cliffs of Zion Canyon, Utah. The *(A)* recesses, *(B* and *C)* collapsed columns, and *(D)* detached columns are the result of weathering of rock cut by conjugate vertical joint sets.

30 degrees. Unless streams transport the debris away from the area it continues to accumulate, and eventually the great platforms of sandstone become interred in the mass of their weathered rubble. In Figure 12-5 we can see the remanent crags of sandstone rising from a massive talus mound. Here the mechanical disintegration of the bedrock is far advanced.

The weathering of the Colorado Plateau sandstone formations has produced a bizarre landscape. But the weathering processes are not unusual. They are active over much of the land surface elsewhere, producing the same effects but in less spectacular proportions.

Geologists are often asked to advise about the possibility of undercutting and collapse for engineering studies concerned with the stability of the landscape. The stability of steep surfaces cut in bedrock for highway construction can be assessed from the standpoint of weathering processes, joint orientations, and lithology.

Figure 12-5 Eagle Crags near the mouth of Zion Canyon are the last remanants of great sandstone platforms now reduced almost entirely to weathered rubble.

Minor Processes

Other factors contributing to the mechanical weathering of rock include growth of plant roots and effects of fire. These phenomena are of relatively minor importance except in unusual situations. Collapse that is caused by dissolution in carbonate rocks is an important aspect that will be discussed in a later section.

CHEMICAL WEATHERING

Decomposition of rock caused by chemical weathering processes is evident almost everywhere near the earth's surface. The most important chemical reactions involve water, oxygen, and carbon dioxide. These substances are much more abundant close to the earth's surface than deeper in the crust where rock is formed. When water, oxygen, and carbon dioxide come in direct contact with rock, chemical reactions commence. Evidence of these reactions is plentiful. Certainly you have seen how old gravestones have become so corroded that lettering can no longer be read. Blocks of building stone in ancient structures have become rounded. Break open a pebble or cobble to see if the rock inside is different than the rock at the outer surface. See how chemical weathering of the specimen illustrated in Figure 12-6 produced the yellow rind that is quite different from the gray crystalline rock inside. Alteration of this rock began with chemical reactions on the outer surface and along surfaces of fractures penetrating the specimen. It is on these surfaces that the rock comes in contact with water or the atmosphere. On a larger scale we can view the regolith that blankets the earth as a rind produced mostly by chemical weathering.

An interesting aspect of chemical weathering is the growth of spheroidal shapes in bedrock masses that are cut by conjugate joint sets. Look at the features illustrated in Figure

Figure 12-6 *(A)* Specimen of gray schist displays chemically weathered rind colored yellow and brown by small amounts of iron oxides. *(B)* Observe how the effects of weathering also penetrate into the rock along a fracture in another specimen.

Figure 12-7 Examples of spheroidal weathering are seen *(A)* in basalt exposed along a roadside in St. Thomas, one of the U.S. Virgin Islands, *(B)* in limestone exposed near Blacksburg, Virginia. Rounding by chemical weathering of originally angular blocks of plutonic rock is seen *(C)* in the Aqueduct of Trajan built in Segovia, Spain, during the period around 100 A.D., and *(D)* the Imperial Palace Wall built in Tokyo, Japan in the sixteenth century (Photograph D by P. H. Ribbe, Virginia Polytechnic Institute and State University.)

Weathering   433

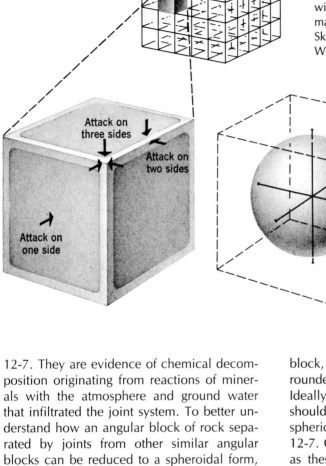

Figure 12-8 A geometrical explanation of spheroidal weathering showing how particles along edges and on corners of angular masses have more exposed surface on which chemical weathering can progress compared with similar particles along the sides of the masses. (Reprinted from R. F. Flint and B. J. Skinner, *Physical Geology*, page 101, John Wiley & Sons, Inc., 1974.)

12-7. They are evidence of chemical decomposition originating from reactions of minerals with the atmosphere and ground water that infiltrated the joint system. To better understand how an angular block of rock separated by joints from other similar angular blocks can be reduced to a spheroidal form, let us examine Figure 12-8. Here we can see that chemical reactions attack three sides of a rock grain occupying a corner of the block. Grains along an edge of the block are attacked on two sides. On a face of the block weathering reactions can occur only on the one exposed side of a grain of rock. Clearly, the grains on the corners will decompose more rapidly than grains along an edge, which in turn decay faster than particles distributed on a face of the block. Chemical weathering destroys the sharp corners and edges faster than it destroys the faces of the block, so that the block becomes more rounded. Eventually a spherical form evolves. Ideally, the effects of chemical weathering should then be the same at all points on the spherical surface. Now look again at Figure 12-7. Chemical analyses of rock masses such as these indicate that mineral grains within the spheroids are relatively unaltered in contrast to minerals in the zones between spheroids. In these zones the original minerals have been replaced largely by compounds that are more stable in the environment at the earth's surface.

The progress of spheroidal weathering, and for chemical weathering in general, depends upon climate as well as the original bedrock composition. The great Aqueduct of Trajan (Figure 12-7C) was constructed during the first century A.D. in the dry climate of central Spain. More than 18 centuries later this structure is not seriously threatened by weathering. In fact it is still used to carry water across the city of Segovia. In comparison, the Imperial Palace wall (Figure 12-7 D), constructed in Tokyo, Japan, in the fifteenth century A.D. has been more severely altered by chemical weathering due to its location in a moist and temperate climate. Both of these structures are exposed to the atmosphere and water from intermittent rain. But the bedrock outcrops seen in Figures 12-7 A and 12-7 B suffer almost continuous exposure to ground water as well as rain and wind. The spheroidal features seen in these exposures have developed in less than 50 years on embankments cut for road construction. Spheroidal weathering is largely a surface phenomenon. There features are not found in new road cuts, but they begin to emerge within a few decades. This suggests the importance of the atmosphere as well as water in the weathering process.

Now let us look more carefully at how certain minerals are affected by chemical weathering. We have learned some things about the decomposition of common rock-forming silicate minerals from a study of the Morton Gneiss in Minnesota that was done by the geologist, S. S. Goldich. From this granitic gneiss he analyzed rock specimens representing different gradations of chemical weathering. The compositions of these specimens are summarized in Figure 12-9. The relatively unweathered specimens contained significant proportions of quartz, orthoclase, and plagioclase. Samples of more severely weathered rock have a much smaller proportion of feldspar and a significantly larger proportion of the clay mineral **kaolinite**. Weathered and unweathered specimens all have important proportions of quartz. The feldspars are clearly more susceptible to weathering than quartz. We can see that plagioclase decomposes more readily than orthoclase. Kaolinite, a hydrous aluminosilicate, is formed from the residue of these feldspars. The diagram in Figure 12-9 A reveals that the other important elements in feldspars, including calcium, sodium, and potassium, have been removed from the rock, presumably in ground water solutions.

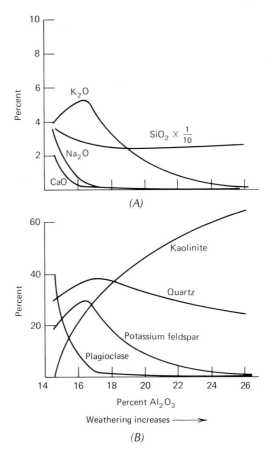

Figure 12-9 Variation diagrams showing differences in (A) chemical and (B) mineralogical composition of specimens from the Morton Gneiss in Minnesota, which display progressively more severe chemical weathering. (After S. S. Goldich, *The Journal of Geology*, page 33, Jan.–Feb. 1938.)

The relative susceptibility of plagioclase, orthoclase, and quartz to chemical weathering is clearly evident from analysis of specimens from the Morton Gneiss. Chemical weathering of other rock-forming silicate minerals has been investigated in similar field studies and by laboratory experiments. From

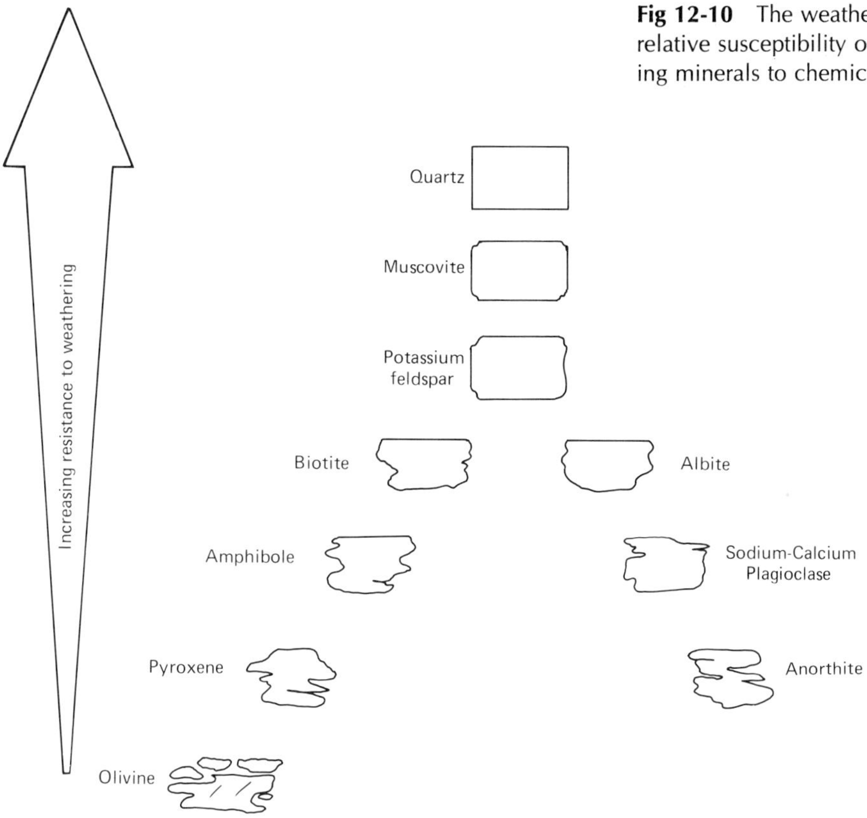

Fig 12-10 The weathering series illustrates relative susceptibility of common rock-forming minerals to chemical weathering.

Increasing resistance to weathering

- Quartz
- Muscovite
- Potassium feldspar
- Biotite — Albite
- Amphibole — Sodium-Calcium Plagioclase
- Pyroxene — Anorthite
- Olivine

this research we have learned about the interesting relationship illustrated in Figure 12-10. Here the important rock-forming silicate minerals are listed according to relative susceptibility to chemical weathering. The arrangement of minerals in this **weathering series** is just the reverse of the order of minerals in Bowen's reaction series. Recall from Chapter 7 that Bowen's reaction series states the order in which the common silicate minerals should crystallize from a cooling magma. We can see from Figure 12-10 that those minerals which crystallize at high temperature are the least stable in the low temperature environment at the earth's surface where weathering takes place. Quartz, the last mineral to crystallize from a cooling magma, is the silicate mineral most resistant to common chemical weathering processes. An important distinction should be made between the weathering series and Bowen's reaction series. Recall from Chapter 7 that in a cooling magma the early formed minerals partially dissolve as temperature decreases, and their constituent elements are subsequently incorporated into other minerals which crystallize at lower temperature. The high temperature minerals, then, must eventually supply some elements needed in the minerals that later crystallize at lower temperature. This is not true in weathering series in Figure 12-10. Substances derived from the decomposition of some minerals of this series are not used to make other minerals in the series. But if the weathering series were extended to include the important clay minerals, then it would be more like Bowen's reaction series because it would have some minerals that are made of substances derived from other minerals in the series.

General Weathering Zone Conditions

The decomposition of bedrock involves the removal of some substances that become dissolved or suspended in ground water. It also involves production of other solid substances, primarily the clay minerals which are relatively stable in the environment of the weathering zone. The varied and complex chemical reactions that occur in this zone almost always involve water. Look at the geometry of the H_2O molecules shown in Figure 12-11. These are electrically polarized molecules because of the way that the hydrogens are grouped close together on one side of the oxygen. A small proportion of these molecules becomes dissociated, that is, a few ions of H^+ and OH^- separate from one another and become distributed in the larger mass of H_2O molecules.

When a substance dissolves in water, the constituents that were formerly joined together to form molecules or crystalline structures become dissociated. Then they exist as ions that are distributed through the fluid. Electrical forces between these charged particles and the polarized water molecules act to keep them evenly distributed. Depending upon the concentration of such dissolved ions, some may combine with the dissociated hydrogen or oxygen, and then precipitate from the solution.

In addition to dissolved ions we also find particles suspended in the water. These particles are not dissociated into ions, but they are broken into very small particles that can possess surface electrical charges along their broken edges. Forces between these charges and polarized water molecules can maintain the particles in suspension almost indefi-

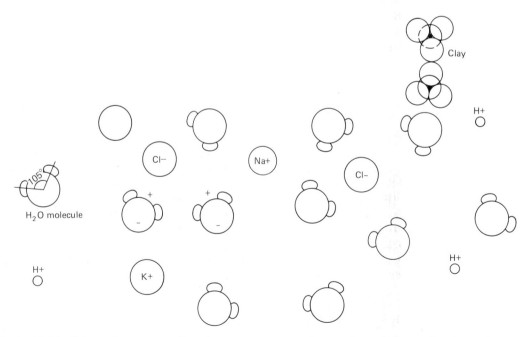

Figure 12-11 Schematic representation of a water mass containing polarized H_2O molecules and dissociated ions of different elements.

nitely. These suspended particles together with dissociated ions are transported by ground water flow. Sometimes, however, the particles react with dissociated ions which adhere to their surfaces and neutralize the electrical charges. If this occurs, the forces maintaining the suspension are greatly diminished. The particles may then flocculate in clusters and settle out of the fluid to form solid deposits.

We can understand from the above discussion that while ground water seeps through a bedrock unit, it maintains substances in solution or suspension. But when these substances are carried by ground water into another bedrock unit, other ions may be encountered. Reactions with these different ions could result in precipitation or flocculation of some of the particles that were formerly dissolved or suspended. The chemical environment of weathering, then, is continually changing because of ground water flow. Chemical reactions in a subsurface zone can change because new substances are introduced and others are removed. Among the most important kinds of chemical weathering reactions occurring in this system are: (a) hydrolysis; (b) ion exchange; (c) hydration; (d) carbonation; and (e) oxidation and reduction reactions.

Hydrolysis

Chemical reactions involving water in such a way that the H_2O molecule does not remain intact are reactions of hydrolysis. These reactions are very important in the production of clay minerals. We can understand the nature of hydrolysis more clearly after we examine some typical chemical reactions that produce the most abundant clay minerals. These minerals include the **kaolinite** group, the **montmorillonite** group, and the **illite** group of clays.

The kaolinite clays are double-layer sheet silicates. Look again at the structure of the mica minerals (Figure 6-23) which are single layer sheet silicates. To get an idea of the structure of a kaolinite clay imagine two of these sheets joined together. Inward facing oxygen ions become bridging oxygens shared by the two layers. A double-layered grain of a kaolinite clay tends to be electrically neutral. The approximate chemical composition for these clays is $Al_2Si_2O_5(OH)_4$. Kaolinite could be derived from orthoclase by the chemical reaction:

$$2KAlSi_3O_8 + 3H_2O \rightarrow \quad (12\text{-}1)$$
$$\text{orthoclase} \quad \text{water}$$
$$Al_2Si_2O_5(OH)_4 + 4SiO_2 + 2K^+ + 2OH^-$$
$$\text{kaolinite} \quad \text{silica}$$

Clays in the montmorillonite group are characterized by three-layer silicate structures. A variety of chemical compositions is found in this group ranging from the aluminosilicate $Al(Si_2O_5)(OH) \, n \, H_2O$ to more complex forms incorporating Ca, Mg, and Na as well. The amount of water in one of these clay minerals is indicated by the value of the integer n. A montmorillonite clay could be produced by the weathering of anorthite in the following way:

$$3CaAl_2Si_2O_8 + 6H_2O \rightarrow \quad (12\text{-}2)$$
$$\text{anorthite} \quad \text{water}$$
$$2H^+ \cdot 2\{Al_2(AlSi_3O_{10})(OH)^-\} + 3Ca(OH)_2$$
$$\text{montmorillonite}$$

Illite is a potassium aluminum silicate with double-sheet structure. It is the most abundant clay in the mica group and is widely distributed in shale and mudstone of marine origin. Illite can be formed by weathering of muscovite or feldspar in the presence of water. It could be obtained from the decomposition of orthoclase:

$$3KAlSi_3O_8 + 2H_2O \rightarrow KAl_2(Al,Si_3)O_{10}(OH)_2$$
$$\text{orthoclase} \quad \text{water} \quad \text{illite}$$
$$+ 6SiO_2 + 2KOH \quad (12\text{-}3)$$
$$\text{silica}$$

These chemical reactions illustrate how feldspars are weathered by hydrolysis to produce important clay minerals. In what way are the reactions similar? In each chemical formula water molecules that react with the feldspars do not remain intact as distinct H_2O units. Rather, the hydrogen and oxygen dissociate, and then become incorporated as separate units into the structure of the clay mineral. This is indicative of a hydrolysis reaction.

Keep in mind that the chemical reactions expressed in Eqs. (12-1), (12-2), and (12-3) are actually simplified representations of the sequence of chemical reactions that probably occur. Actually, the feldspars are almost insoluble in distilled water, and direct reactions such as those expressed in these equations would occur very slowly if at all. But the decomposition of feldspar is greatly accelerated if the ground water is slightly acidic. In ground water we commonly find acids that are produced by reactions with gases in the atmosphere and by biologic processes. For example, we know that small proportions of carbon dioxide can be dissolved in water. In the weathering zone this CO_2 comes from the atmosphere and from plant respiration. It reacts with water to form carbonic acid in the following way:

$$\underset{\text{carbon dioxide}}{CO_2} + \underset{\text{water}}{H_2O} \rightleftarrows \underset{\text{carbonic acid}}{H_2CO_3} \rightleftarrows \underset{\text{dissociated hydrogen}}{H^+} + \underset{\text{bicarbonate ion}}{HCO_3^-} \quad (12\text{-}4)$$

The continual interaction of H_2O and CO_2 causes the ground water to become a weak acid solution. In this environment orthoclase could be decomposed according to the chemical equation:

$$\underset{\text{orthoclase}}{2KAlSi_3O_8} + \underset{\text{carbonic acid}}{2H_2CO_3} + \underset{\text{water}}{9H_2O} \rightarrow \quad (12\text{-}5)$$

$$\underset{\text{kaolinite}}{Al_2Si_2O_5(OH)_4} + \underset{\substack{\text{silicic} \\ \text{acid}}}{4H_4SiO_4} + \underset{\substack{\text{potassium} \\ \text{ion}}}{2k^+} + \underset{\substack{\text{bicarbonate} \\ \text{ion}}}{2HCO_3^-}$$

We can see that both the formation of carbonic acid (Eq. 12-4) and the production of kaolinite (Eq. 12-5) are hydrolysis reactions involving separation of hydrogen and oxygen from the water molecules involved.

Ion Exchange Reactions

Under favorable conditions certain ions in the crystal structure of a particle can become dislodged and replaced by other ions that are present in the surrounding ground water. Where such an exchange takes place, low valence ions tend to be replaced by higher valence ions. Ions found in the common rock-forming minerals and ground water that are susceptible to exchange tend to replace one another in the following order: Na < K < Mg < Ca < Al. In this series ions farther on the right tend to replace those to the left if proper conditions for exchange exist. Thus, a mineral containing primarily magnesium may exchange some Mg^{2+} ions for Ca^{2+} ions in the crystal structure on the surface of a grain. Similarly, some Ca^{2+} ions along the surface could be replaced by Al^{3+} ions obtained from the surrounding ground water. We will not attempt to discuss the conditions favoring such ion exchanges, but you should know that they can occur in the weathering zone if certain ion concentrations exist in the ground water. Ordinarily, we find ground water solutions more enriched in ions on the left side of this exchange series. Do you think that this evidence supports the idea that ion exchange has occurred in the weathering zone?

One important effect of an ion exchange of Al^{3+} for Ca^{2+}, for example, is that it would cause a formerly neutral particle to assume an electrical charge. Such exchange reactions can take place in the clay minerals, especially montmorillonites. Why is this important? One form of montmorillonite has the approximate composition of (Mg, Ca)O·Al_2O_3·$5SiO_2$·nH_2O. Suppose that some Mg^{2+} and Ca^{2+} ions on the surface of a particle are replaced by Al^{3+}. Because of this, the individual particles in a mass of montmorillonite can become positively charged. These charged particles repel one another. In the presence of water the particles tend to draw apart and absorb the water into the mass of clay. In this way ion exchange has an important effect on the capacity of a clay to absorb water.

Ion exchange reactions have been used to increase the fertility of soils for agricultural purposes. Desirable ions are easily introduced into the ground water; eventually they become incorporated into the soil particles by the ion exchange process.

Hydration

Chemical reactions in which H_2O molecules are maintained as distinctive units are hydration reactions. An example is the hydration of anhydrite to form gypsum:

$$\underset{\text{anhydrite}}{CaSO_4} + 2H_2O \rightarrow \underset{\text{gypsum}}{CaSO_4 \cdot 2H_2O} \quad (12\text{-}6)$$

The H_2O molecule exists as a distinctive constituent in the crystal structure of gypsum. This reaction can cause a volume increase of more than 30 percent when an anhydrite crystal is changed into gypsum. Recall our earlier discussion of the mechanical weathering produced by a chemical reaction such as this.

Of much greater importance is the hydration of clay minerals. This is most dramatic in montmorillonite that can expand more than ten times by absorbing water. The capacity of montmorillonite to absorb water depends on the extent to which ion exchange processes have produced a mass of similarly charged particles. Look at the term nH_2O appearing in the chemical formula for montmorillonite $((Mg, Ca)O \cdot Al_2O_3 \cdot 5SiO_2 \cdot nH_2O)$. The value of n is 1 or 2 in a dehydrated mass. But it can have a value above 10 in a water-saturated mass of clay that possesses a large proportion of similarly charged particles. This term also appears in chemical formulas for kaolinite and illite clays. In these clays the value of n ordinarily cannot exceed 3 or 4.

The hydration of kaolinite does not cause such a dramatic change in volume. However, this hydration process has considerable importance in producing bauxite, the primary ore of aluminum. The hydration reaction is:

$$\begin{array}{c}\underset{\text{kaolinite}}{Al_2Si_2O_5(OH)_4} + \underset{\text{water}}{5H_2O} \rightarrow \\ \underset{\substack{\text{hydrated}\\\text{alumina}}}{Al_2O_3 \cdot 3H_2O} + \underset{\substack{\text{silicic}\\\text{acid}}}{2H_4SiO_4}\end{array} \quad (12\text{-}7)$$

The hydrated alumina is an important constituent in bauxite soils. Conditions most favorable to evolution of bauxite soils are found in tropical environments with abundant rainfall.

Hydration is almost always accompanied by volume expansion. Because of this, hydration processes also contribute to mechanical weathering. The possibility of these reactions must be carefully considered in planning engineering projects. Some beds of clay are susceptible to hydration and dehydration in response to intervals of rainy and dry weather. The volume changes can seriously disturb construction foundations and highway road beds.

Carbonation

Carbon dioxide is introduced into ground water by carbonation reactions. The reaction in chemical Eq. (12-4) represents the carbonation of water. Carbonation processes in chemical weathering involve decomposition of minerals in carbonic acid. Other carbonate compounds and the bicarbonate ion (HCO_3^-) are products of these reactions. We can see from the reactions expressed in Eq. (12-4) and (12-5) how carbonation is combined with hydrolysis in the decomposition of silicate minerals. Another chemical weathering reaction is the carbonation of limestone by carbonic acid:

$$CaCO_3 + H_2CO_3 \rightarrow Ca^{++} + 2HCO_3^- \quad (12\text{-}8)$$
calcite   carbonic   calcium   bicarbonate
             acid   ion   ion

Calcite is relatively insoluble in distilled water. This mineral dissolves much more rapidly in ground water which is a weak carbonic acid. The carbonation of limestone can produce some interesting landscape features. We will discuss these features in a later section on karst topography.

Oxidation and Reduction

Reactions in which electrons are removed from ions are called oxidation reactions. The most common of these are reactions that produce oxygen compounds. Oxidation is common in the weathering zone. Here we find oxygen that comes from the atmosphere or is dissolved in ground water. This oxygen is free to react with ions produced by decomposition of the rock-forming minerals.

Reduction processes are the opposite of oxidation; they involve adding electrons to ions. The most common are reactions in which oxygen is removed from a compound. Reducing conditions are not nearly as widespread as oxidizing conditions in the weathering zone. Reduction occurs in an oxygen-deficient environment. This environment can exist in a bog where there are submerged mud deposits that are rich in organic debris. Here the processes are greatly accelerated by the activity of anerobic bacteria.

Iron ores are produced by oxidation and reduction processes in the zone of weathering. Iron can come from the decomposition of silicate minerals. Consider the decomposition of fayalite by hydrolysis:

$$Fe_2SiO_4 + 4H_2O \rightarrow \quad (12\text{-}9)$$
fayalite   water
$$2Fe^{2+} + H_4SiO_4 + 4(OH)^-$$
ferrous   silicic
iron   acid
ion

Conditions most common to the weathering zone favor rapid oxidation of ferrous iron ions (Fe^{2+}) produced by silicate weathering. These ions combine with the oxygen in ground water to form hematite (Fe_2O_3) and hydrated iron oxides such as goethite or limonite ($Fe_2O_3 \cdot H_2O$). These iron oxides are almost insoluble in distilled water. They tend to precipitate from the ground water as soon as they are produced. The hues of red, yellow, and brown that color most weathered silicate rock come from the iron oxides formed in the weathering zone.

Under relatively unusual conditions iron oxides have become concentrated in valuable iron ore deposits. There is little doubt that these ores are derived primarily from rock weathering. But iron occurs only in small proportions in most rock masses. When it is freed by weathering, it ordinarily becomes oxidized and precipitates so rapidly that it cannot be carried away to build up a deposit of ore. So how does nature produce a large iron ore deposit? V. M. Goldschmidt, the father of modern geochemistry, pointed out that (a) oxidizing conditions and alkaline solutions favor precipitation of iron compounds, whereas (b) reducing conditions and acid solutions favor maintaining these compounds in solution. It seems reasonable to suggest that iron is transported out of places where the latter conditions prevail, into an adjacent area where the former conditions exist. One way that this might happen is illustrated in Figure 12-12, where a fresh water lake borders a bog. Reducing conditions exist in the bog which is rich in organic debris. Ground water migrating through the bog becomes charged with CO_2, and carbonic acid is formed by hydrolysis. Iron compounds dis-

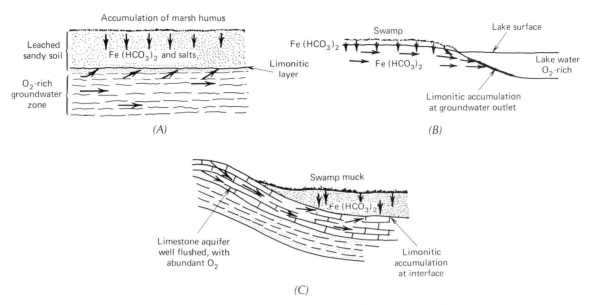

Figure 12-12 A possible explanation of the origin of an iron ore deposit. (A) Under reducing conditions existing in a bog, acidic ground water could dissolve iron compounds. (B) Ground water containing these dissolved iron compounds could then flow into an oxidizing environment. (C) If it encounters other substances that change it from an acidic to an alkaline solution, then the dissolved iron would tend to precipitate and become oxidized. (After H. Borchert, *Bulletin 640*, pages 261–279, Institute of Mining and Metallurgy, 1960, and modified by R. L. Stanton, *Ore Petrology*, page 418, McGraw-Hill Book Co., 1972.)

solve in the acidic ground water in this reducing environment. The highly soluble ferrous bicarbonate (Fe(HCO$_3$)$_2$) is produced by carbonation of the iron oxides. As the ground water migrates from the bog into the lake, conditions change to those of an oxidizing environment. Here the iron from dissolved ferrous bicarbonate is oxidized and precipitated. In this way iron derived from the large mass of the bog could be transported and concentrated in a small zone near the margin of the bog and the lake. A sequence of processes such as this could account for the well-known **bog iron ores.** These ores appear to have formed in temperate and tundra environments of the Northern Hemisphere.

Other kinds of ore deposits including the **Clinton-type ironstone** formations, and the very important **banded Precambrian ironstones** appear to be of different origin than the bog iron ores. The Clinton-type ironstones occur primarily in near-shore sedimentary facies. These facies are often fossiliferous and range from Precambrian to Tertiary in age. The banded Precambrian ironstones include zones of hematite, magnetite, and taconite, together with other bands that do not possess iron. These ores appear to have originated during Precambrian time when oxygen may have been absent from the atmosphere. Many questions remain unanswered concerning the origin of these ores of iron.

KARST LANDSCAPES AND CAVES

In regions where thick layers of limestone are found, ground water and rock weathering processes can play a dominant role in forming the landscape. Perhaps the most exotic features in these regions are labyrinths of underground passages that make up great cave systems. Prominent landscape features are produced by collapse of rock in these caverns. This kind of landscape is particularly well developed in the Karst region of Yugoslavia, so it has been named **karst landscape.** Similar karst landscapes are found in areas of southeastern United States as well as other parts of the world where carbonate terranes exist.

Most caves are formed by the dissolving of limestone in ground water. This involves the carbonation reactions stated in chemical Eq. (12-4) and (12-8). Caves commonly develop along a regional joint system (Figure 12-13). The joints are channels through which the ground water can circulate. Geologists have argued about whether caves are originally formed above or below the water table. The passages that we can explore most easily are, of course, drained. But were they formerly filled by water?

The walls of many passages are decorated by depositional features. Among the bizarre structures are **stalactites** which hang in icicle-like form from cave ceilings, **stalagmites** which stand like inverted icicles from cave floors, and columns presumably formed by the joining of stalactites and stalagmites. Look at the examples in Figure 12-14. What do these features tell us about the origin of a cave? When droplets of ground water seep into a cave, slight changes in pressure and temperature as well as evaporation can cause dissolved CO_2 to be released from the water. This diminishes the concentration of carbonic acid so that less Ca^{2+} and HCO_3^- can be maintained in solution. The result is precipitation of $CaCO_3$ as **dripstone,** the common name for the limestone that is deposited in caves. We find, then, that deposition rather than dissolution of $CaCO_3$ is the primary pro-

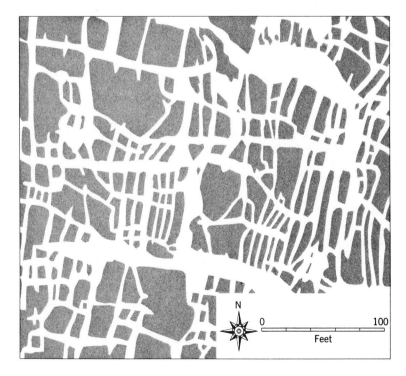

Figure 12-13 Map of passages in Anvil Cave near Decatur, Alabama, prepared by explorers from the National Speleological Society. The pattern illustrates the control of regional joint sets on cave development. (Reprinted from R. F. Flint and B. J. Skinner, *Physical Geology*, page 167, John Wiley & Sons, Inc., 1974, adapted from map by W. W. Varnedoe.)

Figure 12-14 Dripstone deposits in Carlsbad Cavern in New Mexico form stalactites, stalagmites, and columns. (R. V. Davis, U.S. Geological Survey.)

cess in caves that are not filled with water.

Arguments about cave formation tend to favor the idea that $CaCO_3$ is dissolved beneath but close to the water table. The shapes of some passages support this idea. These passages pinch out in upward and downward directions alike in a way that suggests that they were originally filled by water. However, relatively rapid circulation of ground water is required because of the limited concentrations of Ca^{2+} and HCO_3^- which can be maintained in solution without saturating the water. It would appear that caves begin to form when $CaCO_3$ beneath the water table is dissolved. Passages are later drained by lowering of the water table. Then deposition of the bizarre dripstone formations commences.

Labyrinths of cave passages are mostly horizontal even where the limestone strata are inclined. This suggests that development may be controlled by the position of the water table and takes place close beneath it. The water table may then be lowered perhaps by changes in the landscape caused by stream erosion (Figure 12-15) or by a change in climate. The cave passages are then drained.

A few caves such as those in the Black Hills of South Dakota are partially lined with crystals rather than dripstone. Conditions favoring the growth of such crystals exist in passages filled with water that is oversaturated with Ca^{2+} and HCO_3^-. These caves appear to have formed beneath the water table where $CaCO_3$ was being dissolved. Condi-

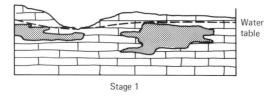

Stage 1

Stage 2

Figure 12-15 A possible explanation of the origin of caves. Caverns first develop by dissolution of limestone in the phreatic zone (Stage 1). At a later time the water table was lowered and the caverns were drained. Then dripstone precipitates from vadose water seeping into the caverns to produce stalactites and stalagmites (Stage 2).

tions subsequently changed so that crystals began to form on the walls of the water-filled passages by precipitation of $CaCO_3$.

Much remains to be learned about the evolution of caves. Careful exploration of drained passages is now being supplemented by the discoveries of still more daring cave explorers, equipped with aqualungs, who probe the submerged portions of cave systems.

In regions that are extensively undermined by cave systems peculiar landscape features develop where the land surface collapses. **Sinkholes** such as the one shown in Figure 12-16 can form suddenly when the ground falls. With the passing of time additional rubble will wash into this depression and vegetation will take root on its slopes. It will then become another conical depression similar to the ones that mark the landscape seen in Figure 12-17. This is a characteristic karst landscape.

Figure 12-16 Sinkhole more than 130 m wide and 45 m deep formed in central Alabama in 1972 by sudden collapse of limestone into a subsurface cavern. (U.S. Geological Survey.)

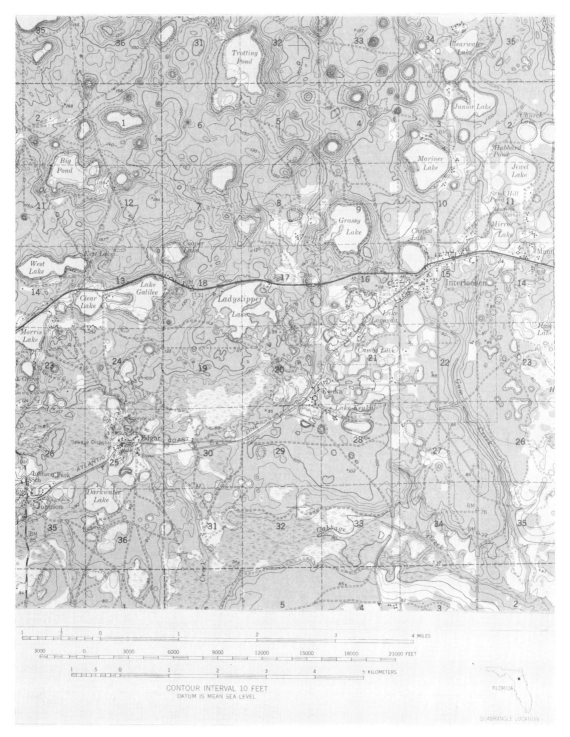

Figure 12-17 Topographic map showing circular and oval depressions, some filled by water, which typify karst topography seen near Interlochen, Florida. (U.S. Geological Survey.)

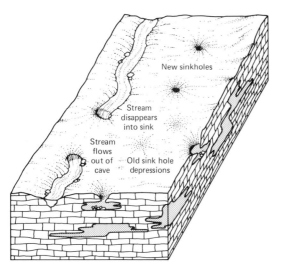

Figure 12-18 Typical features of Karst landscape.

Stream drainage can be altered by caves and karst features. A stream valley may terminate at a point where the stream plunges beneath the surface to follow a cave passage, only to emerge elsewhere to flow along the surface. These features appear on the diagram in Figure 12-18, which shows stream drainage and typical karst features related to a cave system.

It is a good idea to probe construction sites in karst areas for hidden caverns, because failure to learn about such caverns can have serious consequences (Figure 12-19). Less spectacular, but eventually more costly to urban and industrial development, is slow subsidence and tilting of unstable ground caused by the weight of buildings.

Figure 12-19 Edge of sinkhole that formed in 1967 by sudden collapse of unstable limestone beneath a part of Bartow, Florida, with destructive consequences. (U.S. Geological Survey.)

SOILS

Soil could be viewed as our most basic and important natural resource along with fresh water. Agriculture is sustained by soil and water. The processes of soil formation and weathering are very much the same. But soil is a material that has been modified by weathering processes and biological processes in such a way that it is capable of supporting rooted plants. Most soils tend to be arranged in stratified profiles, and they continue to evolve by chemical, biological, and physical processes that are active in the regolith. In this brief discussion we will examine the principal soil-forming processes and the relationships between different soils and climate.

Soils evolve by chemical reactions involving ground water that infiltrates the vadose zone. The important processes described below all contribute to the transport and deposition of soil constituents in ways that produce a soil zone that can be divided into layers with different properties.

Leaching and Precipitation

Leaching is the process by which substances are dissolved and removed from a soil layer by circulation of ground water. The process is most active in transporting constituents from upper soil zones into lower zones. In these deeper zones where there is a different environment, other chemical reactions cause these dissolved substances to precipitate. Where changes in climate or excessive irrigation cause the water table to rise, substances can be leached from lower zones and precipitated in upper zones. Sometimes this degrades soil fertility. A process known as **cheluviation** represents a particular kind of leaching caused by ion exchange reactions involving organic compounds.

Eluviation and Illuviation. The mechanical dislodging and transport of clay particles and other grains small enough to be suspended in ground water is the process of **eluviation.** We might think of this as washing a soil zone. Ordinarily material is transported from upper to lower soil zones where it is deposited. The deposition of such mechanically transported material is called **illuviation.** Upper soil layers are sometimes described as zones of eluviation and the lower layers are illuviation zones.

Organic Accumulation and Sorting. Surface vegetation contributes important substances to the soil. Subsurface organic processes as well as the mechanical activity of burrowing organisms can separate the decayed vegetation into particles of different composition and grain size.

As a result of these soil-forming processes more or less horizontal soil layers are developed. A sequence of soil layers is called a **soil profile.** Three zones are distinguished in most soil profiles; these are illustrated in Figure 12-20. The upper zone labeled the A-horizon is an eluviated zone that is ordinarily rich in organic compounds. The gray organic **humus** produced by decay of vegetation is usually an important constituent of the A-horizon. The B-horizon is an illuviated zone that is almost devoid of humus. It is rich in the constituents leached or eluviated from the overlying A-horizon. The soil zone that we call the C-horizon consists of parent material, either bedrock or regolith, which has not become stratified by the action of soil-forming processes.

Residual soils form where the products of rock weathering are not transported from the area and sedimentary deposits are not accumulating in the area. Here the weathering processes first produce a heterogeneous blanket of regolith by fragmentation and decompositon of the bedrock. The soil zones then form from this unconsolidated debris by activity of the soil-forming processes. This sequence of events in soil evolution is simplified. In nature it is usually interrupted by

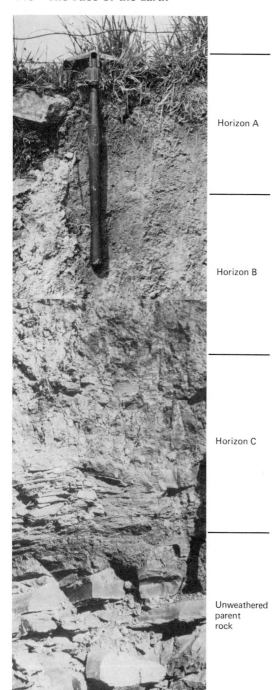

Figure 12-20 Profile illustrating the major zones in a mature soil.

erosion of partially formed soil or by deposition of sediment from other places. In some instances this is beneficial. Sediment that is periodically deposited by flooding rivers may enrich soil fertility, but rapid erosion can greatly diminish the agricultural potential of the land. The thicknesses of different soil zones depend upon these factors as well as the conditions of climate, bedrock composition, and the time available for soil evolution. Soil-forming processes have been active throughout geologic history, and we can find evidence of fossil soil profiles in layers of ancient rock.

Soils are classified according to geographical location and local and regional conditions of climate. **Zonal** soils are related mostly to regional climate and vegetation. Local soil-forming conditions which differ from prevailing regional conditions cause **intrazonal** soils to form. **Azonal** soils are characterized by partially developed profiles because of factors that interrupt or inhibit the soil-forming processes. The principal kinds of zonal soil are described in Table 12-1.

The soils that have been described above are related both to temperature and moisture conditions. Those soils found in relatively dry environments tend to retain proportionally more calcium in the A-horizon. These soils are generally classified as **pedocals.** In more humid environments the calcium tends to be leached from the A-horizon leaving higher concentrations of the less soluble iron and aluminum enriched minerals. These soils are called **pedalfer** soils. The map in Figure 12-21 shows the worldwide distribution of soils. This map together with Table 12-1 and your knowledge of global topography can be used to identify conditions of climate in different parts of the world.

PERSPECTIVES

The blanket of regolith, a weathered rind on the surface of the earth, consists of those frag-

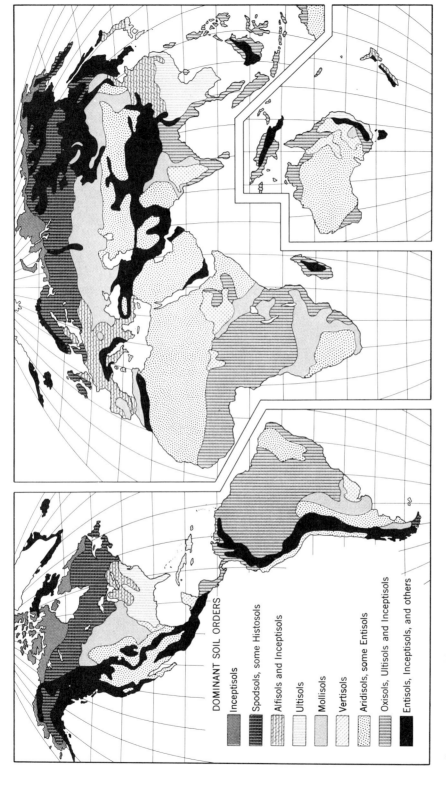

Figure 12-21 Worldwide distribution of soil. (From Soil Conservation Service, U.S. Department of Agriculture.)

Table 12-1
Distinguishing Characteristics, Climatic Range, and Vegetation of the Ten Soil Orders According to the United States Department of Agriculture Soil Classification System*

Soil Order	Climatic Range	Natural Vegetation	Distinguishing Characteristics
Entisol	All climates, from polar regions to subtropics, arid to humid. No climatic signficance.	Highly variable, from forests and grass, to desert shrubs.	Little or no horizonation with no more than A and C horizons. Common in recently transported materials such as river floodplain deposits, in regoliths highly resistant to weathering such as quartz sands, and on steep slopes subject to rapid removal of weathering products. Occur on flat to steep slopes.
Vertisol	Humid to arid. Hot dry seasons followed by season of higher precipitation.	Mostly grass and woody shrubs.	Soils of high clay content that form cracks in dry season and expand during wet season. B horizon commonly absent. Occur on flat surfaces as well as sloping uplands. Parent material widely variable but always characterized by high content of clays that expand when wet.
Inceptisol	Humid and subhumid areas from Arctic to the tropics. Do not occur in desert regions.	Mostly forests but some grass.	Weak horizon differentiation with no significant eluviation or illuviation. Parent material not weathered. Appreciable accumulation of organic material in A horizon. Occur on land surfaces ranging from flat to moderately steep slopes.
Aridisol	Arid, semiarid, steppe.	Sparse grasses, shrubs, cactus, and other desert plants.	Calcium carbonate (lime), calcium sulfate (gypsum), or sodium chloride concentrated in various parts of the profile. Lime may cement soil particles to form *caliche*. Horizons not well differentiated, generally. Colors range from gray in steppe climates to red in true deserts. Little organic matter in A horizon.
Mollisol	Humid, subhumid, semiarid. Mostly cold winters, hot summers.	Dense grass.	Dark soils due to heavy concentration of organic matter in the A horizon well mixed with mineral constituents. Lower B or upper C horizon may have secondary lime accumulation.

From Zumberge and Nelson, John Wiley & Sons, Inc., Copyright © , pp. 154-155.
*Prepared with the help of Roy W. Simonsen, Soil Conservation Service, U.S. Dept of Agriculture.

ments of rock that are relatively stable in a low temperature-low pressure aqueous environment. Many rock-forming minerals which were produced in a deeper plutonic or metamorphic environment decompose in the weathering zone to yield the more stable clay minerals. The weathering processes serve to concentrate deposits of clay that are widely used in the ceramics industry. Bauxite and iron ores are also products of weathering. Because weathering processes alter the strength of bedrock and regolith, they must be consid-

Soil Order	Climatic Range	Natural Vegetation	Distinguishing Characteristics
Spodosol	Humid continental to subtropical	Coniferous forest, heather, some deciduous forest.	An A_2 horizon of organic matter over a mineral A_2 horizon with pale colors due to lack of iron oxides and organic matter. B horizon illuviated with organic matter, iron oxides, and aluminum oxides; generally brown or black in color. Soils do not develop in parent material high in clay.
Alfisol	Cool humid to subhumid.	Generally deciduous forest.	Thin A_1 horizon rich in humus and thicker light brownish-gray A_2 horizons over B horizon illuviated with very fine clay particles and with small amounts of organic matter and iron oxides. Parent materials relatively young (geologically speaking) and usually calcareous.
Ultisol	Humid, subtropical, and tropical.	Mixed forest to rain forest.	Thin A_1 horizon with some humus and thicker A_2 horizon of pale to moderate color over B horizon illuviated with very fine clay particles and some iron and aluminum oxides. More strongly weathered and leached than Alfisols. Parent materials generally old and strongly weathered but some unweathered feldspars and micas remain.
Oxisol	Tropical and subtropical.	Rain forest.	Somewhat darkened and fairly thick A_1 horizon over B and C horizons of great thickness. Aluminum and iron oxides are residual concentrates through loss of other components, including silica. High in clay but porous, permitting ready downward movement of water. Parent material is deeply weathered with no original minerals remaining.
Histosol	Moist to wet.	Swamp, marsh, and bog vegetation.	Solum made up chiefly of organic materials under conditions of very poor drainage. No visible horizons. Commonly called bog soils, muck, or peat.

ered when planning industrial and urban development.

We have concentrated our discussion on the prosaic natural weathering processes. Since the advent of the Industrial Revolution during the mid-nineteenth century, the scope of weathering has broadened. Exotic pollutants added to the atmosphere and ground water have markedly accelerated the corrosion of ancient buildings and perhaps the underlying bedrock as well. The long-term effect of these new agents of weathering remains to be assessed.

STUDY EXERCISES

1. Suppose that monuments for display in New Orleans, Louisiana, and Minneapolis, Minnesota, can be constructed from either granite or quartz arenite. Discuss the advantages of using one or the other of these rocks with regard to its resistance to weathering in the climates of these two cities. Why might one of these kinds of rock be more resistant to mechanical weathering than the other, which might be more resistant to chemical weathering?

2. Consider the weathering of a large mass of rock that is cut by two intersecting joint sets. Is there reason to believe that a smaller and smaller proportion of the remaining fresh rock will be destroyed by weathering with each passing year? Explain!

3. Describe two weathering reactions that could introduce silicic acid into the ground water.

4. What influence does the interaction of water with the earth's atmosphere have on the process of leaching calcium from the A-horizon of the soil profile? What differences in the rate of leaching explain why pedocal soils develop in dry environments, whereas pedalfer soils develop in moist environments?

SELECTED READINGS

Birkeland, Peter W., *Pedology, Weathering, and Geomorphology,* Oxford University Press, New York, 1974.

Goldich, S. S., A Study in Rock Weathering, *Journal of Geology,* vol. 46, pp. 17–58, 1938.

Reiche, P., *A Survey of Weathering Processes and Products,* University of New Mexico, Publications in Geology, No. 3, Albuquerque, 1950.

Sweeting, Marjorie M., *Karst Landforms,* Columbia University Press, New York, 1973.

13

RUNNING WATER, MASS WASTAGE, WIND, AND CONTINENTAL LANDSCAPES

Surface Drainage Systems
  Drainage Patterns
  Drainage Geometry
Streams and Rivers
  Channel Geometry
  Hydraulic Geometry
  Stream Flow and Sediment Transport
  Fluvial Landscape Features
    Stream Valleys in Mountainous or Hilly Terrain
    Meandering Streams on Alluvial Plains
    Rejuvenation
    Floods
  Base Level and Deltas
  The Graded Stream
Mass Wastage
  Creep
  Earthflow and Solifluction
  Landslides
Wind and Eolian Landforms
  Global Air Circulation
  Eolian Landscape Features
Evolution of Landscape
  The "Cycle of Erosion" of W. M. Davis
  The Concept of Dynamic Equilibrium

The Willow River used to follow a winding path over the countryside of southwestern Iowa. Water flowed into it from a branching network of smaller streams. Frequently the river overflowed its banks causing floods that often interrupted spring planting and sometimes destroyed crops. People living nearby welcomed the idea of a flood control project. Work began in 1906, and by 1920 a drainage channel more than 28 miles long was excavated along the Willow River Valley. Unlike the winding river, this channel was nearly straight. It was wide enough and deep enough to carry even the large amounts of water that formerly would overflow the Willow River. After the river was diverted into this new channel floods were no longer a serious problem.

Soon after the new drainage channel was put into operation peculiar things began to happen. It rapidly became choked with sediment near the place where it entered the Missouri Valley. Here frequent dredging and rebuilding were needed to keep it open. But farther upstream the channel was being enlarged by erosion. In less than 40 years its width increased from less than 50 ft to more than 100 ft. Its original depth of about 15 ft increased to between 30 and 40 ft. Similar things occurred in the branching network of streams that flowed into the channel. These streams began to enlarge their channels. It became necessary to build new bridges to span the widened streams. Erosion of cropland became a more serious problem. New gulleys were eroded, and former gulleys rapidly grew wider and deeper. Why did the streams and landscape begin to change so dramatically after the Willow River was diverted into the newly excavated channel? We now realize that the Willow River flood control project was planned without a clear understanding of the delicate balance of a landscape and the network of streams flowing through it. The Willow River is only one of many interdependent components in an intricate system. We have learned that the effects of changing one of these components are felt throughout the system.

Streams are the most important tools used by nature to shape the landscape. Erosion near the Willow River and deposition of sediment farther downstream are examples of how a stream can carve and build landscape features. Processes such as these in which streams play the dominant role are called **fluvial** processes. Streams can have an indirect effect on the landscape in other ways. Slopes produced by erosion can become so steep that rock debris breaks loose and moves downhill by gravity. Such downslope movement or rock debris is called **mass wastage.** Gulleys near the Willow River became widened by mass wastage processes as well as by stream erosion. Wind, too, has played a role in the evolution of the landscape in southwestern Iowa. This area is blanketed by beds of wind-transported silt that were deposited earlier during the Pleistocene epoch. These silt beds are products of **eolian** processes which are processes that depend on wind.

The continental landscape presents us with a panorama of mountains, rolling hills, valleys, plains, and canyons. This panorama is continually changing. Even the most enduring landforms are eventually modified by processes that work on the earth's surface. The science of landscape is called **geomorphology.** It is concerned with the description of landforms and analysis of the processes that produce them. In this chapter we will learn how fluvial processes, mass wastage, and eolian processes shape the land surface.

SURFACE DRAINAGE SYSTEMS

There are different ways to look at a river. When standing on the bank or crossing a bridge, we can watch the swirling water and see how it cuts into the muddy banks. From an airplane we get another view. Here we can observe the pattern of the river. We can see that it is one of a branching network of rivers and streams that joins with one another to make a **drainage system.** The major drainage systems in the United States are illustrated in Figure 13-1. Here we see only the most important rivers. Water from a myriad of smaller streams flows into each of these rivers. Do these drainage systems form distinctive patterns? Is there a systematic way to describe them?

On the land surface water from rain, melting snow, and ice either evaporates, seeps into the ground water system, or flows overland in a stream drainage system. Most of the ground water eventually returns to the atmosphere by transpiration or is discharged into the network of streams. Columns in Figure 13-1 compare the proportions of surface water discharged by surface drainage in streams and by evaporation and transpiration in different areas of the United States.

Surface drainage commences with the trickle of water droplets down the slope of the land surface. These droplets flow along a multitude of tiny temporary streams called **rills.** These threads of water join to form larger rills, which in turn merge into streams. Where two streams come together, the large one is called the **master** stream and the smaller one is its tributary. Farther along, that same master stream can become the tributary of a still larger stream or river.

A drainage system consisting of a master

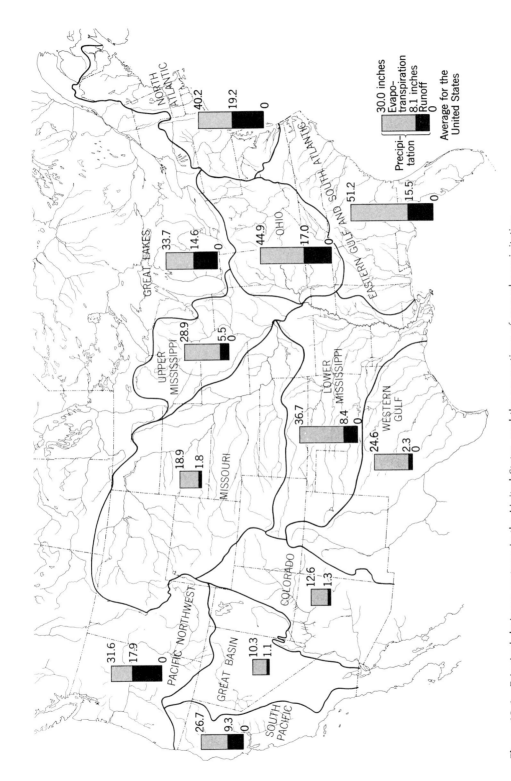

Figure 13-1 Principal drainage systems in the United States, and the proportions of annual precipitation that are dispersed by evaporation and by streams. (After *The Physical Basis of Water Supply and Its Principal Uses*, The Physical and Economic Fundamentals of National Resources, Vol. II, House of Representatives, U.S. Congress, 1952.)

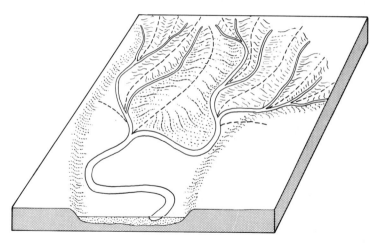

Figure 13-2 Subdivision of the landscape into drainage basins. Dotted lines outline areas drained by different streams.

- - - - - Drainage divide

stream and its tributaries receives water from an area called a drainage basin (Figure 13-2). Observe how a large drainage area contains several smaller drainage areas that are separated by **drainage divides.** A drainage divide is a line situated so that land on one side slopes downward into one drainage basin, whereas land on the other side slopes downward into the other basin.

Drainage Patterns

If we took the time to examine a great many drainage systems, we would find that most of the streams are arranged in one or another of the patterns illustrated in Figure 13-3. Most common is the **dendritic** pattern, which is an arrangement of irregularly branching streams. Nothing in the dendritic pattern indicates any systematic regional control of the direction of stream flow. Here the land is more or less uniformly resistant to erosion regardless of stream direction. Compare this with the **parallel** drainage pattern in which the streams tend to flow in the same general direction despite some obvious irregularity. Parallel drainage can occur where sedimentary rock layers covering an area all dip gently in the same direction. Other drainage patterns also indicate effects of geology on stream directions. **Angular** drainage occurs where streams are situated along linear zones of weakness produced by intersecting regional joint sets. Streams flowing over a terrane of parallel anticlines and synclines can produce the **trellis** pattern. Here we observe that the larger master streams are situated in parallel valleys, whereas the tributaries flow down the slopes of adjacent ridges. Sedimentary strata that form the ridges are more resistant to erosion than the beds exposed in the valleys. There are some drainage systems that are seemingly unaffected by obvious differences in resistance to erosion. Geologists have proposed that such drainage systems may have originated at an earlier time when the area was covered by rock now worn away by erosion (Figure 13-4). The drainage pattern that was established in an older different terrane has been preserved as the streams eroded down into another geologic environment. We call this a **superposed** drainage system.

Drainage Geometry

It is surprising to discover that a seemingly irregular drainage system possesses distinct relationships between streams of different size

Running Water, Mass Wastage, Wind, and Continental Landscapes 457

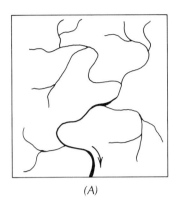
(A)

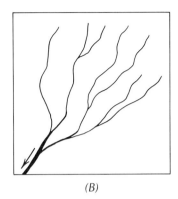
(B)

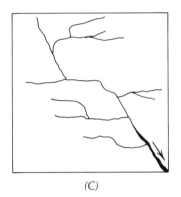
(C)

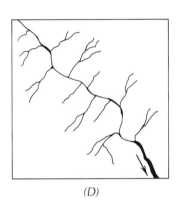
(D)

Figure 13-3 Principal drainage patterns: *(A)* dendritic drainage, *(B)* parallel drainage, *(C)* angular drainage, and *(D)* trellis drainage.

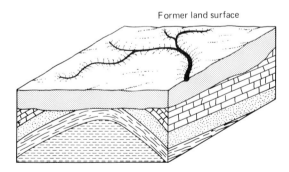

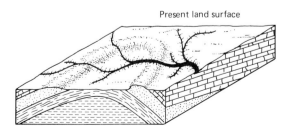

and the areas of drainage basins. These relationships were first pointed out in 1945 by R. E. Horton, a civil engineer. Let us examine Horton's method of analysis. Streams in a drainage system are assigned an **order number** according to their positions in the network of tributaries. A **first-order** stream has no tributaries. A **second-order** stream has only first-order tributaries. A **third-order** stream may have second-order and first-order streams as tributaries, and so on. The application of this scheme is illustrated in Figure

Figure 13-4 Development of superposed drainage. In this example the same drainage pattern initially adapted to a terrane that was uniformly subject to erosion is shifted downward onto a terrane of variable durability after complete erosion of the original terrane.

13-5A, where we find 32 first-order streams, 10 second-order streams, 4 third-order streams, and 1 fourth-order stream. If we plot the number of streams of different order on a semilogarithmic graph, the points are seen (Figure 13-5B) to lie along a straight line.

Let us next examine the lengths of streams of different order. For this purpose we include as part of a second-order stream the longest first-order tributary merging into it, and similarly for higher order streams the length includes lower order tributaries. This is evident from the way numbers are assigned in Figure 13-5A. Having measured stream lengths, the sum of all first-order stream lengths, all second-order stream lengths, and so on, are divided by the number of streams of the particular order to obtain values of average stream length for the different orders. Now observe that these data (Figure 13-5C) plot along a straight line on a semilogarithmic graph.

Drainage area can also be compared with stream order. The area of the drainage basin in Figure 13-5A is 17.2 km² (6.65 sq miles). This is the area drained by the fourth-order stream and all of its tributaries. Lower order streams drain smaller basins within this area. The average areas of the drainage basins for streams of different order plot along the straight line on the semilogarithmic graph in Figure 13-5D. The **drainage density** for the area is the ratio L/A, where L is the sum of the lengths of all streams in a drainage basin and A is the area of the basin.

This analysis of a drainage system reveals the simple geometrical relationships that are part of the **drainage geometry** of the area. Similar analyses for a large number of drainage systems show the same results. Although the numbers may differ from one system to another, the straight-line plots on semilogarithmic graphs are always observed. If this consistency can be assumed to hold for very large areas, it is possible by extrapolation to make estimates of continental drainage density. Look again at the map in Figure 13-1 that shows the major river systems for continental United States. These very large rivers are few enough in number to be analyzed according to Horton's method. Streams of enough different orders are evident in Figure 13-1 to determine the positions of straight lines on semilogarithmic graphs. By assuming the largest river, the Mississippi, to be, say, a tenth-order stream, the graphs can be constructed so that straight lines determined from higher order streams can be extended to yield information about the numerous lower order streams that drain into them. Results of such an analysis, summarized in Table 13-1, are extended to include small streams less than 2 km long. The study indicates that more than 2 million streams between 1 and 2900 km in length probably exist in the continental United States. Clearly, it would be impractical to attempt to compile data for each of these streams. We can see how the general relationships found from study of selected drainage basins provide a basis for estimating average stream characteristics for large regions.

Now we can begin to understand why all of the streams that drained into the Willow River began to change when water was di-

Table 13-1

Number and Length of River Channels of Various Sizes in the United States

Order	Number of Streams	Average Stream Length (km)
10	1	2900
9	8	1250
8	41	545
7	200	235
6	950	100
5	4,200	45
4	18,000	19
3	80,000	8.5
2	350,000	3.7
1	1,570,000	1.6

Modified from L. B. Leopold, M. G. Wolman, and J. A. Miller, Fluvial Processes in Geomorphology, W. H. Freeman and Co., San Francisco, p. 142, Table 5-2, 1969.

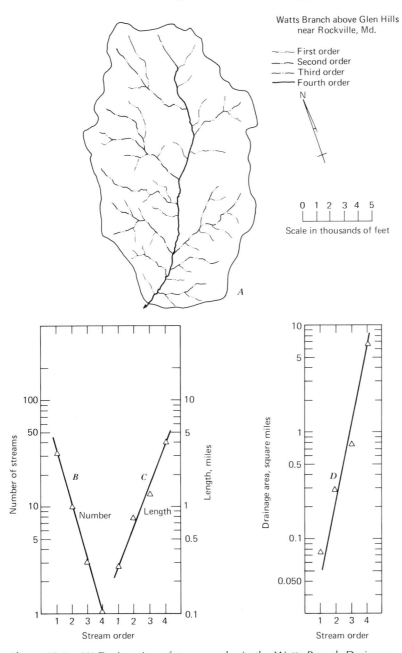

Figure 13-5 (A) Designation of stream order in the Watts Branch Drainage Basin near Glen Hills, Maryland. Relationships in this drainage basin between order and (B) number of streams, (C) average stream length, and (D) average drainage area are displayed on semilogarithmic graphs. (After L. B. Leopold, M. G. Wolman, and J. P. Miller, *Fluvial Processes in Geomorphology*, pages 136–137, W. H. Freeman and Co., 1964.)

verted into the new drainage channel. Before the channel was opened the number of streams and the length and drainage area of each stream were adjusted to fit a pattern of drainage geometry. When the master stream, in this case the Willow River, was forced to change, then all of its tributaries would also be forced to adjust so that a new pattern of drainage geometry could be established with features similar to those seen in Figure 13-5.

STREAMS AND RIVERS

Now let us look more closely at the individual streams in a drainage system. Each of these streams possesses distinctive features that indicate a delicate balance between the shape of the channel and the water that flows in it. Landscape features are produced by the stream as it erodes its channel in some places and deposits sediment elsewhere in its effort to maintain this balance.

Channel Geometry

A stream threads an irregular course over the landscape, following a straight path in some places, and elsewhere a curved and sinuous channel or web of intertwining channels. The shape of this path when displayed on a map illustrates the **channel pattern** of the stream. Typical patterns (Figure 13-6) include **straight** channels, **meandering** channels, and **braided** channels.

Straight channel patterns ordinarily are restricted to short segments of a stream and do not extend over long distances. The form of the channel cross section along a straight segment tends to be asymmetrical and different from one location to another. Observe on the cross sections in Figure 13-6 how the location of deepest water can be described by a curving line. Along a straight channel this line tends to cross the center at enough places so that the lengths of channel asymmetric toward one side or the other are approximately equal. With the passing of time the position of this line shifts because of the continual redistribution of sediment along the bottom of the channel.

A stream that flows along a winding path of curves has a meandering channel pattern. A single S-shaped segment of the channel is a **meander** whose length and amplitude are described in Figure 13-7. Measurements of numerous meandering streams, large and small, reveal that meander length and radius of curvature are related to channel width. These relationships are indicated in Figure 13-7 by the straight line plots on logarithmic graphs.

The irregularity of a meandering stream channel can be stated numerically in terms of the ratio x/L, where x is the average channel length and L is the average meander length. This ratio expresses the **sinuosity** of the stream. There are different ways to use sinuosity to describe channel geometry. According to one classification scheme, channels having sinuosity of less than 1.5 are called straight, and those displaying a sinuosity of more than 1.5 are meandering channels. Values of sinuosity range from 1.0 to in excess of 5.0.

The form of the channel cross section in a meandering stream tends to vary systematically with position in the meander. We can see on the cross sections in Figure 13-6 how a meandering channel is deepest near the outside bank of a bend in the stream. The channel remains slightly asymmetrical at the inflection of the meander and becomes symmetrical at a downstream location where the point of deepest water crosses from one side of the channel to the other. Keep in mind that these diagrams indicate idealized shapes, and that actual channel geometry in some particular river may not be exactly the same. Stream channels in unconsolidated sediment that is more or less uniformly susceptible to erosion come closest to the patterns shown in Figure 13-6. Meanders and channel cross

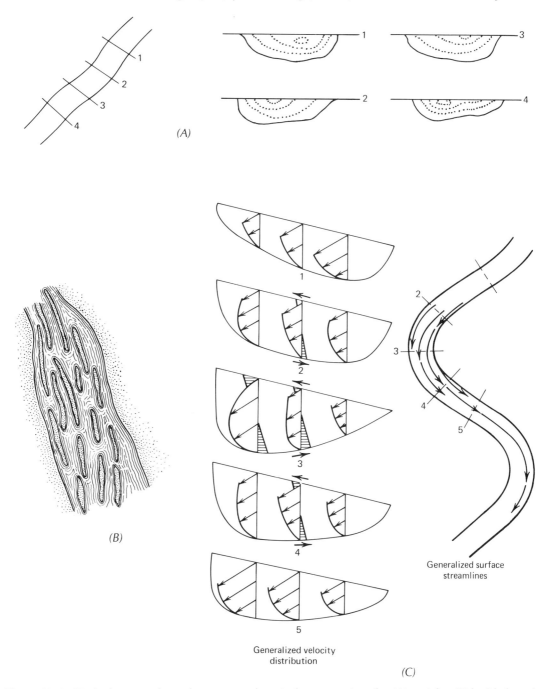

Figure 13-6 Typical stream channel patterns and vertical cross sections for (A) straight, (B) braided, and (C) meandering channels. Dotted contours in (A) represent current velocity. (Diagrams in B after L. B. Leopold and M. G. Wolman, *Bulletin of the Geological Society of America,* page 780, June 1960.)

sections can have more complicated shapes in places where channels cross a terrane of mixed bedrock and regolith, because of variations in the resistance of the stream bed to erosion.

An interwoven network of stream channels separated by narrow islands and bars of sediment produces a braided channel pattern. The individual channels are not tributary to one another. Instead, they should be thought of as parallel but alternate paths that intersect in numerous places. The locations and shapes of islands and bars of loose sediment can change markedly in a short time as the flowing water shifts and redeposits the loose material. During seasons of high water they become submerged and reappear in different configurations in drier times of the year. Typically, a braided channel is produced by a river that flows over a relatively flat plain or broad valley and does not have enough water to carry all of the loose sediment which is available to be transported. A braided channel indicates a particular balance of several factors, including shape and inclination of the channel, discharge of water, and supply of sediment. You will understand more clearly the interdependence of these factors after studying the following sections on hydraulic geometry and sediment transport.

An individual stream can display more than one channel pattern; it is straight in some places and changes to a winding pattern of meanders elsewhere. If it encounters a susceptible terrane, it can become braided. These changes are not altogether random. River channels tend to become more sinuous and less braided in the downstream direction.

Figure 13-7 Geometrical properties of a meandering channel. Width of the channel and radius of curvature of a meander are related to meander length. (After L. B. Leopold and M. G. Wolman, *Bulletin of the Geological Society of America*, pages 772 and 773, June 1960.)

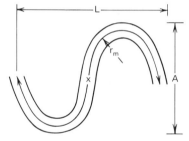

L = Meander length (wave length)
A = Amplitude
r_m = Mean radius of curvature

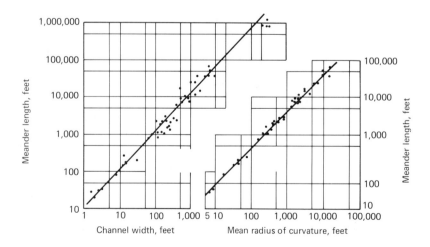

Hydraulic Geometry

The relationships between the shape of a stream channel and the discharge of water through it can be used to describe the **hydraulic geometry** of the stream. At a particular location along a stream we can measure the shape of the channel cross section and the average velocity (v) at which water passes this cross section. The discharge (Q) of water past this location is given by:

$$Q = vA \qquad (13\text{-}2)$$

where Q is expressed in units of volume/time, for example, cubic meters/second. The area (A) of the channel cross section is the product of the width (w) and the average depth (d). Ordinarily the flow of water depends partly on the weather and the season. Following a period of rain or melting snow, more water reaches the stream, so that increases in the width and depth of the channel as well as an increase in the velocity of the flowing water can be observed. This means that the discharge will be larger. The reverse is observed during dry periods when both the size of the stream and the discharge of water diminish.

Suppose that we measured stream width, depth, and velocity and calculated the discharge at the same place several times during a year. Would we observe any systematic changes from time to time? Scientists have discovered from these observations that different values of Q corresponding to different values of v, w, and d at a particular location on a stream tend to plot along straight lines on a logarithmic graph. Examples of these patterns, which describe the hydraulic geometry of the stream at that location, are illustrated in Figure 13-8. We know from analytic geometry that equations of the following form represent straight lines on logarithmic graphs:

$$w = aQ^b, \qquad d = cQ^f, \qquad v = kQ^m \qquad (13\text{-}3)$$

Relationships of this form appear to exist for cross sections on small streams and large rivers alike. The numerical values of the coefficients a, c, and k, and the exponents b, f, and m that we find for one particular cross section are not necessarily the same as values for another cross section where conditions are different. But for a particular location where $A = wd$, the expressions (13-3) can be substituted into Eq. (13-2):

$$Q = aQ^b \times cQ^f \times kQ^m$$

which requires that

$$b + f + m = a \times c \times k = 1.0 \qquad (13\text{-}4)$$

This tells us that the shape of a stream channel must adjust in a systematic way to changes in the amount of water flowing through it.

Let us examine the physical significance of these coefficients and exponents. Look at the results of measurements on the San Juan River at a location near Bluff, Utah (Figure 13-8). Here an increase in Q corresponds to increases in w, d, and v. The stream normally flows in a bedrock channel lined with loose sediment. We can see from the shape of the channel that the stream will become wider as water level rises. Stream depth increases in two ways: (1) a rise in water level and (2) deepening of the channel as loose sediment is removed from the bottom and taken into suspension. When discharge and average velocity increase, the hydraulic geometrical relationships expressed in Eq. (13-2), (13-3), and (13-4) dictate the way that width and depth will also increase.

The hydraulic geometry that we have been discussing thus far refers to a particular location along a stream. Let us next consider hydraulic geometry along the length of a channel. A stream has a point of origin and a point of termination. The channel between these points receives water from numerous rills and tributaries and perhaps from ground water seepage. This water flows downstream along an inclined channel. We can prepare a graph

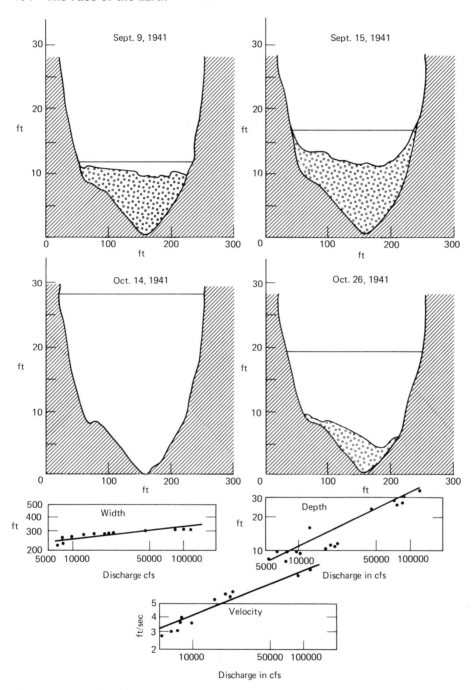

Figure 13-8 Hydraulic geometry and changes in channel shape of the San Juan River near Bluff, Utah. Observe that channel shape depends on water level and the amount of sediment that remains on the bottom or suspended in the water. (After L. B. Leopold and T. Maddock, *Professional Paper 252,* page 32, U.S. Geological Survey, 1953.)

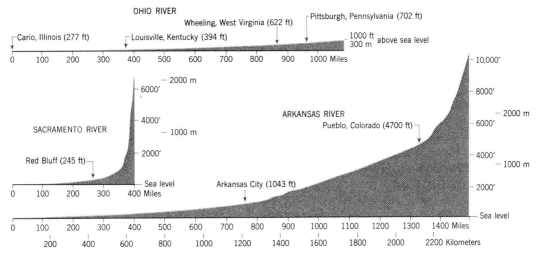

Figure 13-9 Longitudinal profiles of three rivers in the United States. (After H. Gannett, *Paper 44*, U.S. Geological Survey Water Supply Papers, 1901.)

that shows channel elevation at different distances from the termination point (Figure 13-9). Such a graph illustrates the **longitudinal profile** of the stream. Observe that the profile tends to be concave upward so that the average inclination of the channel decreases in a downstream direction.

The way that the flow of water is related to the shape of the longitudinal profile is called the longitudinal hydraulic geometry of the stream. We can make various measurements to determine the longitudinal hydraulic geometry. For example, we can measure the channel inclination and the discharge at different places along the stream. We will observe that discharge increases in a downstream direction. This is what could be expected in a river that is receiving water from more and more tributaries at locations farther and farther downstream. But we have also observed that channel inclination tends to decrease in a downstream direction (Figure 13-9). When we plot discharge and channel inclination on a logarithmic graph, we discover that the points lie close to a straight line (Figure 13-10). This indicates that the slope of the longitudinal profile is related in a partic-

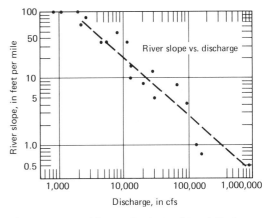

Figure 13-10 Observed relationship of discharge and the inclination of a stream channel. (After L. B. Leopold, M. G. Wolman, and J. P. Miller, *Fluvial Processes in Geomorphology*, page 246, W. H. Freeman and Co., 1964.)

ular way to the amount of water carried by a stream.

Stream velocity tends to increase in the downstream direction along the longitudinal profile. We may find this surprising in view of the fact that the channel inclination decreases in that direction. But other factors as well as inclination influence velocity. Most important is discharge. Increasing discharge is accompanied by increasing velocity. Another factor is the roughness of the channel that acts to retard the flow of water. Ordinarily there is more roughness in the narrow tortuous channel where a river begins than exists farther downstream. Therefore both discharge and roughness favor an increase in velocity in a downstream direction. These factors combine to counteract the effect of inclination. The result in most streams and rivers is a downstream increase in velocity. Does this mean that a rushing mountain stream flows more slowly than a river winding its lazy way over a broad valley? This is often true.

As the seasons and the weather change, the amount of water carried by a stream also changes. But the hydraulic geometry of the stream is maintained by delicate adjustments in the velocity, channel dimensions, and channel inclination. The logarithmic graphs in Figure 13-11 summarize all of the important features of the hydraulic geometry of a typical stream. The straight lines on these graphs indicate how the channel dimensions, longitudinal profile, and stream velocity will adjust to a change in discharge. Look first at the lines A_0C_0, A_1C_1, and A_2C_2. They illustrate how channel shape adjusts to changes in discharge at a cross section near the headwaters of the stream. The lines B_0D_0, B_1D_1, and B_2D_2 show how a different channel cross section farther downstream adjusts to changes in discharge. These features are similar to those seen in Figure 13-8. They tell us how the channel changes from one time to another because changes in discharge related to weather and season.

But what do lines A_0B_0, A_1B_1, and A_2B_2 indicate? They tell us how the channel shape along the longitudinal profile is adjusted to downstream differences in discharge at some particular moment in time. Lines C_0D_0, C_1D_1, and C_2D_2 illustrate these adjustments at another moment in time when the stream is carrying more water.

In Figure 13-10 we see data that indicate how discharge and channel inclination are related at one moment in time. Similar information is shown by line A_5B_5 in Figure 13-11 for a particular time when the stream is not carrying very much water. Line C_5D_5 shows the relationship at a different time when much more water is flowing in the stream. But what can we learn from line A_5C_5? It tells us that the channel inclination at a particular location adjusts from one time to another in response to changes in discharge.

Think again about how the Willow River drainage system changed after the flood control channel was opened. The longitudinal profile along this channel was different from the profile of the old Willow River channel. Therefore, the relationship between channel inclination and discharge in the master stream was changed. The new channel was not originally designed with a shape that was properly adjusted to discharge. All of the tributaries flowed into the channel at locations different than where they had entered the old river. Therefore, their longitudinal profiles were changed. Clearly, adjustments were required everywhere in the drainage system to reestablish the patterns of hydraulic geometry similar to those that seen in Figure 13-11. Such adjustments certainly would involve erosion, transportation, and deposition of sediment. We can see in Figure 13-11 that relationships involving suspended sediment are part of the hydraulic geometry of a stream. Let us now examine the factors that influence the capacity of a stream to erode and carry sediment.

Stream Flow and Sediment Transport

The water in a river usually looks cloudy or muddy because of the small particles of sediment suspended in it. This is not the only sediment carried by the river. It also rolls and bounces larger fragments along the bottom. These larger fragments make up the **bed load** of the river. The bed load and the load of sus-

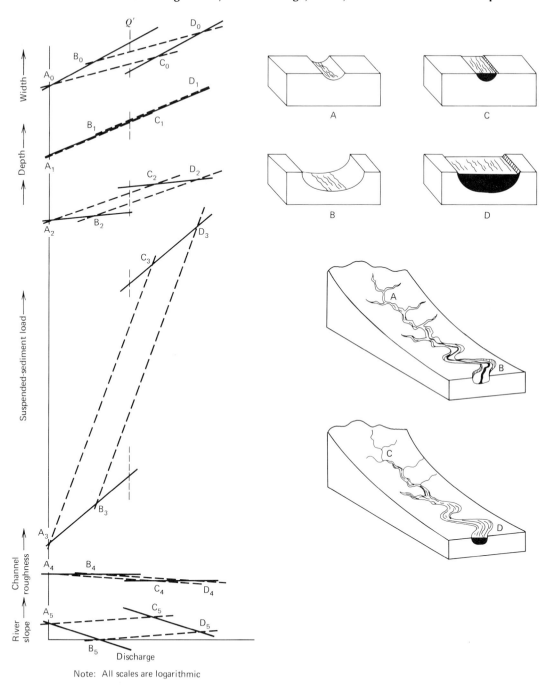

Figure 13-11 Diagrams summarizing the hydraulic geometry of a stream. Sketches show channel cross sections at two locations, labeled A and B at a time of low water, and at the same locations, labeled C and D at a time of high water. Graphs display the complete hydraulic geometry of the stream. (After L. B. Leopold and T. Maddock, *Professional Paper 252*, page 27, U.S. Geological Survey, 1953.)

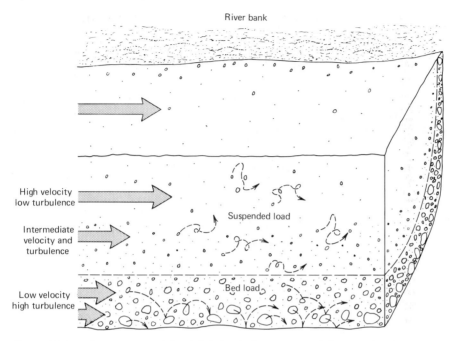

Figure 13-12 Components of the debris load of a stream and their mode of transport. Suspended load consists of small particles maintained in suspension, and the bed load consists of larger particles that roll and bounce along bottom.

pended sediment together are the **debris load** of the river (Figure 13-12). We can describe it in terms of the average mass of sediment in a standard volume of water, for example, grams of sediment per cubic meter of water.

The turbulence of water flowing in a river enables it to lift particles of sediment and to keep some of them suspended. Turbulence is indicated by the irregularity of motion in the water. Suppose that we could observe a particle of water. Its downstream component of motion contributes to the average downstream current of the river. Its motion in directions other than downstream helps to produce **eddy** currents. These are the swirls and gyres that are called turbulence. In some places large, semipermanent eddy currents exist because of irregularities in the channel (Figure 13-13). But most eddies are only temporary swirls in a turbulent river.

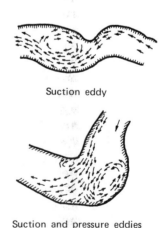

Figure 13-13 Semipermanent eddy currents in irregular stream channels. (After D. O. Elliott, *The Improvement of the Lower Mississippi River for Flood Control and Navigation*, U.S. Waterways Experiment Station, Vicksburg, Mississippi, 1932.)

The amount of turbulence in a river is related to the downstream current velocity and the roughness of the channel. We can see in Figure 13-6 that velocity is slowest along the walls of the channel. It is here that turbulence is greatest because the rough walls interfere with the movement of water. Closer to the center of the channel the water flows most swiftly and has the least turbulence. You may wonder why the fastest velocity is not at the surface. There appears to be a surface tension effect which retards the flow so that the maximum velocity exists in a zone beneath the surface. This is also evident in Figure 13-12. Now suppose that more water was added to the river so that discharge increased. Ordinarily this would result in an increase in downstream current velocity and an increase in turbulence as well. This is useful to know because current velocity can be measured much more easily than turbulence. But what can it tell us about the debris load of a river?

We know that an increase in turbulence favors erosion, which will add sediment to the debris load. A decrease in turbulence favors deposition of part of the debris load. The fact that a change in turbulence accompanies a change in velocity means that we can look for relationships between current velocity in a river and conditions favoring erosion or deposition of sediment. One such relationship (Figure 13-14) is based upon laboratory experiments and observations of typical rivers. Lines on the graph indicate velocity ranges for erosion, suspension, and deposition of different sized grains. Look first at the lower line. It tells us that water flowing 1 cm/sec can keep in suspension all particles smaller than 0.15 mm in diameter. However, particles larger than this are too heavy to stay suspended in a 1 cm/sec current. But water flowing 10 cm/sec can keep particles up to 1.5 mm in diameter in suspension.

Now look at the upper line in Figure 13-14. It tells us that water must be flowing at least 100 cm/sec to have the capacity to erode 0.01 mm particles from the walls of the

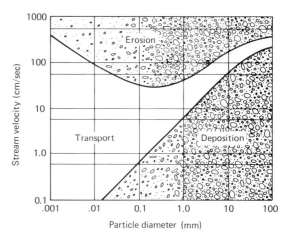

Figure 13-14 Conditions of stream velocity and associated turbulence required for erosion, suspension, or deposition of particles of different average diameter. (After F. Hjulstrom, *Bulletin 25*, pages 221–527, University of Uppsala Geological Institute, 1935.)

channel. But a current of only 40 cm/sec is sufficient to erode 0.1 mm particles. Why is a stronger current needed to erode smaller particles? The very small 0.01 mm particles are mostly flat clay grains that pack together more firmly than the larger 0.1 mm particles which tend to be more rounded in shape. Therefore a swifter and more turbulent current is required to break loose the very small particles. This is not true for particles larger than 0.5 mm in diameter. Weight rather than shape is the critical factor controlling the current necessary for erosion of sediment composed of these large grains. A current of almost 200 cm/sec is needed to erode sediment composed 10 mm particles, but water flowing only 45 cm/sec can lift 1.0 mm particles because they weigh so much less.

What do these conditions for erosion and deposition (Figure 13-14) tell us about transport of sediment by a river? They indicate that the addition of sediment or its removal from the debris load is influenced by the current velocity. We already know that velocity and discharge are related. Therefore, we can un-

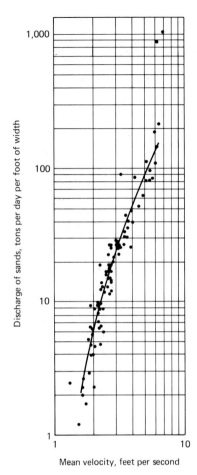

Figure 13-15 Observed relationship of average velocity and the debris load transported by the Rio Grande River bordering New Mexico. (After B. R. Colby, *Paper 1498-D,* pages 1–10, U.S. Geological Survey Water Supply Papers, 1961.)

derstand that the size of the debris load (grams of sediment per cubic meter of water) must be related to discharge. This relationship is part of the hydraulic geometry of the river seen in Figure 13-11. Observe that line A_3C_3 shows how the debris load at some particular location changes from one time to another in response to changes in discharge. Line B_3D_3 indicates a similar relationship at another location. We can tell from lines A_3B_3 and C_3D_3 how the debris load is adjusted to discharge along the entire length of the river at two different times. Now, if we multiply the average size of the debris load and the average current velocity at some location, we obtain a value called the **sediment discharge.** Observe in Figure 13-15 how sediment discharge tends to be adjusted to the current of a river.

Recall that where the Willow River enters the Missouri Valley the drainage channel became choked with sediment shortly after it was opened. This sediment was produced by intensified erosion farther upstream. These effects indicate the adjustment of the debris load of the Willow River to a new hydraulic geometry.

The Willow River and many other small rivers carry water and sediment into the Missouri River which is one of the large tributaries of the Mississippi River. Each day the Mississippi discharges approximately 2 million metric tons of sediment into the Gulf of Mexico. Other large rivers carry similar amounts of sediment to the ocean. To account for this tremendous debris load which is transported through the drainage systems of North America, sediment must be eroded from the land surface at an average rate of 6mm/century. Suppose that tectonism and plutonism did not act more or less continuously to restore the continent. Would streams eventually wear it down to a sea level plain? If this were to happen, how long would it take?

FLUVIAL LANDSCAPE FEATURES

Think about the hills, valleys, plains, and other features of the landscape. To what extent are they formed by stream erosion or by deposition of sediment carried by streams? Clearly, any adjustment in the shape of a stream channel is also an adjustment of the landscape. But tectonism and volcanism, too, play an obvious role in shaping the land. So do the processes involving mass wastage, wind, and glaciers. Let us begin our study of landscape features by looking at the particular effects of fluvial processes.

Stream Valleys in Mountainous or Hilly Terrain

One of the scenic features of Yellowstone National Park in Wyoming is the Grand Canyon of the Yellowstone River (Figure 13-16). Observe that it is a valley with a V-shaped cross section. This shape is typical of the small valleys in mountainous or hilly countryside. Perhaps you can recall seeing steep mountainsides that are cut by V-shaped valleys (Figure 13-17). These valleys are carved by stream erosion and associated mass wastage. Here a turbulent stream tends to follow a straight channel that has a relatively steep inclination. It carries a bedload of large rock fragments as well as smaller particles. The principal effort of the stream is directed toward deepening the channel. Nature uses it as a saw to cut downward into the land. As a consequence the sloping sides of the valley become unstable, and debris begins to slide into the stream. At the same time the runoff of rain and melting snow washes sediment down from the valley sides. The combined effects of runoff and mass wastage act to widen the valley while the stream is deepening it (Figure 13-18).

The action of many turbulent mountain streams produces the landscape of merging V-shaped valleys as seen in Figure 13-17. Observe that some of the streams flow out of the rugged mountainous area onto a plain. At the border the turbulence of the stream diminishes abruptly so that it can no longer move a bedload of large rock fragments and coarse particles of sediment. Here much of the bedload is deposited. Sediment that has been transported and deposited by streams is called **alluvium**. Look at the fan-shaped deposits that have been built up along the border of the plain (Figure 13-17). We call these

Figure 13-16 The Yellowstone River has eroded a straight channel in a deep V-shaped valley in Yellowstone National Park.

depositional landscape features **alluvial fans.** Recall from Chapter 10 that geologists can reconstruct ancient environments of deposition from distinctive combinations of sedimentary facies. Lithification of alluvial fan deposits could produce the combination of coarse-grained sandstone and conglomerate layers that might tell us of the existence of an ancient landscape much like the one shown in Figure 13-17.

Meandering Streams on Alluvial Plains

The Mississippi River and its major tributaries meander over the plains and broad valleys of central United States. Because these plains

Figure 13-17 Streams have eroded the V-shaped gullies into the steep mountain sides. The sediment is deposited on alluvial fans bordering an upfaulted mountain range in the Mojave Desert of California. (J. R. Balsley, U.S. Geological Survey.)

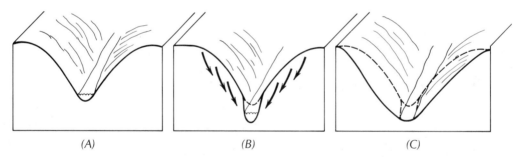

Figure 13-18 Enlargement of a V-shaped valley *(A)* resulting from channel deepening followed by slumping of the steepened and unstable valley sides *(B)*. In addition, weathered debris is continually washed into the stream from the valley slopes by rain showers *(C)*.

are largely covered by a blanket of alluvium, they are called **alluvial plains.** The rivers gradually wear down the land, but they work differently than the steep mountain streams that we have just discussed. If the mountain stream is like a saw that cuts into the land, then the meandering river is more like a carpenter's plane; nature uses it to shave sediment from the land surface. We can begin to understand how this is accomplished by looking at certain features produced by a meandering river. In some places the former paths of a river are clearly etched in the landscape (Figure 13-19). We can easily see that this river has been shifting its channel over the entire area of the plain. In doing this it has

Figure 13-19 *(A)* Landscape features indicate the former channel positions of the Laramie River in Wyoming that meanders over a broad plain of subdued relief. *(B)* Note the oxbow lakes and dry remnants of cut-off meanders seen along the White River in Arkansas. (J. R. Balsley, U.S. Geological Survey.)

474 The Face of the Earth

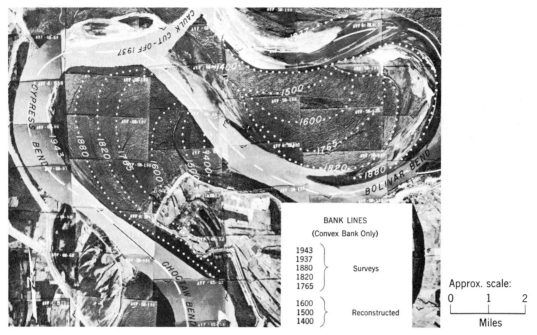

Figure 13-20 Comparison of former and present channel positions of the Mississippi River near Greenville, Mississippi. Dotted lines indicates outer bank positions dating from about 1400 A.D. Positions prior to 1765 were estimated from landscape features. Beginning in 1765 the channel was periodically surveyed. (After H. N. Fisk, *American Society of Civil Engineers Transactions,* pages 667–682, 1952.)

distributed a blanket of alluvium over the plain.

How rapidly can a river shift its channel? Near Greenville, Mississippi, we can use landscape features together with old maps and archaeological relics to document changes in the Mississippi River (Figure 13-20). Observe that the amplitudes of the two meanders, Boliver Bend and Cypress Bend, increased by about 4 km/century for more than five centuries. Then, during the flood of 1937 water surged over the bank and quickly eroded a new segment of channel called the Caulk Cutoff. It was gradually enlarged to carry the full flow of water directly into Cypress Bend. The river abandoned Boliver Bend, which now remains as a **cutoff meander.** Parts of the cutoff meander still hold ponds of water. Ponds like these are called **oxbow lakes.** Look again at Figure 13-19; everywhere on the plain we can see evidence of the steady growth and then abrupt cutoff of meanders.

We can construct laboratory models to learn more about the way that a meandering river moves its channel. Experimental results in Figure 13-21 illustrate typical features. Here we see that erosion occurs along the outer bank of a meander, whereas sediment is deposited along the inner bank to produce a feature called a **point bar.** This is what would be expected according to Figures 13-6 and 13-14. The high-current velocity that favors erosion tends to be close to the outer bank of a meander, and the low velocity favoring deposition of sediment is near the inner bank. Be sure to observe that the point bars are slightly lower than the original surface of the plain (Figure 13-21). We can see

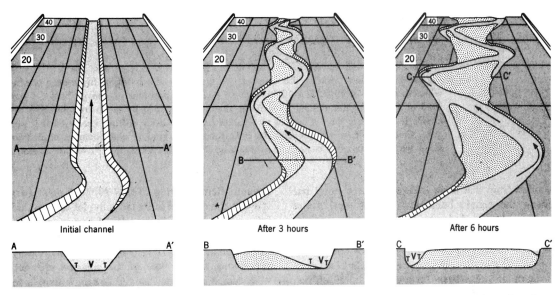

Figure 13-21 Migration of meanders in a laboratory model during a 6-hour experiment conducted at the U.S. Army Engineers Waterways Experiment Station in Vicksburg, Mississippi. Trough length is about 15 m (50 ft). (After J. F. Friedkin, U.S. Waterways Experiment Station.)

how a meander increases in amplitude and at the same time migrates slowly in a downstream direction. As the river gradually moves its channel over the entire area of the plain, point bars grow and coalesce to produce a new surface at a slightly lower elevation. Some sediment has been removed from the plain, carried farther downstream, and eventually into the sea. Like the carpenter's plane, the river has shaved the land surface. Look again at Figure 13-19 to see the features produced in the process of shaving an alluvial plain.

Let us examine more closely the way that sediment is eroded and redeposited by a meandering river. Cross sections in Figure 13-22 illustrate the position of a river at different times. Observe how the river has been cutting laterally into the outer bank, whereas at the same time the inner bank is extended by deposition of sediment which was produced by erosion farther upstream. See how the increments of sediment are deposited as inclined laminations on the inner bank of the channel. Each lamina adds to the horizontal extension of a layer of cross-bedded sediment. The meandering river slowly wears down the land into a plain covered by a blanket of alluvium consisting of these cross-bedded layers of sediment. These strata may also display ripple marks formed by the flow of water over alluvium. Alluvial plain sediments ordinarily consist of smaller and better-sorted particles than alluvial fan deposits.

The continuing change of the channel position of a river flowing over a broad valley covered by alluvium leads to the development of distinctive topographic features. As the river cuts laterally into older alluvial deposits and redistributes the sediment in point bar deposits at lower elevation, a landscape of alluvial terraces may develop (Figure 13-23). The different terrace levels indicate former levels of the alluvial plain over which the river flowed in earlier times. Alluvial terraces are particularly well developed in the valley pictured in Figure 13-24.

Meandering streams are not only active in

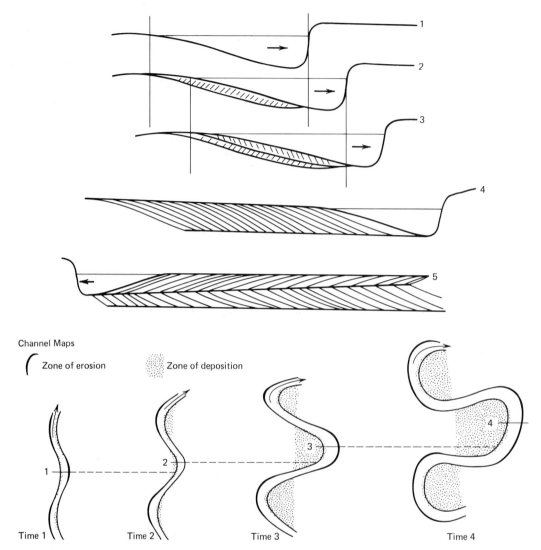

Figure 13-22 Sedimentation associated with outward channel migration contributes to the growth of a cross-bedded stratum of alluvium. Numbers indicate channel at a succession of times.

redistributing unconsolidated sediment; they also continue to erode the topographic remnants of bedrock that project above an alluvial plain or border a broad valley. Bedrock is considerably more resistant to erosion than alluvium. Nevertheless, it can be carved into distinctive landscape features by running water. The most durable rock remains to form these features, whereas the less-resistant bedrock is reduced to alluvium and distributed over the surrounding countryside. Examples of bedrock topographic features are illustrated in Figure 13-25. Prominent isolated bedrock masses rising from alluvial plains are called **inselbergs.** Other features carved from a terrane of horizontal sedimentary rock stand as

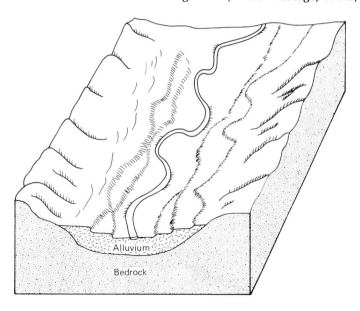

Figure 13-23 Terraces cut in the alluvial valley deposits by migrating stream meanders. Successively higher terraces indicate the level of the alluvial plain at successively earlier times.

Figure 13-24 Alluvial terraces cut by the Waimakariri River located on the South Island of New Zealand.

478 The Face of the Earth

(A)

(B)

(C)

Figure 13-25 *(A)* Broad flat-topped mesas and smaller buttes capped by horizontal strata relatively resistant to erosion are typical landscape features in Monument Valley, Arizona. (I. J. Witkind, U.S. Geological Survey.) *(B)* Hogbacks in central Colorado are formed from steeply inclined strata that are relatively resistant to erosion. (T. S. Lovering, U.S. Geological Survey.) *(C)* Isolated remanants of relatively durable bedrock rise as inselbergs from the surrounding plain in Pima County, Arizona. (J. Gilluly, U.S. Geological Survey.)

broad, flat-topped **mesas** or smaller **buttes.** Where sedimentary strata are tilted, the more durable layers can form steep ridges called **hogbacks** that are bordered by valleys covered by alluvium.

Rejuvenation

The balance between a stream and the landscape can be upset by changes that are independent of stream activity. A change in climate can alter the amount of water carried by a stream. Tectonic tilting of the land surface can change the stream channel inclination. An increase in channel inclination ordinarily increases the capacity of the stream to erode downward and deepen its channel. When this happens, we say that the stream has been **rejuvenated.** It then commences to erode downward rather than outward, tending to maintain or straighten its channel pattern rather than to increase meander amplitude. One result of rejuvenation is the development of entrenched meanders (Figure 13-26).

Eventually a rejuvenated stream ceases further entrenchment and redirects its effort to outward erosion that increases the meander amplitude. We do not fully understand how this redirection of effort occurs, but we know that it is influenced by two factors. First is the manner in which channel inclination is changed by continued downward erosion. Second is the amount of sediment being delivered to the stream from the steep sides of the gorge by runoff and mass wastage. These factors can create a situation in which the stream cannot continuously transport all of the sediment delivered to the channel. This leads to channel pattern adjustments that redirect most of the stream energy from downward to outward erosion. Once again the stream seeks to plane off the land rather than to continue sawing into it. When this occurs, a new valley is gradually widened. It is bordered by the topographic remnants of the former alluvial plain that existed before rejuvenation (Figure 13-27). Observe how these topographic remnants form terraces that stand at about the same elevation on both sides of the new valley. We call these features **paired terraces.** Their origin is different than the ori-

Figure 13-26 The Goosenecks of the San Juan River in southern Utah are entrenched meanders eroded by the rejuvenated stream during tectonic uplift of the Colorado Platau. (E. C. LaRue, U.S. Geological Survey.)

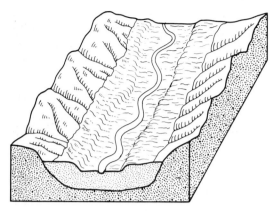

Figure 13-27 Formation of paired alluvial terraces by a stream as it develops a new lower alluvial floodplain after an episode of rejuvenation. Terraces on both sides of the valley stand at the same elevation.

gin of the irregular terraces (Figure 13-23) produced by migrating meanders uninterrupted by episodes of rejuvenation.

Tectonic tilting that decreases the inclination of a river channel tends to cause a larger amount of sediment to be deposited in a region. This can lead to a rise in the level of an alluvial plain. The alluvial deposits that we would expect to find covering a broad valley or plain probably represent several depositional episodes separated by periods of rejuvenation and erosion. Sediments from former episodes of deposition could be partially eroded first, then buried beneath later deposits.

Floods

From time to time water discharged into rivers exceeds the capacity of the channel and spills over the banks flooding the surrounding countryside. In some drainage systems the flood is an annual event, occurring during a season of excessive rain, or when melting of winter snow contributes to a rapid increase of discharge. Although velocity and turbulence increase as discharge builds up, the river channel cannot be enlarged rapidly enough to contain the larger volume of water. The low countryside inundated by the flood is called the **floodplain** of the river.

When water can no longer be confined to the channel and proceeds to spread over the flood plain, velocity and turbulence decrease. Part of the suspended load then settles from the water, adding a covering of sediment to the flood plain. In some regions an annual flood is important for agricultural purposes insofar as it enriches and replenishes soil. Annual floods have helped to shape the economy and religion over many centuries in such areas as the Nile Valley in Egypt and along the Ganges River of India. Floods are not always beneficial, and depending upon the nature of the suspended load, they may degrade or erode the soil cover of the floodplain. They are clearly undesirable in urban, industrialized areas.

At the onset of a flood the most abrupt decrease in velocity and turbulence takes place immediately adjacent to the channel. Here the greatest thickness of flood-transported sediment tends to accumulate. This sediment builds up the banks of flood-prone rivers, forming **natural levees.** These features add to the future channel capacity. In some places they are supplemented by manufactured levees or dikes constructed for purposes of flood control. Other flood-control practices include alterations of channel size and shape and construction of diverting channels that direct the spill of flood water into specified areas.

Base Level and Deltas

Thus far we have been discussing the ways that streams in a drainage system wear down the land surface and carve landscape features. But what happens where the master stream flows into a lake or the sea? Here the stream reaches the **base level,** which is the critical level below which the streams in that drainage system cannot further wear down the land. Ideally, the base level surface for a drainage area is a surface which is flat or so gently inclined that streams flowing on it cannot possess the capacity to erode it. A lake or the sea determines the base level at the mouth of a river. Upon reaching such a body of standing water, a river unloads its freight. The sediment that is deposited at the mouth of a river produces an alluvial structure called a **delta.** The large deltas that exist where major rivers empty into the sea are amazingly complicated structures. Here many fluvial and shallow marine processes work together to produce a variety of interfingering sedimentary facies. The fertile lands on some deltas, such as the Nile Delta, have nurtured civilization since ancient times. But the hostile environments of disease-ridden swamps and treacherous marshes of other deltas have rebuffed the growth of civilization.

What are the important features of a delta? We can observe many of these features on the Mississippi Delta (Figure 13-28). Here we find the river winding over a nearly flat countryside of swamps, marshes, and bayous, and irregular bays reaching inland from the Gulf of Mexico. We use the term **bayou** to describe lakes and ponds occupying abandoned meanders of the river, as well as the wide sluggish streams that thread their irregular paths through the swamps and marshes. But look at the very end of the Mississippi River itself, and notice how it branches into many channels that flow into the sea. They look like tributaries, but they do not flow into the master stream. Instead, they flow out of it. For this reason they are called **distributaries** rather than tributaries.

Typical of many large rivers is a system of distributary channels, which all together makes up the mouth of the river. The present Mississippi distributaries are building a **birdfoot delta** (Figure 13-28), so named because of the narrow fingers of land that extend seaward like bird claws. Between these fingers of land are marshes and shallow bays that cover submerged alluvial deposits. Farther offshore at the delta front these alluvial deposits slope downward into deep water. How does the river build narrow fingers of land? By what processes do sediments accumulate in the bays between them?

Let us consider a greatly simplified example that helps to explain how a river builds a delta (Figure 13-29). Consider first a stream flowing into a pond. At the mouth of the stream velocity and turbulence decrease. The coarse particles settle most rapidly and accumulate where the bottom of the pond is sloping downward from the shore. Therefore, these coarse particles collect to produce an inclined lamination of sediment that we call a **foreset bed** (Figure 13-29 A). The momentum of the water carries the fine-grained part of the debris load farther from shore. Here these small particles settle in the deep water to produce a **bottom-set bed** (Figure 13-29 A). As this process continues, a fan-shaped submerged delta platform is built outward from the shore of the pond. Now let us add a complication to this simple example. Suppose that floods occur during the season of highest discharge. As water spills over the stream banks, sediment is added to the natural levees which not only grow higher, but begin to grow outward over the submerged delta platform. When the flood recedes, the natural levees remain as narrow arms of land (Figure 13-29 B). Each flood extends them farther from the original shore of the pond. As the natural levees grow outward over the shallow platform they extend the stream channel out to the steep delta front. Sediment accumulates most rapidly in channel mouth bars directly in front of the extended stream channel so that the delta tends to grow outward but does not become appreciably wider. During floods, the stream with its debris load spills over the natural levees onto the shallow delta platform. The coarser sediments accumulate on the natural levees, but the finer particles spread farther away, where they accumulate

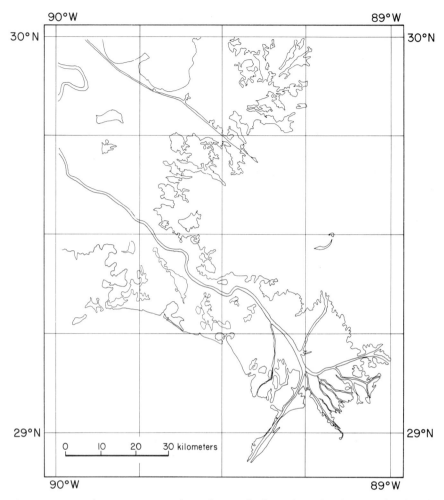

Figure 13-28 The Mississippi Delta. Observe the branching distributaries bordered by natural levees. (Modified from *Breton Sound, Louisiana*, 1:250,000 Topographic Map, U.S. Geological Survey.)

to form a topset bed on the delta platform (Figure 13-29A).

Gradually a limit is reached. The stream has been building the delta outward but has not widened it very much. The channel that extends over the delta platform is only slightly inclined. Then, during one of the floods, the swirling waters cut through a natural levee at a place closer to the shore and open another mouth into the pond. Now a new distributary channel begins to form (Figure 13-29C). The delta commences to grow outward from the mouth of this distributary. Eventually other distributary channels form in the same way. Ordinarily, water continues to flow in several of these channels. Now the delta grows wider as distributaries build natural levees in several directions and add sediment to the foreset and bottom-set beds. Flood waters continue to spread sediment on the topset beds in the

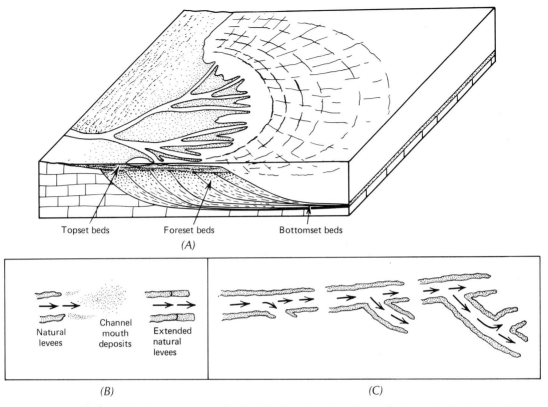

Figure 13-29 *(A)* Simplified structure of a small delta showing river distributaries and principal deposits of sediment. *(B)* Accumulation of sediment on natural levees and in channel mouth deposits. *(C)* Development of distributaries by growth and subsequent breaching of natural levees.

bays between the natural levees. Finally, some bays become so shallow that marsh grasses begin to grow. Sediments deposited by repeated flooding may eventually raise the level of these areas slightly higher than water level in the pond.

Many other processes contribute to the growth of a delta. The pond in our example (Figure 13-29) is undisturbed by the waves, tides, and currents that exist in the ocean. The marine processes that influence erosion, transport, and deposition of sediment are discussed in Chapter 15. These processes can have profound effects on a delta. Ordinarily these marine processes act to disperse the sediment carried to the ocean by fluvial processes. The balance between these competing processes dictates the size and shape of the delta that can exist at the mouth of a river. Conditions in the Gulf of Mexico near the mouth of the Mississippi River favor growth of the long arm of land that forms the birdfoot delta illustrated in Figure 13-28. This is very different from the Nile Delta which is more or less triangular in shape. In fact the word **delta** was chosen to describe the land at the mouth of the Nile that is shaped like the Greek letter Δ. But, because of the different ways that fluvial and marine processes achieve a balance, there is no truly typical shape for a delta.

A delta must adjust to seasonal changes in

the debris load and discharge of the river. In temperate and humid tropical environments where there is a continuous flow of water throughout the year the distributaries ordinarily make continual adjustments to maintain their hydraulic geometry. It is more difficult for a river in an arid or arctic environment to make these adjustments because of intermittent flow of water in the distributaries. For example, the arctic climate of northern Canada keeps distributaries on the MacKenzie Delta frozen for much of the year. Then heavy summer rain together with thawing ice and snow cause a rapid and dramatic increase in discharge. The distributaries adjust only imperfectly to the rapid influx of water and sediment. Therefore, the channels tend to be somewhat unstable. Yearly changes in the size and direction of these channels are more pronounced than the changes occurring in delta distributaries in which the discharge is more regular and continuous during the year. Distributaries on arid deltas also tend to be unstable and changeable where most of the discharge comes during infrequent torrential rains.

The inclination of a submerged delta front is influenced by the density differences between sea water and the stream water. Pure water (density of 1.0 gm/cm^3) tends to float temporarily on sea water (approximate density of 1.027 gm/cm^3). However, suspended sediment increases the density of the stream water. Because the stream water has the lower density, its momentum tends to carry it farther from the mouth of the stream before it becomes completely mixed with the sea water. Its debris load, too, is distributed over a broader area of the sea floor, so that a gently sloping delta front develops. A steeper delta front is produced where the stream water density equals or is greater than the sea water density.

Over a long period of time certain distributaries contribute most to the growth of a delta. Then conditions change so that sedimentation increases at the mouths of other distributaries, and marine processes erode and disperse the alluvial lobe produced by the formerly active channels. Subsidence of parts of the delta caused by compaction of sediment or by tectonic tilting and isostatic adjustments can change the areas where sediment collects. Look again at Figure 9-28 to see different stages in the growth of the Mississippi Delta. Observe the interfingering facies of alluvial and marine sediments that were produced by changes in the balance of fluvial and marine processes over a long period of time. At present the growth is most active on the Plaquemines-Modern birdfoot (Figure 13-28).

Over the vast interior of a continent rivers wear down the land. But where a river flows into the sea it changes its role to builder. Here its delta is an extension of land that enlarges the area of the continent.

The Graded Stream

Much of the foregoing discussion has been about the adjustments that establish a balance between streams and landscape. Does this mean that a stream and the land surface tend to exist in a condition of equilibrium? Professor William Morris Davis (1850–1934), a pioneer geomorphologist, pointed out what he thought was a tendency for a river to attain a "balance between erosion and deposition" given sufficient time for it to adjust to the landscape. Davis called a stream that had achieved such a balance a **graded stream.** The late J. Hoover Mackin, elaborating on this concept, considered the graded stream as:

. . . one in which, over a period of years, slope is delicately adjusted to provide, with available discharge and with prevailing channel characteristics, just the velocity required for the transportation of the load supplied from the drainage basin. The graded stream is a system in equilibrium: its

diagnostic characteristic is that any change in any of the controlling factors will cause a displacement of the equilibrium in a direction that will tend to absorb the effect of the change.

All that we have learned about stream hydrology supports the idea that streams can become graded. But how can we tell when this condition is reached? According to Mackin's definition we would have to determine the amount of debris supplied from a drainage basin. Also, we would have to devise some way to find out if the stream velocity was sufficient to transport precisely the amount of debris. In practice these factors can be very difficult if not impossible to measure. So, although the concept of a graded stream is attractive, the practical application of this concept is probably quite limited. This is why some geomorphologists have tried to redefine in measurable ways the equilibrium that may exist between a stream and the landscape.

One idea is the concept of the **poised** stream. This a stream that has the capacity to move all of the loose debris in its channel only when the amount of water in the channel reaches flood stage. What does this mean? We know that as water rises in the channel, velocity and discharge increase so that the debris load also increases. But only when the water is about to spill over the banks in a flood does the stream have the capacity to carry all of the loose debris in the channel. According to this definition, we can make measurements to determine if a stream is poised. We can collect samples of debris from its channel while the water is rising. In this way we can find out if all of the loose debris is picked up just when the stream reaches flood stage.

The concept of the poised stream does not take into consideration all possible factors that contribute to the balance between a stream and the landscape. But this concept does recognize some important features that indicate a condition of equilibrium, and these features can be measured. There are practical reasons for making such measurements. For example, what happens when a dam is constructed across a river? The effects are not restricted to the land covered by the reservoir; changes can occur throughout the entire drainage system. The reservoir becomes the new base level of streams draining into it. Initially, the streams draining into and out of the reservoir are not properly adjusted to the landscape. To once again reach a condition of equilibrium, streams must reshape their longitudinal profiles by eroding some areas and depositing the sediment in other places. We cannot predict all of the changes that will eventually occur as a result of dam construction. But with careful planning some of these adjustments can be anticipated.

MASS WASTAGE

The landscape is a complex arrangement of surfaces, gently or steeply inclined, merging in a varied pattern of valleys and ridges. An inclined surface tends to be an unstable feature that is subject to modification by downslope movement of rock debris under the direct effect of gravity. The rate at which this debris is transported ranges from almost imperceptibly slow movement to the sudden and sometimes catastrophic advance of a massive landslide. The downslope movement of rock debris in direct response to gravity is **mass wastage.**

We describe the important kinds of mass wastage according to the rate of movement and the condition of the material involved.

Creep

The downslope migration of regolith and broken-up rock at a very slow but more or less continuous rate is called **creep.** We can see the effects of creep on the appearance of the landscape only after the processes have been

Figure 13-30 Evidence of creep in regolith and bedrock in Washington County, Maryland. (G. W. Stose, U.S. Geological Survey.)

active for several years. Look at the distorted bedding surfaces of weathered sedimentary strata (Figure 13-30) that indicate creep.

Creep is a near-surface phenomenon and diminishes with depth to where movement is no longer discernible. There are different mechanisms that might cause creep. By analyzing how gravity produces stress in an inclined layer of loose grains, Richard Haefeli discovered that individual particles would gradually move downward along the trajectories illustrated in Figure 13-31A. Another mechanism for creep is associated with alternate expansion and contraction of an inclined soil layer (Figure 13-31B). Here we see that a loose particle projected outward during expansion of the layer does not draw back along the same path when the layer contracts at a later time. Rather, it shifts to a new position slightly farther downhill. Movement caused by creep is usually less than 1 cm/yr at the land surface.

Earthflow and Solifluction

A porous mass of regolith that becomes sufficiently saturated by water may begin to flow like a viscous fluid. This sometimes occurs in arctic soils that contain interstitial ice rather than ground water. Seasonal thawing of interstitial ice occurs close to the surface, whereas the deeper zone remains permanently frozen. During the summer thaw, water is unable to percolate into this deeper zone that is called **permafrost.** Instead, the water stays in the

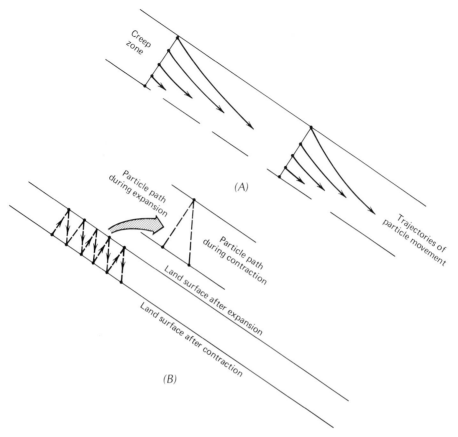

Figure 13-31 Particle movement associated with creep caused by (A) flow of regolith and (B) alternate expansion and contraction of regolith.

near-surface layer. As melting continues, the soil may become so charged with water that even gently sloping layers become unstable masses of mud, and begin to flow. The phenomenon is known as **solifluction,** which means soil flow. Solifluction can also occur in more temperate regions where water can collect in a surface layer of loose debris because it is underlain by impermeable clay or bedrock. During a period of intense rainfall the strength of the porous layer is reduced. Movement of a considerably larger mass of regolith resulting from a condition of water saturation is described as an **earthflow.** Rates of movement associated with near-surface solifluction and larger earthflows range from a few centimeters per year to in excess of several meters per year. A landscape modified by these processes is illustrated in Figure 13-32 A.

Mudflows are rapid movements of water-saturated regolith which move faster than earthflows. Rates of several meters per day or more have resulted in steeply inclined masses of unconsolidated debris. A large mudflow occurred on December 9, 1950, near Oakland, California, when debris on a hillside became unstable after a period of unusually

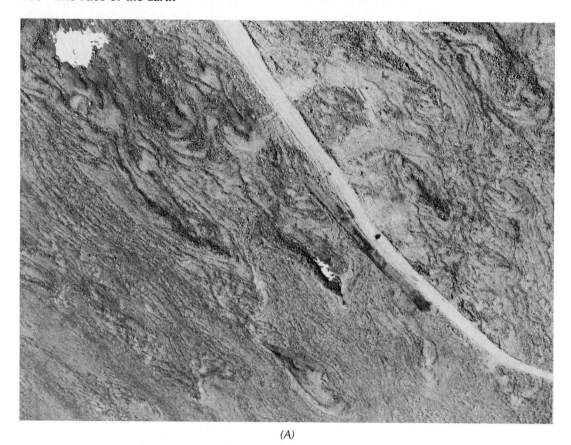

(A)

Figure 13-32 *(A)* A hillside in Alaska modified by solifluction displays a succession of lobes formed in mobile regolith. (R. B. Colton, U.S. Geological Survey.) *(B, opposite)* A modern mudflow that occurred on December 9, 1950 near Oakland, California. (Bill Young, *San Francisco Chronicle*.)

frequent rain. In Figure 13-32B this mudflow is seen encroaching on the highway.

The difference between solifluction and earthflow is gradational with regard to depth. The former term refers to a relatively thin near-surface layer compared with the larger mass of an earthflow. Mudflow and earthflow are transitional with regard to rate of movement, the latter term describing slower mass wastage. All of these terms refer to mass transport of water-charged masses of unconsolidated sediment which behave more or less as viscous fluids.

Landslides

The downslope movement of a coherent mass of regolith or rock over a distinct surface of failure is characteristic of a landslide. Commencing as a more or less unbroken block, the mass may become broken apart as the landslide progresses.

Landslides occur in different forms. **Rock falls** and **rock slides** describe the collapse of sheets of rock from cliff faces or bedrock slopes. The remains of a 5000 metric ton rock fall which occurred in August 1968, in Zion

(B)

(A)   (B)

Figure 13-33 *(A)* Debris from a 5000 metric ton rock fall which occurred in The Narrows of Zion Canyon in August 1968. *(B)* Scar of light-colored relatively unweathered rock is evident on the canyon wall following the rock fall.

Canyon, Utah, are illustrated in Figure 13-33. The debris from rock slides or rock falls such as this collects at the base of the slope or cliff to form deposits of broken-up rock fragments called *talus*.

Rock slides can occur along a more or less planar surface where initial failure occurs in an incompetent bed that lies beneath other more durable layers. Let us consider two situations where a rock slide might be expected. In the first example (Figure 13-34A) an unstable condition develops after a gully has been eroded downward into the zone of weak rock. Failure will occur if the shearing stress parallel to the surface of failure exceeds the shear strength of the material along that surface. In this example we can calculate the shear stress (τ) from the weight of the overlying rock above a point on the surface of failure:

$$\tau = \rho g h \sin \theta$$

where ρ is density of overlying rock, h is the depth to a point of failure, θ is the dip of the surface of failure, and g is the acceleration of gravity.

Now look at the second example (Figure 13-34B). Here a zone of weakness is initially exposed near the base of the hillside. But let us suppose that the material is strong enough to maintain the overlying burden provided it

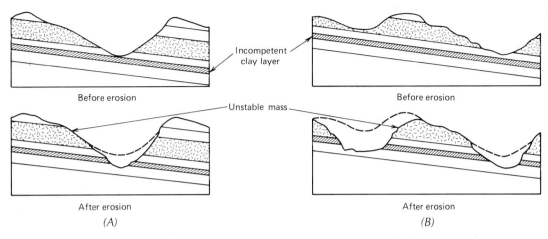

Figure 13-34 Two situations where stream erosion causes instability in a rock mass and a subsequent rock slide. (A) Gully eroded near base of a slope removes support of rock mass overlying a zone of weakness. (B) Gully eroded near the top of a slope exposes a porous zone causing an increase in the infiltration of water.

remains dry. What happens when a gully is eroded into the incompetent bed at a location farther uphill? Now water can seep into this zone causing a reduction in its shear strength to the extent that failure occurs.

Slump is another form of landslide. It describes an initially coherent mass that is dislodged along a curved surface of failure. Slumps commonly occur in deposits of unconsolidated sediment where the shear strength is more or less uniform throughout the mass. Slump can involve only a single coherent block or a series of blocks displaced along en-echelon surfaces of failure (Figure 13-35 A). Slumping can progress very slowly at rates of less than a few centimeters per year, or at increasingly faster rates including sudden failure and large displacement that take place in a few moments.

The sudden downslope movement of a large and heterogeneous mass of regolith and rock is an **avalanche.** We also use this term to describe sudden movement of masses of snow and ice in mountainous regions. The processes of slump, rock sliding, and earthflow may all be active during an avalanche.

The great Madison Landslide occurred on August 17, 1959, along the Madison River west of Yellowstone National Park in Montana. You can see some details of this avalanche in Figure 13-35B. A mass of dolomite and foliated metamorphic rock broke from the mountain during a severe earthquake. Almost 40 million cubic meters fell into the valley damming the river and covering an adjacent campground. A lake has since formed upstream from the slide area. Twenty-six persons were killed in this massive landslide. Evidence of other similar landslides is abundant in the geologic record.

The various kinds of mass wastage can have a marked effect on the landscape. Practically all sloping surfaces are modified to some extent by downslope mass transport. Together with weathering the processes of mass wastage act to reduce bedrock to fragments that can be eroded and transported by streams.

492 The Face of the Earth

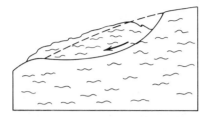
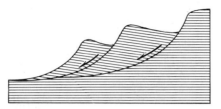

(A)

Figure 13-35 Slumping may involve (A) detachment of a single coherent mass of regolith, or a series of en echelon slump blocks, or a large mass originating as a coherent block then disintegrating and progressing in a manner similar to a mudflow. The Madison Landslide (B) occurred August 17, 1959 in southern Montana. Photo shows the surface of failure as a massive scar on the mountain side. The debris lodged in the lower valley forms a dam on the Madison River behind which a lake has formed. (Photograph by R. W. Bayley, U.S. Geological Survey.)

(B)

WIND AND EOLIAN LANDFORMS

The atmosphere of the earth is a turbulent fluid. The shifting air masses produce currents of wind that can move sediment and mold **eolian** landscape features. In the following discussion we will examine the role of wind in landscape evolution.

Global Air Circulation

Atmospheric circulation on a global scale displays persistent patterns of prevailing wind currents. Near the surface of the earth we observe local erratic winds that are temporary eddy currents in the larger and more regular system of shifting air masses. The directions of the permanent wind currents near the earth's surface are shown in Figure 13-36. These wind patterns are produced by the interaction of several thermal and mechanical phenomena including solar heating, the rotation of the earth, and differences in the specific heat of oceans and continents.

Wind is a turbulent current of air possessing power to erode the earth and to transport and deposit sediment. It is most effective in redistributing the unconsolidated sediment that exists in regions of sparse vegetation. Therefore, most eolian landforms are found in arid and semiarid areas (Figure 13-36). Elsewhere in more moist environments these features are found mainly along coasts where there are abundant beach sands. Eolian processes are relatively ineffective in regions where the land surface is protected by a cover of vegetation.

Eolian Landscape Features

Under what conditions can wind lift and carry sediment? Eolian transport of sediment is possible where upward moving eddy currents flow more swiftly than the settling velocity of the particles in air. We can learn from Figure 13-37 how rapidly different sized particles will fall through quiet air. We know that the average velocity of updrafts of air is usually about 10 percent of the average wind velocity. Therefore, we can tell from Figure 13-37 that an updraft velocity of approximately 5×10^{-2} m/sec (0.1 mile/hr) could lift a very fine-grained sand particle (diameter of 0.0625 mm on the Wentworth Scale), because the particle falls at a slower velocity than the updraft. Such sediment could be moved by a wind blowing at the approximate velocity of 0.5 m/sec (1.2 miles/hr). Coarser sand grains of, say, 1 mm diameter could be lifted by an updraft of more than 5.0 m/sec (11 miles/hr). These larger particles could be maintained in suspension only by a strong wind of almost 25 m/sec (56 miles/hr). Such winds are infrequent, so ordinarily we could not expect sand grains of intermediate size to remain airborne for more than a few moments. Smaller silt- and clay-sized particles, on the other hand, could remain suspended for much longer periods of time in regions where light to moderate prevailing winds are common. These very fine-grained particles sometimes collect in **dust clouds** (Figure 13-38) over semiarid areas.

Eolian transport of sand particles takes place mostly by **saltation.** This is the process of bouncing the particles over the land surface. After each bounce the particles move along asymmetrical trajectories (Figure 13-39). Ordinarily they do not rise more than a few meters above the land surface. In the atmosphere the saltation zone is equivalent to the bedload zone of a stream. Above this zone only fine-grained particles can remain suspended. They are borne along in the suspended load of the turbulent wind.

The principal landforms produced by eolian transport and deposition of sand are **dunes.** The shape of a sand dune depends upon the volume of sediment available, the nature of the prevailing winds, and the topography and

494   The Face of the Earth

Figure 13-36 Maps showing global patterns of prevailing wind (after A. N. Strahler, *Physical Geography*, 2nd edition, page 126, John Wiley and Sons, Inc., 1965), and regions having an arid or a semi-arid climate. (After R. F. Flint and B. J. Skinner, *Physical Geology*, page 174, John Wiley & Sons, Inc., 1974.)

Running Water, Mass Wastage, Wind, and Continental Landscapes 495

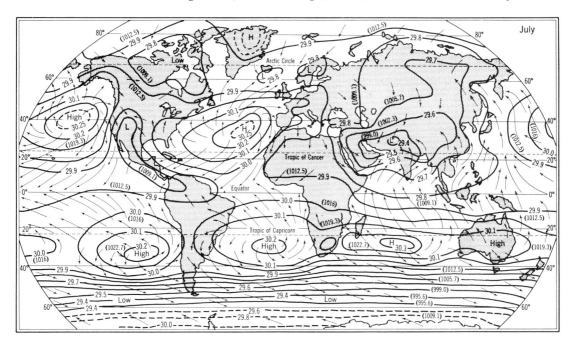

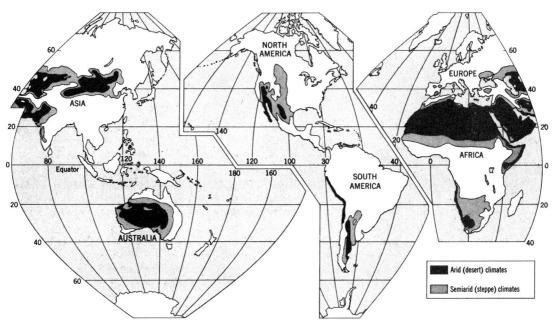

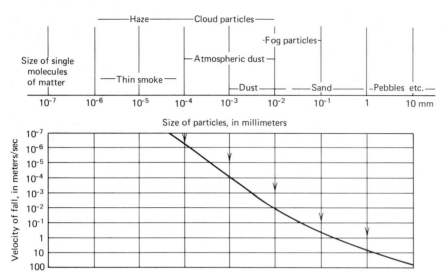

Figure 13-37 Experimentally determined relationship of particle size and its settling velocity through quiet air. (After R. A. Bagnold, *The Physics of Blown Sand and Desert Dunes*, page 1, Methuen and Co., Ltd., 1941.)

Figure 13-38 Dust cloud over an area in New Mexico was created by air turbulence. (U.S. Soil Conservation Service.)

vegetation in the region. A **barchan dune** is a cresent-shaped deposit with an asymmetric cross section in the direction of prevailing wind (Figure 13-40). The surface of the dune facing the wind is inclined gradually to an arcuate crest where it meets the steep backslope. As wind blows along the gradual slope, sand is moved by saltation. Upon passing the crest, wind velocity diminishes and its sediment load is dropped abruptly. In this process a lamination of sand is deposited on the backslope. As sand continues to be moved from the wind-facing slope to the backslope, the position of the dune changes. Barchan dunes ordinarily do not grow more than 50 m high. They are most ideally formed where the landscape is flat and the supply of sand is sparse. Where sand is more abundant, dunes coalesce to form other patterns.

In areas covered by an abundant supply of sand and little vegetation, persistent wind produces **transverse dunes.** These dunes display more or less continuous crests that are oriented perpendicular to the prevailing wind direction. The surface facing the wind is gently inclined in comparison with the steeper backslope of the dune. The dune crest moves in the downwind direction as sand is transferred from the wind-facing slope to the backslope. Transverse dunes form best in flat terrain with an abundance of sand.

The largest dunes that exist anywhere are

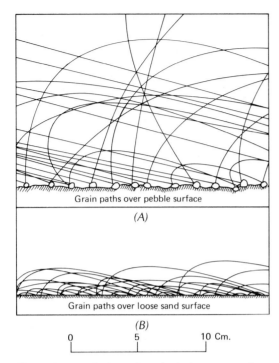

Figure 13-39 Trajectories of wind-transported sand grains moving by saltation over (A) a pebble covered surface, and (B) a sand covered surface. Note that the particles bounce considerably higher above the surface when migrating over a cover of pebbles compared to a cover of sand. (After R. A. Bagnold, *The Physics of Blown Sand and Desert Dunes,* page 36, Methuen and Co., Ltd., 1941.)

great sand ridges oriented parallel to the direction of the prevailing wind. These **longitudinal** or **seif dunes** have been known to grow to more than 200 m in height and extend to lengths of over 40 km. In Iran there are vast dune complexes consisting of closely spaced seif dunes (Figure 13-41). We do not fully understand the origin of these eolian deposits. Some geologists believe that they begin with the coalescence of barchan dunes caused by a change in the prevailing wind direction. Thereafter they exert a control on the direction of near-surface wind. Some asymmetry may develop from small fluctuations of the prevailing wind, shifting sand from one side of the dune crest to the other. Seif dunes are best developed by strong prevailing winds in regions where sand is not abundant.

Eolian transport of sand is restricted by vegetation. **Parabolic** dunes can develop where gaps in the cover of vegetation permit sand to be scoured from a local **blowout** area and subsequently deposited in a U-shaped ridge along its margin. Irregular but relatively abundant vegatation found near beaches can control the movement of sand so that it becomes deposited in small **hummocks**. These hummocks or **beach dunes** are common to some coastal areas in relatively moist climates.

Deposits of very fine-grained eolian sediment have accumulated in some regions. A small proportion of wind-transported silt and clay is found almost everywhere in the mixture of regolith that comes mostly from other sources. However, thick beds of unstratified silt and clay almost entirely of eolian origin blanket large areas of central United States, southwestern China, and parts of central Asia. These deposits are known as **loess.** They consist of very fine-grained particles that appear to have been produced by glacial erosion and transported in suspension by winds blowing off the former continental glaciers onto the bordering landscape. Because the clay-sized particles tend to have flat platelike shapes, the loess deposits usually consist of well-packed interlocking grains. Where a blanket of loess is eroded by a stream, the walls of the erosion gully can stand as nearly vertical cliffs because the grains of sediment are so tightly packed. Look at the loess bluffs near the city of Vicksburg, Mississippi (Figure 13-42).

Erosional landforms of eolian origin include broad **deflation basins** which are arid areas where almost all loose sediment has been removed by the wind. Because the wind

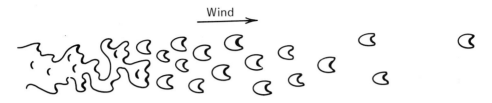

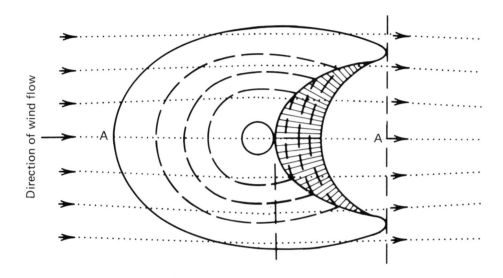

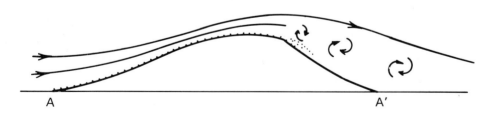

Figure 13-40 Barchan dunes are crescent shaped, and develop with a steep side facing away from the wind direction. Arrows indicate flow of the wind. (After R. A. Bagnold, *The Physics of Blown Sand and Desert Dunes*, pages 202, 209, and 218, Methuen and Co., Ltd., 1941.)

Running Water, Mass Wastage, Wind, and Continental Landscapes 499

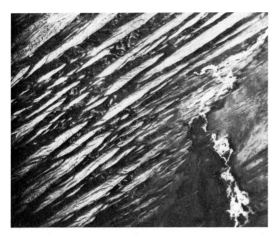

Figure 13-41 Longitudinal dunes, or seifs, extend as long ridges over part of the desert in Iran. (U.S. Geological Survey.)

selectively removes the fine-grained sands, silts, and clays, a concentration of larger fragments eventually accumulates on the surface of a deflation basin. Thereafter it yields virtually no sediment to the wind. The surfaces of these larger fragments become worn as they shift against one another. Eventually the fragments become closely fitted into a hard and relatively smooth surface that is called a **desert pavement.**

A strong prevailing wind charged with fine sand and silt grains is an abrasive fluid with the capacity to carve exotic bedrock feature. Where found, the wind-carved features add a unique and bizarre character to the landscape.

EVOLUTION OF LANDSCAPE

A landscape of plains and low hills reaches across the Canadian Shield from the Great

Figure 13-42 Bluffs eroded in massive loess deposits near Vicksburg, Mississippi. (E. W. Shaw, U.S. Geological Survey.)

Lakes to the Arctic Circle. But the rocks exposed here tell us that mountain chains once crossed this area. Clearly, these mountains could not forever resist the effects of erosion. The landscape must change with time, but what is the sequence of change? Can we recognize different steps in the evolution of landscape?

The Cycle of Erosion of W. M. Davis

William Morris Davis (1850-1934) thought that features could be recognized that represented different stages in a process he called the "cycle of erosion." Davis proposed that the landscape evolved through a **youthful** stage into a **mature** stage and finally reached a stage of **old age.** He chose a simple example to illustrate the "cycle of erosion" (Figure 13-43). It begins with the emergence of a featureless land surface from beneath the sea that rises to form a high plateau. Sparsely distributed streams then begin to carve narrow V-shaped valleys into this plateau. At this stage the landscape is youthful. The effort of a stream on the youthful landscape is concentrated on downcutting. This effort continues until the stream and landscape become adjusted so that the stream can transport just the amount of sediment that it receives from tributaries, sheet wash, and mass wastage from the valley sides. After this balance is established Davis believed that stream effort becomes redirected to the process of lateral erosion. This marks the transition from the youthful stage to the mature stage. He believed that a stream became graded when its activity changed from downward to lateral erosion, but we now know that this is not necessarily true. Thereafter, the stream continues to widen rather than deepen its valley by cutting laterally into the bordering uplands and redistributing the eroded sediment on the ever-broadening valley floor. As the stream merges with other similar streams, a mature landscape of broad valleys bordered by steep uplands develops.

Eventually the coalescence of ever-broader mature valleys produces a wide plain of subdued relief. Here only scattered upland remnants of the original plateau remain. This nearly featureless surface, called a **peneplain,** is a landscape in old age. Sluggish streams follow broadly meandering channels over the peneplain during this final stage in the "cycle of erosion."

The idea of the "cycle of erosion" was a singularly important scientific contribution. It provided geomorphologists with a scheme that could be used for analyzing landscape and for comparing stages of landscape development from different regions. Davis realized that tectonic events and changes in climate would certainly interrupt the ideal "cycle of erosion." The actual evolution of landscape would be a sequence of incomplete cycles. But he believed that vestiges of ancient landscapes could be recognized and used as evidence for reconstructing a sequence of incomplete cycles.

Is it actually possible to identify the remnants of old landforms? Can this information be used to determine how the landscape has changed with time? Now we have much better information about drainage geometry, hydraulic geometry, and rates of erosion than was available when W. M. Davis proposed his ideas. Judging from the debris loads of streams and their rates of sediment discharge geomorphologists now estimate that the land surface of the United States is being worn down at an average rate of 6 cm/1000 years. Higher rates of more than 1 m/1000 years are estimated for rugged mountainous terrain in some regions. These erosion rates suggest that uninterrupted fluvial processes could reduce the continent to a sea level plain in less than 10 million years. If these rates are correct, how could the features that W. M. Davis identified as vestiges of Mesozoic and early Cenozoic landscapes still exist? We know that fluvial processes can level a formerly rug-

Running Water, Mass Wastage, Wind, and Continental Landscapes 501

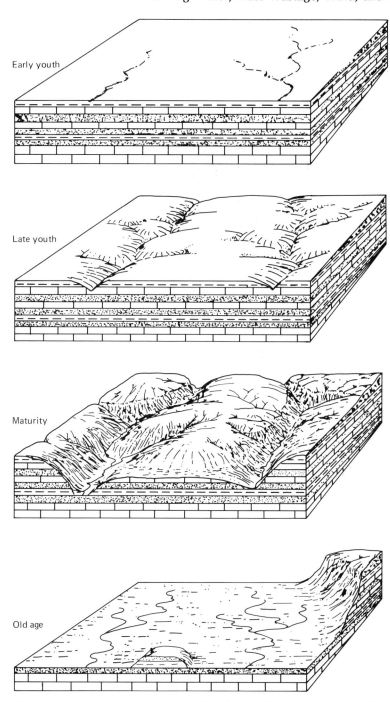

Figure 13-43 The principal stages of landscape development during the "cycle of erosion" proposed by W. M. Davis. (After E. J. Raisz and the Kentucky Geological Survey.)

ged terrain. The Canadian Shield and the Eastern Piedmont of North America display the cores of ancient orogenic belts on a landscape of low relief. The important question is whether or not the remnants of any ancient landscapes still exist. Modern evidence suggests that this is an unlikely possibility. Without this evidence we cannot be certain about what features distinguish different stages of landscape evolution. But if the landscape does not evolve according to the ideas of W. M. Davis, in what other way might it be evolving?

The Concept of Dynamic Equilibrium

A newer idea about the way landscape evolves has been described by the well-known geomorphologist John T. Hack. This idea is called the **dynamic equilibrium concept of landscape evolution.** It represents a significant departure from the earlier ideas of W. M. Davis. According to the dynamic equilibrium concept, a drainage system and the surrrounding landscape adjust quickly to differences in the resistance to erosion of the rocks and regolith in the drainage basin. During this brief period of adjustment, valleys develop in the least resistant material, hills and ridges remain where the material is more durable, and the streams become graded. Thereafter, a balance is achieved, so that there will be no further change in the shape of the land surface. The steeply inclined hillside streams have the additional power to wear down the durable rock at the same rate as the more gently inclined streams wear down the valley floors. As long as this condition of equilibrium continues, all the land, hills and valleys alike, is worn down at the same rate. An equilibrium topography is maintained. Landforms produced when the streams were becoming graded continue to exist as long as the system stays in balance. But we have no way of knowing how long the balance has lasted. Therefore we cannot distinguish an ancient feature from one produced recently.

Events of tectonic uplift and climate change will disturb the balance between streams and landscape. So do changes in the rocks and regolith that are encountered as the streams wear away the land. Adjustments to these events and different materials produce almost immediate changes in the landscape. But no vestiges of the former features are preserved.

Clearly, the dynamic equilibrum concept of landscape evolution is very different from W. M. Davis' "cycle of erosion" in which landforms are worn down in a way that diminishes topographic relief. A landscape in dynamic equilibrium experiences no change in topographic relief even though the absolute level of the land surface is being lowered everywhere. Only when the lower parts of the area are reduced to the base level does topographic relief begin to diminish. At this point the valleys at base level can be lowered no further, but erosion of higher landforms can continue.

The dynamic equilibrium concept proposes that no stages in the evolution of landscape can be distinguished from one another. Accordingly, the shape of the land surface depends only on the present balance of stress that acts on it. We do not know how the land looked formerly or how it will look in the future. The diversity of landforms in some regions is more clearly understood in terms of the dynamic equilibrium concept rather than the "cycle of erosion." But what about features such as a superposed drainage system (Figure 13-4)? Clearly, it is the imprint of a former landscape that is difficult to interpret in terms of the dynamic equilibrium concept. A landscape produced by streams in an **internal** drainage system contradicts the idea of maintaining equilibrium topographic relief in the drainage area. An internal drainage system is one in which no master stream flows out of the area. In some arid regions stream water evaporates or seeps into the ground water system before it can flow out of the

Running Water, Mass Wastage, Wind, and Continental Landscapes 503

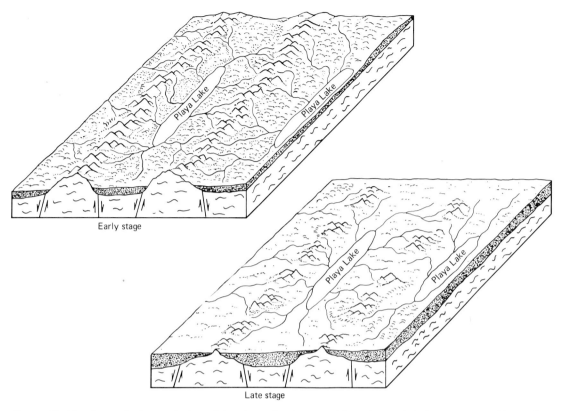

Figure 13-44 Stages in the evolution of the desert landscape with an internal drainage system. Mountains are reduced by erosion and bordering valley floors are raised by the deposition of sediment. Gradually the mountains become buried in their own debris. Topographic relief diminishes with time.

drainage area. Here sediment washed from mountains by intermittent streams accumulates in nearby arid valleys. As the mountains are worn down the valley floors are raised as sediment continues to accumulate there (Figure 13-44). In this way an internal drainage system causes continual change in topographic relief. But the consensus of geomorphologists is that the dynamic equilibrium concept, despite its inadequacies, provides us with the best understanding of landscape evolution in most regions. So perhaps there is no way to tell how the land once looked, or to predict the future details of its appearance.

STUDY EXERCISES

1. On the map of a drainage system (Figure 13-45) points A, B, C, and D all lie on the master stream, and points E and F lie on one of its tributaries.

   (a) What drainage pattern does this system display?

   (b) What channel pattern is displayed by the master stream between

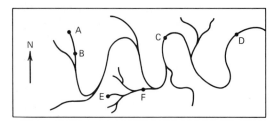

Figure 13-45

points B and D?

(c) What is the order number of the master stream?

(d) At point C is the stream most likely to have (1) a higher velocity and a lower gradient; (2) a lower gradient and higher turbulence; (3) higher gradient and lower turbulence; or (4) higher velocity and higher turbulence than at point B? Explain!

2. According to Figure 13-14,
   (a) What are the largest particles that will remain suspended in a stream with a velocity of 0.5 cm/sec?
   (b) If this stream velocity increased to 5 cm/sec, what will be the diameter of the largest particles that the stream can erode from the bed and take into suspension?
   (c) What is the minimum velocity at which any kind of erosion can occur assuming that a wide range of grain sizes is available to be eroded?

3. Suppose that your measurements of velocity (v), width (w), depth (d), and discharge (Q) for a cross section of a stream show that : $d = 0.5Q^{0.3}$, $v = kQ^{0.5}$, $w = 4Q^b$.
   (a) What particular kind of stream geometry do these equations describe?
   (b) At a time when the stream width is 5 m, what is its discharge?
   (c) At a time when the stream is discharging 4 m³/sec, what is its width, depth, and velocity?

4. Are alluvial floodplain deposits coarser grained and more poorly sorted or finer grained and better sorted than alluvial fan deposits? Explain!

5. Suppose that a river draining an area of 500 km² has an average width of 30 m and an average depth of 3 m. The average sediment load is 20 gm/kg. Precipitation in the area is 40 cm/year.
   (a) What is the average discharge of this river?
   (b) How much time would be needed for the level of the land surface to be worn down by an average of 10 cm?

   Note: Use data from Figure 13-1 in your calculations.

6. Discuss what kinds of information you would need to prove that the concept of dynamic equilibrium explains evolution of the landscape during Paleozoic time.

SELECTED READINGS

Bloom, Arthur L., *The Surface of the Earth*, Prentice-Hall, Inc., Englewood Cliffs, N.J., 1969.

Davis, William Morris, The Geographical Cycle, *Geographical Journal*, vol. 14, pp. 481–504, 1899 (reprinted in *Geographical Essays*, Dover Publications, 1957).

Hack, John T., Interpretation of Erosional Topography in Humid Temperate Regions, *American Journal of Science*, vol. 258-A, pp. 80–97, 1960.

Leopold, L. B., M. G. Wolman, and J. P. Miller, *Fluvial Processes in Geomorphology*, W. H. Freeman and Co., San Francisco, 1964.

Tuttle, Sherwood D., *Landforms and Landscapes*, Wm. C. Brown Co., Publishers, Dubuque, Iowa, 1970.

14

GLACIERS AND THE CONTINENTAL LANDSCAPE

Continental Glaciers
   The Antarctic Ice Sheet
   The Greenland Ice Sheet
   Other Modern Ice Sheets
Ice Streams in Mountainous Terrain
Mass Balance in a Glacier
   Accumulation and Ablation
   Ice Movement
   Advance and Retreat of Glaciers
Glaciation

The Debris Load of a Glacier
Erosional Features of Glaciation
Depositional Features of Glaciation
Frozen Ground
Ice Ages
   The Late Cenozoic Ice Age
   The Paleozoic Ice Age
   Causes of Ice Ages
Perspective

Antarctica is a vast continent extending over the southernmost reaches of the earth. It is a unique continent because of the cover of glacial ice that blankets all but the highest mountains. The antarctic landscape (Figure 14-1) is almost everywhere a nearly featureless expanse of ice, punctuated in a few locations by rocky peaks that protrude through the blanket of ice. Beneath this icy landscape is another bedrock landscape with hills, plains, plateaus, and lowland areas not unlike other continents. Antarctica remained virtually unknown until the twentieth century when it has become a frontier for modern exploration. In the northern hemisphere the island subcontinent of Greenland is similarly covered by ice. Like Antarctica, it was a mysterious uncharted land that was crossed by Fridtjof Nansen for the first time near the end of the nineteenth century. Taken together Antarctica and Greenland comprise approximately 10 percent of the earth's continental surface. Glaciers in other parts of the world are located mostly in mountainous regions.

Large areas of northern North America and northern Eurasia are covered by an irregular and unstratified mixture of gravel, boulders, clay, and sand. During the eighteenth and nineteenth centuries these deposits were described by naturalists who were perplexed as to their origin. Sedimentary structures indicative of a fluvial or marine environment were absent. The geologic processes familiar to these early observers could not account for such deposits. A convincing explanation in terms of uniformitarianism was not evident.

The **diluvian theory,** widely accepted early in the nineteenth century, attributed these irregular deposits to a stupendous deluge of water that washed over the continents. This

Figure 14-1 The south polar landscape in West Antarctica. The nunataks are peaks of subglacial mountains otherwise buried by the continental ice sheet that covers more than 98 percent of Antarctica. (U.S. Navy photograph.)

explanation was consistent with the widely held belief in Noah's flood. The idea of gradual inundation by ice had not occurred to anyone. At this time the continental glaciers of Antarctica and Greenland were unknown to scientists in Europe and North America. It was difficult to believe that ice could have any important effect on the landscape of a temperate continental lowland. However, a few scientists and engineers began to recognize relationships between the unstratified deposits and nearby glaciers occupying valleys in the European Alps. The more imaginative observers speculated about the former existence of massive continental glaciers. The best-known proponent of this idea was the French scientist, Louis Agassiz (1807–1873), who in 1837 proposed a theory of continental glaciation. He argued that ice, not water, had spread over large areas of Europe. Upon melting, this glacier left widespread deposits of unstratified sediment. At first his ideas

were received with interest. The older diluvian theory, however, was too firmly implanted in the minds of many naturalists to yield readily to the new glaciation theory. Argument persisted for several decades during which time the deposits of sediment in question were most carefully examined, and explorers discovered more about the north polar regions. Evidence favoring former continental glaciation became convincing by the end of the nineteenth century. By then the concept of an ice age was generally accepted.

By studying the distribution of glacial deposits we have learned that several times during late Tertiary and Quaternary time vast ice sheets have spread from the polar regions to cover more than 30 percent of earth's continental area. During those times of glaciation the typical landscape over much of northern North America and northern Eurasia was probably similar to the scene in Figure 14-1. There are more ancient deposits that point to extensive glaciation during parts of the Paleozoic era. Still older sediments of probable glacial origin suggest the possibility of ice ages during intervals of Precambrian time. The fact that more than 10 percent of the earth's continental area is now covered by ice indicates that the late Tertiary-Quaternary ice age has not ended.

Our interest in the processes of glaciation is stimulated by the realization that we now dwell in an ice age. The activities of the industrialized civilization of northern North America and northern Eurasia are related to an ice age climate. Sediments of recent glacial deposition sustain agriculture, supply the sand and gravel for construction purposes, and store important reserves of fresh water. The landscape is molded in a terrane of glacial deposits.

We will begin discussion of the processes of glaciation by describing the continental glaciers and mountain glacier systems that exist today. Then we will examine conditions for maintaining glaciers and the nature of glacier movement. We will look into glacial geomorphology and the typical landscape features and deposits produced by glaciers. The chapter concludes with a review of some aspects of glacial history and discussion of theories concerning origin and evolution of ice ages.

CONTINENTAL GLACIERS

The most inhospitable environment on earth for human activity is found on the vast continental ice sheets of Antarctica and Greenland. The harsh and frigid climate provides neither the warmth nor the food needed for survival. The polar traveler must carry everything needed to sustain life. Attempts to explore the interior of Greenland began about 1875, and inland Antarctic journeys began after 1900. The first explorers transported all supplies on sledges hauled by themselves or with the aid of dogs or ponies. They found the interior regions of these ice sheets to be immense windswept plateaus reaching to altitudes of more than 3000 m (about 10,000 ft). Motor vehicles and aircraft became increasingly important in polar exploration after 1920. Exploration continued on an expanded scale, reaching a peak effort in 1957 to 1959 during the International Geophysical Year (IGY) when several nations participated in a cooperative program of polar research.

An important part of the IGY program was to continue and expand upon earlier efforts to systematically survey the large continental ice sheets. Elevation on the vast ice plateaus was measured by methods described in Chapter 2. The ice thickness was determined by the techniques of exploration seismology that we learned about in Chapter 3. Physical characteristics of the glacier ice were studied from surface samples and cores obtained from boreholes, and various conditions of climate were recorded. These studies were carried out by exploration parties traveling overland in motor vehicles or transported by aircraft. Some scenes in Figure 14-2 illustrate modern exploration activities on a continental glacier.

The Antarctic Ice Sheet

Glacier ice in Antarctica extends almost everywhere to the edge of the continent. In some places the inland ice spills into the sea through narrow valleys in coastal mountain ranges and breaks up into icebergs (Figure 14-3). Elsewhere in low coastal regions the ice moves off the continent as a broad sheet which disintegrates into large floating "ice islands" and flat tabular ice bergs (Figure 14-4).

Antarctica is a nearly circular continent more or less centered on the South Pole (Figure 14-5). In the interior the ice plateau rises to above 4000 m (over 13,000 ft). Elevation contours reveal the relief on this surface. Two large marine embayments, the Ross and Weddell Seas, extend onto the interior of the

Figure 14-2 Modern exploration of Antarctica: *(A)* Sno-Cat used on overland exploration journeys; *(B)* air-transported exploration party lands near the Thiel Mts. in central Antarctica; *(C)* base camps are maintained beneath the ice surface in Marie Byrd Land; *(D)* detonating an explosive charge for seismic echo sounding of the thickness of ice; *(E)* glaciologists measuring snow hardness with a ramsonde penetrometer.

Glaciers and the Continental Landscape 509

Figure 14-3 Outlet glacier in the Transantarctic Mountains near Cape Hallett on the Ross Sea coast of Antarctica.

Figure 14-4 The ice sheet in West Antarctica flows into the Amundsen Sea over a coastal region of subdued relief and disintegrates into tabular icebergs.

510 The Face of the Earth

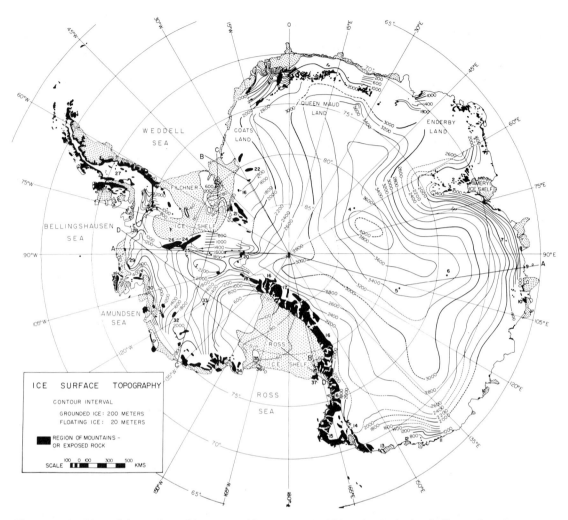

Figure 14-5 Map of Antarctica with topographic contours at 200-m intervals that indicate elevations on the surface of the ice sheet. Areas of exposed bedrock are shown in black. Stippled areas are floating ice shelves. (Courtesy of Charles R. Bentley, University of Wisconsin.)

continental platform. The southern reaches of these seas are permanently covered by floating **ice shelves.** Seismic reflections indicate that the shelf ice is between 300 and 600 m thick.

The Antarctic ice sheet averages more than 3000 m in thickness over a large part of the continent. Profiles in Figure 14-6 illustrate the

Figure 14-6 *(opposite)* Cross sections illustrating the shape of the Antarctic ice sheet and the underlying bedrock surface along lines indicated in Figure 14-5. (Courtesy of Charles R. Bentley, University of Wisconsin.)

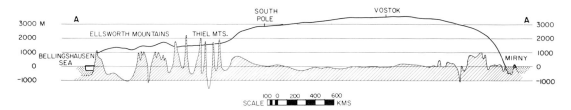

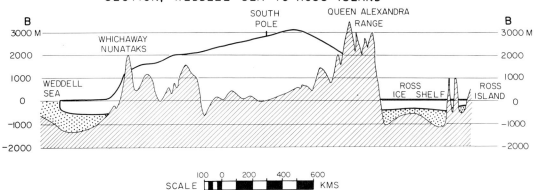

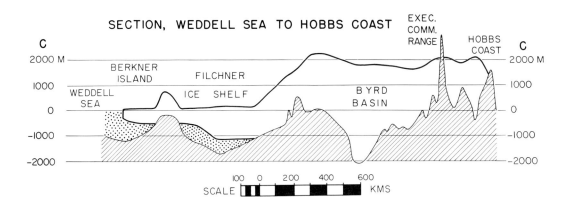

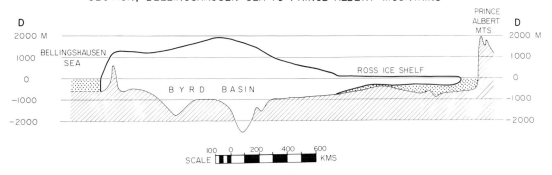

form and thickness of the ice sheet and the topography of the underlying bedrock surface. Without its ice cover Antarctica would appear much like other continents. The rock exposed in the mountain ranges that reach above the ice and seismic velocity data from beneath the glacier indicate a typically continental structure.

To the polar traveler the Antarctic ice plateau is almost featureless. Small grooves and ridges that we call **sastrugi** are carved in the wind-roughened surface (Figure 14-2A). In some areas this surface is a hard crust of snow, and elsewhere it is soft owing to a less severe prevailing wind. Snow falling on the plateau produces a surface layer with a density that is mostly between 0.3 and 0.4 gm/cm^3. As this layer becomes buried by additional snowfall, it becomes compressed by the overburden. Because of this compaction, density increases in the successivly deeper layers. Original snowflakes recrystallize into an increasingly compact mass of ice grains called **firn** or **névé**. At a depth of approximately 200 m pore space is virtually absent, and the mass of interlocking crystals reaches the density of about 0.91 gm/cm^3. Here the metamorphism of firn into pure ice is complete. Snow which is eventually incorporated into this ice mass builds up over the Antarctic continent at annual rates of less than 5 gm/cm^2 over a large inland area. This is equivalent to an annual rainfall of 5 cm (1.97 in.). It is obvious that compared with other continents these accumulation rates are so low that Antarctica must be considered an arid desert, albeit a frigid one.

Temperature conditions on the surface of the ice sheet remain perpetually below freezing except along the northernmost coastal margins. In the continental interior the temperature ranges from about $-15°$ C ($5°$ F) during the austral summer to below $-80°$ C ($-112°$ F) in the coldest times of winter. Conditions along the coasts are generally milder, usually remaining warmer than $-50°$ C during the winter months. Deep within the ice mass itself the temperature increases and is probably close to the freezing point at the base. This temperature increase is related to geothermal heat flow from within the earth and friction of ice sliding upon the bedrock base.

The interior of Antarctica is devoid of life except for occasional dormant spores and bacteria carried by wind, and those foolhardy travelers who venture away from the coasts. But marine life abounds in the seas bordering the continent. Various species of seals, penguins, and other birds such as skuas dwell part of the time on the coastal margin, never migrating far inland.

Practically all of the continent lies south of the antarctic circle. Consequently, there is an extended period of darkness during the austral winter during which time the sun does not rise above the horizon. The sea surface surrounding the continent freezes during this cold time to a distance of as much as 1000 km from the coast. This sheet of **pack ice** reaches a thickness of 1 to 2 m, and creates a formidable barrier to ship transport. The pack ice disintegrates in the austral summer, when for several weeks the sun remains perpetually above the horizon.

Such is the environment of the largest continental glacier. From this description we can tell how Canada and northern Eurasia probably appeared during former times of glaciation.

The Greenland Ice Sheet

The large island of Greenland is bordered mostly by mountains. The irregular coast is indented by numerous fjords. Except for these coastal regions the island is covered by an ice sheet. Like Antarctica, the interior of Greenland is an immense ice plateau reaching an elevation above 3200 m (almost 10,500 ft) near the center of the island. Contours in Figure 14-7 illustrate relief on this ice plateau.

Beneath the ice sheet the bedrock surface

Figure 14-7 Map of Greenland and other Arctic Islands where ice sheets exist. Topographic contours at 500-m intervals indicate elevations on the surface of the Greenland Ice Sheet. Areas of exposed bedrock are shown in black. (After H. Bader, The Greenland Ice Sheet, in *Cold Regions Science and Engineering* [F. J. Sanger, editor], U.S. Army Cold Regions Research and Engineering Laboratory Monograph, Part I, Section B2, 1961.)

of the interior of Greenland appears to be largely a lowland plain. Profiles prepared from seismic reflection surveys illustrate the thickness of ice and the character of the underlying bedrock surface (Figure 14-8). The continental glacier is seen to reach a maximum thickness of more than 3200 m. For the most part its base is near sea level.

The surface of the ice plateau is a wind-roughened plain of sastrugi similar to the antarctic plateau. Snow accumulating on this surface adds to the layer of firn, which becomes metamorphosed by compaction into a solid ice mass at depths of about 200 m. Mean annual snow accumulation exceeds 80 gm/cm^2 (equivalent to 80 cm of annual rainfall) near the southern margin of the ice cap. Accumulation decreases northward to a value as low as 15 gm/cm^2 over the northern interior part of the continental glacier. Here Greenland, like Antarctica, is a virtual desert.

Ice from the interior is discharged through relatively narrow coastal valleys into the sea, where it disintegrates into icebergs, some of which display exotic shapes. Large tabular ice islands similar to those found near the low coastal margins of Antarctica are not produced along the mountainous coast of Greenland.

The climate of Greenland is harsh, although not quite as severe as in Antarctica. Mean annual temperature on the ice plateau is near −30°C, but extreme winter temperatures of more than −65°C (−85°F) have been recorded. During the summer average temperatures of about −10°C are found in the interior. Within the ice mass, the temperature increases with depth reaching a value near the freezing point at the base.

Arctic pack ice 1 to 3 m thick develops around Greenland except along a small extent of the southwestern coast. This pack recedes during summer so that only the northernmost coast is ice locked.

The coastal margins of Greenland support a much more abundant and diversified fauna and flora than are found in Antarctica. Human settlements were established during the tenth century by Scandinavian colonists. From still earlier times nomadic Eskimo groups lived along the coastal margins of the island where the climate is somewhat milder than in the interior.

Other Modern Ice Sheets

Small ice sheets are found on some of the larger arctic islands. The distribution of these glaciers is shown in Figure 14-7. Glacier ice

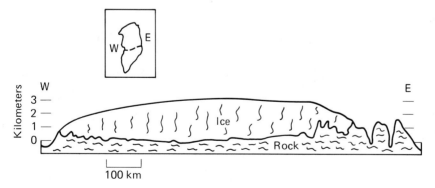

Figure 14-8 Cross section illustrating the shape of the ice sheet and the underlying bedrock surface along a line crossing Greenland. (Modified from M. Alain Joset and Jean Jacques Holtzscherer, *Annales de Geophysique,* Tome 10, page 365, 1954.)

flowing off Ellesmere Island forms a significant floating ice shelf, although not nearly as large as the Ross Shelf and the Filchner-Ronne Shelf in Antarctica. Large tabular ice islands more than 10 km long break from this shelf and drift with the currents of the Arctic Ocean. The best known of these are Fletcher's Ice Island (T3) and Arlis on which expeditionary groups can live and monitor conditions in the Arctic basin.

ICE STREAMS IN MOUNTAINOUS TERRAIN

Ice-filled valleys are part of the rugged landscape in many of the most spectacular mountain systems. Glaciers in these regions make up about 4 percent of the total volume of ice on the earth. The other 96 percent is contained in the vast ice sheets of the polar regions. Although the glaciers in mountainous terrain account for this relatively small proportion of the global ice mass, they have proven to be remarkably effective in shaping unique and scenic landscapes.

These glaciers are mostly streams of ice contained within valleys, so they are called **valley glaciers.** Some of them, for example the Argentiere Glacier in the Alps of France (Figure 14-9), are individual tongues of ice terminating within a valley. Others merge where valleys intersect and form still larger ice streams. The glacier shown in Figure 14-10 is confined to a large valley in Alaska and receives the flow of ice from bordering tributary valley glaciers. In some regions several valley glaciers flow onto a plain adjacent to the mountains and coalesce to form a larger **piedmont glacier.** An example is the Malaspina Piedmont Glacier (Figure 14-11) that borders the Saint Elias Range in Alaska.

Glaciologists have made detailed studies of many individual valley glaciers to discover the typical characteristics of these ice streams. They measure ice thickness by the seismic methods described in Chapter 3. The thickness of glacier ice can also be estimated from Bouguer gravity anomalies by the same methods used to estimate the thickness of unconsolidated sediment covering a valley (Figures 2-30 and 2-31). The smallest active valley glaciers are only a few hundred meters long and less than a few tens of meters thick. The largest ones are the great outlet glaciers of Antarctica and Greenland that extend more than 100 km and exceed 1000 m in thickness.

On most valley glaciers we can see surface features which indicate that the ice is moving. Observe in Figure 14-10 the linear patterns produced by variations in the volume and color of sediment carried by the glacier.

Figure 14-9 Argentiere Glacier is a valley glacier near Chamonix in the Alps of France. (From R. Vivian, *Les Glaciers des Alpes Occidentales,* page 49, Imprimerie Allier, Grenoble, 1975.)

516 The Face of the Earth

Figure 14-10 Valley glacier in the Chigmit Mountains of the Cook Inlet region of Alaska. Patterns on the glacier surface are the result of rock fragments frozen in the ice. (A. Post, U.S. Geological Survey.)

Look at the sets of **crevasses** that cut the surface of the Argentiere Glacier (Figure 14-9). These are vertical fractures produced by stresses that are related to ice movement. They can extend to depths of more than 50 m. Some crevasses become hidden by a thin cover of snow that can collapse if it is stepped on.

How fast do glaciers move? We can find out by placing markers on a glacier surface. Then we survey these markers repeatedly from fixed positions on land bordering the ice. In some areas aerial photo surveys are repeated periodically to detect changes that reveal movement. These kinds of measurements indicate that the rate of movement varies considerably from one glacier to another.

The flow of some is almost imperceptible, whereas others flow faster than 10 m/day. Most glaciers experience seasonal changes in flow velocity, but ordinarily the average annual velocity changes only slowly in response to changes in climate.

The motion that is measured on the upper surface is produced by the glacier sliding over its bed and by plastic flow within ice mass. In most glaciers bottom sliding accounts for a large part of the movement. Glaciologists can measure bottom sliding of the Argentiere Glacier in an interesting way. A cavity not filled by ice was discovered beneath the glacier. Here apparatus is operated to record ice movement continuously (Figure 14-12). The results indicate that this glacier is sliding

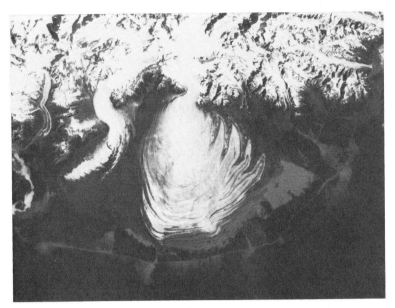

Figure 14-11 The Malaspina piedmont glacier bordering the Saint Elias Range in Alaska. Patterns on the ice surface are the result of rock debris frozen in the glacier. (LANDSAT Photograph, U.S. Geological Survey.)

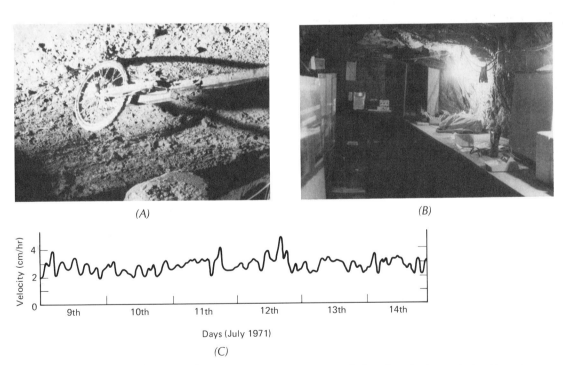

Figure 14-12 Velocity of bottom sliding can be measured in a subglacial cavity beneath the Argentiere Glacier in France. The equipment includes (A) a frame mounted on bedrock that holds a wheel against the base of the glacier ice. (B) Movement of the wheel is recorded electronically in a nearby tunnel. (C) Record of the movement of the Argentiere Glacier by bottom sliding. (Data from R. Vivian, *Les Glaciers des Alpes Occidentales,* page 50, Imprimerie Allier, Grenoble, 1975.)

over its bed at a rate of a few centimeters per hour.

Glacier movement caused by plastic flow depends upon the inclination of the glacier surface, ice temperature, and other physical properties of the ice. Ordinarily the velocity of movement related to plastic flow is greatest at the surface near the center and decreases toward the sides and bottom of the glacier (Figure 14-13). In most glaciers the movement related to plastic flow does not exceed a few tens of meters per year. The nature of plastic flow is discussed in more detail in another part of this chapter.

Most glaciers seem to have reliable habits. From year to year the flow of ice is quite steady, but this is not true for a relatively small number of glaciers. These glaciers undergo sudden and unexpected paroxysms of movement that are called **glacial surges.** Prior to surging some of these glaciers remained almost quiescent for several decades. Then during brief periods of a few weeks they can accelerate to velocities of perhaps 10 m/day or even as fast as 100 m/day. A surge can continue for a year or more. Then the exhausted glacier lapses into quiescence. Its surface becomes a chaotic jumble of ice blocks and twisted crevasses that give it a shattered appearance. What triggers a glacial surge? When can one be expected? During a period of apparent quiescence stresses are building up until a threshold of imbalance is reached. Johannes Weertman, a noted glaciologist, suggests that surges are caused by an increase in the amount of water along the bed of a glacier that provides lubrication needed to accelerate bottom sliding. But so far we have no comprehensive explanation of this phenomenon, and we cannot predict when a surge will begin or how long it will last.

Valley glaciers are an important part of the mountainous landscape in many parts of the world, but they are by no means enduring features. The photographs in Figure 14-14 illustrate dramatically how rapidly glaciers can

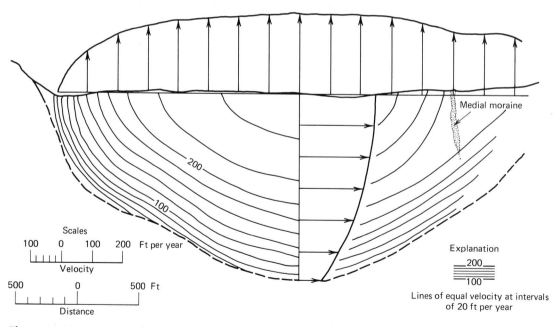

Figure 14-13 Average velocity of ice movement in the Saskatchewan Glacier based upon measurements of markers on the glacier surface and calculations of plastic flow in the ice. (After M. F. Meier, *Professional Paper 351*, page 48, U.S. Geological Survey, 1960.)

Figure 14-14 Retreat of the Hintereisferner and its tributary the Kesselwandferner, two ice streams in the Austrian Alps near the Italian border. Photographs (A–D) made at four different times from the same location indicate changes. (Courtesy of H. Hoinkes, University of Innsbruck, Austria.)

waste away and perhaps eventually disappear. During the past century a large number of valley glaciers have grown noticeably smaller. However, some evidence suggests that this trend is changing. What are the factors that affect the size and equilibrium of a glacier?

MASS BALANCE IN A GLACIER

Glaciers grow or decline in size because of imbalances between the accumulation and loss of mass. The factors bearing on mass balance include accumulation and loss at the surface, and transfer of ice at depth by plastic flow and bottom sliding. Under ideal conditions a glacier presumably can maintain its size and shape in a state of dynamic equilibrium. Alteration of these conditions causes instability and subsequent changes in the size of the glacier.

Accumulation and Ablation

All processes by which mass is added to a glacier contribute to **accumulation.** Snowfall, perhaps rain, avalanches of snow from higher on the sides of a valley, and **rime** formed by crystallization of water vapor from the atmosphere upon contact with the glacier surface all add to accumulation. **Ablation** includes all processes by which mass is removed from a glacier. Melting and runoff, sublimation, wind erosion, and iceberg **calving** are the primary ablation processes. Calving occurs when a mass of floating ice splits away from the glacier to form a drifting ice berg.

At a particular location on a glacier accumulation may exceed ablation during part of the year. During this time the thickness of snow and firn increases. At other times of the year ablation predominates and the glacier thickness decreases.

During a complete year the accumulated mass over part of the glacier exceeds mass loss by ablation. This part of the glacier is the **accumulation zone.** Elsewhere over the **ablation zone** of the glacier annual ablation is larger than annual accumulation. The boundary separating the accumulation and ablation zones is the **equilibrium line.** Zones of accumulation and ablation can be clearly recognized on valley glaciers in temperate regions. The distinction is less obvious on polar glaciers. On a typical valley glacier the accumulation zone lies upstream where the weather is cooler than the weather on the lower downstream ablation zone. There are no ablation areas of significant size on the Antarctic continent. Here the equilibrium line is close to the coast, and calving of icebergs from the margins of the continental glacier is the important ablation process.

Ice Movement

Mass is added annually to the surface of the accumulation zone, and it is removed from the surface of the ablation zone. Consequently, even if accumulation and ablation are in balance and total mass is preserved, the shape of the glacier would change markedly if the ice were otherwise static. The surface of the accumulation zone would continue to rise while the ablation surface was being lowered. This change is counteracted by mass transport of ice by plastic flow and bottom sliding. Under ideal conditions the glacier could maintain its total mass and shape in dynamic equilibrium if the ice mass flowing from beneath the accumulation zone into the ablation zone was equal to the net mass added to the accumulation zone.

A simple example will help us to understand how a glacier flows. Imagine a sheet of ice of uniform thickness resting on a bed that is inclined at an angle α (Figure 14-15). Plastic flow within the ice sheet can be represented by the movement of infinitesimally thin, parallel laminations. This is a laminar flow process in which each lamination slides over another lamination directly beneath it. Observe how these movements displace volume elements in a column of ice. Each displacement is produced by a shear stress (τ) that acts in the direction of downward inclination along the surface between two laminations. The shear stress is:

$$\tau = \rho g z \sin \alpha \qquad (14\text{-}1)$$

where ρ is ice density, g is gravity, and z is depth below the ice surface. If each lamination has a thickness Δz, then the displacement (Δd) can be expressed in terms of the shear stress, the effective viscosity of ice (η), and the increment of time (Δt) required to produce the displacement:

$$\Delta d = \Delta z \Delta t \, (\tau/\eta)^m \qquad (14\text{-}2)$$

where $m \simeq 3$ is a constant determined in laboratory experiments. This formula is called a *flow law for ice.* Recall from our discussion in Chapter 5 that effective viscosity is a physical property used to describe the flow of fluids and plastic substances. In ice this property changes with temperature.

Observe in Figure 14-15 that the size of an individual displacement (Δd) increases with depth. We can understand this from (a) Eq.

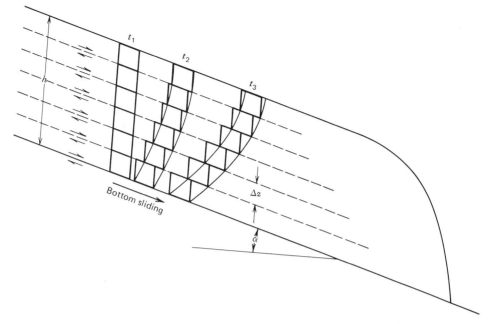

Figure 14-15 Movement of an inclined layer of ice by bottom sliding and plastic shear strain. Position of an ice column, vertical at time t_1, is shown to be displaced and deformed at later times t_2 and t_3. Movement is represented by displacements of infinitesimally thin laminations in the ice.

(14-2) which indicates that Δd will increase if τ increases, and (b) Eq. (14-1) which shows that τ increases with depth (z). But what do these equations tell us about the velocity of glacier flow? By rearranging Eq. (14-2) we find the relative velocity (Δv) at which one lamination slides over another directly beneath it: $\Delta v = \Delta d/\Delta t = \Delta z(\tau/\eta)^m$. But the velocity ($v$) of that lamination will be the sum of the relative velocities between all of the laminations lying beneath it. This summation is done by the mathematical process of integration to obtain an equation for the velocity (v_z) at depth z in the glacier that is produced by plastic flow:

$$v_z \simeq K(h-z)^4 \sin^3 \alpha \qquad (14\text{-}3)$$

where h is depth to the glacier bed, and $K \simeq 0.68 \times 10^{-18}$ is a constant related to gravity, ice density, and ice viscosity. Now, suppose that a glacier 300 m thick is flowing in a valley inclined at a 3° angle. We can calculate that at the surface where $z = 0$ the velocity $v_z = 8 \times 10^{-5}$ cm/sec or about 25 m/year.

A glacier moves by bottom sliding as well as by plastic flow. The velocity of bottom sliding depends on the balance between the shear stress at the base of the glacier and the stress of friction that acts to retard movement of ice over the bed. Shear stress, $\tau_b = \rho g h \sin \alpha$, is not difficult to calculate. But ordinarily we cannot obtain an accurate estimate of friction on the rough irregular bed of a glacier. Therefore the velocity of bottom sliding can only be inferred. This is done by subtracting the contribution of plastic flow from the velocity measured at the glacier surface by surveying markers.

Laminar flow lines in the simple example in Figure 14-15 are all parallel to the plane of the base. In nature it is unlikely that the glacier will rest on a plane surface. Rather, the

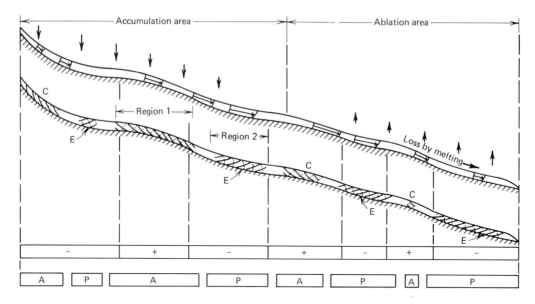

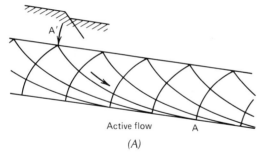

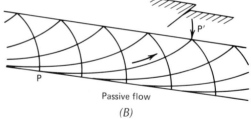

E - extension
C - compression

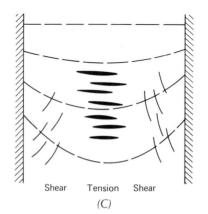

Figure 14-16 Laminar flow trajectories corresponding to *(A)* active flow and *(B)* passive flow, and *(C)* crevasse patterns in an idealized ice stream. (After J. F. Nye, *Journal of Glaciology*, page 88, November 1952.)

base will have some irregularities. Professor John F. Nye analyzed the stresses that would theoretically be expected in a glacier flowing over an undulating and inclined surface. He described the patterns of laminar flow that are summarized in Figure 14-16. Where the glacier bed is convex upward the ice undergoes **active flow.** Here the upper laminations of the ice layer slide downward over the lower laminations along infinitesimally close slip surfaces that are similar to normal faults. Active flow can produce crevasses close to the glacier surface. Here the ice is brittle. But at depth below a few tens of meters the ice becomes plastic. Elsewhere on the glacier where the bed is concave upward, the ice undergoes **passive flow.** Upper laminations are pushed over underlying laminations along slip surfaces that are similar to thrust faults. Crevasses do not form from passive flow.

The diagrams in Figures 14-15 and 14-16 illustrate the important features of laminar flow in a plastic ice mass and the movement of a glacier by bottom sliding. The plastic flow component of velocity is most important in slowly moving ice streams. From our knowledge of the physical properties of ice, however, we know that plastic flow is unlikely to produce movement of more than a few tens of meters per year. Bottom sliding is the most important kind of motion in more rapidly advancing ice streams. Both of these kinds of movement must be considered when calculating the flow of ice from beneath the accumulation zone to the ablation zone of a glacier.

Laminar flow processes are also active in continental ice sheets such as those now covering Antarctica and Greenland. To get some idea about movement within the ice sheets Professor Nye calculated the effects of plastic flow in a hypothetical circular ice sheet resting on a horizontal bed (Figure 14-17). Without going into the details of these calculations we can examine the results. If the ice sheet is

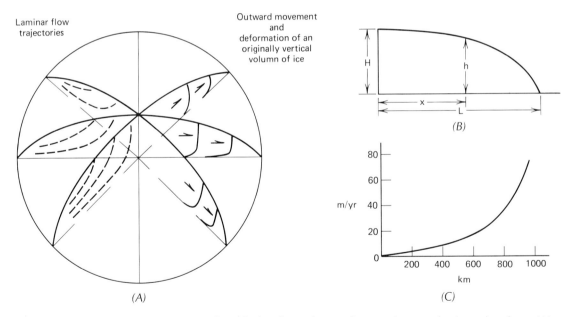

Figure 14-17 Shape and movement of an idealized circular ice sheet resting on a horizontal surface. *(A)* Dashed lines are flow paths of individual particles of ice. Solid lines followed by short arrows indicate velocity variation with depth. *(B)* Cross section illustrating the equilibrium shape of an ice sheet according to eq. (14-4). The shape is close to being a parabola. *(C)* Curve showing the velocity of ice at the surface of the ice sheet as a function of distance from the center. (Prepared from information in J. F. Nye, *Journal of Glaciology,* pages 493–507, October 1959.)

524 The Face of the Earth

in equilibrium so that its shape and its mass remain constant, then the thickness (*h*) at distance (*x*) from the center is related to the radius (*L*) of the glacier, and the maximum thickness (*H*) at its center:

$$\left(\frac{h}{H}\right)^{2+1/n} + \left(\frac{x}{L}\right)^{1+1/n} = 1 \qquad (14\text{-}4)$$

where the constant (*n*) appears to have a value slightly smaller than 2. This indicates that the shape of the glacier cross section (Figure 14-17 B) is not too different from a parabola. Particles of ice should move along the curved paths indicated by dotted lines in Figure 14-17 A. The horizontal component of velocity of such a particle depends upon its distance from the center of the glacier. Values are given in Figure 14-17 C for a glacier with a 2000 km diameter. They indicate that a particle of ice beginning at a location 50 km from the center would move to the outer edge of the ice sheet in about 140,000 years.

We have learned from recent studies in Antarctica that the slow, steady plastic flow illustrated in Figure 14-17 is not the only movement occurring in a large ice sheet. From time to time parts of the ice sheet can surge. This happens where accumulations of water apparently produced by pressure melting lubricate the base of the ice sheet. This greatly enhances bottom sliding. Here streams of ice a few tens of kilometers wide and hundreds of kilometers long flow at rates of more than one meter per day. On either side of such a surging stream the ice remains practically motionless. Only recently have glaciologists realized that surging processes are an important factor in the movement of ice in continental glaciers. We do not have a clear understanding of the mechanism that causes a surge. Flow processes within a large ice sheet appear to be complicated indeed!

Look again at Figure 14-4. Observe what happens where an ice sheet reaches the ocean. Here there is no friction at the base of the floating ice. Once afloat the ice rapidly thins by plastic flow. It spreads out over the water forming a floating ice shelf. Upon thinning to less than a few hundred meters the ice shelf begins to disintegrate under stress from wind and ocean currents. It breaks into large drifting ice islands and smaller icebergs. Ice shelves tend to be unstable except where they are contained within bays and small seas. Because of the plasticity and low breaking strength of ice, glaciers are restricted almost entirely to the continents. They can extend for only a short distance over the bordering sea before becoming unstable. There are no glaciers in the ocean basins.

Advance and Retreat of Glaciers

The size of a glacier is almost always changing. If the ideal balance of accumulation, ablation, and ice movement is ever realized, it probably persists only for a short period of time. When mass is not preserved, a glacier either expands over a larger area or retreats by melting as the example in Figure 14-14 illustrates. Under certain conditions these adjustments cause the glacier to become more stable in its environment. In other circumstances the adjustments can make the glacier increasingly unstable.

Factors controlling the location of the equilibrium line on a glacier affect the stability of the ice mass. Mean annual temperature and the seasonal variation of temperature influence its position. Mean annual temperature tends to decrease with increasing latitude and increasing elevation. Atmospheric precipitation rates, which tend to be higher on coastal areas than on continental interiors, also influence the position of the equilibrium line on a glacier. Bearing in mind all of these factors, we can understand why equilibrium lines are high in equatorial continental interiors and are lower in areas of higher latitude and nearer to the sea (Table 14-1).

Let us consider the response of a glacier to a change in mass balance. Suppose first that

Table 14-1
Latitude, Precipitation, and Equilibrium Line Elevation for a Worldwide Distribution of Glaciers

Location	Latitude	Precipitation (water equivalent in cm)	Equilibrium Line Elevation in m
Equador	0°	150	5800
East Africa	3°N	200	5200
New Guinea	4°S	150	4900
Columbia	11°N	200	4550 to 4900
Tibet	29°N	150	5500 to 6100
Pyrennes, France	41°N	200	2680 to 2800
Wyoming, USA	45°N	50	3350
Alps, Switzerland	47°N	200	2450 to 2900
Scandinavia	71°N	25	900
Franz Josef Land	80°N	25	300

Data taken from R. F. Flint, Glacial Geology and the Pleistocene Epoch. John Wiley & Sons, 1947.

factors causing this change do not affect the location of the equilibrium line, which remains the same distance from the sea or at the same latitude as before the change in mass balance. This implies that the area of the accumulation zone remains constant. The size of the ablation zone will then increase by ice flow from beneath the accumulation zone, or it will decrease by melting until a size is reached wherein ablation balances accumulation. This kind of adjustment tends to make the glacier more stable in its environment.

Suppose next that a change in the environment alters the area of the accumulation zone by changing the position of the equilibrium line to a different latitude or a different distance from the coast. The glacier will respond to such a change by expanding or decreasing the area of the ablation zone so as to reestablish the balance of accumulation and ablation. The adjustment will tend to make the glacier more stable in the changed environment. Adjustments tending to stabilize a glacier in a changing environment are illustrated in Figure 14-18.

Let us now consider the possibility that elevation is the dominant factor controlling the position of the equilibrium line (Figure 14-19). Assume that latitude and distance from the sea are unimportant. Suppose that a change in environment causes increased precipitation of snow. This additional snow raises the elevation of the glacier surface and thereby expands the area of the accumulation zone (Figure 14-19A). The glacier will then flow out onto a larger area. For the glacier to maintain an equilibrium cross-sectional shape

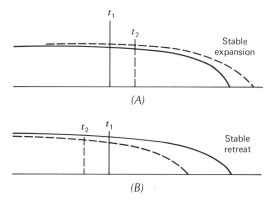

Figure 14-18 Adjustments of a glacier in response to change in location of the equilibrium line caused by factors independent of the glacier, which (A) expands or (B) recedes toward a position of equilibrium.

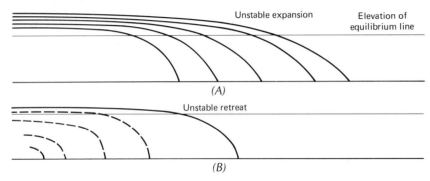

Figure 14-19 Adjustments of a glacier to change in mass balance where the equilibrium line is determined by elevation. *(A)* Change in the glacier shape causes change in position of the equilibrium lines. *(B)* Continued adjustment of the ice mass makes the glacier progressively more unstable.

such as the one illustrated in Figure 14-17, however, it must become thicker as it expands. If it becomes thicker, its accumulation area grows still larger. This in turn causes it to expand still farther. It becomes impossible for the glacier both to dispose of the ever-increasing accumulation and to establish an equilibrium shape. It becomes more and more unstable until it grows large enough to alter the climate that initiated its growth.

A climate change causing a decrease in the rate of snow precipitation will cause a glacier to become increasingly unstable if the equilibrium line remains at a fixed elevation. The glacier will retreat in response to the decrease in accumulation. As it retreats it must also become thinner (Figure 14-19*B*). But as it becomes thinner, the surface elevation is lowered and the accumulation area grows still smaller. This in turn causes further retreat and additional thinning. The glacier becomes more and more unstable and retreats at an ever increasing rate until it completely wastes away.

GLACIATION

Alteration of the land surface caused by glaciers is called **glaciation**. This includes erosion of bedrock and regolith and deposition of the debris transported by glaciers. We can recognize distinctive landscape features and sedimentary deposits that are characteristic of a **glaciated landscape** where the land is no longer covered by ice. The following discussion is concerned with the processes of glaciation that produce these features and deposits.

The Debris Load of a Glacier

All glaciers carry a debris load of rock fragments. In some valley glaciers these fragments are distributed throughout the ice. They tend to be concentrated nearer the base where the debris load can account for more than 50 percent of the moving mass. Observe the sediment near the glacier surface that is clearly evident in Figures 14-10 and 14-11. On large ice sheets the debris load is almost entirely confined within a few hundred meters of the base. The typical distribution of the debris load in a valley glacier and an ice sheet is illustrated in Figure 14-20.

Fragments making up the debris load range in dimensions from clay-sized particles to boulders several meters in diameter. Glaciers move primarily by sliding at the base. The moving ice mass exerts tremendous stress on

Glaciers and the Continental Landscape 527

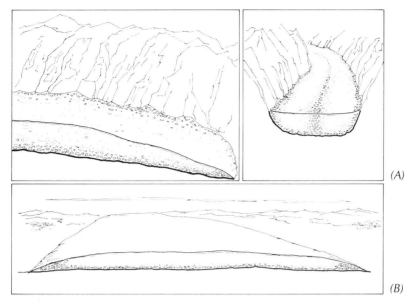

Figure 14-20 Distribution of the debris load in (A) a valley glacier and (B) an ice sheet.

the bed and dislodges some fragments of bedrock which become imbedded in the glacier. Other fragments are ground into a fine-grained powder. Part of this material is drawn farther into the glacier by mechanisms such as **passive flow** in the plastic mass (Figure 14-16). The process by which rock is lifted from the bed and incorporated into the debris load is called **glacial plucking.** Some fragments that have become frozen to the base continue to scrape the bed and make the moving glacier even more abrasive.

The debris load is ordinarily most concentrated in ice nearest the contact of the glacier and its bed. There are ways, however, by which it can accumulate elsewhere in an ice stream. Where valley glaciers join (Figure 14-10), fragments along the merging sides of each ice stream form a concentration of debris nearer the center of the combined ice stream. On other glaciers debris from rock slides and wind-transported sediment accumulates on the surface. The extent to which debris from the base and surface is further distributed within the glacier by plastic flow is uncertain, because the mechanisms involved are not clearly understood.

Boulders, cobbles, and finer sediment from almost everywhere along the bed of a glacier are steadily added to the moving debris load (Figure 14-21). This mass is continuously transported toward the terminus regardless of whether the glacier is advancing or retreating. The **terminus** is the outer margin of an ice sheet or the lower end of a valley glacier. In an advancing glacier the terminus itself is moving outward because the discharge of flowing ice exceeds the rate of melting. Retreat of the terminus results when the rate of melting exceeds the discharge of flowing ice. In either situation the ice itself always flows toward the terminus. Here it melts and the debris load is deposited. The glacier, then, acts as a conveyor belt. This effect is illustrated in Figure 14-21. Sediment accumulated along the base of the glacier or on its surface is moved to the terminus where it is discharged from the melting ice.

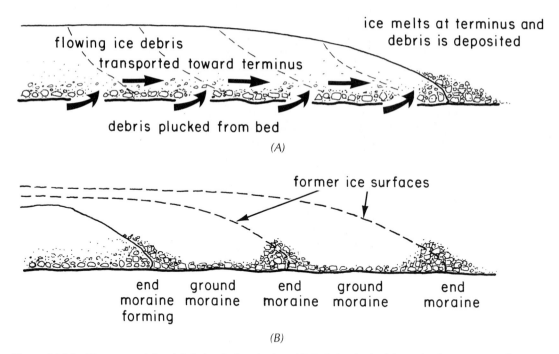

Figure 14-21 Transport of the debris load of a glacier. (A) Debris plucked from the base is moved toward the terminus where it is deposited as the ice melts. (B) Formation of end moraines and ground moraine as the terminus remains first in one position then retreats to other positions where it remains temporarily stable.

Associated with ice transport of sediment is the fluvial transport by melt-water streams. Some streams flow over the surface of the ablation zone, dislodging and transporting sediment that was formerly frozen in the ice. Other melt-water streams form at the base of the ablation zone and flow out from beneath the glacier along its terminus. These streams remove sediment that was formerly frozen in the basal zone of the glacier, and they further erode the bed. As a result of these melt-water streams, the fluvial deposits become interbedded with the debris load released directly by melting ice near the terminus of the glacier.

Deposits formed from the debris load of glaciers and associated melt-water streams are referred to as **glacial drift.** One kind of glacial drift is a completely unsorted and heterogeneous mixture that is called **till.** The other variety of drift possesses stratification that is revealed by sedimentary structures such as lamination, ripple bedding, cross bedding, and imbrication. Recall that these sedimentary structures are described in Chapter 9.

Erosional Features of Glaciation

Sliding glaciers made more abrasive by the rock debris frozen in the basal ice can eventually carve a distinctive erosional landscape. Local effects of glacial erosion are grooves and striations cut in glaciated bedrock surfaces. Larger knolls of bedrock become streamlined in the direction of ice movement (Figure 14-22). We can use orientation of features such as these to surmise the former di-

(A)

(B)

Figure 14-22 (A) A knoll of bedrock rounded and streamlined by glacial erosion, and (B) striations carved in a bedrock surface by glacial erosion on Isle Royale National Park in Michigan. (N. K. Huber, U.S. Geological Survey.)

rection of ice movement over a glaciated landscape.

Several landscape features are typical of mountainous terrain that has been subjected to prolonged valley glacier erosion. Valleys become U-shaped in cross section (Figure 14-23). A thick ice stream which almost fills a valley can receive the flow of ice from tribu-

(A)   (B)

Figure 14-23 Valleys of U-shaped cross section are carved by glacial erosion: (A) Little Cottonwood Canyon in the Wasatch Mountains near Salt Lake City, Utah; (B) hanging glacial valley in Glacier National Park, Montana.

Figure 14-24 Landforms of glacial origin in a mountainous terrain near Mount Stuart in the Wenatchee Range of Washington. Cirques, horns, aretes, and U-shape valleys are all part of the scenery here. (A. Post, U.S. Geological Survey.)

Glaciers and the Continental Landscape   531

Figure 14-25 A drumlin in Wayne County, New York. (G. K. Gilbert, U.S. Geological Survey.)

tary glaciers that merge at high elevations along the larger valley side. After retreat of the ice these higher glaciated valleys appear truncated where they merge with the deeper valley carved by the main ice stream. The trunctated tributary valleys are described as **hanging valleys** and are also typically U-shaped in cross section (figure 14-23 B).

Large bowl-shaped features that open into U-shaped valleys are characteristic of the glaciated mountainous landscape. These landforms are called **cirques.** As glacial erosion continues in a cirque basin, the steep headwall is cut deeper into a mountain mass. As growing cirques on opposite sides of a mountain gradually merge, a steep ridge known as an **arête** is formed. Where more than two cirques expand toward a common point, the eventual remnant of rock separating them is described as a **horn.** These different erosional features combine to form the spectacular scenery of a glaciated mountainous region (Figure 14-24).

Glaciated landscapes of subdued relief are marked by erosional features carved by ice sheets. These features are formed in bedrock terrane and in areas covered by drift which were subsequently overridden by advancing ice. Widely distributed in northern North America and Europe are **drumlins.** These are streamlined hills that are elongate in the direction of ice movement. An individual drumlin pictured in Figure 14-25. A large number of these hills are grouped in central Wisconsin. Drumlins are carved from a combination of drift and bedrock, or entirely from drift.

On lowland areas such as the Canadian

532   The Face of the Earth

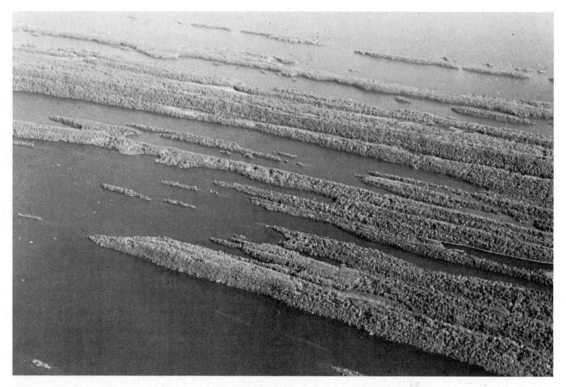

Figure 14-26 Bedrock terrain on Isle Royale National Park shows effects of glacial erosion. Upturned edges of Precambrian basalt layers were carved into parallel ridges by the movement of the North American ice sheet of late Quaternary time. (N. K. Huber, U.S. Geological Survey.)

Shield where bedrock exposures are abundant, the locally striated and rounded knobs and knolls of rock indicate erosion by ice. We can see the effects of glacial erosion on the broad bedrock surface pictured in Figure 14-26.

Depositional Features of Glaciation

The important depositional landforms resulting from glaciation are **moraines** and **glaciofluvial** features. Moraines are formed from glacial till, and glaciofluvial features consist primarily of stratified drift.

Moraine ridges form along the margins of a glacier. When the terminus of the glacier remains in a more or less stationary position for an extended period of time, the debris load conveyed by moving ice continues to be discharged along this margin. Eventually a ridge of till called an **end moraine** is formed (Figure 14-21). An end moraine can also be formed by an advancing glacier which plows and moves a ridge of regolith and bedrock fragments along its terminus. Here the glacier is similar to a bullldozer in its capacity to move rock and soil. During a succession of advances and retreats a glacier can form several end moraines. The ridge indicating the farthest advance is the **terminal moraine.**

A relatively thin and uniform layer of till covers some parts of a glaciated terrain. This is a **ground moraine.** It can form during the

Glaciers and the Continental Landscape 533

withdrawal of a glacier when the terminus retreats too rapidly for a ridge of till to accumulate. Rather, the debris load is spread in a thin layer over the area from which the ice is receding (Figure 14-21). The distribution of end moraines and intervening ground moraine in Ohio is displayed on the glacial geologic map in Figure 14-27.

Ridges called **lateral moraines** and **medial moraines** are found in glaciated valleys. The lateral moraines form along the valley sides during the retreat of the ice stream. Medial moraines are deposited by receding glaciers in which debris carried from the margins of merging ice streams has become concentrated nearer the center of the larger glacier. The streams of debris so obvious in Figure 14-10 would form medial and lateral ridges if the glacier eventually melts.

Glaciers transport large boulders that are eventually deposited as glacial **erratics.** Boulders such as the one shown in Figure 14-28 are common in glaciated areas. Continental ice sheets have transported large erratics many tens or even hundreds of kilometers from their original bedrock locations.

Glaciofluvial landscape features are formed by glacial melt-water streams. These streams form on the ice surface and along the bed beneath the glacier. Some of the sediment transported by these streams is deposited in ponds on the glacier surface, along the bed underneath, and against the margins of the glacier. **Ice-contact** features are eventually formed from these deposits which accumulated partly against a surface of ice. Some sediment is transported away from the ice mass by melt-water streams. **Outwash** features are formed by deposition of this debris. Ice-contact features and outwash features display sedimentary structures which indicate that the sediment was deposited in small lakes and along stream beds.

Sinuous ridges of stratified drift are found in glaciated regions. These ridges are ice-contact features called **eskers.** A typical esker is pictured in Figure 14-29. We believe that these features were formed from subglacial stream deposits. They stand higher than the surrounding landscape because the stream flowed in an ice tunnel on the glacier bed. Ice formed the sides of the stream channel which therefore rose above the bed.

Mounds of stratified drift known as **kames** are another kind of ice-contact feature. Kames appear to have been produced from sediment that was deposited in ponds on the surface of the ablation zone or as alluvial cones along the glacier margin. After retreat of the ice these localized accumulations of sediment remained as isolated mounds. Long ridges of stratified drift along the sides of glaciated valleys are called **kame terraces.** These are supposedly the deposits of streams flowing along the sides of a valley glacier where the channel is against rock on one side and ice on the other side.

Local pitlike depressions in a terrane of till and stratified drift are called **kettles**. They appear to have developed during the final stages of glacial retreat. When melting was almost complete, isolated patches of ice were all that remained of a former ice sheet. Some of these isolated ice masses became covered by drift. As they eventually melted and produced ground water that seeped away through the porous debris, the drift surface settled into a depression. In some places **kettle lakes** remain in these depressions.

Merging with glacial moraines and ice-contact features are broad alluvial outwash plains. Much of the debris originally transported by ice is carried much further by the streams flowing well beyond the limits of the glacier. Some of the outwash debris was removed directly from the ice, whereas other portions came from stream erosion of bordering moraines and ice-contact landforms. The various landforms illustrated in Figure 14-30 combine to make up the typical glaciated landscape.

After repeated advance and retreat of an ice sheet the landscape becomes blanketed by a heterogeneous assortment of interbed-

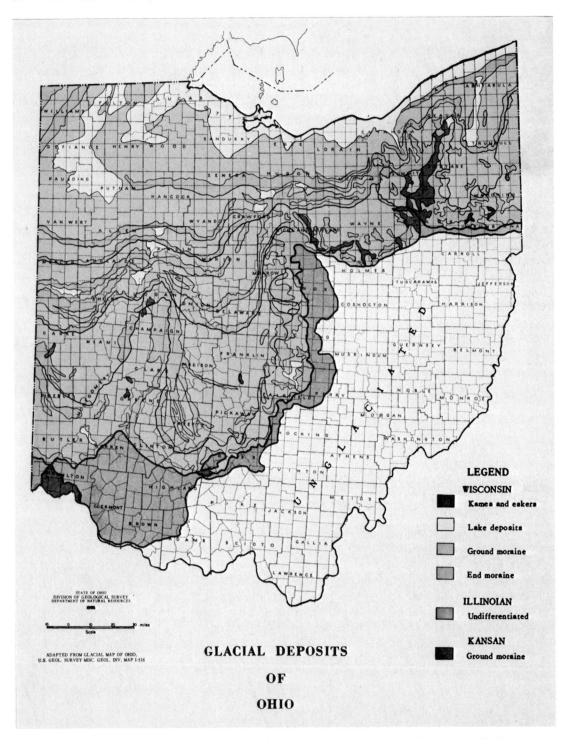

Figure 14-27 Map showing the distribution of glacial deposits in Ohio. (*Glacial Deposits of Ohio*, Division of Geological Survey, State of Ohio, 1966.)

Glaciers and the Continental Landscape   535

Figure 14-28 Large glacial erratic near Hardwick, Massachusetts, rests on a surface of glacial drift. (W. C. Alden, U.S. Geological Survey.)

Figure 14-29 An esker in Dodge County, Wisconsin, appears to have been deposited by a melt-water stream flowing in a tunnel at the base of a continental glacier. (W. C. Alden, U.S. Geological Survey.)

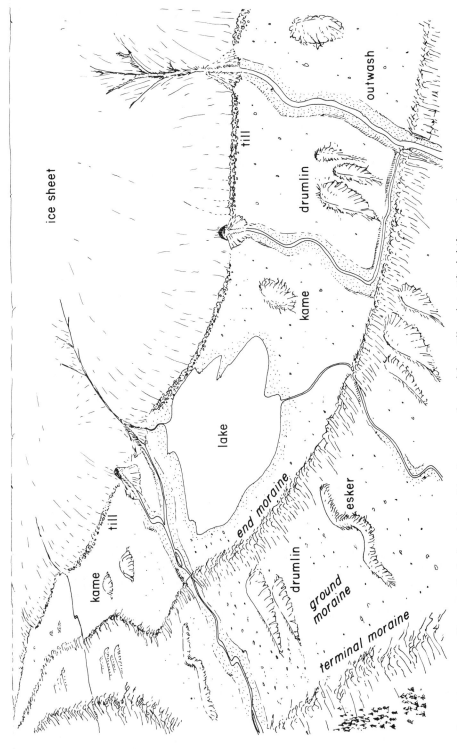

Figure 14-30 Landforms typical of a glaciated landscape are formed from till and stratified drift as an ice sheet recedes.

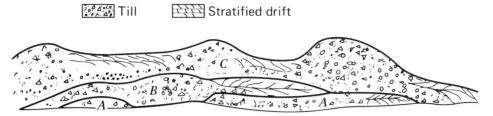

Figure 14-31 Hypothetical cross section illustrating interbedded glacial till and stratified drift. Deposits *(A)* of the earliest glacial stage were later overridden by ice and subsequently covered by drift *(B)* from a later glacial stage. Still later ice expanded over the region again, and as the ice of this final glacial stage receded, the deposits *(C)* were laid down.

ded glacial sediments. The moraines and stratified drift features produced by retreat are partially, but not completely, obliterated by subsequent ice advance. Another assortment of landforms develops above the remnants of the former glaciated landscape when the ice once again recedes. The cross section in Figure 14-31 illustrates a hypothetical assemblage of interbedded glacial deposits which could accumulate during multiple glaciations of a region. These kinds of deposits are distributed over northern North America and Eurasia.

Frozen Ground

Porous regolith and bedrock close to the earth's surface in polar regions is commonly saturated with ice rather than ground water. In some places this frozen ground is seasonal, and some summer melting occurs. Elsewhere it remains frozen the year around and is called **permafrost.** Where some seasonal freezing and thawing occurs irregular hummocks can be formed by frost heaving. Recall from Chapter 13 the effect of this process on mass wastage. Perhaps the most dramatic landscape features formed in frozen ground are **pingos.** They form either irregular mounds and ridges, or in more regular conical shapes (Figure 14-32). The development of pingos is not clearly understood. Some are believed to result from hydraulic pressure in ground water caused by encroaching zones of frozen ground. The tops of some pingos are craterlike and yield water from fissures during seasonal thaws.

Roadbeds and building foundations on frozen ground must be carefully designed to resist damage from seasonal changes. Local heating and pressure from these structures may partially melt the otherwise frozen permafrost. This can stimulate worse frost heaving than is observed on the undisturbed countryside. Scientists and engineers have been very concerned about the effect of the Alaskan pipeline on arctic permafrost. If heat from the pipeline were to cause melting, the support structures might become unstable and allow breaks in the pipeline. Development of polar regions poses engineering problems not encountered in more temperate climates.

ICE AGES

The present ice sheets of Antarctica and Greenland together with smaller glacier systems elsewhere cover more than 10 percent of earth's land surface. But glacial drift of late Cenozoic age covers a much larger area. These deposits indicate the former advance of

538 The Face of the Earth

Figure 14-32 Pingo located in the Yukon region of Alaska. (G. W. Holmes, U.S. Geological Survey.)

ice over the regions shown in Figure 14-33 which make up more than 30 percent of the present global land area. Evidence from glaciated areas and other regions as well reveal not one, but a succession of continental ice sheets which advanced, then receded, during late Cenozoic time. The different advances of ice represent stages in the present ice age which probably began late in the Tertiary period. Prior to this, evidence of widespread glaciation is found in tillites of Paleozoic age and of late Precambrian age. **Tillite** is a sedimentary rock produced by compaction and cementation of glacial till.

The Late Cenozoic Ice Age

The limits of ice advance during different stages of the late Cenozoic ice age are marked by moraines. Successions of end moraines such as those indicated in Figure 14-27 were formed during substages of the final retreat of the most recent North American ice sheet. When this continental glacier spread over the land, overriding a formerly glaciated surface, it partially erased the evidence of earlier ice sheet advances. The deposits of these previous stages become more and more fragmentary as we look farther back in time. Nevertheless, tills and stratified drift from several stages can be identified beneath and bordering the more recent deposits. These different drift deposits can be distinguished from one another partly on the basis of the ages of buried plant material that are obtained by carbon 14 dating. In many places buried soil profiles and other weathering features help the geologist to distinguish the deposits of drift that were produced during different ice advances.

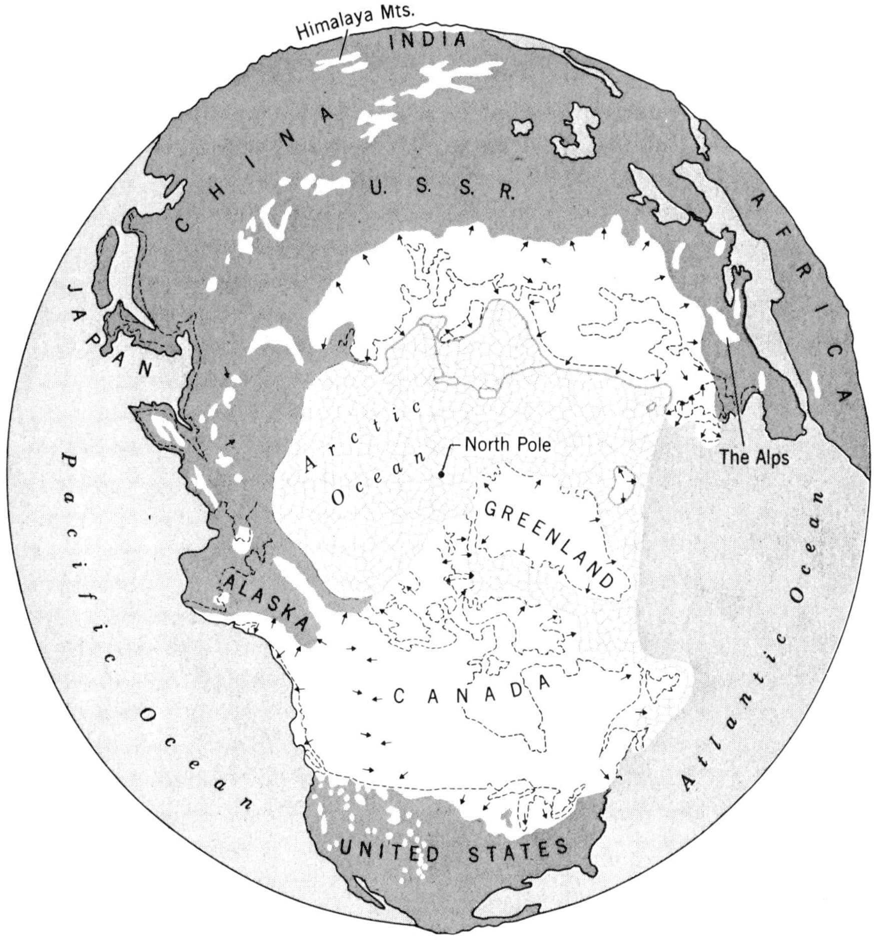

Figure 14-33 Maximum extent of the glacial cover during stages of the Quaternary ice age. Grounded ice is shown in white, and floating shelf ice and pack ice are displayed by the irregular pattern of gray and white. Continental coast lines are drawn for a sea level 100 m lower than at present. Reprinted from R. F. Flint and B. J. Skinner, *Physical Geology*, page 214, Jon Wiley & Sons, Inc., 1974.)

Interbedded with the Quaternary glacial deposits are fluvial and lake bed sediments indicative of milder climate. These sediments were laid down during **interglacial** stages when the ice had receded at least to the extent of the glacial cover at the present time. They contain a floral and faunal fossil record which reveals former interglacial climates as mild as we experience at the present time.

The stratigraphic assemblages of glacial and interglacial deposits in North America and Europe indicate four major ice advances during Quaternary time. The generally accepted sequences of glacial and interglacial stages are given in Table 14-2. This history of ice advances has been established independently for the two continental areas. To what extent did the North American and European ice advances occur at the same time? We are not completely certain. But stratigraphic cor-

Table 14-2
Principal Stages in the Quaternary Ice Age

North America	Northern Europe	European Alps
Wisconsin glacial	Weichel glacial	Würm glacial
Sangamon interglacial	Eem interglacial	Riss-Würm interglacial
	Saale glacial	Riss glacial
Illinoian glacial	Holstein interglacial	Mindel-Riss intergalcial
Yarmouth interglacial	Elster glacial	Mindel glacial
Kansan glacial		Günz-Mindel interglacial
Aftonian interglacial		Günz glacial
Nebraskan glacial		

relation suggests that the major stages of North American and European glaciations were nearly simultaneous. Correlation of substages is considerably more uncertain.

The growth and retreat of glaciers has a dramatic effect upon sea level. So much H_2O was taken into the ice mass when the continental glaciers reached maximum size that sea level fell by almost 140 m below the present level. A much broader expanse of continental shelf was above the sea during these times. Melting of the present mass of ice in Antarctica and Greenland would raise sea level by approximately 65 m. Former stands of sea level during Quaternary time are marked by elevated beaches and terraces eroded by waves and submerged terraces as well. These kinds of features are described in Chapter 15. The position of sea level relative to a continent is changed not only by melting ice; the effects of isostasy must also be considered.

Sea level during the Sangamon interglacial stage was approximately 6 m higher than at present. The glaciologist John Mercer pointed out that a volume of water sufficient to raise global sea level by 6 m would be produced by melting of the present West Antarctic icecap. This is the portion of the Antarctic ice sheet between South America and the Transantarctic Mountains (Figure 14-5). Many glaciologists believe that this ice sheet as well as those on North America and Eurasia melted during the Sangamon stage. There is evidence that the West Antarctic icecap is currently receding.

The various kinds of evidence found in the Quaternary glacial deposits and the related data pertaining to climate and sea level suggest the following ice age chronology. Massive ice sheets spread over northern North America and Eurasia almost 3 million years ago. The duration of this early Nebraskan stage remains uncertain, and we cannot identify substages with confidence. The Kansan stage, too, is obscure, but fragmentary evidence suggests that it ended about 700,000 years ago. Three substages of the Illinoian glacial stage have been tentatively recognized during the interval of time from 350,000 years b.p. to 250,000 years b.p. The abbreviation b.p. means "before the present time." The Sangamon interglacial stage lasted until approximately 70,000 years b.p. At this time the ice sheets of the Wisconsin stage began advancing over North America and Eurasia. Six substages marked by partial retreat and advance are recognized. During an interglacial substage from 40,000 years b.p. until 30,000 years b.p. the ice sheets receded almost completely except in Antarctica and Greenland. The ice sheets again spread to the approximate limits seen in Figure 14-33 between 30,000 years b.p. and 25,000 years b.p. This position was maintained more or less continuously until 10,000 to 12,000 years ago when major recession commenced. The ice sheets had nearly disappeared from North America and Eurasia 8000 years ago.

We believe that continental glaciers formed much earlier in Antarctica and Greenland. Although diminished in size during interglacial stages these ice sheets have remained large during late Tertiary and Quaternary time. The East Antarctic ice sheet appears to have existed for at least 20 million years as an extensive continental glacier.

The Paleozoic Ice Age

Evidence of extensive glaciation during the Paleozoic Era is found in regions of Africa, South America, Australia, India, and Antarctica. Geologists have interpreted sediments resembling conglomerate and breccia to be lithified glacial till. Rock fragments in these **tillite** deposits and underlying strata display striations and grooves typical of glaciation. Areas where these deposits occur and the directions of ice movement inferred from them are shown in Figure 14-34. Also shown is a line indicating the relative position of the south pole surmised from paleomagnetic measurements. Tillite deposits are usually hard to distinguish from other poorly sorted sediments deposited by nonglacial processes. The origin of any individual deposit can be questioned. When viewed collectively, however, these sediments indicate more convincingly the probablility of continental glaciation during Paleozoic time. Similarly, the paleomagnetic evidence is fragmentary and of varying quality. The line showing relative pole position represents a reasoned speculation compatible with the incomplete paleomagnetic record.

We can tell from the information in Figure 14-34 that the region of the continent covered by ice changed as the position of the continent changed relative to the earth's axis during the Paleozoic era. Tillites in North Africa were deposited in early Paleozoic time

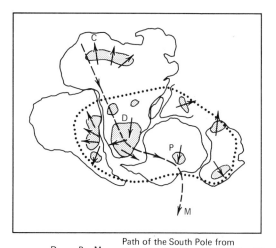

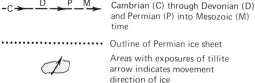

Figure 14-34 Distribution of Paleozoic tillite deposits and orientation of Paleozoic glacial striations in Africa, South America, India, Australia, and Antarctica. Line indicates the position of the earth's axis of rotation relative to the drifting continent based on paleomagnetic data. Age of glacial deposits suggests early and middle Paleozoic glaciation centered over what is now North Africa. The late Paleozoic glaciation in regions which now appear farther south is outlined. Reconstruction of the continents is based upon their present shapes, paleomagnetic pole positions, and alignment of orogenic belts. (Prepared from data in M. W. McElhinny, *Paleomagnetism and Plate Tectonics,* page 260, Cambridge University Press, 1973; H. J. Harrington, *American Association of Petroleum Geologists Bulletin,* pages 1773–1814, Vol. 46, No. 10; R. W. Fairbridge, *Natural History,* pages 66–73, Vol. 80, No. 6, 1971.)

when the pole was situated in that region. But by Permian time the large icecap was supposedly located over the area centered on South Africa. Observe that evidence of this Permian icecap found along the margins of South America and Australia indicate ice movement toward the interiors of these continents. We know that ice sheets cannot spread from oceanic areas onto continents. Therefore it seems probable that these present margins of South America and Australia were then part of a larger continental platform. The reconstruction of the super continent illustrated in Figure 14-34 brings the worldwide Permian glacial deposits into an arrangement which outlines a large and coherent continental glacier. Other kinds of geologic evidence that we learned about in Chapter 5 support this former grouping of the continents prior to Cretaceous time. The concept of continental drift offers a reasonable explanation for the Paleozoic glacial sediments that are now widely separated on the present continents.

Causes of Ice Ages

Growth of a large ice sheet requires: (a) a continental platform, (b) a cold climate, and (c) a source of moisture. These conditions are most likely to be found where continents occupy polar regions. Because solar radiation diminishes with increasing latitude, polar regions tend to be relatively cold. During the present ice age Antarctica has been located almost entirely south of the antarctic circle and is surrounded by ocean. In the northern hemisphere large continental regions extend north of the arctic circle, more or less enclosing the Arctic Ocean. Evidence in Figure 14-34 tells us that a Paleozoic ice sheet existed on a former continent in a polar setting.

Glaciers form and expand when accumulation of snow and ice on a continent exceeds ablation. We are not certain if conditions necessary for an ice age are always produced when a continent migrates into a polar location. Possibly there are other external factors that also contribute to a colder climate. We know from astronomical observations that the orbit of the earth about the sun is periodically perturbed. The earth moves slowly farther from, then nearer the sun, reaching a maximum distance every 92,000 years. A colder climate could be expected at these times because of diminished solar heating of the earth. Changes in the rate of solar radiation have been suggested. This hypothesis is questionable, but possibly could contribute to climatic variations. We are not certain whether or not continental migration into polar regions must occur at the same time as a temperature decline related to these other factors. We have not yet learned enough about the distribution of continents relative to the earth's axis during geologic history to know if any land masses have existed for long periods of time in polar regions without ice sheets.

Quaternary glacial deposits indicate the repeated advance and retreat of ice sheets. Why does an ice sheet fluctuate in this manner? What factors affect its stability? The late Maurice Ewing and W. L. Donn discussed the interaction of a polar continental ice sheet and the bordering seas from which moisture is derived. They suggested a glacial cycle involving alternate covering and uncovering of the Arctic Ocean by ice shelves and pack ice. When free of ice this ocean would supply moisture to ice sheets, which would then expand over the bordering continents of Eurasia and North America. Continued growth of these ice sheets would eventually lead to expansion of large ice shelves over the Arctic Ocean. The glaciers would also contribute by their vast cold and white surface to a lowering of temperature. In this colder climate perpetual pack ice would form over the parts of the polar ocean not already covered by ice shelves. A lowering of sea level would further restrict the circulation of warmer ocean currents from lower latitude into the polar ocean. In this ice-choked condition the Arctic Ocean would cease to supply moisture for nourishment of the ice sheets. Because of the altered climate and moisture supply the over-extended glaciers could become unstable and commence to recede. The processes contributing to this recession are not clearly understood. Perhaps it is triggered in some way by the balance of north polar and south polar glacial cycles.

There is the possibility that a mechanism exists whereby early glacial recession in one hemisphere can stimulate recession in the other hemisphere. We know that large portions of the West Antarctic ice sheet are now grounded below sea level. Because of isostatic subsidence, we know that portions of northern North America and Eurasia were depressed below sea level under former ice loads. Johannes Weertman has demonstrated from theoretical rheological considerations that ice sheets grounded below sea level on horizontal beds are inherently unstable. This means that such ice sheets must always be either expanding toward the margin of the continental platform or receding. Raising sea level causes formerly grounded ice along the edges of the sheet to come afloat and surge so that ice shelves are produced. These shelves disintegrate into icebergs which melt and further increase sea level. This causes still more grounded ice to come afloat and to break up into icebergs until the ice sheet eventually disintegrates. The idea has been suggested that the earlier melting of northern hemisphere glaciers has raised sea level to the point of accelerating the disintegration of the West Antarctic ice sheet. Eventually the glaciers recede to the extent that formerly ice-choked seas become open once again. With moisture supplied by these open seas the ice retreat is halted and glaciers again begin to advance.

Let us briefly summarize our understanding of ice ages. The movement by continental drift of a large land mass into a polar region is probably the single most important factor in initiating an ice age. Because large ice-covered areas of a land mass will probably be depressed below sea level by isostatic adjustment, we can expect an ice sheet to undergo a series of advances and retreats during the ice age.

PERSPECTIVE

Human culture has evolved during the Quaternary ice age. Our early forebears adjusted their activities and wanderings in accordance with the advance and retreat of the ice sheets and related changes in sea level. In recent centuries our civilization has become more permanently centered in large metropolitan complexes, many of which are situated along coastal and lowland areas. We now realize that ice sheets can advance over or recede from large areas in a few thousand years or less. Significant changes in sea level and climate are associated with these fluctuations. What adjustments will we be able to make when the ice once again advances?

STUDY EXERCISES

1. Suppose that a continental icecap centered on James Bay extended as far as Columbus, Ohio. If its maximum thickness was 3 km, estimate its thickness over Minneapolis, Minnesota.

2. A large valley glacier 400 m thick moves over a bed inclined at an angle of 2 deg. Surveys indicate that a stake on the glacier surface moves at a speed of 300 m/year. Estimate the bottom sliding velocity.

3. Estimate the volume of ice in the West Antarctic icecap.

4. Discuss how deposits of sediment and landscape features produced by continental glaciation can be used to reconstruct the dimensions of a large ice sheet.

5. Tillite deposits tell us that a large ice sheet existed in what is now Africa during much of Paleozoic time. Discuss the factors that contributed to ending this Paleozoic ice age.

SELECTED READINGS

Flint, Richard Foster, *Glacial and Quaternary Geology,* John Wiley & Sons, Inc., New York, 1971.

Hughes, T., The West Antarctic Ice Sheet: Instability, Disintegration, and Initiation of Ice Ages, *Reviews of Geophysics and Space Physics,* vol. 13, pp. 502–526, 1975.

Paterson, W. S. B., *The Physics of Glaciers,* Pergamon Press, New York, 1969.

Price, R. J., *Glacial and Fluvioglacial Landforms,* Hagner Publishing Co., New York, 1973.

Weertman, J., Stability of the Junction of an Ice Sheet and an Ice Shelf, *Journal of Glaciology,* vol. 13, pp. 3–11, 1974.

15

GEOLOGIC PROCESSES IN THE MARINE ENVIRONMENT

Deep Sea Sediments
  Dredging and Coring
  Marine Seismic Profiling
  Ocean Basin Sediments
  Ocean Basin Stratigraphy
The Oceanic Water Mass
  Dissolved Constituents in Sea Water
  Ocean Current Measurements
  Surface Currents
Deep Sea Circulation
The Ocean Tide
Ocean Waves
Long-Term Changes in Sea Level
Coastal Landforms
  The Coastal Landscape
  Evolution of Coasts
  Maintenance of Coasts

The fascination of the sea is evident in the accounts of ancient and modern seafarers alike. For the most part their wanderings over the oceans have been a search for trade, geographic discovery, and good fishing. Oceanography, which is the scientific investigation of the marine environment, has become an important activity in modern times. The principal divisions of this subject include marine and coastline geology, physical and chemical oceanography, and marine biology.

Information used in the scientific study of the oceanic regions is obtained from oceanographic vessels and at coastal observatories. A typical ship equipped for investigation of many aspects of the marine environment is pictured in Figure 15-1. Activities include collection of rock and sediment specimens from the ocean floor and surveys of submarine topography and sediment thickness. Oceanographers make temperature measurements at many representative locations in the oceanic water mass and collect sea water samples for analysis of dissolved and suspended constituents. They measure motion of the water mass including waves, tides, currents, and long-term changes in sea level.

Oceanographic expeditions in vessels provided with measuring and sampling devices have traversed much of the global ocean. Some scientific information is obtained while a ship is moving. Other measurements and observations require a period of time in a fixed position. Certain studies entail many months or years of work at permanent or semipermanent coastal installations.

The mass of sea water which forms the world ocean is spread over more than 70 percent of the earth's surface. The water is perpetually in motion, circulating in a global system of currents and perturbed by waves and tides. The nature of the water mass is the con-

Figure 15-1 The oceanographic vessel *Eastward* is equipped with permanent deck winches, booms, and rigging required to lower and tow an assortment of measuring devices. (Courtesy of Duke University Marine Laboratory, Beaufort, North Carolina.)

cern of physical and chemical oceanographers. Interactions of the diverse assortment of organisms in this environment are the subjects of marine biology.

The ocean floor is largely covered by a veneer of organic and inorganic sediment. Waves and currents produce distinctive landscape features along the coastal borders of the sea. These features and processes are important aspects of marine and coastal geology. Other topics such as volcanism, seismicity, and tectonism in the lithosphere beneath the oceans are discussed elsewhere in the book. Our principal interests in this chapter are the geologic processes directly related to the oceanic water mass. Recall that topography in the ocean basins is discussed in Chapter 2. You might find it useful to review the features of the different submarine topographic provinces. In the present chapter we will examine ocean floor sedimentary deposits, chemical and physical oceanographic features, and coastal geomorphology.

DEEP SEA SEDIMENTS

In recent decades oceanographers have developed practical methods for investigating sediments on the ocean floor. We now have instruments for mapping submarine stratigraphy and for sampling the bedrock and unconsolidated sediment beneath the sea. But compared with knowledge of the continental land surface, our comprehension of the ocean floor is rudimentary. Nevertheless, we now have enough data to begin to piece together a coherent picture of the submarine geologic environment.

Dredging and Coring

Specimens of unconsolidated sediment and bedrock from the marine environment are collected by devices lowered to the ocean floor from a moving or stationary ship. The principal methods are dredging, free fall coring, and core drilling. Rock fragments and soft sediment can be retrieved from the sea floor with dredges (Figure 15-2). The heavy dredge is dragged along the bottom and fastened by cable to a moving ship. Its sharp lower edge can break fragments from bedrock ledges and scoup sediment and loose cobbles. Ordinarily another heavy mass is attached ahead of the dredge to insure that it remains on the bottom. Specimens can also be obtained by means of a grab-sampler (Figure 15-2) which bites into the bottom debris when the jaws are suddenly closed.

Soft sediment on the ocean floor can be penetrated by the free-fall coring devices (Figure 15-3). After being lowered to a prescribed height above the bottom the corer is released to fall under its own weight into the sediment. The simple **gravity corer** can ordinarily penetrate 1 or 2 m of sediment. In the more complicated **piston corer** a piston is drawn upward in the core chamber as the device penetrates. The core of sediment is forced into the chamber by hydrostatic pressure as well as from the momentum of the falling mass. Under optimum operating conditions cores more than 15 m long have been retrieved in piston corers.

In recent decades we have used drilling ships to obtain cores of ocean floor sediment and bedrock several hundred meters in length. The *Glomar Challenger* (Figure 5-18)

Figure 15-2 Devices used for collecting rock samples from the ocean floor. *(A)* Dredge is dragged along the sea flooor. *(B)* Grab sampler bites into the ocean bottom. (After P. K. Weyl, *Oceanography*, page 265, John Wiley & Sons, Inc., 1970.)

(A) (B)

is perhaps the best-known vessel equipped for such work. The expanding search for offshore petroleum and gas reservoirs has spurred the development of improved marine drilling techniques. Although these valuable commodities have been found only in continental margin deposits, the same drilling methods are useful for obtaining data elsewhere in the ocean basins.

Dredging and coring efforts are not always successful. Often an empty dredge or broken core barrel is all that is obtained from many hours of difficult work. After several decades of persistent effort, however, we have obtained thousands of cores and dredge samples from a broad distribution of locations in the world ocean.

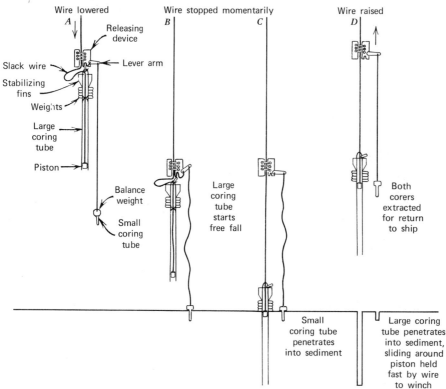

Figure 15-3 Free fall soft sediment coring devices. (After P. K. Weyl, *Oceanography,* page 265, John Wiley & Sons, Inc., 1970, and C. R. Longwell, R. F. Flint, and J. E. Sanders, *Physical Geology,* page 351, John Wiley & Sons, Inc., 1969.)

Figure 15-4 Apparatus used for seismic sparker profiling: (A) frame with electrodes for generating sparks; (B) "snake" containing pressure sensitive signal detectors; (C) chart recorder.

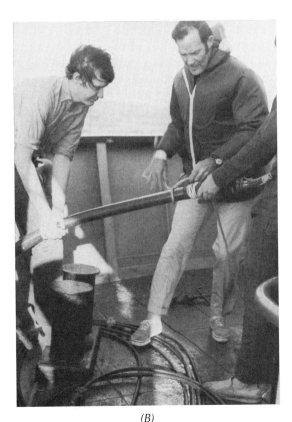

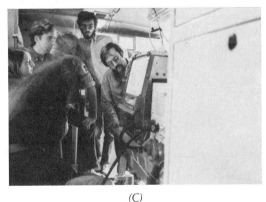

Marine Seismic Profiling

Submarine stratigraphy can be mapped from shipboard seismic reflection and refraction measurements. One method of seismic profiling is similar in principle to echo sounding (described in Chapter 2), but utilizes a low-frequency pulse commonly in the 100 to 1000 hertz range. The ship's engines and other shipboard activities may also produce low-frequency vibrations. For this reason the seismic profiling apparatus must be trailed some distance behind where these other vibrations will not interfere with the seismic waves. Typical equipment is illustrated in Figure 15-4. The sound pulse is generated by a powerful electrical spark emitted at regular intervals. Echos are returned not only from the ocean floor, but also from deeper bedding surfaces within the underlying submarine sediments. Reflection travel times are recorded in much the same way as echo sounder data. The apparatus is called a **sparker-profiler.**

Charts produced by a sparker-profiler on a moving ship indicate patterns of layering in the ocean bottom sediment (Figure 15-13), but these patterns are distorted because of the different P-wave velocities in sea water and the individual layers of sediment. Velocity data must be incorporated into later analysis to obtain correct thickness values.

Marine reflection and refraction surveys are also carried out using large explosions to generate seismic waves. These waves penetrate deeper than those from a sparker source. They are recorded on arrays of hydrophones. Sometimes two or more ships are required to maintain long lines of hydrophones and to place explosives at the proper positions. These methods of seismic surveying are described more fully in Chapter 3.

Sparker profiles and other seismic measurements have been obtained from many representative locations in the oceans of the world. These measurements together with core and dredge specimens provide the basic data that is needed to determine the type and distribution of oceanic sediments.

Ocean Basin Sediments

Unconsolidated sedimentary deposits cover large areas of the ocean basin floor. The sediment is irregularly distributed on the oceanic ridges, being concentrated in the valleys and absent on the more precipitous slopes. The deposits are much more uniform over the abyssal plains of the ocean basin floor. There are three principal kinds of deep sea sediments: **biogenic** deposits, **inorganic detritus,** and **hydrogenic precipitates.** Almost everywhere these different types of sediment are mixed together, but one or another of the constituents is dominant.

The remains of marine organisms that have settled to the sea bottom form **biogenic** deposits. Calcium carbonate and silica in tests and other shell fragments are the most abundant chemical constituents. This debris comes almost entirely from floating (**planktonic**) and swimming (**nektonic**) **pelagic** organisms. A pelagic organism is one that is free to move in the water and is not attached to the ocean floor.

Calcite secreted by **foraminifera** and **coccolithophores,** which are microscopic marine organisms, and aragonite produced by organisms called **pteropods** make up most of the calcium carbonate deposits. Most of these sediments are called **calcareous ooze** or mud. The clay- and silt-sized grains, which are the abundant constituents, are fragments of tests or particles from decomposed tests and other shells. A few examples of the numerous organisms that produce this sediment are pictured in Figure 15-5. In mixtures of deep sea sediment calcareous ooze is the principal constituent in regions of the global ocean basin shown in Figure 15-6.

We know that the rate at which calcareous ooze accumulates on the ocean floor depends upon water depth. To a lesser extent the rate is affected by the amount of calcium carbonate-producing organisms living in the sea. Almost all of these deposits accumulate originally at depths of less than 4500 m. Chemical analysis of sea water from depths of more than 500 m indicates that dissolved calcium carbonate is below the saturation concentration. Therefore, calcareous particles are soluble in the water through which they must move to reach the ocean floor. The extent to which a particle is dissolved depends upon the dissolution rate and the time required to settle to the bottom and to become buried by subsequent sedimentation. Factors influencing the dissolution of calcareous material in sea water are not well understood. Chemical oceanographers are now studying the effects of pressure, temperature, and the presence of other dissolved ions on this process. We can tell from the global distribution of calcareous oozes that in water less than 4500 m deep many particles do reach the bottom. But we find calcareous ooze in places where the water is more than 4500 m deep. How did it get

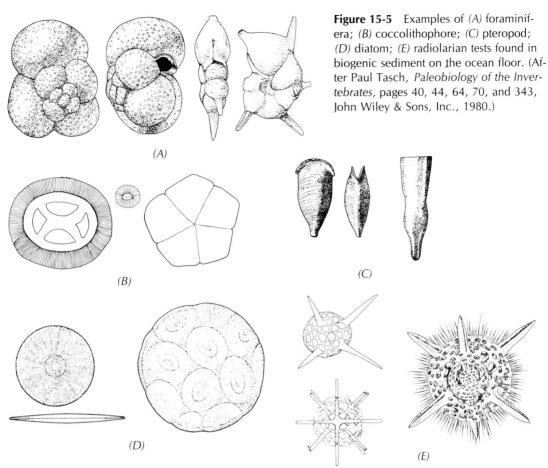

Figure 15-5 Examples of (A) foraminifera; (B) coccolithophore; (C) pteropod; (D) diatom; (E) radiolarian tests found in biogenic sediment on the ocean floor. (After Paul Tasch, *Paleobiology of the Invertebrates*, pages 40, 44, 64, 70, and 343, John Wiley & Sons, Inc., 1980.)

there? Perhaps sea floor spreading plays a role; we will look into this possibility later in the chapter.

The principal sources of biogenic silica in deep sea sediments are pelagic **diatoms** and **radiolaria.** A few of the large variety of these single-cell organisms are pictured in Figure 15-5. Fragments of their silica tests accumulate to form **siliceous ooze.** The distribution of this kind of deep sea sediment is shown in Figure 15-6. Most of it lies in the southern oceans between 45 deg S and 60 deg S latitude. Elsewhere it occurs along the northern margin and in equatorial parts of the Pacific Ocean Basin.

Sea water is undersaturated in silica and has the capacity to dissolve the siliceous tests of pelagic organisms. Large tests which may be partly shielded by other less soluble organic material adhering to their surfaces are most likely to settle to the ocean floor before becoming completely dissolved. Summer conditions at high latitudes appear to favor growth of large diatoms. Recycled nutrients carried by upwelling currents in the equatorial Pacific Ocean also favor growth of silica-producing organisms. These are the regions where siliceous oozes are the dominant deep sea sediments.

Other biogenic sediments include phosphate compounds in bone and teeth fragments and incompletely decomposed organic materials that have accumulated in oxygen-deficient parts of the ocean basin. These deposits account for a very small proportion of the total mass of deep sea biogenic sediment.

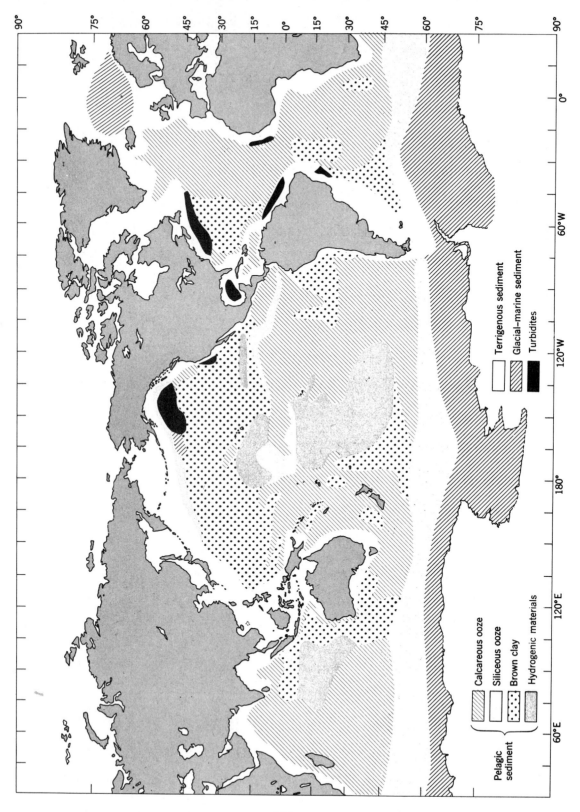

Figure 15-6 *(opposite)* Global distribution of deep sea sediment. (Compiled from several sources.)

Reducing conditions in a few small and locally isolated basins, for example the Black Sea, favor accumulation of carbonaceous biogenic deposits.

Inorganic detritus makes up a significant proportion of the mass of deep sea sediments. Most of this material consists of weathered rock fragments transported from the continents into the ocean by streams, wind, and ice. Here it accumulates on the sea floor with other biogenic and hydrogenic material. **Brown clay** is the most abundant inorganic detrital constituent. Kaolinite, illite, and chlorite are important minerals in brown clay. We believe that the relative proportions of kaolinite and chlorite depend on the climate of the continental source areas. In the equatorial Atlantic Basin the kaolinite/chlorite ratio exceeds 5/1. But at higher latitudes in the Atlantic Ocean the chlorite fraction increases. How can this be explained? A moist tropical climate is favorable for producing kaolinate by silicate rock weathering. Chlorite tends to be unstable in this environment. Therefore we would expect inorganic sediments derived from nearby moist equatorial continents to be relatively enriched in kaolinite and deficient in chlorite. The reverse is true for colder temperature climates in which chlorite is more stable and production of kaolinite proceeds more slowly. These findings suggest that the brown clay constituents in the Atlantic Ocean are derived for the most part from nearby continents. But a similar pattern is not so clearly evident in the Pacific Ocean. Brown clay is the dominant constituent of ocean floor sediment in regions shown in Figure 15-6.

Illite is derived from the weathering of micas. The illite fraction of inorganic material in deep sea sediments from the Pacific Ocean is highest near continents, but diminishes toward the central part of the ocean. This supports the argument that this brown clay constituent has been transported from continental areas.

Other inorganic constituents in brown clay include silt- and clay-sized quartz grains. These are most abundant in oceanic regions near the large tropical deserts, from where they most likely have been transported by wind and rivers. Volcanic ash is also a significant consituent in some brown clay deposits. In a few places larger volcanic fragments perhaps produced by submarine eruptions are found in deep sea sediments. Small amounts of montmorillonite may have been produced by submarine weathering of volcanic debris.

The Arctic Ocean floor and the region of the southern ocean basin bordering Antarctica are covered predominantly by a particular variety of inorganic detrital sediment. These deposits consist of clay- and silt-sized particles that show relatively little oxidation compared with brown clay constituents elsewhere in the ocean. Occasional boulders and other large fragments are found in the otherwise fine-grained sediment. We believe that this material has been transported from glacierized continents by icebergs, which release it when they melt. These deposits are called **glacial-marine** sediments, and their distribution on the ocean floor is shown in Figure 15-6.

We can recognize sedimentary structures in cores of deep sea deposits. Many deposits display features indicative of uniform sedimentation in quiet water. Some deposits, however, contain structures such as rippled bedding which suggest deposition from flowing water. In still other cores these primary bedding structures are absent, presumably destroyed by burrowing organisms. But we have learned from a worldwide distribution of cores and from seismic profiles that in most parts of the ocean deep sea sediment has accumulated in a quiet water environment, as particles of organic and inorganic debris set-

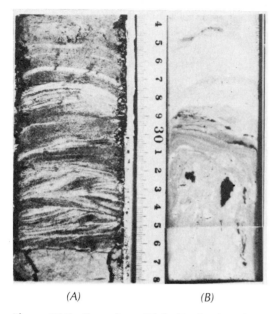

Figure 15-7 Cores from *(A)* the Tyrrhenian Abyssal Plain, and *(B)* the Mediterranean Ridge display the contorted sedimentary structures and graded bedding typical of turbidites. (From W. B. F. Ryan and others, in *The Sea* [A. E. Maxwell, editor], Vol. 4, Part II, page 435, Wiley Interscience, 1970.)

tle to the ocean floor. These particles form **pelagic** sediments. Here the term **pelagic** refers to both inorganic and organic particles which had become suspended and mixed in the water mass and eventually settle to the bottom. Almost all pelagic deep sea sediments are a mixture of organic and inorganic composition. Ordinarily we classify them as biogenic or inorganic detrital according to which constituent is dominant.

A smaller amount of deep sea sediment has been deposited from turbulent and flowing water. These deposits are most abundant in continental margin provinces, but have also been found in the deeper ocean basins. They are characterized by grading, cross-bedding, ripple marks, and distorted laminations. These current-deposited deep sea sediments are called **turbidites.** Cores illustrated in Figure 15-7 display sedimentary structures typical of turbidites.

We attribute the transport of deep sea turbidites to submarine **turbidity currents.** The occurrence of such currents has been surmised from reports of broken submarine cables and the nature of bottom sediment in the vicinity of these disruptions. A well-documented example was described by B. C. Heezen and M. Ewing in 1952. They examined evidence of a rapid sequence of breaks in communications cables on the continental margin bordering Newfoundland in the North Atlantic Ocean following the Grand Banks earthquake of 1929. The locations of these cables are illustrated in Figure 15-8. Also shown are the times at which they were broken in sequence by a current generated on the continental shelf by the earthquake. This current is judged to have been a dense mixture of turbulent sea water and sediment discharged by earthquake-related slumping on the shelf. The dense mass flowed rapidly down the slope, then spread more slowly over the continental rise, finally diminishing on the abyssal plain. According to the times at which cables were broken by this submarine torrent, it reached velocity of more than 70 km/hour (over 40 miles/hour) on the continental slope. Its speed over parts of the continental rise exceeded 20 km/hour. Cores of

Figure 15-8 *(opposite)* *(A)* Map of the Atlantic continental margin near Newfoundland showing location of submarine cables broken following the Great Banks earthquake of 1929. *(B)* Times of successive cable breaks are shown on the cross section. (Modified from B. C. Heezen and C. L. Drake, *American Association of Petroleum Geologists Bulletin*, pages 221–233, Feb. 1964, by C. R. Longwell, R. F. Flint, and J. E. Sanders, *Physical Geology*, page 363, John Wiley & Sons, Inc., 1969).

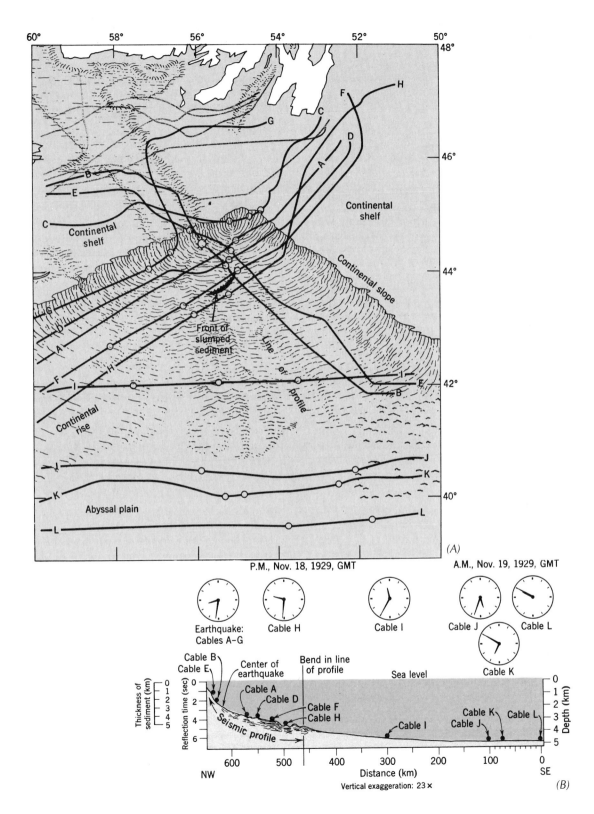

Figure 15-9 *(opposite)* Sand suspended in water flows down a submerged embankment in San Lucas Bay near Baja, California. (R. F. Dill, permission of U.S. Navy Electronics Laboratory).

bottom sediment were later obtained from the path of this proposed current. A surface layer approximately 1 m thick contained typical shallow water fossils in graded beds of silt and sand particles. Progressively farther down the continental rise the proportion of silt and clay particles increased, and the sand fraction declined. These data support the argument for an earthquake-induced turbidity current. It seems reasonable to explain the occurrence of other deep sea turbidites by similar currents. They probably also contribute to erosion of canyons on the ocean basin floor.

Sand and silt particles derived from continental erosion are widely distributed over the continental rise areas of the ocean basin. This sediment also makes up an important fraction of the deposits covering continental shelf and slope areas. These inorganic detrital accumulations are called **terrigenous marine sediments** and are most abundant along the continental margins (Figure 15-6).

Turbidity currents are of major importance in transporting terrigenous sediment into the ocean basin. Cores from the continental rise near Newfoundland reveal that these sediments account for a large fraction of the deposits left in the wake of the current triggered by the Grand Banks earthquake in 1929. Elsewhere in the oceans broad fan-shaped deposits rich in terrigenous sediment spread out from submarine canyons onto the bordering continental rise. Scientists using scuba equipment and small research submarines have observed sediment transport in the upper parts of these canyons. They have photographed sand falls (Figure 15-9) that occur when sediment slumps from steep slopes. They have also watched fine-grained sediment being steadily transported by slower currents that flow more or less continuously at speeds of a few tenths of a kilometer per hour down some submarine canyons. Marker poles have been placed on the ocean floor in these canyons to aid in detecting erosion and deposition. They indicate the downslope transport of sediment by creep as well as by suspension in flowing water. But turbidity currents rather than these slower, steadier currents appear to play the dominant role in eroding submarine canyons. Most terrigenous sediment and shallow water organic sediment found in the continental rise turbidites appear to have been transported into the ocean basin by currents flowing from the continental shelves through submarine canyons.

The **hydrogenic** sediments of the deep ocean floor are formed by chemical alteration and precipitation processes involving sea water. Nodules of manganese and iron oxides are found in many parts of the ocean basin. Examples of **manganese nodules** are illustrated in Figure 15-10. The oxides form a shell of concentric layers about various kinds of fragments including volcanic rock, bones, and teeth. Manganese is not an abundant dissolved constituent in sea water. Small amounts may come from submarine volcanism and from weathering of continental rock. The origin of the nodules is not clearly understood and is an interesting topic of chemical oceanographic research. Some iron-manganese nodules on the Pacific Ocean basin floor are sufficiently rich in nickel, copper, and cobalt to be classified as commercial grade ores. Hydraulic suction apparatus has been developed to retrieve these nodules so that deep sea mining is technologically feasible.

Other hydrogenic deposits include montmorillonite and a zeolite mineral **phillipsite.** Submarine chemical alteration of basalt could yield phillipsite, which is most abundant on the oceanic ridge system in the central Pacific

Geologic Processes in the Marine Environment 557

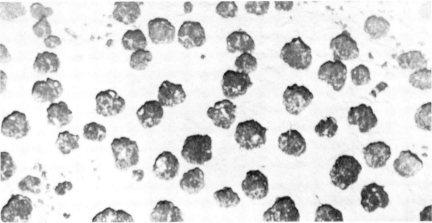

Figure 15-10 Nodules of manganese oxide on the South Pacific ocean floor. (National Science Foundation Photograph.)

Ocean. Here bedrock outcrops of basalt are widespread and can react with sea water. Chemical alteration of ocean floor volcanic ash is a probable source of hydrogenic montmorillonite. The areas where these different kinds of hydrogenic sediment form the important fraction of ocean floor deposits are shown in Figure 15-6.

We find carbonate reefs and detrital deposits derived from them in warm and shallow marine environments. They occur on the tropical and subtropical continental shelves and as coastal fringes bordering island-arc volcanoes and emerged seamounts. Reef carbonates also form ringlike coral islands known as **atolls** which rise only a few meters above sea level. Typically a ribbon of coral (Figure 15-11) almost surrounds a shallow lagoon. We have learned from core drilling and seismic surveys of atolls that volcanic rock lies beneath the reef structure and carbonate detritus which fills the lagoon. The atoll, it seems, forms a cap on a seamount or submerged island-arc volcano.

Coral reefs grow in water less than 20 m deep (about 65 ft) and warmer than 18.5° C (65° F). These organic structures could not have grown upward from deeply submerged seamounts to form atolls. The most widely accepted hypothesis of atoll formation was originally proposed during the nineteenth century by the well-known naturalist Charles Darwin. He suggested that coral initially fringing a volcanic island continued to grow upward during subsidence of the original island. This idea was later elaborated and modified by W. M. Davis and other geomorphologists. The proposed sequence is illustrated in Figure 15-12. The atoll continues to develop as long as the subsidence rate does not exceed the rate of upward growth of reef-building corals. The reef forms the sides of a cup-like container which becomes nearly filled with the fragments eroded and washed into the lagoon by wave action.

Elsewhere in the ocean basin reef structures and associated carbonate fragments have been found on the tops of some guyots. These deposits suggest that the flat-topped seamounts may be former atolls which eventually subsided too rapidly for reef growth to continue. Prior to becoming completely submerged a lagoon is filled with sediment almost to the level of the bordering reef. The nearly level surface formed by the reef and lagoon bottom then subsided to greater depth where it forms the top of a guyot.

Carbonate reef fragments are found in

Figure 15-11 Lake Atoll in the Marshall Island Group is formed from a ribbon of reef coral surrounding lagoon. (D. B. Doan.)

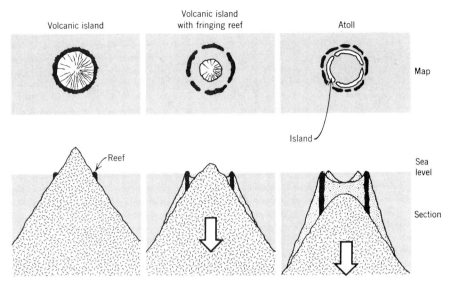

Figure 15-12 Stages in the evolution of a coral atoll on a subsiding volcano proposed by Charles Darwin. (After P. K. Weyl, *Oceanography*, page 452, John Wiley & Sons, Inc., 1970.)

some deep sea turbidites, but these fragments were not produced in this deep environment. The particles were transported along with inorganic detritus from shallow marine source areas into the ocean basins. These and other abundant shallow marine sediments, which are widely distributed in epicontinental seas and on the continental shelves, are discussed in some detail in Chapter 9. Transport and deposition of these sediments in coastal regions are discussed in another section of this chapter.

Ocean Basin Stratigraphy

Sediments in the ocean basin cover a rugged basaltic bedrock surface. These deposits are thin and irregular in the rift mountains and fractured plateau areas of the oceanic ridge system. They form more continuous sequences over the outer ridge steps and the ocean basin floor where thicknesses of several hundred meters or more are common.

Some submarine trenches have only thin accumulations of sediment, whereas others have been filled to a greater extent. The thicknesses of several hundred meters or more are common. The thickest deposits of marine sediment have accumulated on the continental margins during the evolution of orogenic belts and in epicontinental seas in response to epeirogeny. These terrigenous and shallow marine carbonate deposits are described more fully in Chapters 9 and 10.

The ages of deep sea sedimentary beds, determined principally from fossil foraminifera, indicate that almost all ocean basin deposits have accumulated since middle Mesozoic time. But the distribution of the older deposits is somewhat different from the present pattern of accumulation for deep sea sediment seen in Figure 15-6. For example, some areas now receiving inorganic detritus formerly received mostly biogenic sediment. Global variations in climate, the extent of glaciation, sea floor spreading, and changes in the shape of the ocean basin must be considered in attempting to explain these differences. Let us examine in more detail the stratigraphy and age of these deposits and the processes affecting their accumulation.

Sediment is more or less symmetrically distributed relative to oceanic ridge axes. Deposits are sparse, if not absent, from the rugged central area along a ridge axis and become thicker and more abundant farther and farther from the axis on both sides. This pattern of symmetry is evident in Figure 15-13A which shows a typical Mid-Atlantic Ridge cross section. Observe that most of the sediment is confined to the valleys and depressions in the rugged bedrock surface. The seismic profile reveals some horizontal bedding surfaces and other irregular sedimentary structure. We have learned from core samples that these deposits are interbedded pelagic sediment and turbidites.

The few deposits that are found in the central rift mountains and fractured plateau regions consist mostly of Quaternary age sediment. But the deeper layers that lie beneath these Quaternary deposits on the outer steps of the ridge system are of middle- and late-Tertiary age. Recall that these age relationships are described in Figure 5-22 and the accompanying discussion in Chapter 5. Sea floor spreading outward from the ridge axis appears to be the principal reason for this age distribution of sediment. The older bedrock surface on the outer ridge steps has been present since middle Tertiary time, and therefore has been receiving sediment longer than the newly formed sea floor along the ridge axis. This process of sedimentation on a spreading sea floor is illustrated in Figure 15-14.

Sedimentary deposits on the ocean basin floor are generally thicker and more continuous than on the oceanic ridges. Seismic profiler records (Figure 15-13B) indicate that the rough bedrock surface found on ridges also exists beneath the ocean basin floor where it is more completely covered by sediment. In some places the bedrock surface rises above the sediment to form abyssal hills. But along other profiles we can trace continuous horizontal sedimentary beds for tens or even hundreds of kilometers. Cores reveal mostly pelagic sediment interbedded with some turbidites. Most of the strata are of Tertiary and Quaternary age. However, smaller amounts of Cretaceous and perhaps late Jurassic sediment have been discovered. Almost all ocean basin sediment is less than 100 million years old.

The different varieties of sediment in a core from the ocean basin can tell us about the changing conditions of sedimentation. We have discovered that the siliceous ooze now accumulating in the southern parts of the Pacific, Atlantic, and Indian Oceans (Figure 15-6) is underlain by glacial marine deposits. These glacial sediments were transported farther northward during former times of more extensive glaciation. During those times icebergs from Antarctica drifted farther north carrying glacier debris into parts of the ocean that are now too warm. Similarly in the North Pacific and North Atlantic Oceans, modern biogenic and terrigenous sediment is underlain by beds of glacial marine sediment.

Elsewhere in the ocean basins where brown clays are now accumulating on deep abyssal plains, we find biogenic sediment underneath. The present sea floor in these regions is too deep to expect calcareous or siliceous fragments to settle without completely dissolving. Some older biogenic deposits probably accumulated in shallower water, perhaps closer to an oceanic ridge. These beds were then moved by sea floor spreading to deeper parts of the basin where they are now being covered by inorganic detritus. Variations in climate could also contribute to these changes in the kind of sediment being deposited. Shells of former organisms less susceptible to dissolution may have accumulated, whereas organic material produced at the present time cannot.

Turbidites displaying sedimentary structures related to current deposition are dominant on the continental rise areas. Wedges of

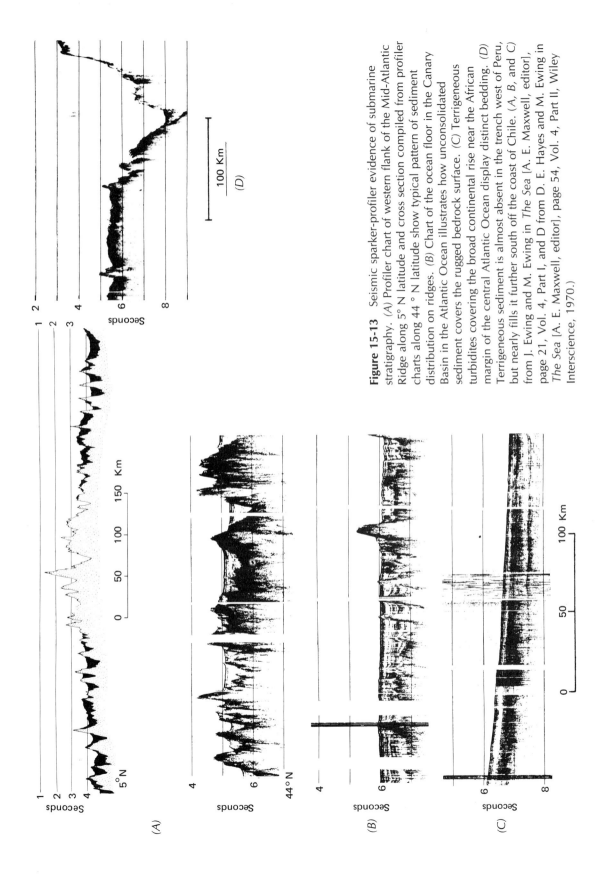

Figure 15-13 Seismic sparker-profiler evidence of submarine stratigraphy. (A) Profiler chart of western flank of the Mid-Atlantic Ridge along 5° N latitude and cross section compiled from profiler charts along 44° N latitude show typical pattern of sediment distribution on ridges. (B) Chart of the ocean floor in the Canary Basin in the Atlantic Ocean illustrates how unconsolidated sediment covers the rugged bedrock surface. (C) Terrigeneous turbidites covering the broad continental rise near the African margin of the central Atlantic Ocean display distinct bedding. (D) Terrigeneous sediment is almost absent in the trench west of Peru, but nearly fills it further south off the coast of Chile. (A, B, and C from J. Ewing and M. Ewing in *The Sea* [A. E. Maxwell, editor], page 21, Vol. 4, Part I, and D from D. E. Hayes and M. Ewing in *The Sea* [A. E. Maxwell, editor], page 54, Vol. 4, Part II, Wiley Interscience, 1970.)

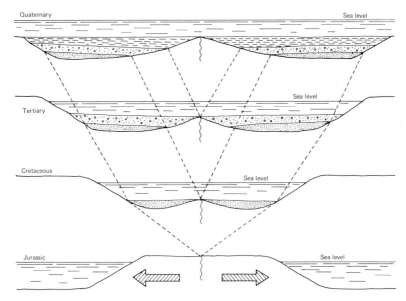

Figure 15-14 Stages in the continuous process of sedimentation on a hypothetical ocean floor spreading outward from a ridge. Observe how older sediment is transported farther from the ridge with the passing of time.

terrigenous turbidites with some biogenic constituents (Figure 15-13C) are typical of broad continental rises. Observe how these deposits thicken closer to the continental slope. Currently these sediments are being intensively surveyed for possible petroleum and natural gas accumulations. The apron of continental rise turbidites, so broad in some regions, is much smaller where continents are bordered by submarine trenches. In some parts of the Pacific Basin older continental margin turbidites are found on the seaward side of a trench which presumably formed after they were deposited. These turbidites were later covered by pelagic deposits. Here we find that changes in the configuration of the ocean basin have affected the distribution of sediment.

Most submarine trenches contain very little sediment, but there are some exceptions where trenches are nearly filled by terrigenous turbidites. Pelagic sediment is almost absent in these deposits. Examine the seismic profiler records (Figure 15-13D) that reveal deposits of sediment in trenches. Observe that most of these beds are undeformed and possess almost no tectonic structure. This is difficult to explain in such a tectonically active zone.

There are four important mechanical processes of deep sea sedimentation. These include: (1) transport by deep ocean currents that are part of the global system of circulation; (2) turbidity currents; (3) low-velocity density currents; and (4) pelagic settling. The low-velocity density currents operate because of density differences related to the mass of suspended sediment, but they flow much more slowly than turbidity currents. One or another of these processes predominates in a particular area, but all the others are probably active to a lesser extent.

Rates of sediment accumulation differ with location and lithology. Thus coarse-grained terrigenous deposits accumulate rapidly along continental margins. But the finer-grained debris remains suspended and is transported to the deeper ocean basin. More biogenic sediment reaches a shallow sea floor than reaches the bottom in deeper water where it has more time to dissolve while settling. The melting rate and drifting range of icebergs affect the accumulation of glacial marine sediments. Oceanographers estimate average sedimentation rates for these different kinds of marine sediment from age and thickness data. Biogenic oozes accumulate at an average rate of between 1 and 5 cm per 1000 years. A much slower rate of 0.1 to 1.0 mm per 1000 years is found for brown clay. Intermittently high accumulation rates occur during turbidity current flows that sometimes exceed several centimeters in a few hours. These currents together with the slower density currents add sediment to large continental rise areas at average rates exceeding 10 cm per 1000 years.

The pattern of distribution of ocean basin sediments of different ages emphasizes dramatically the process of sea floor spreading. The presence of sediment that is everywhere no older than middle Mesozoic age resting on a basaltic bedrock surface points to the relative youth of the deep ocean floor compared with the continental platforms. Only the youngest sediments are found near the ridge axes, but farther away they are underlain by increasingly older deposits. Sea floor spreading is the only process that has been proposed to explain these age distribution patterns of ocean basin sediments.

THE OCEANIC WATER MASS

Knowledge of the chemical and physical properties of the water in the ocean will give us a clearer understanding of its importance in geologic processes. Oceanographers have collected samples of sea water from most parts of the world's oceans. They have been able to recognize global patterns of variation in temperature, density, and the principal dissolved constituents in the ocean from analyses of these sea water samples. Measurements of currents and water level fluctuations indicate local and worldwide systems of water movement and circulation. We have learned that the oceanic water mass possesses a dynamic structure, and that large sea water zones can be distinguished from one another by subtle differences of temperature and composition. In this section we will examine some of these important aspects of chemical and physical oceanography.

Dissolved Constituents in Sea Water

Sea water is a saline fluid containing the dissociated ions of several dissolved salts. The principal ions listed in Table 15-1 have been found in all parts of the world ocean from chemical analyses of water samples. Collectively these ions make up the **salinity** of sea water. Salinity is specified by the total mass of dissolved constituents in a standard mass

Table 15-1
Principal Ions Dissolved in Sea Water

Ion	‰ in Ocean	‰ in Rivers	Ratio $\frac{‰\ Oceans}{‰\ Rivers}$	Residence Time
Cl^-	19	.00176	10800	475 m.y
Na^+	11	.00186	5900	260
SO_4^{2-}	2.7	.0108	250	11
Mg^{2+}	1.3	.0036	360	16
K^+	0.4	.00235	170	7.5
Ca^{2+}	0.4	.0148	27	1.2
HCO_3^-	0.14	.0583	2.4	0.11

of water. Almost everywhere in the ocean the salinity is between 33‰ and 37‰, where the symbol ‰ indicates grams of dissolved constituents per kilogram of water. The average salinity is close to 35‰.

We have discovered a very interesting fact about the relative proportions of the principal dissociated ions in sea water. These proportions are constant throughout the ocean. Chemical analyses of numerous samples reveal that regardless of salinity the ion fractions are everywhere identical. This means that in a sea water sample from anywhere in the ocean chloride comprises 55 percent of its total mass of dissolved constituents. And there will be 1.798 times more chloride than sodium in that sample. We can verify these numbers from the data in Table 15-1. Recognition of this feature of sea water simplifies the effort required to measure the salinity of a water sample. We have used precise and time-consuming chemical studies of a relatively small number of samples to find the individual ion fractions given in Table 15-1. Then, further tests indicate that electrical conductivity is directly proportional to sea water salinity. Therefore, we can routinely measure the electrical conductivity of a water sample and then calculate salinity and the concentrations of the principal ions from the predetermined constant ratios.

The principal ions in sea water are also found in the water of rivers draining into the ocean. The concentrations of these ions (Table 15-1) are much smaller in rivers, and their relative proportions are different. The mass of dissolved constituents in a kilogram of sea water is approximately 500 times greater than the mass in a kilogram of river water. Nevertheless, a very large proportion of the ions in sea water appears to have been transported by river drainage from the continents where they were produced from rock weathering. Much smaller amounts probably come from submarine volcanism and eolian transport. Chloride is the important exception; apparently it is not derived from the continents.

The occurrence of metallic chlorides on the walls of volcanic vents, mentioned in Chapter 7, suggests that erupting magma may be a source of chloride.

Let us assume that all of the principal ions in sea water except chloride were transported from the continents by river drainage. We know that at the present time worldwide river drainage discharges water into the ocean at the approximate rate of 3.15×10^{16} kg/year. The mass of water now in the world ocean is about 1.4×10^{21} kg. We can then calculate that at the present rate of river drainage it would take a little more than 44,000 years to discharge the amount of water now in the world ocean. We designate this value as the **residence time** of the oceanic water mass. We know that water is continually circulating in the hydrologic cycle through ocean, atmosphere, and continent. The residence time of the oceanic water mass might be viewed as the average time required for a water droplet to pass through the hydrologic cycle.

As water drains into the sea, ions are accumulated at rates proportional to their fractions in river water. Ratios given in Table 15-1 indicate the differences of individual ion concentrations in sea water and average river water. For example, the concentration of magnesium in a kilogram of sea water is 360 times greater than the average concentration in river water. Consider the assumption that all of the magnesium in a kilogram of sea water was transported by river drainage. This would require that a kilogram of water has passed 360 times through the hydrologic cycle, each time adding another increment of magnesium to the oceans. Since each cycle takes about 44,000 years, a total of 16 million years would be needed to accumulate the magnesium concentration now found in the sea. This duration is the residence time for magnesium. The values of residence time for other abundant ions given in Table 15-1 are obtained in a similar way.

Because ions are continuously transported by rivers into the ocean, we might think that sea water salinity and the relative proportions of these ions should change with time. It would appear from data in Table 15-1 that some should be accumulating much more rapidly than others. However, analyses of fossil sea water obtained from the pore space of ancient impermeable sediments indicate similar ion concentrations for former times. This and other evidence suggest that sea water composition probably has not changed very much at least since Paleozoic time. Certainly if ions were simply added and retained in the oceans, the sea water would have become saturated long ago. Because the ocean remains undersaturated, there must be other processes that remove these ions from the water. What do we know about these processes?

Large concentrations of the biocarbonate ion HCO_3^- and the calcium ion Ca^{2+} would rapidly develop if these ions were not otherwise removed. Their importance in marine biologic processes is well known. They are synthesized by numerous organisms to form shells of calcium carbonate which eventually sink to the ocean floor and become incorporated into the mass of deep sea sediment. Insofar as these ions are removed by such biological processes at the same rate as they are discharged into the sea, their concentration in sea water will remain unchanged. Apparently such a balance has existed for a very long time.

Chemical processes must also be removing other ions from the sea. The nature of these processes to a large extent remains unknown. According to Table 15-1 sodium is added at a very slow rate, and therefore can be balanced by a slow removal rate. We know that sodium exists in some deep sea sediments. How does it get there? Is it removed from sea water by some process yet to be discovered? Or does it come from decomposition of minerals such as plagioclase that have not interacted with sea water? It will be very difficult to make an accurate determination of the rates at which these and other sources add sodium to the deep sea sediments. Similarly for other ions, much research is necessary to learn about the different processes affecting them and rates at which they are added or removed from the sea.

Almost all of the natural chemical elements are found in minute concentrations in sea water. Relative proportions of many of these are significantly different from one location to another. Constant ion fractions such as those given in Table 15-1 are found only for the most abundant constituents.

Industrial processes have been developed to extract some constituents from sea water. Since ancient times salts have been collected from evaporation ponds. Sodium chloride (NaCl) makes up more than 95 percent of the mass that can be crystallized in this way. The important salts that can be precipitated are listed in Table 15-2. Electrochemical processes have been used to extract chlorine and small concentrations of bromine from sea water. These methods are most widely applied in processing ancient brines pumped from subsurface reservoirs in continental areas. Some of these brines were concentrated by

Table 15-2

Compounds that Would Yield the Principal Ions if Dissolved in Water

Compound	Uses
Sodium chloride NaCl	Table salt
Sodium bicarbonate $NaHCO_3$	Settles the stomach (antacid)
Magnesium sulfate $MgSO_4 \cdot 7H_2O$	Epsom salts
Sodium tetraborate $NaB_4O_7 \cdot 10H_2O$	Borax
Calcium chloride $CaCl_2$	Absorbs moisture—dust control on roads
Potassium chloride KCl	Electrolyte
Calcium sulfate $CaSO_4 \cdot 2H_2O$	Gypsum for plaster of paris
Magnesium chloride $MgCl_2$	Silkscreen etching agent

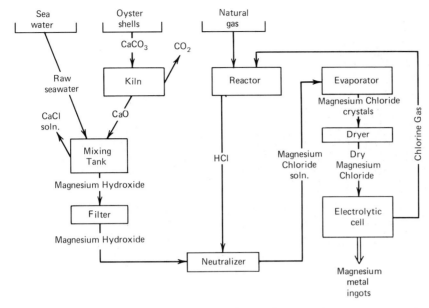

Figure 15-15 Diagram of the Dow electrochemical process for extracting metallic magnesium from sea water.

partial evaporation of former epicontinental seas, and therefore have significantly higher salinity than modern sea water.

Sea water is the principal source of metallic magnesium, which is a useful industrial metal. The electrochemical procedure for processing raw sea water is illustrated in Figure 15-15. This diagram indicates the complicated series of steps in the industrial process for extracting this useful commodity from a very abundant raw material.

Ocean Current Measurements

Several techniques are used to estimate the direction and speed of currents in the ocean. Surface currents can be inferred from the drift of floating objects. Drift bottles can be placed at designated locations in the ocean to be retrieved by chance wherever found by other seafarers. The message in the bottle instructs the finder to report the time and location where the bottle was discovered. This seemingly crude method has provided much data about the general pattern of currents near the ocean surface.

More sophisticated techniques involve current meters that measure the flow of water past a fixed point. One such device is the current meter illustrated in Figure 15-16. It can be anchored in a fixed location and maintained at a certain depth by a submerged buoy. Flowing water then causes the impeller to rotate at a rate proportional to the current speed. Fins orient the device in line with the current direction. Electrical signals indicating speed and direction are transmitted by radio or cable to the recording vessel. Currents of less than 1 cm/sec (0.036 km/hr) can be detected. Other electromagnetic devices have been designed to measure currents to even greater precision under proper operating conditions.

Extremely slow deep sea circulation can be inferred from analysis of carbon 14 dissolved

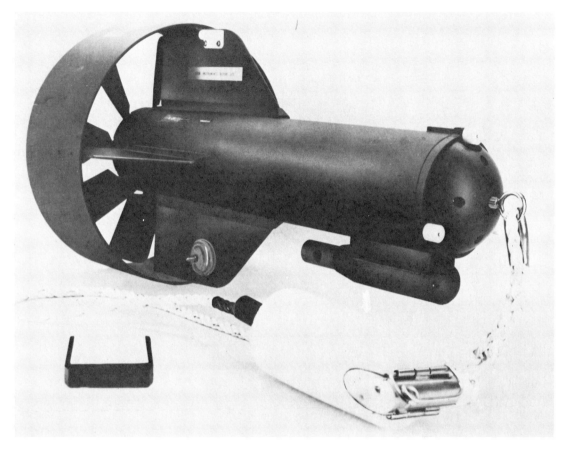

Figure 15-16 A modern current meter measures speed of ocean current by impeller blades turned by flowing water. (Courtesy of Environmental Devices Corp., Marion, Massachusetts.)

in sea water. The ratio of C^{14}/C^{12} that we learned about in Chapter 1 indicates the time elapsed since the water was at the sea surface where interaction with the atmosphere could take place.

Surface Currents

We can tell from historical accounts that early seafarers knew about ocean currents along established trades routes. They describe the differences in the time required to sail to and from different ports and the drift of ships in different parts of the ocean. Their navigation records provided the information used in the first attempts to compile a global chart of ocean currents. In modern times systematic studies making use of drift bottles, current meters, and other devices have greatly clarified this circulation pattern. The principal currents measured on the ocean surface are shown in Figure 15-17. We know that these currents diminish below the surface and become negligible at depths of approximately 100 m. For this reason they are called **surface currents**. They follow a circulation pattern much different from that of the slow deep sea currents.

Surface currents flowing along the western

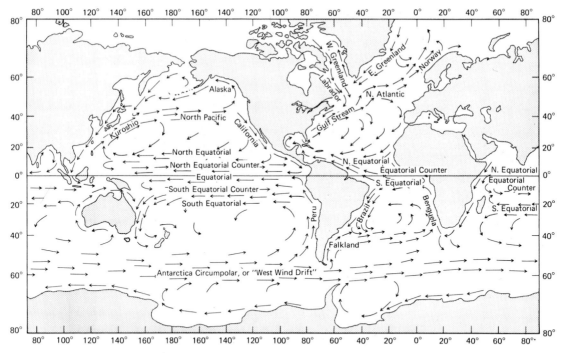

Figure 15-17 Surface currents in the world ocean. (Reprinted from K. S. Stowe, *Ocean Science*, page 284, John Wiley & Sons, Inc., 1979.)

margins of the oceans are most swift. The Gulf Stream moves northeastward near the Atlantic border of North America at speeds of between 2 km/hr and 9 km/hr depending upon location and season. Similar velocities are recorded in the Kuroshio Current that flows along the Asian margin of the Pacific Ocean. In the western Indian Ocean the Agulhas Current moves southwestward along the coast of Africa at speeds exceeding 2 km/hr. Elsewhere in the world ocean, surface currents move more slowly in a system of gyres. Between latitudes 55 and 65°S the West Wind Drift (or Antarctic Circumpolar Current) flows in an uninterrupted path around the earth. Narrow surface streams called the equatorial countercurrents flow opposite the prevailing water movement in the central Atlantic and the central Pacific Oceans. In this global system of surface currents water draining from a particular continental area can be transported great distances in periods of a few years or less. These currents play an important role in mixing dissolved and suspended constituents so that the differences in sea water composition from one place to another remain small.

Atmospheric circulation is the principal cause of surface currents in the ocean. These currents are the result of friction between the ocean surface and prevailing winds. Insofar as the atmospheric circulation pattern persists during the year with only secondary seasonal variations, so do the surface currents in the sea. The diagrams in Figure 13-36 show the circulation of those air currents in direct contact with the ocean. The largest seasonal changes in surface currents occur in the eastern Indian Ocean in response to seasonal changes in wind direction.

Density differences in the water near the ocean surface can also contribute to surface currents. These density differences are related to temperature and salinity. Near the sea surface water tends to flow to establish uniform density and hydrostatic equilibrium.

Deep Sea Circulation

The flow of water deep in the ocean basin is primarily the result of subtle density differences in the water mass related to temperature and salinity. Examine the pattern of circulation that is believed to exist in the Atlantic Ocean (Figure 15-18). Water cooled near the surface in the polar latitudes sinks, then migrates horizontally into lower latitudes. Water nearer the surface moves from intermediate and equatorial latitudes into the north polar latitudes. Otherwise the pattern is somewhat complicated by circulation within individual zones which can be distinguished from one another by subtle temperature and salinity differences.

We can calculate estimates of the speed of deep sea circulation by determining the length of time water is away from the surface. Analysis of C^{14}/C^{12} ratios indicates that water sinking to the ocean bottom near Antarctica remains about 750 years away from the surface in the Atlantic Ocean and approximately 1500 years away from the surface in the Pacific Ocean. During these periods of time the water follows a path of perhaps 5000 km distance. These deep sea currents, then, flow at the rate of a few kilometers per year. Although this is much slower than the movement of surface currents, it is fast enough to insure mixing of dissolved constituents in the ocean.

Turbidity currents are known to flow from time to time in the deep ocean basin. These currents are not part of the steady circulation system in the ocean. When they occur, they have an important effect upon ocean bottom topography and distribution of sediment. Recall that these currents are described earlier in this chapter.

The Ocean Tide

Periodic rise and fall of sea level is observed everywhere in the ocean. The cycles of water level variation lasting several hours or more make up the **ocean tide.** Superposed upon the tidal cycle are short-term fluctuations caused by wind-generated waves on the sea surface.

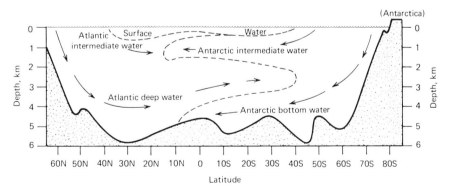

Figure 15-18 Inferred deep sea circulation in the Atlantic Ocean. Arrows indicate flow of water. Dotted lines separate the principal water masses that are distinguished from one another by subtle differences in salinity and temperature. (Reprinted from K. S. Stowe, *Ocean Science,* page 309, John Wiley & Sons, Inc., 1979.)

570 **The Face of the Earth**

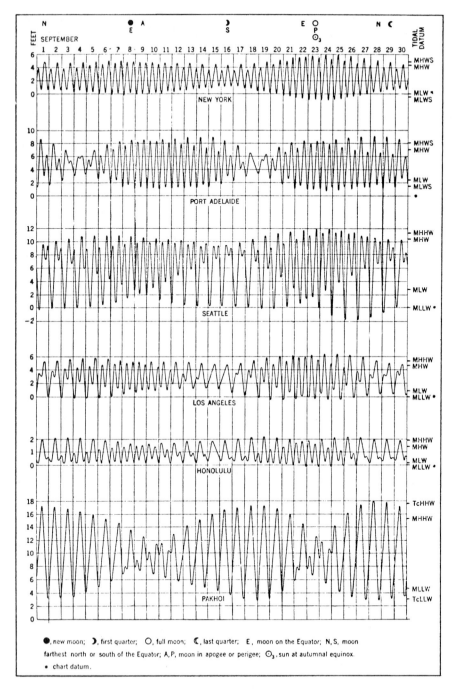

Figure 15-19 One-month records of the ocean tide showing typical semidiurnal tides, mixed tides, and diurnal tides. (U.S. Navy Oceanographic Office.)

Oceanographers have measured tidal water level fluctuations and associated currents at a large number of coastal sites near continents and islands. Large differences are found from one location to another. These are evident in the examples shown in Figure 15-19. The difference between high water level and low water level during a cycle is called the **range** of the tide. In some parts of the ocean the range is less than a few centimeters and in a few places it is between 10 and 20 m. Along the Atlantic margin of North America ranges of 1 or 2 m are common. We know from Roman history that tidal fluctuations of between 5 and 10 m in the English Channel were at first unknown to Caesar's invaders and seriously upset their first efforts to transport an army into Britain. The tide on their familiar Mediterranean coast is very small.

A feature of all tidal records is the change in range during a period of approximately one month. During times of new moon and full moon the range is higher than average. These few days of abnormally high range which occur twice each month are the times of **spring tide.** At alternate times the tide is abnormally low for a few days. These days are the times of **neap tide.** The change in range between spring tide and neap tide is quite different from one place to another. Look at the record from Pakhoi (Figure 15-19) that shows a spring tidal range more than three times larger than the neap tidal range.

At most locations the tide rises and falls twice each day. Successive times of high water occur at intervals of approximately 12.5 hours. The water level at successive times of high tide may be quite different, however. Consider the examples in Figure 15-19. Near New York the ranges of two successive cycles vary by less than 10 percent. The tide at this location can be described as principally semidiurnal. In contrast, the times of high water near Pakhoi occur at intervals of approximately 25 hours. Here the tide is principally diurnal. Elsewhere, as near Los Angeles, the ranges of successive cycles differ markedly.

This pattern is called a **mixed tide.** Tidal records from many regions indicate that mixed tides and semidiurnal tides are much more common than diurnal tides.

Tidal currents in the ocean basin are usually too slow to have any appreciable direct effect on geologic processes. This is not true along the margins of continents and islands, where tidal currents can be effective in modifying the landscape. These currents are most swift in narrow channels connecting coastal bays with the ocean. Water must flow into the bay as the tide rises and back out into the sea as the tide falls. Consider the idealized example in Figure 15-20. Assume that a semidiurnal tide with a range of 3 m is recorded in the bay that covers an area of 5 km $\times$ 5 km = 25 km^2. The bay is connected to the sea by a channel 200 m wide and 10 m deep at mean sea level. In the 6-$\frac{1}{4}$ hour time between low tide and high tide a water volume of 75×10^6 m^3 must flow into the bay through the channel, which has a cross-sectional area of 2000 m^2. The average current velocity would have to be 6 km/hr in order to transport that volume of water. Observe that at the instants of high water and low water levels the current changes direction and momentarily has no velocity. It then increases, reaching a maximum of more than 10 km/hr at a time midway between high and low water times. We see from this numerical example that the tidal current could reach a velocity more swift than currents in many major rivers. The current would be slower if the channel were wider and deeper or if there were several channels connecting the bay with the sea. As the tide swirls in and out of a bay with a broad opening to the sea, the current pattern can be more complicated.

Because of the rise and fall of the tide, there is a coastal fringe of land that is alternately flooded and drained during the cycle of water level fluctuation. This is the **intertidal zone.** It is very narrow along steep coasts where the land rises abruptly from the sea. It can be much broader in coastal re-

572   The Face of the Earth

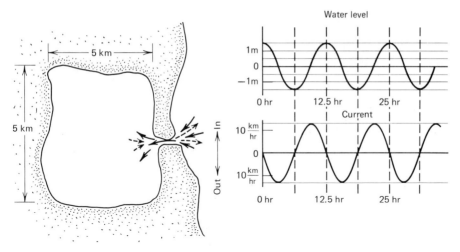

Figure 15-20 Semi-diurnal tide in an idealized rectangular bay connected to the sea by a narrow channel. Graphs show water level and current fluctuation within the channel. Note that current is most swift at times when water level is changing most rapidly, that is, where the water level graph crosses mean sea level. Current velocity diminishes to zero at times of high and low water. All water contributing to tidal fluctuations in the bay must flow in or out through the channel.

gions of subdued relief. Tidal marshes several kilometers wide are found in some places. An example is the marsh along the coast of South Carolina (Figure 15-21).

Isaac Newton was able to explain several important features of the tide by analyzing the gravitational and centrifugal forces associated with planetary motion. In terms of these forces it is possible to calculate the relative motions of pairs of masses. Consider for example the earth and the sun. Both of these bodies follow orbits about a common center of mass that lies close to the sun because of its greater individual mass. The earth remains in its orbit because at its center the outward centrifugal force is exactly balanced by the inward gravitational force due to the sun's mass. On the surface of the earth, however,

(A)

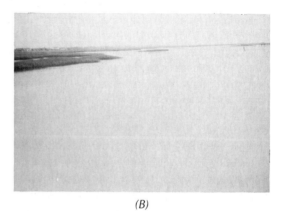

(B)

Figure 15-21 Marsh near Surfside Beach, South Carolina, during (A) low tide and (B) high tide.

these two forces are not equal in magnitude and opposite in direction. Instead, we find that the centrifugal force has the same magnitude and direction everywhere in the earth, whereas the magnitude and direction of the gravitational force due to the sun are different from one point to another in the earth. Consider in Figure 15-22A the line of length r connecting the centers of the earth and the sun, and the line of length r_p connecting the center of the sun and the point p in the earth. The centrifugal force $\vec{F}_c$ on an element of mass m_p located at any point in the earth is in the direction of r and has the magnitude:

$$F_c = (GM_s/r^2)m_p \qquad (15\text{-}1)$$

where M_s is the mass of the sun and G is the universal gravitational constant. At any particular point p, the gravitational force $\vec{F}_g$ is in the direction of r_p and has the magnitude:

$$F_g = (GM_s/r_p^2)m_p \qquad (15\text{-}2)$$

The solar tidal force at p is

$$\vec{F}_T = \vec{F}_g - \vec{F}_c \qquad (15\text{-}3)$$

Vectors in Figure 15-22B indicate the field of tidal force which is symmetrical about the line r.

If we examined the earth-moon system independently, we would find a similar distribution of lunar tidal force vectors relative to a line passing through the centers of the earth and the moon. The moon is much closer to the earth so that it causes a tidal force more than twice as large as the solar force.

The direction and relative magnitude of the net luni-solar force at different points on the earth's surface, referred to a line between the centers of the earth and the moon, is illustrated in Figure 15-23. An outward component of force exists over surface regions facing toward and away from the moon. Ocean water would be drawn outward to form a high tide simultaneously in these regions on

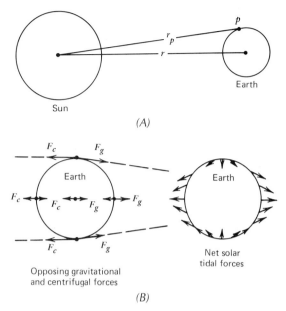

Figure 15-22 Balance of gravitational (F_g) and centrifugal (F_c) forces on the earth caused by the sun. See text for detailed explanation of these diagrams.

opposite sides of the earth. Between these zones of outward force is a region banding the earth where the net luni-solar force is inwardly directed. In this region ocean water would be drawn down to form a low tide. During any given day the earth rotates about its axis, whereas the changes in the positions of the sun and moon are relatively small. Therefore, during one day a point on the ocean surface will alternately experience outwardly and inwardly directed forces. Because of the form of the force field the water will be drawn outward to form a high tide during two intervals of time each day. During two alternative intervals it will be shifted to form low tide.

The alignment of the force vectors relative to the position of the moon persists throughout the month; however, the absolute strength of the force field fluctuates with the changing positions of the moon and the sun. This is illustrated in Figure 15-24. The force

574   **The Face of the Earth**

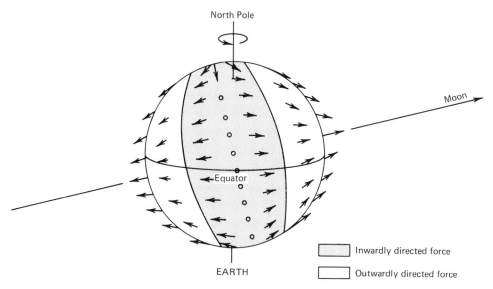

Figure 15-23 Vectors illustrating the variation of the tide-generating force on the earth's surface due to the moon. Line connecting the centers of the earth and moon is different from axis of earth's rotation. Therefore, the force at a point on the surface of the earth changes with time because of axial rotation.

is strongest during times of new moon and full moon when the moon, sun, and earth are approximately in line. The higher than average tides at these times are spring tides. When the moon revolves to positions where a line between its center and the earth is approximately perpendicular to the line between the sun and earth, the luni-solar force field on the earth is weakest. These are times of neap tide.

Because the luni-solar tidal force field illustrated in Figure 15-24 is of simple form we might expect the ocean tide itself to be equally simple and predictable. But observations at many different places show that it varies in a much more complicated way. This complexity is related to the irregular shape of the ocean. Tidal currents tend to shift water toward regions where the luni-solar force is outwardly directed. These currents are deflected by continental margins and ocean basin topography. Therefore, the water is transported in paths quite different from those that would be predicted simply from consideration of the luni-solar force field. The observed ocean tide is the result of water flowing in a basin of irregular shape in response to a known force field.

A water mass in a basin of simple shape will oscillate in a predictable way if it is disturbed. The oscillation can be described in terms of waves with periods and wavelengths related to the dimensions of the basin. We can observe this by shaking a small pan partly filled with water. The shaking causes the water to slosh around the pan. But by shaking the pan first rapidly, then more and more slowly, we discover one particular period of shaking that causes unusually violent sloshing of the water. This period is the resonant period, and it depends on the size and shape of the pan.

The luni-solar tidal force disturbs the ocean water mass. This disturbance is periodically repeated as the earth rotates. In effect it gives the ocean a shake every $12 \frac{1}{2}$ hours. Differ-

Geologic Processes in the Marine Environment 575

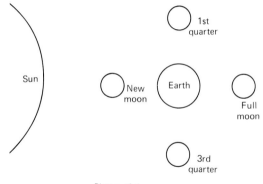

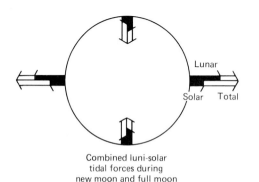

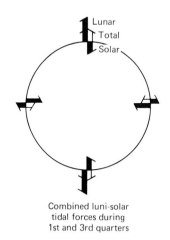

Figure 15-24 Tidal force field caused by combined lunar and solar effects for different relative positions of the earth, moon, and sun. See text for a more detailed discussion of these diagrams.

ent parts of the ocean can be viewed as adjoining pans or basins. In each of these the water will slosh about, if distrubed, at periods related to the dimensions of that particular basin. For some basins these periods of oscillation correspond to the repetition period of the luni-solar tidal force. High tidal ranges are measured in such parts of the ocean owing to the resonance effect. Elsewhere resonance is less perfectly developed and lower tidal ranges are observed. Depending upon the region, semidiurnal resonance may be better or more poorly developed than diurnal resonance. Because of the large differences in resonance effects from one part of the irregular ocean basin to another, the features such as the range, and the semidiurnal, diurnal, or mixed nature of the observed ocean tide change with location.

Ocean Waves

The surface of the ocean is continually disturbed by waves. Most of these waves are generated by wind. Everywhere over the ocean wind blows in light to moderate breezes or in more violent gales. The ocean waves produced by this turbulent atmosphere propagate along the sea surface, interfering with one another to shape a constantly changing, and seemingly confused pattern of crests and troughs (Figure 15-25). Ordinarily the turbulence related directly to the wind-generated waves extends only to depths of less than 100 m. These waves, then, are called **surface waves** on the ocean. The sea level fluctuation caused by waves can be recorded with water level recording devices (Figure 15-26). Observe that oscillations recur at periods of a few tens of seconds or less.

We can learn about the principal features of surface waves on the ocean from experiments and from the theory of fluid dynamics. If we disturb the water surface in a pan in several places, we can make it look as irregular and confused as the ocean surface. The

576   The Face of the Earth

Figure 15-25 Wave patterns on the surface of the Atlantic Ocean near North Carolina.

complicated shape of the surface results from superposing several series of waves. But the waves originating from a particular disturbance have a simple form. They could be generated by dropping an object into the water or by directing a jet of air at some point on the surface. As the energy propagates away from the point of disturbance a series of concentric wave crests and troughs is formed. Profiles in Figure 15-27 illustrate the shape of these waves, how they change position with time, and the oscillatory movement of water particles that they produce. The distance between successive crests is the wave length L, and the vertical difference between crest and trough is the wave height H. Observe that the position of a wave crest changes with time, advancing outward from the point of disturbance. The velocity at which the crest moves is called the wave **celerity** (C), and the wave period (T) is the time required for the crest to advance the distance of one wave length. Therefore, we have the following relationship:

$$C = L/T \qquad (15\text{-}4)$$

There is no similarly simple relationship be-

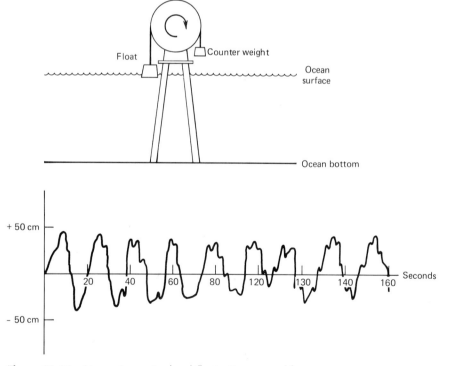

Figure 15-26 Measuring water level fluctuation caused by waves.

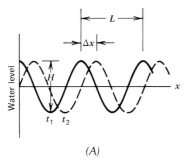 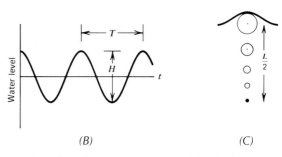

Figure 15-27 The approximate shape, dimensions, and motion of a wave sequence of simple form on the surface of a liquid. (A) Wave profile at two different times t_1 and t_2. (B) Time varying water level variation at one location. (C) Paths of orbital motion of water particles at different depths.

tween H and these other features of a wave.

The celerity of a wave also depends on gravity (g), which acts to restore the disturbed surface to a level surface and the water depth (d) in the following equation that we will not attempt to derive here:

$$C^2 = \frac{gL}{2\pi} \tanh \frac{2\pi d}{L} \qquad (15\text{-}5)$$

This equation can be simplified in different ways by making certain assumptions.

First let us consider a place where water depth is much greater than wave length: $d \gg L$. This is ordinarily true in the ocean basin where depths are several thousands of meters and wave lengths are less than a few hundred meters. Here we see that the ratio d/L assumes a large value so that: $\tanh \frac{2\pi d}{L} \approx 1$; and

$$C^2 \approx \frac{gL}{2\pi} \qquad (15\text{-}6)$$

Consider next the propagation of waves in shallow water where depth is much smaller than the wave length: $d \ll L$. For this case the ratio d/L has a very small value; therefore, $\tanh \frac{2\pi d}{L} \approx \frac{2\pi d}{L}$, so that

$$C^2 \approx gd \qquad (15\text{-}7)$$

We also know that the agitation of water by surface waves diminishes with depth. In water that is much deeper than the wave length the path of a particle of water is nearly circular (Figure 15-27C). The radius of this circular path is largest at the surface and decreases to a negligible size at the approximate depth: $h \approx L/2$. Theoretically, wave agitation extends everywhere in the water, but for practical purposes the agitation is assumed to be restricted to depths of less than one-half wave length.

We can tell from Eq. (15-5), (15-6), and (15-7) that the celerity of a wave will decrease as it moves from deep water into shallow water. But the wave period does not change. This means that the wavelength must decrease, as can be seen by substituting the right-hand side of Eq. (15-4) into Eq. (15-7). Wave height increases as the wave moves into shallower water. These changes in wave length and wave height are illustrated in Figure 15-28. Observe that these effects become most obvious where water depth diminishes to less than $L/2$, and the wave is said to "feel" the bottom. In this zone the water agitated by the waves is capable of eroding the bottom.

When wave crests approach the coast at an angle, an interesting thing happens. The near shore part of the crest encounters shallow water earlier than the part which is farther out in

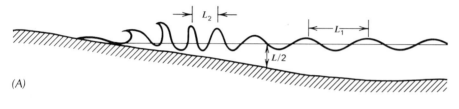

(A)

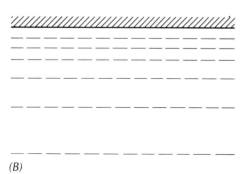

(B)

Figure 15-28 Change in wave length and celerity related to movement of waves from deep water into shallow water along a coast. As celerity C and wavelength L decrease, wave height increases. Very near the coast the wave becomes unstable and overturns in a breaker. (A) Cross section showing wave length and height variation with distance from the shore. (B) Map showing positions of a wave crest (dashed lines) at successively later times.

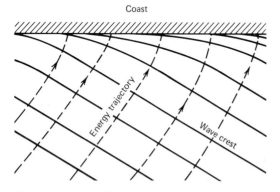

Figure 15-29 Refraction of waves moving obliquely from deep water into shallow water along a coast. Solid lines indicate wave crests. Dashed lines showing trajectories of movement of individual points on a wave crest represent directions of energy propagation.

the ocean. The celerity of this inshore part becomes slower than the part still in deep water, which causes the wave crest to bend (Figure 15-29). We can see that as celerity decreases, the direction of the advancing wave crest changes. This is the phenomenon of **wave refraction.** The trajectories shown in Figure 15-29 illustrate the direction at which wave energy approaches the coast. Observe that wave refraction causes the energy to be concentrated more directly upon the coast.

As waves advance into very shallow water, wave height increases, whereas wave length decreases. Graphs in Figure 15-30 illustrate how these dimensions change. Here the waves become unstable, overturn, and collapse as **breakers** (Figure 15-28). During collapse of a breaker, water is thrown forward. When a wave breaks upon a sloping coast, water is first thrown up on the beach, then runs back down the slope where it is picked up by the next breaking wave. Where waves advance obliquely on the coast, water from a breaker is thrown forward in the direction of the wave movement, but then recedes directly downslope. Therefore, particles of water cycled through a succession of breaking waves migrate along the coast in a zigzag path. This path of water flow is shown in Figure 15-31. The water that washes along the coast in this way makes up a **longshore current.** This current includes the **swash zone** of breaking waves and water that extends a short distance beyond the swash zone which is pulled along by the momentum of the more turbulent water. Depending upon the size of the waves and the angle at which they break upon the coast, the width of the longshore current ranges from a few meters to several tens of meters. This is a turbulent stream that

Geologic Processes in the Marine Environment 579

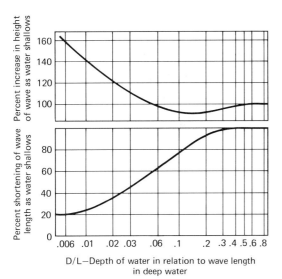

Figure 15-30 Graphs showing decrease of wavelength and corresponding increase in wave height as a wave moves into water of depth less than one-half its wavelength in deep water. (After R. C. H. Russell and D. H. MacMillan, *Waves and Tides,* page 76, Greenwood Press, 1970.)

Figure 15-31 Generation of a longshore current by waves breaking obliquely upon a coast. A succession of breakers causes a particle of water to be alternately thrown at an angle upward on the coast then recede directly downslope, therefore traversing a zigzag course along the slope. This irregular flow of water in the swash zone constitutes a turbulent longshore current.

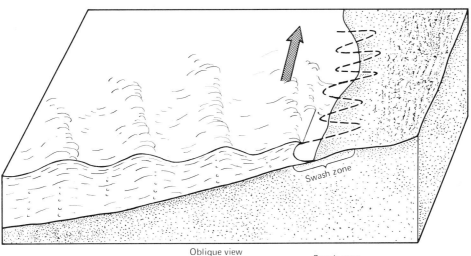

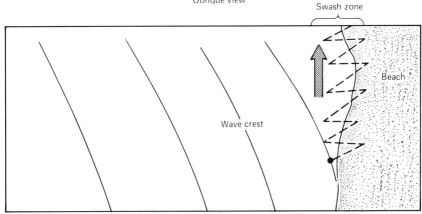

can flow along the coast at velocities of up to a few kilometers per hour. Longshore currents can have a very important effect upon coastal landforms. This effect, and related effects of wave erosion, are discussed in the following section of this chapter.

How high can waves become? Because the interaction of two turbulent fluids, the ocean and the atmosphere, is complicated, there are no exact equations for predicting the wave heights that will be produced by a certain wind speed. Instead we rely on useful empirical relationships that are based on observations and experiments. Following this approach we have learned that wave celerity is less than the speed of the wind that is generating the waves. After sufficient time has elapsed for a steady wind of velocity W to raise waves of maximum size, observations suggest the approximate relation:

$$C = 0.8 W \qquad (15\text{-}8)$$

Using this expression we find that a moderate wind with a velocity of 10 m/sec (36 km/hr or 22 miles/hr) could eventually generate waves of celerity $C = 8$ m/sec (or 29 km/hr). Knowing this, we can use Eq. (15-6) to calculate that the wave length will be approximately 65 m.

An empirical relationship for estimating the height of waves raised by light to moderate winds has been developed from observations:

$$H = .031 W^2 \qquad (15\text{-}9)$$

where W is given in m/sec and H is given in meters. A wind velocity of 10 m/sec would raise waves to the approximate height of 3.1 m according to this formula.

We know from observations of the ocean surface during severe storms that the largest wind-generated waves may exceed 300 m in wave length and reach heights of between 10 m and 15 m. Such waves are very infrequent. The size of waves generated by more moderate winds can be estimated from the empirical formulas (15-8) and (15-9) and Eq. (15-6). From these formulas we can determine the dimensions of surface waves which we could reasonably expect in the ocean.

Very large waves known as **tsunami** are occasionally generated by tectonic disturbances of the ocean floor. Earthquakes, volcanic explosions, and large submarine sediment slumps can produce these waves. Large tsunamis are known to have wave lengths of more than 200 km and move at celerities of over 700 km/hour. In midocean the wave height is ordinarily smaller than a few meters. Because of the long wavelength and low height, a tsunami is imperceptible to an observer on a ship. As it advances into shallow coastal waters, however, the height can rise to more than 20 m. Such a wave produces an enormous breaker that can devastate low coastal areas situated several thousand kilometers from the origin of the tsunami.

There are many accounts of destruction caused by tsunami. One recent occurrence followed a large earthquake in Chile on May 22, 1960. This shock generated a tsunami which swept across the Pacific. In Honshu, Japan, more than 14,000 km away, the tsunami broke upon the coast almost 24 hours after the earthquake. It brought death to 180 people and caused damage estimated at more than $50 million. This tsunami also moved up the coast of North America causing much devastation. It is one example of many such large waves generated by tectonic disturbances.

A tsunami warning system is now maintained through international cooperation to diminish the destruction caused by these waves. Because their wave lengths greatly exceed the depth of water in the ocean, celerity can be predicted from Eq. (15-7) and bathymetric charts. The epicenter of an earthquake and the predicted travel time of a tsunami that might be produced by it can be communicated within minutes to numerous coastal warning stations, where the alert is announced. This provides ample time for ships to move to open sea and people to abandon beaches and low coastal areas.

Long-Term Changes in Sea Level

When we speak of sea level as the surface from which elevation is measured, we are referring to **mean sea level.** This would be the ocean surface in the absence of tides, waves, and currents. Its shape corresponds to the geoid which we learned about in Chapter 2.

There is abundant evidence that the mean sea level surface has changed during the history of the earth. Sedimentary rocks of marine origin are widespread on the continents. Ancient beach deposits are found on inland areas. In evaluating evidence of mean sea level changes, we must be careful not to confuse indications of localized uplift or subsidence of the land surface with indications of worldwide changes. Tectonic activity in a particular region can elevate a former coastline. In contrast, water added to the ocean by melting glaciers contributes to a global rise in sea level. We refer to worldwide rise or fall of the ocean surface as a **eustatic** change in mean sea level.

Eustatic changes in sea level can result in two ways: (1) a change in the global water volume, and (2) a change in the volume of the ocean basin. What are the principal causes of change in water volume? One cause is the advance and retreat of glaciers. Eustatic sea level fluctuations of approximately 200 m accompanied changes in the size of later Tertiary and Quaternary glaciers described in Chapter 14. Glacial processes do not alter the total volume of H_2O near the earth's surface, but only its physical state, that is, ice or water.

Volcanic processes, in contrast, can add to the volume of H_2O. Erupted volcanic gases that are described in Chapter 7 consist mostly of H_2O. However, less than 5 percent of this H_2O is believed to come directly from the magma; most of it is recycled surface water. Nevertheless, the continued addition of water from the earth's interior at a slow rate causes a eustatic change in sea level. This has been estimated to be as much as 3 m in one million years. At this rate all of the H_2O near the earth's surface could have been produced by volcanism in less than 2×10^9 years, well within the time since the origin of this planet.

The volume of the ocean basin can be diminished by the addition of sediment derived from the continents. This sediment may accumulate in deltas or on the ocean basin floor, partly filling the basin and contributing to the rise of mean sea level. We can use the sedimentation rates given earlier in this chapter together with the area of the ocean to compute an estimate of this effect on eustatic rise of sea level.

Solid material is added to the ocean basin by volcanic eruptions. But at the same time other material is removed as the oceanic crust is subducted along destructive plate margins. A temporary imbalance of these effects could contribute to eustatic sea level changes.

We would expect the slow addition of water to the oceans to eventually submerge the continents unless they too are growing. There is evidence to argue that the growth of continental crust from volcanism is approximately in balance with the increase in water volume. Perhaps slight imbalances in the past have contributed to encroachment of epicontinental seas, which are otherwise controlled by tectonic subsidence.

COASTAL LANDFORMS

Waves pound relentlessly on coasts. The roiling and turbulent water loosens fragments from bedrock headlands and dashes them against the shore. As these fragments grind together, they become reduced to pebbles and sand from which beach deposits are formed. Wave action is effective in modifying the landscape along a narrow coastal fringe which advances and recedes with the tide. In some places tidal currents supplement wave action in the processes of coastal erosion and sediment transport. Long-term changes in the position of the zone of wave action are

582 The Face of the Earth

caused by vertical tectonic movements and by eustatic variation of sea level.

The Coastal Landscape

The landscape bordering the ocean varies from rugged mountainous terrain rising steeply from the sea to low marshy coastal plains. The shoreline is almost straight along some stretches of coast and elsewhere is extremely irregular. Typical coastal landforms caused by wave erosion are steep **bluffs** and **sea cliffs.** Waves from the stormy North Sea have produced the bluffs along the eastern coast of Scotland, as pictured in Figure 15-32. This illustration also shows a low rocky **wave-cut terrace** projecting from the base of the steep bluff. In some places bedrock terraces are submerged in shallow water and perhaps covered by a veneer of sediment. Elsewhere they are emergent during low tide, and the sediment has been washed from the surface. Remnants of durable bedrock, which have resisted the erosion of coastal waves and currents, stand above wave-cut terraces to form **sea stacks** in some locations. A large sea stack pictured in Figure 15-33 is situated along the coast of New Zealand near the city of Christchurch. The cliffs and bluffs, wave-cut terraces, and occasional sea stacks are the principal erosional landforms shaped by wave action. Exotic sea caves and arches found in a few localities have been carved

Figure 15-32 The coast of the North Sea near Berwick, Scotland. Low, rocky wave-cut terraces project from the base of the sea cliffs along the headlands. Sand has accumulated along the beach on inner margin of the small bay.

Geologic Processes in the Marine Environment 583

Figure 15-33 Sea stack and sea cliff along the South Pacific coast near Christchurch, New Zealand.

Figure 15-34 Cobble beach and sea cliff along the North Sea coast near Berwick, Scotland.

in sea cliffs by unusual focusing of wave energy.

The persistent pounding of the surf on a steep coast undercuts the slope until it becomes unstable and collapses. This is the process by which bluffs and sea cliffs are formed. The debris that accumulates at the base is further fragmented by wave action and becomes broken into smaller and smaller cobbles, pebbles, and grains of sand. Depending upon the irregularity of the coast and the strength and direction of longshore currents, these sediments are transported along the beach to zones of less turbulent water where they form distinctive depositional features. Mixed with the sediment derived from wave erosion is the debris load transported into the sea by streams and wind. Figure 15-34 shows a narrow beach covered by cobbles and pebbles that are still close to the bedrock from which they were derived. Farther along this coast is the small bay seen in Figure 15-32 where the smaller sand grains, transported by longshore currents, have accumulated along a broader beach. This pattern of sediment distribution is typical of irregular coasts where bays are separated by rocky headlands. Sediment eroded from bordering headlands has accumulated in the bay seen in Figure 15-35 and forms a low flat plain on which a small town is now situated.

The typical zones of erosion and sediment deposition along an irregular and rocky coast are illustrated in Figure 15-36. Here we see the tendency for sea cliffs and wave-cut terraces to form along projecting headlands, and beach deposits to accumulate in bays. We can understand the process of erosion and deposition by considering the refraction of waves and the local longshore currents they produce along an irregular coast. As a long wave crest approaches land, the segments directed toward headlands encounter shallow water, and they decrease in celerity, whereas other segments proceed with unreduced celerity toward coastal bays. Observe in Figure 15-37 how wave crests become warped because of the change in celerity. Refraction causes the wave to move obliquely along the sides of the bay and to produce local longshore currents that flow from headlands into the bays.

Trajectories in Figure 15-37 show the movement of points on a wave crest. They indicate the way in which wave energy is directed on the coast. The curved trajectories show how this energy is focused upon the headlands. Here the bluffs and terraces are formed by erosion. The sediment produced by this erosion is then transported by local longshore currents and deposited in the bays.

Other depositional features are produced

Figure 15-35 Low flat plain of sedimentary deposits fills a bay between projecting headlands along the coast of the Banks Peninsula near Christchurch, New Zealand.

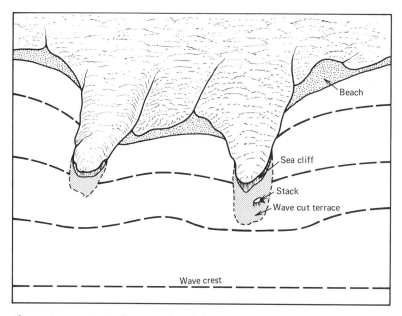

Figure 15-36 Typical erosional and depositional landforms resulting from wave action on an irregular coast.

Geologic Processes in the Marine Environment 585

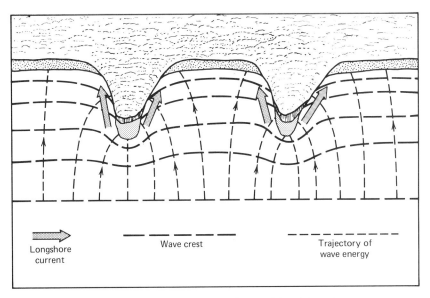

Figure 15-37 Wave crests, energy trajectories, and local longshore currents related to wave refraction along an irregular coast.

Figure 15-38 This spit separates a shallow bay from the South Pacific Ocean near Christchurch, New Zealand.

from sediment carried by longshore currents. Arms of land called **spits** are built seaward by deposition of sand and pebbles. Narrow **bars** of sand and gravel accumulate to form low causeways between the mainland and nearby islands and between headlands across the openings of bays. Examine the example of a spit that is pictured in Figure 15-38. We find that these features develop where there is an abundance of sediment. A spit is produced by a longshore current which moves along a stretch of coast to a place where the direction of the coast changes abruptly. Here the momentum of the current carries the water and its sediment load seaward beyond the coast. There the current diminishes quickly and the sediment load is deposited to form a projecting arm of land. The process is illustrated in Figure 15-39.

In some locations the continued growth of a spit can eventually form a bar reaching entirely across the mouth of a bay. If the longshore current is relatively strong and sediment is abundant, the bay may become entirely separated from the ocean by such a bar. Usually, however, channels cutting through the bar are maintained by tidal currents. As these channels grow narrower by further deposition, the tidal currents become more swift and turbulent. In many locations a condition is reached where the tidal current in the channel is just sufficient to remove the sediment that has been carried to that location by the longshore current.

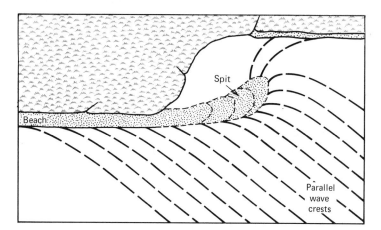

Figure 15-39 Patterns of wave refraction near a spit.

We observe the extensive development of spits and bars in some regions, but in other places we find that most sediment accumulates along the inner shores of bays. How can we explain these differences? Whether we find growth of spits and bars or inshore beaches depends upon the direction of prevailing waves. Consider, for example, a coast that extends in an eastwest direction for several hundred kilometers, along which are found many local bays and headlands. If waves approach this coast from the south (Figure 15-37), only local longshore currents will develop. These will transport sediment from headlands into the bays. If the waves approach the coast obliquely (Figure 15-40) from the southeast, however, a westward-flowing longshore current will develop along the headlands, as well as the local longshore currents leading into the bays. Here some of the sediment eroded from headlands will be deposited on westward extending spits, whereas other sediment will be transported to the inner margins of the bays.

It is important to note that as a spit develops, it alters the shape of the coast and changes the longshore current pattern. The change in current in turn alters the development of the spit. Because of wave refraction on the end of the spit, deposition of sediment tends to be greatest along the inshore margin of the end of the spit (Figure 15-39). This

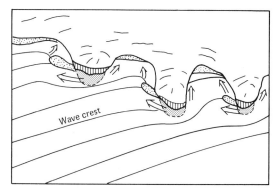

Figure 15-40 Patterns of growth of coastal depositional landforms related to waves moving obliquely on an irregular coast. Here longshore currents flowing parallel to sea cliffs along headlands transport sediment to form bars and spits that extend in the direction of the current.

causes it to grow in a curved hooklike form. Observe in Figure 15-41 the growth of a spit along the New Jersey coast during a 23-year period. This sequence of photographs reveals how dramatically a local stretch of coastline can be altered in a few decades.

Figure 15-41 *(opposite)* Growth of a spit at Little Egg Harbor, New Jersey. Dates of the photographs are 1940, 1957, and 1963. (C. D. Denny et al., U.S. Geological Survey Prof. Paper 590, 1968.)

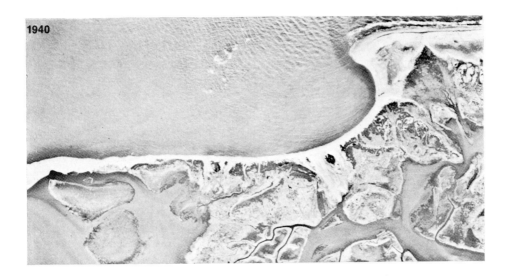

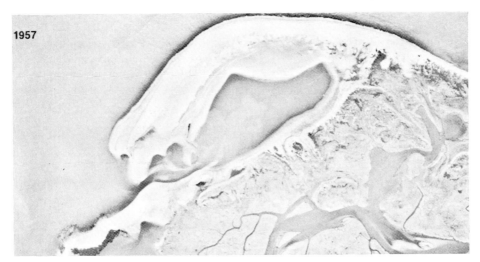

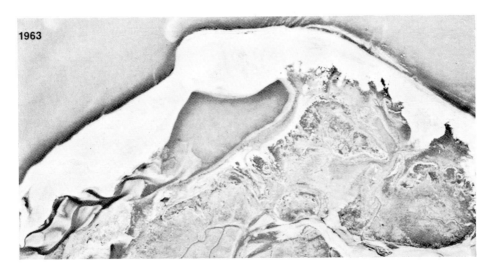

588   The Face of the Earth

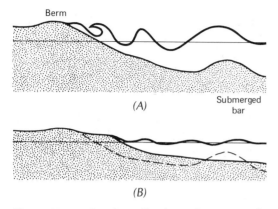

Figure 15-42 Beach profiles formed in unconsolidated sediment by wave action. (A) Shape of the bottom formed by large waves of relatively long wave length and high steepness. (B) Bottom formed by waves of relatively short wave length and low steepness.

Seasonal changes in the distribution of sediment along beaches are widely observed. The cross section in Figure 15-42 compares a beach profile typical of the relatively calm summer season with the profile developed during the stormy winter season. During both seasons a low ridge called the **berm** borders the swash zone of the coast. The position of the crest of the berm follows a seasonal cycle, migrating landward with the onset of winter and seaward during spring and summer. The landward shift of the berm during winter is typically accompanied by growth of a submerged offshore bar. With return of summer the bar disappears and the sediment is redeposited on the berm that is now building seaward. Seasonal changes of the beach near Carmel, California, (Figure 15-43) cause a shoreline fluctuation of more than 50 m.

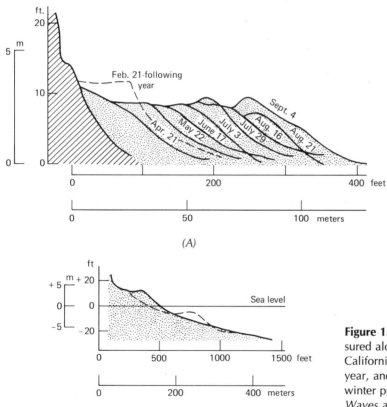

Figure 15-43 (A) Beach profiles measured along the same line near Carmel, California, at different times of the year, and (B) average summer and winter profiles. (After Willard Bascom, *Waves and Beaches*, page 197, Anchor Books, Doubleday and Co., 1964.)

Geologic Processes in the Marine Environment 589

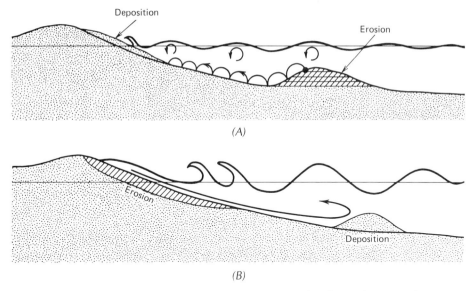

Figure 15-44 Water circulation and sediment transport related to a change in the prevailing length and steepness of waves. (A) Change from long, steep waves to short waves of low steepness causes erosion of submerged bar and deposition of sediment along the shore. Berm position moves seaward. (B) Change from short waves to long, steep waves. Erosion of the shore causes berm position to shift landward. Sediment is deposited to form a submerged bar.

We can correlate short-term, day-to-day or week-to-week changes in a beach profile with changes in the intensity of wave action. Submerged bars develop during severe summer storms, only to disappear later as the waves diminish. Fluctuations of this kind are evident in Figure 15-43 B.

The landward or seaward shift of beach sediments appears to be related to the **steepness** of waves, which is the ratio of wave height and length *(H/L)*. Laboratory studies indicate that berm erosion and growth of submerged bars begins when wave steepness exceeds the approximate value of 0.03. Bars do not form when steepness is less than 0.025. Erosion of bars and deposition of sediment on the seaward slope of a berm commences when wave steepness diminishes below a critical value. This value is related to the average slope of the coast and the grain size of beach sediment.

The differences in water circulation and sediment transport processes for waves of low steepness and high steepness are compared in Figure 15-44. When steepness is small, gentle orbital motion of water caused by low waves lifts a particle from the bottom at one point and sets it on the bottom slightly nearer the shore. Its seaward return to its original position is retarded by friction. Consequently, the particle advances slowly toward the shore. In this way low waves erode submerged bars and build the berm out in a seaward direction. In constrast, steep waves breaking upon the berm wash sediment from its slope. The sediment is carried by the receding swash into deeper water where it is deposited to form a bar.

Evolution of Coasts

The persistent action of waves on a coast tends to reduce the irregularity of the shoreline. In the previous discussion we have noted how wave energy is concentrated on the erosion of headlands. Sediment tends to be redistributed along inner margins of bays, causing the position of the shore to migrate seaward. Spits and bars form as extensions of a coast which otherwise would be more irregular. Beginning with an initially irregular rocky coast, progressive changes illustrated in Figure 15-45 could be expected to result from wave action alone.

Wave action independent of other processes of geomorphology would probably not eliminate entirely irregularities along a rocky coast. The final stage illustrated in Figure 15-45 reveals a modified but still irregular shoreline. Initially a large proportion of wave energy is directed upon the headlands. A wave-cut terrace begins to evolve. As it develops, this terrace forms the bottom of a shallow water zone. Waves must subsequently move over a broader zone of shallow water before expanding energy on the receding shore. A limit is reached when the terrace is so broad that waves moving over it decrease in celerity to the extent that no appreciable energy remains at the shore. Along the coastline illustrated in Figure 15-33 waves wash gently on the shore because most energy has been dissipated farther out as the wave drags over the shallow terrace. Even though some irregularity remains along the shore, the effectiveness of waves for further modification is negligible.

Other processes interact with waves to alter the form of the coast. Streams discharge sediment into the sea which can be transported by longshore currents even where waves are no longer effective in producing sediment by erosion. Deltas formed by these streams extend the margin of land: As the delta develops, introducing irregularity along the shore, we observe a highly complicated interplay of coastal marine processes and fluvial processes.

The modification of a coast by wave action can be interrupted by eustatic sea level changes and vertical tectonic movements of

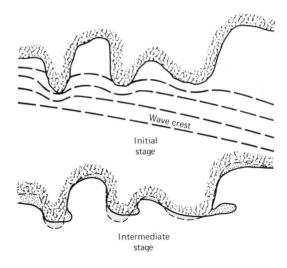

Figure 15-45 Stages in the evolution of a rocky, irregular coast caused by wave action alone. Because the effectiveness of waves for erosion is diminished by progressively broader wave-cut terraces, the coast will retain some irregularity after prolonged wave action.

the coastal land. An **emergent coastal landscape** develops when the sea recedes. If the sea level change relative to the coast was intermittent, then former beach lines, bluffs, and terraces are seen at elevated positions near the coast. Typical features of an emergent coastal area are shown in Figure 15-46.

The submergent coastal landscape is found where the sea is progressively advancing over a landscape that has been molded by subaerial processes. Submergent coasts tend to be irregular, the shoreline following the contour of the land as it had been shaped by streams, wind, and glaciers. Deep fjords and bays indent the land. Below the surface of the

Geologic Processes in the Marine Environment 591

Figure 15-46 Features of an emergent coast. Note the raised shorelines and wave-cut terraces. (Photograph by John S. Shelton.)

Figure 15-47 Features of a submergent coast north of Portland, Maine. Irregular coast follows a contour on a land surface originally shaped by stream erosion. (J. R. Balsley, U.S. Geological Survey.)

encroaching sea, echo sounder measurements reveal a bottom topography that retains characteristics of stream drainage patterns. Wave action has not had sufficient time to modify the shore. Beaches tend to be narrow, and wave-cut terraces are in the initial stages of development. Examine the features of a submergent coastal landscape illustrated in Figure 15-47. We know that the recent retreat of northern hemisphere continental glaciers

592 The Face of the Earth

has caused an eustatic sea level rise of almost 140 m. As a consequence, we now observe typically submergent coastal landscapes in many parts of the world.

Maintenance of Coasts

Communities have developed along seacoasts since ancient times. Traditionally, sheltered coastal bays have been the harbors about which commerce has flourished. More recently, the sandy beach itself has become the important natural resource for expanding resort activities. Because of the dynamic nature of the coastal landscape, the form of harbors and beaches can become altered significantly in a few years or decades. Consider the prospect of establishing a resort along the beach of a sheltered cove such as the one seen in Figure 15-41 A. Within 20 years the beach becomes the border of a brackish marsh. Think about the community that was built along the shore pictured in Figure 15-48. The consequences of continuing shoreline erosion are obvious.

Control of coastal erosion and sedimentation can be accomplished by construction of breakwaters, sea walls, groins, and jetties, and by dredging operations. A **breakwater** is a wall placed some distance from shore. It absorbs and reflects wave energy and shelters a zone of quiet water from the open sea. The breakwater pictured in Figure 15-49 A extends the area of a harbor.

Seawalls are constructed directly along the shoreline to protect the land from erosion. They must be more durable than the land and designed to reflect and absorb wave energy.

Walls projecting perpendicularly from the shore into the sea are called **groins.** These structures interfere with the longshore currents causing sediment to be deposited in the manner shown in Figure 15-49 B. Fields of groins have been constructed for the purpose of expanding beaches.

Figure 15-48 Two views showing erosion and collapse of a cliff formed by wave action on the Lake Michigan shore near St. Joseph, Michigan. (Hann Photo Service, Hartford, Michigan.)

(A)

Figure 15-49 Structures designed to control erosion and sediment transport along coasts. *(A)* Breakwaters constructed to form a harbor at Santa Barbara, California. Note growth of spit following completion of breakwater. A dredge located near the end of the spit operates to prevent further growth of the spit by removing sand to a location further along the coast in the downstream direction of the longshore current. *(B on next page)* Groins along the Atlantic Coast near Willoughby Spit, Virginia, cause deposition of sand carried by the longshore current. *(C on next page)* Jetty at the entrance of Cape May Harbor, New Jersey, insures sufficient tidal current to prevent sedimentation in the inlet. (*A* is from Mark Hurd Aerial Surveys, and *B* and *C* are U.S. Army Corps of Engineers photographs.)

(B)   (C)

Figure 15-49 *(continued)*

A harbor channel can be protected from sedimentation by construction of a **jetty.** This structure consists of walls arranged to maintain the channel at a proper width so that the tidal current will be sufficient to flush the sediment that would otherwise be deposited. A simple jetty is illustrated in Figure 15-49C.

The presence of structures designed to control coastal erosion and sedimentation alters the pattern of wave action. Sediment deflected from one zone begins to accumulate elsewhere, perhaps with undesirable effects. Observe in Figure 15-49A the spit that developed after construction of the breakwater.

Dredges equipped with powerful pumps are sometimes operated to suck the accumulating sediment from the bottom and move it via pipelines or barges to still other locations.

Most attempts to control the process of coastal geomorphology are only partially successful. Combinations of structures and dredging operations, however carefully designed, may produce unanticipated perturbations in the powerful natural system of waves and currents. Nevertheless, many useful coastline engineering systems have been designed.

STUDY EXERCISES

1. The distribution of sediment illustrated in Figure 15-6 shows that calcareous ooze is the dominant sediment in the Mid-Atlantic Ridge Province, but that inorganic detrital sediment is more abundant on the bordering ocean basin floor prov-

inces. Discuss two reasons for this difference in the abundance of these two kinds of sediment.

2. The concentrations of potassium and calcium in sea water are almost equal. But the concentration of potassium is much smaller than the concentration of calcium in rivers that drain into the ocean. Explain why the residence times of these two ions are different. How do we know that calcium is being removed from sea water at a much faster rate than potassium is being removed from sea water?

3. In a bay connected to the ocean by a narrow channel high tide will occur at 10:30 A.M. and 11:06 P.M. This implies that the current flowing into the bay is most swift at which one of the following times: 6:00 A.M.; 10:30 A.M.; noon; 2:00 P.M.?

4. Waves from the southwest break on the sandy beach shown on the map.
    (a) What erosional and depositional features will develop at locations A

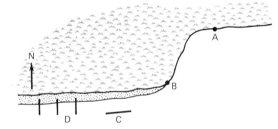

Figure 15-50

and B as a result of wave action?

(b) Discuss how the transport and deposition of sand will change if a breakwater is constructed in location C, and groins are built at location D.

5. Suppose that waves that have a wave length of 50 m and a wave height of 1 m where the water is more than 30 m deep approach a sandy beach. According to the information in Figure 15-30 will these waves cause the berm to move shoreward and build an offshore bar, or will they cause the berm to shift seaward and erode an offshore bar?

SELECTED READINGS

Bascom, W., *Waves and Beaches,* Anchor Books (Doubleday and Co.), New York, 1964.

Bird, E. C. F., *Coasts,* M.I.T. Press, Cambridge, Mass., 1968.

King, Cucklaine A. M., *Beaches and Coasts,* 2nd ed., St. Martin's Press, New York, 1972.

Turekian, K. K., *Oceans,* Prentice-Hall, Inc., Englewood Cliffs, N.J., 1968.

Weihaupt, John G., *Exploration of the Oceans,* Macmillan Publishing Co., Inc., New York, 1979.

SECTION FIVE
OTHER PLANETS

16

The Earth in the Solar System

Motions and Dimensions of the Solar Planets
The Terrestrial Planets
  Moon
  Venus
  Mars
  Mercury
  Pluto

Development of the Terrestrial Planets
The Jovian Planets
  Jupiter
  Saturn
  Uranus
  Neptune
Is the Earth Unique?

What can we expect to learn about geologic processes by studying other planets? We can begin looking for an answer to this question by discovering how individual planets are similar and how they differ from one another. Our nearest neighbors are the celestial bodies of the solar system. The earth which we have been studying is one in a group of nine principal planets in orbit about the sun. These principal planets and their smaller satellite planets make up the solar system.

The dimensions and masses of the solar planets and their orbital motions can be measured from the earth. The sizes of these planets are compared in Figure 16-1. We can get some information about chemical composition and temperature from electromagnetic waves emitted and reflected from their surfaces. Telescopic photographs and radar images reveal some landscape features. Earth-based observations provide almost no information about planetary interiors and the tectonic and geomorphic processes that shape the solar planets. This information can be obtained only by traveling to these celestial bodies.

Space travel became a reality on October 4, 1957, when *Sputnik I* was lifted into orbit from the plains of Russia. Two years later *Lunik II* was launched in Russia and became the first vehicle to land on the moon. On July 21, 1969, the *Apollo 11* spacecraft was guided in a lunar orbit by Michael Collins while his companions, Neil Armstrong and Edwin Aldrin, descended to the surface of the moon in the lunar landing vehicle *Eagle*. They were the first human visitors to another planet. In the decades following launch of *Sputnik I*, numerous satellites have been placed in orbits around the earth. Several spacecraft have traveled to other planets carrying instruments for photographing the landscape, detecting gravity anomalies, temperature, magnetism, seismicity, and electromagnetic fields. Data

600 Other Planets

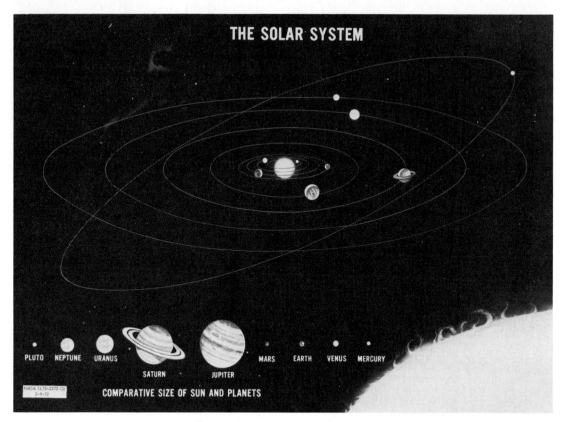

Figure 16-1 Comparison of the sizes of the sun and the principal solar planets and the orbits of these planets. (Courtesy of NASA.)

from these space probes are beginning to yield information about the interior structure and the processes that form the landscape of these planets.

Information about other nearby celestial bodies plus knowledge of the earth's structure and geologic processes provide us with a basis for speculating about aspects of development common to several planets. The fact that the earth has a liquid outer core, whereas the moon is apparently solid throughout might be the result of these two planets being in different stages of a similar pattern of development. An apparent global pattern of tectonic structures seen on photographs of **Mercury** suggests a different balance of crustal geologic processes in this small planet compared with the larger earth. Progress in this kind of **comparative planetology** may lead to better understanding of how the present balance of geologic processes in the earth will change with time. We will explore evidence for these ideas in this chapter. Discussion of the motions and dimensions of the solar planets will be followed by observations and inferences about their internal structure. We will conclude with thoughts about the extent to which the earth is a unique planet.

MOTIONS AND DIMENSIONS OF THE SOLAR PLANETS

Ancient astronomers observed that the locations of stars relative to one another appeared to remain unchanged. The fixed pattern

The Earth in the Solar System

Table 16-1
Dimensions and Geometry of the Orbits of the Principal Solar Planets

Planet	Mean Distance from Sun	Eccentricity of Orbit	Inclination of Orbital Plane	Period of Revolution
Mercury	57.7×10^6 km	0.2056	7.004 degrees	88 earth days
Venus	107×10^6	0.0068	3.394	225
Earth	149×10^6	0.0167	—	365
Mars	226×10^6	0.0934	1.850	687
Jupiter	775×10^6	0.0484	1.306	11.99 earth years
Saturn	1421×10^6	0.0557	2.490	29.5
Uranus	2861×10^6	0.0472	0.773	84.0
Neptune	4485×10^6	0.0086	1.774	165
Pluto	5886×10^6	0.2486	17.144	248

moved across the sky as the night passed, and its orientation shifted from one day to the next, finally returning to a similar position in the course of a year. Groups of stars appearing to outline familiar forms were called constellations. They proved to be helpful in describing the night sky.

Long ago the sun and moon and five other objects visible to the naked eye were observed to move differently from the pattern of stars. The sun and moon followed regular paths, whereas those five celestial bodies, which became known as planets, seemed to move erratically across the sky. Eventually geometrical schemes were developed to find the movement of a planet from measurements of its position at a succession of times. After careful study of such measurements Johannes Kepler (1571–1630) devised three empirical laws to summarize the principal motions of the solar planets.

1. The orbit of a planet is an ellipse with the sun located at one focus.

2. A straight line between a planet and the sun sweeps across equal areas in equal lengths of time.

3. The square of the period of orbit is proportional to the cube of the mean distance between the planet and the sun.

This important work and the later development of Newton's laws of motion and gravitation provide us with ways for calculating the masses of the planets and the manner in which the mass and motion of one planet introduces subtle irregularities in the otherwise elliptical orbits of other planets. Three additional solar planets were then discovered after their existence was predicted from orbital irregularities of other planets.

The periodic reappearance of surface markings that we can observe through telescopes indicates the rate of rotation of a solar planet. The size of a planet and the extent of flattening caused by rotation can be found from the angles between lines of sight to different points on the rim of the illuminated disc that appears in the sky.

At the present time nine principal planets are known to be in orbit about the sun. Their orbit dimensions and rates of revolution are summarized in Table 16-1. Smaller satellite planets orbit six of these principal planets. In a zone between the orbits of Mars and Jupiter at least 100,000 small asteroids can be seen from telescope observations to move around the sun. The largest asteroid is **Ceres** which has a radius of 350 km. If all the asteroids were grouped in a single celestial body, its radius would be less than 400 km.

The principal solar planets can be separated into two groups on the basis of their di-

Table 16-2
Factors related to Size and Mass of the Principal Solar Planets

Planet	Radius	Mass	Mean Density	Surface Gravity	Rotation Period	Flattening
Mercury	2418 km	3.3×10^{26} gm	5.58 gm/cm^3	376 gals	59 earth days	<1/300
Venus	6097	4.87×10^{27}	5.16	874	243	<1/300
Earth	6370	5.97×10^{27}	5.51	980	1	1/1/298
Moon	1738	7.34×10^{25}	3.3	162	27.3	<1/300
Mars	3380	6.45×10^{26}	4.12	372	1.03	1/191
Jupiter	71372	1.90×10^{30}	1.33	2587	9.83 earth hours	1/15
Saturn	60400	5.68×10^{29}	0.62	1146	10.23	1/10
Uranus	23550	8.66×10^{28}	1.6	1009	10.80	1/15
Neptune	22300	1.01×10^{29}	2.2	1470	15.00	1/50
Pluto	2980	6.57×10^{26}	4?	493?	6.4 earth days	?

mensions and masses. Information related to the sizes, shapes, and masses of these planets is presented in Table 16-2. The **terrestrial** planets include Mercury, Venus, Earth, and Mars. These are relatively small planets, Earth being the largest and most massive. They are characterized by relatively high values of mean density and low values of flattening. The size and mean density of Pluto suggest that it may be similar to the terrestrial planets. The **Jovian** planet group includes Jupiter, Saturn, Uranus, and Neptune. These are the giant planets of the solar system, much larger and more massive than the terrestrial planets. They are characterized by relatively low values of mean density and large values of flattening.

THE TERRESTRIAL PLANETS

The celestial neighbors passing closest to the earth are the moon and the terrestrial planets Venus, Mars, and Mercury. Optical and radio telescopes and spacecraft passing near or landing upon their surfaces have been used to learn about these planets. The obscure Pluto is tentatively grouped with the terrestrial planets on the basis of its size and mean density.

Landscapes of the nearby terrestrial planets have been mapped from spacecraft photographs and radio telescope observations. The large parabolic dish of the Jodrell Bank radio telescope is pictured in Figure 16-2. Features less than 0.5 km wide can be discerned on

Figure 16-2 The large parabolic dish antenna 76 m (250 ft) in diameter of the radio telescope located at Jodrell Bank in England. This is the largest fully steerable telescope on earth.

the lunar surface with instruments such as this. Images of the surface of Venus prepared from radar reflections indicate topographic features less than 40 km across.

Several space probes have traveled to terrestrial planets. In Figure 16-3 the Mariner 4 vehicle is shown undergoing tests prior to its journey to Mars in 1965. Spacecraft such as this are designed to fly close to a planet to obtain photographs and measure magnetism, temperature, electromagnetic radiation, and other phenomena. More recently, spacecraft operated by remote control have been designed to land upon the surface of a planet

Figure 16-3 The Mariner 4 spacecraft undergoing prelaunch tests. Extended arms of the vehicle contain panels for absorbing solar energy to operate some instruments. (Courtesy of NASA.)

604 Other Planets

and carry out x-ray measurements of rock composition, record seismicity, and determine levels of radioactivity.

It is from these kinds of observations that we can begin to understand the nature of a nearby planet. Let us now consider some of the important features of individual terrestrial planets.

Moon

The moon is a companion planet to the earth. Because its period of rotation is equal to its orbital period, the same hemisphere of the moon always faces the earth. Prominent regions of its surface are evident even without the aid of a telescope. The rugged **lunar highlands** extend over about 60 percent of the hemisphere facing the earth. Plains of subdued relief known as the **lunar maria** occupy the remaining 40 percent. Spacecraft photographs show that highland terrain is found almost everywhere on the lunar surface facing away from the earth. These features of lunar geography are evident in Figure 16-4. The average level of the maria is about 5 km lower than the average elevation of the highlands.

The lunar landscape is marked everywhere by craters. The largest craters are more than

(A)

250 km in diameter and the smallest are less than a few centimeters wide. There are more than 200,000 craters larger than 2 km in diameter. Some of their different features are illustrated in Figure 16-5.

The very large lunar craters are circular elevated rims surrounding broad floors of subdued relief. The floors of some of these large craters are nearly flat, and marked only by the pockmarks of a few small craters (Figure 16-5 A). Other large craters have more features. For example, concentric slump terraces mark the rim of **Tycho** (Figure 16-5 B), and an irregular mountain group is seen near its center. Low and disconnected ridges show the circular outlines of ghost rings (Figure 16-5 C). Here the **Mare Humorum** floor within a ghost ring appears to be continuous with the surrounding mare surface.

Many craters appear to be pits in the lunar surface surrounded by rims of deformed rock and rubble. They are similar in form to meteorite craters and bomb craters formed on the earth's surface. Dimensions of these lunar

Figure 16-4 The surface of the moon: *(A) (opposite)* earth-facing side; *(B)* side facing away from the earth. (U.S. Air Force Mosaic LEM-1, courtesy of NASA.)

(B)

Figure 16-5 Large lunar craters; *(A)* a flat floored crater approximately 50 km wide, *(B)* the crater Tycho, which is approximately 80 km wide, and *(C)* discontinuous circular rim of a ghost ring in Mare Humorum. (Courtesy of NASA.)

craters reveal that the volume of the pit is about the same as the volume of material forming the elevated rim. Very few craters display a form similar to the conical volcanic structures found on the earth.

Most lunar craters appear to have originated with the impact of meteoroids. These meteoroids are fragments of material traveling in interplanetary space. They can strike the lunar surface at speeds of between 10 and 20 km/sec. They can range in size from tiny dust particles to bodies the size of the larger asteroids. Impact of large meteoroids in the regions of the maria apparently stimulated volcanism. Lava flowed into the vast impact pits and spread out to the crater rims, congealing to form the broad crater floors. The nature of the volcanic activity is not understood. Conversion of the kinetic energy of the moving meteoroid into heat energy upon impact could cause some rock melting. Would this mechanism produce the vast quantities of lava that apparently filled the large craters? Perhaps early in its history the moon had a molten interior contained in a thin solid crust which could be penetrated by the more massive meteoroids. Could lava from this reservoir then have welled out over the lowland regions to form the maria? We do not have answers to these questions.

Lunar craters are found in various conditions of disrepair. The rims of some have been broken by later impacts. Others, seen now as ghost rings, have been almost submerged by the maria lava floor. The rugged features of a newly formed crater eventually become softened by a cover of small fragments and dust thrown out by subsequent impacts that formed nearby craters. Radiating from some craters are rays that extend over other presumably older craters. Observe the particularly obvious ray patterns in Figure 16-4. These various features of the lunar surface aid in distinguishing the relative ages of craters, linear structures such as the great Alpine Rift Valley (Figure 16-6), and other long escarpments. This information has been used to

Figure 16-6 The large Alpine Valley extends 130 km along the lunar surface and is as much as 10 km wide. (Courtesy of NASA.)

prepare tentative bedrock geologic maps for portions of the moon.

The Sea of Tranquility (Mare Tranquillitatis) was visited by the *Apollo 11* astronauts, the first men to reach the moon. Materials returned from this site included specimens of basalt and gabbro, volcanic breccia, and unconsolidated regolith. Plagioclase and pyroxene minerals were found to be the most abundant constituents. The lunar regolith consists principally of rock fragments and glassy spherules, transported in the spray of debris caused by meteoroid impact from surrounding craters. The glassy spherules and a small proportion of iron-nickel spherules presumed to be meteoroid remnants apparently congealed from the fluid spray that was melted by the impact.

Specimens from the *Apollo 12* landing site in the Sea of Storms include similar breccia and lunar regolith. In addition, olivine- and pyroxene-bearing peridotite samples were recovered, as were some fine particles similar to volcanic ash. This site was located on a ray projecting from **Copernicus**. Dust and small fragments of a light gray color contribute to the visibility of the ray on high elevation photographs of the lunar surface.

Lunar highland terrain is found near the **Fra Mauro** landing site of the Apollo 14 mission. The **Apennine Mountains** are close to the Apollo 15 landing site, and the Apollo 16 and Apollo 17 journeys reached highland areas near **Descartes** and the **Littrow-Taurus** region. Bedrock exposures of anorthosite were found to be abundant in these highland settings. This rock is richer in calcium plagioclase and aluminosilicates and has a much lower iron content than the typical mare basalts.

The lunar rocks are similar to their counterparts on earth in most aspects, but some differences are apparent. The lunar rocks are comparatively enriched in refractory elements such as titanium, and they are almost totally devoid of water. None of the hydrated forms common on the earth occur on the moon.

The ages of lunar rock specimens have been determined from ratios of isotopes in the radioactive decay series. Anorthosite specimens from the lunar highlands crystallized as long ago as 4.5×10^9 years. Outpourings of mare basalts and flooding of the large craters occurred between 4.1×10^9 and 3.1×10^9 years ago. There is no evidence of any significant lunar tectonism since that time. Apparently the numbers of large meteoroids striking the moon were much greater during these early stages of its development.

The moon has no detectable atmosphere, and no water or other fluids exist near its surface. Erosion of the landscape may occur at almost imperceptible rates by abrasion from interplanetary dust and meteoroids. The temperature near the lunar equator varies between the extremes of $+130°$ C ($+266°$ F) and $-200°$ C ($-328°$ F) during the 29-day rotation period. Perhaps this large temperature change contributes to mechanical weathering, but at a very slow rate. The lunar regolith consists of fragments formed by meteoroid impact rather than by weathering. It accumulates at rates estimated to be less than 1 mm in one million years. Lacking the geomorphic processes which alter the earth's surface, the lunar landscape has probably not changed in any important way for more than three billion years.

The seismicity of the moon has been measured by seismographs placed in four locations by the *Apollo 12, 14, 15,* and *16* landing parties. Their instruments have been operating simultaneously and transmit information to observatories on the earth. Seismic

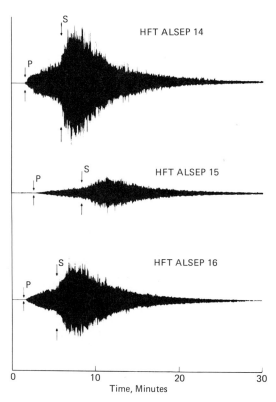

Figure 16-7 Seismograms recorded at three lunar stations indicate a moonquake that occurred on March 13, 1973. (From M. Nafi Toksoz, A. M. Dainty, S. C. Solomon, and K. R. Anderson, *Reviews of Geophysics and Space Physics*, page 556, November 1974.)

waves are generated by moonquakes in the lunar interior and by meteoroids striking the surface. Examples of moonquake records are shown in Figure 16-7. Moonquakes occur much less frequently than earthquakes. The number of shocks during an interval of time changes in a cyclic way as the moon traverses its orbit. A tidal force field due to the earth and the sun acts on the moon. Its form is similar to the earth's tidal force field illustrated in Figure 15-23. Because the earth has a much larger mass than the moon, the tidal force on the moon is much stronger than on the earth. This force field causes elastic deformation of the moon, pulling it into an ellipsoid. The average extension of its radius is approximately 10 m. This value varies by about 33 percent as the distance between the earth and moon changes from 356,404 km at orbital perigee to 406,680 km at apogee. Perigee is the point on the orbit closest to the earth, and apogee is the most distant point.

We believe that adjustment of the moon to periodically changing tidal stress is the principal cause of moonquakes.

Travel time curves for P-waves and S-waves have been prepared from the records of moonquakes and meteoroid impacts. These have been used to calculate the seismic velocities shown in Figure 16-8 for the lunar interior. Three concentric zones are found, including a crust about 60 km thick, a mantle extending to about 1040 km below the surface, and a core with a 700 km radius (Figure 16-8). Layers of mare basalt reaching thicknesses of about 20 km form the upper part of the crust in the lunar maria regions. Otherwise the crust consists principally of anorthosite. A thin high-velocity layer directly beneath the crust is indicated in some places. There is no convincing evidence of any liquid zone; the moon appears to be entirely solid.

The variation of density in the moon has been estimated from knowledge of its gravity

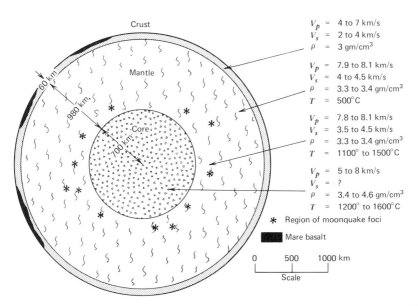

Figure 16-8 Diagram illustrating interior zones of the moon, and estimates of density (ρ), temperature (T), and values of seismic P-wave (V_p) and S-wave (V_s) velocities. (Prepared from information in M. Nafi Toksoz and others, *Reviews of Geophysics and Space Physics*, November 1974.)

field and interior seismic velocity characteristics. The gravity field is determined from perturbations of the orbits of spacecraft which have followed several different lunar orbits. Estimates of the density variation with depth are shown in Figure 16-8 and illustrate the range of uncertainty about interior mass distribution.

Satellites passing over the moon are drawn closer to the surface in some regions where the gravity field is anomalously strong. The cause of these anomalies is not known. It has been suggested that local mass concentrations called **mascons** increase the strength of the local gravity field. One idea is that mascons are high-density buried meteorites. There is no convincing evidence to confirm or refute this interesting possibility. The mascon regions of the moon are not in isostatic equilibrium. Presumably, this condition of disequilibrium has existed for a very long time. Perhaps in the weak lunar gravity field the rocks are sufficiently rigid to maintain mass irregularities without adjusting to the isostatic stresses.

Inferences about temperature in the lunar interior are based upon magnetism and electromagnetic conductivity. Remanent magnetism has been detected in specimens of lunar rock, and orbiting spacecraft have detected very weak magnetic anomaly patterns. The strength of this field is weaker than the earth's field by a factor of about 10^{-4}. Nevertheless, there is sufficient continuity to this weak and irregular field to suggest that the moon may have once had a liquid interior. At that time convection of the ionized fluid generated the field which is now only fossilized in the remanent magnetism of the rock. An estimate of the present temperature distribution in the moon is shown in Figure 16-8. The moon is now cool enough for solid rock to form its core.

The origin of the moon more than 4.6 billion years ago is unknown. One hypothesis has it amassing from the accumulation of solid fragments clustering because of their mutual gravitational attraction. Another supposes condensation from a gaseous cloud. A very different idea is that the moon was torn from the earth at a very early stage in the history of these planets. Whatever the origin, it is unlikely that the present zoned structure of the moon existed from the beginning. Rather, it developed after a large mass had accumulated. Presumably, the first stage of this development involved radioactive buildup of heat in the interior which was insulated by the outer shell of the planet. Eventually temperature was raised to the point that a large part of the mass was melted. Heavier fractions of the fluid migrated to the core, and the low-density constituents moved to the surface where they congealed to form a crust of anorthosite. Continued radioactive heating maintained a fluid core that was somewhat enriched in iron, and a mantle sufficiently hot for episodes of melting to be stimulated, perhaps by meteoroid impact. Basaltic lavas were erupted and filled in the lowland regions of the maria. As radioactivity diminished, the planet cooled to the point that volcanism ceased more than three billion years ago.

The stages of active development of the moon were during the first 1.5 billion years of its existence. Since then it appears to have remained a passive zoned planet. There is still sufficient radioactivity to maintain elevated internal temperatures, so the planet is not altogether internally inert.

Venus. The nature of Venus remains obscure. The dimensions and mass of this planet show that it is only slightly smaller than the earth. Venus has a dense and clouded atmosphere that hides completely from the optical telescope the features of its solid surface. Distinctive cloud patterns are evident in photographs made from approaching spacecraft. The appearance of Venus at a distance of 720,000 km is seen in Figure 16-9. Commencing with the Mariner 2 mission, nine

The Earth in the Solar System   611

Figure 16-9 The disc of Venus photographed from the Mariner 10 space vehicle at a distance of 720,000 km reveals cloud patterns in the atmosphere. (Courtesy of NASA.)

space probes have been launched between 1962 and 1978 to investigate this planet. Two of these vehicles, *Venera 7* and *Venera 8,* descended through the atmosphere to the solid surface and telemetered information about the soil at the landing site. In December 1978, the Pioneer Venus spacecraft was put into orbit around the planet to perform several experiments, including radar surveying of the surface topography.

The atmosphere of Venus is a dense and turbulent gaseous blanket. Carbon dioxide makes up more than 90 percent of its volume, and traces of water vapor have been detected. Wind speeds of more than 48 m/sec (110 miles/hour) were encountered about 40 km above the surface by *Venera 8*. At its landing site, however, the wind speed was less than 1.8 m/sec (4 miles/hour). Atmospheric patterns were observed from Mariner 10 photographs to move at speeds of over 100 m/sec (223 miles/hour). Clearly the atmosphere of Venus is a dynamic and turbulent mass. Atmospheric pressure on the surface of the planet is about 100 times greater than the pressure on the earth's surface. This is as large as the pressure experienced by a submarine at a depth of 1000 m in the ocean.

Temperature on the solid surface of Venus reaches 425° C (800° F) in equatorial regions on the day side (side facing the sun). Night side (side facing away from the sun) temperature is not appreciably lower, which indicates the effect of heat transfer by the turbulent atmosphere. One explanation of high surface temperature on Venus is sometimes called the "greenhouse effect." Short wavelength solar radiation easily penetrates the atmosphere to warm the solid surface. The dense mass of carbon dioxide is opaque to the long wavelength infrared energy radiated back from the solid surface. Therefore, outgoing heat is trapped by the atmosphere and maintains a high temperature near the surface.

Radar mapping from the Pioneer Venus orbiter spacecraft has revealed striking topographic features on Venus. The largest canyon yet discovered in the solar system reaches more than 15,000 km along the surface. It is about 5 km deep and almost 300 km wide. Radar images reveal more than 10 km of relief on the surface of Venus. An artistic conception of the Venus landscape (Figure 16-10) shows a continent-sized plateau more than 3 km higher than the bordering lowland plain. From careful examination of the radar images, scientists have recognized smaller features suggestive of shield volcanoes. Some are aligned in a way that could be related to faulting. These features together with the broad ridges indicate that tectonic processes have had an important effect on the topography of Venus. The gravity field over large regions of the planet has been calculated from irregularities in the spacecraft orbit. We can tell from variations in gravity that the major

Figure 16-10 Artistic conception of the Ishtar Terra region of Venus based on radar images. This continent-size feature stands about 3 km higher than the bordering lowland, and the mountains rise about 6 km higher than the bordering lowland plateau. (Courtesy of NASA.)

topographic features are in isostatic equilibrium.

We have very little information for basing speculations about the internal structure of Venus. No seismicity measurements have been attempted, and a very weak magnetism has been detected. Because Venus rotates so slowly, it cannot be determined if this is a remanent field or if it is caused by slow convection in a liquid core active at the present time. We know that the mean density of the planet is approximately 5.2 gm/cm^3. This is higher than the value of 1.5 gm/cm^3 found for the surface regolith. It would be higher than the density of granitic rock from which regolith may have been derived. We might then speculate that a relatively low-density shell surrounds the higher density interior forming a zoned planet. Other considerations of radioactive heating in a planet of this size argue that it must consist of concentric zones. The existence of a liquid core cannot yet be proven or refuted.

Mars

The *Mariner 4* spacecraft traveled as close as 10,000 km to Mars in 1965. It transmitted the first photographs to reveal clearly recognizable landscape features. These photographs put to rest any remaining thoughts that the vague markings seen through optical telescopes could be canals, perhaps constructed by intelligent creatures. They revealed a rugged and varied terrain marked by impact craters, volcanic features, and tectonic structures. More detail was added by the *Mariner 6* and *Mariner 7* space probes that passed the

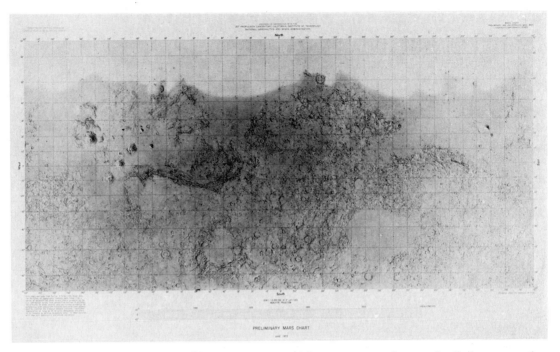

Figure 16-11 Relief map of the surface of Mars prepared from Mariner 9 photographs displays cratered terrain, plains, and mountains of volcanic and tectonic origin. (Courtesy of NASA.)

planet in 1969. The very productive *Mariner 9* vehicle was launched on May 30, 1971 and reached Mars 5-½ months later. For more than 11 months it followed a changing orbit around the planet transmitting photographs and other data. The map of the Martian surface shown in Figure 16-11 was compiled from *Mariner 9* photographs. Coverage of almost the entire surface was obtained on a scale that revealed features as small as 1 km wide. High resolution photographs of about 2 percent of the surface area showed features only a few hundred meters across. Then in 1976 two vehicles, *Viking Landers I* and *II*, reached the surface of Mars. Equipment on these vehicles was used to analyze the surface materials, make weather observations, and to search for evidence of life.

A thin atmosphere of carbon dioxide with traces of water vapor blankets Mars. Atmospheric pressure on the surface varies from one region to another between 0.002 and 0.008 bar, much lower than the pressure of about 1 bar at the earth's surface. Storms of tremendous turbulence occur from time to time in the Martian atmosphere. Mariner 9 photography was interrupted for several weeks because the large mass of dust suspended in the turbulent atmosphere obscured surface features.

Mars experiences changing seasons. Icecaps presumed to be principally frozen H_2O and with a covering of frozen CO_2 fluctuate over the polar regions. Evidence of these icecaps is seen in Figure 16-12. As Mars traverses its orbit, the north pole is inclined toward the sun at apogee, where it is farthest from the sun, and is therefore colder than the south pole which is illuminated at perigee, where it is nearest to the sun. Frost cover persists throughout the year at both poles. The larger icecap in the north is about 1000 km wide. Winter precipitation of frost greatly extends the icecap margins which then recede

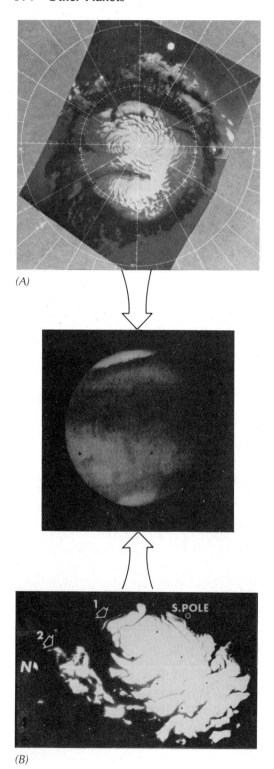

Figure 16-12 The polar icecaps of Mars. (A) North Polar cap, (B) South Polar cap. (Courtesy of NASA.)

by summer sublimation or melting. Polar temperatures are about $-125°$ C ($-193°$ F), much colder than the equatorial temperature which may reach $+25°$ C ($+77°$ F) on a warm summer day.

Large areas of the Martian surface are marked by impact craters similar to those seen on the moon. Some of these are broad circular rims surrounding nearly flat floors. Others are smaller pits in the surface. Impact craters with the fresh appearance of lunar craters are rare on Mars. Most look weathered and eroded and seem to represent the most ancient of Martian landscapes.

Volcanic structures similar to those formed by central eruptions on the earth are widespread on Mars. The larger ones exceed the dimensions of the largest volcanic piles on the earth. Olympus Mons is a shield volcano (Figure 16-13) which is 600 km wide at the base and rises 23 km higher than the surrounding plain. Three high conical volcanoes on the Tharsis Ridge stand more than 17 km above the floor of the adjacent Amazonis Basin. The lobed surface features typical of volcanic flows and elongate gulleys which closely resemble lava channels are further evidence of volcanism. Many volcanic structures display relatively little evidence of weathering and erosion, suggesting recent eruptive activity. Broad plains and basins resemble the surfaces formed by plateau basalts in western North America and India.

Tectonic structures on Mars are evident in the block-faulted terrain and long escarpments seen in some regions. Faults separating grabens between 1 and 5 km wide are seen (Figure 16-14) in the Tharsis region. Elsewhere topographic domes and elevated plateaus apparently of tectonic origin are observed. There is no evidence of folding and thrusting typical of the orogenic belts on the

The Earth in the Solar System 615

Figure 16-13 Olympus Mons is a conical volcanic structure on Mars which rises about 23 km higher than the surrounding plain. It is the largest Martian shield volcano with a diameter of approximately 400 km. (Courtesy of NASA.)

Figure 16-14 Graben separated by parallel fractures on the surface of Mars on the side of Alba Patera which is located near the Tharsis region. The graben illustrated here are as much as 5 km wide. (Courtesy of NASA.)

earth. The tectonic and volcanic features on Mars resemble structures on the earth, but there is no evidence of plate tectonic processes. Displacement along faults appears to be principally vertical.

Age relationships can be established for Martian landforms by the ways in which they appear to overlap one another. Different landscapes that can be seen in Figure 16-11 are believed to have developed during different times in the history of Mars. About one-half of the surface is a densely cratered terrain. This region extends over much of the

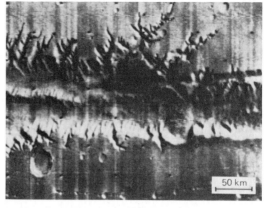

(A)

(B)

Figure 16-15 *(A)* Large canyon on Mars near 7.7° S lat., 84.3° W longitude is more than 100 km wide and 3 or 4 km deep in places. Dendritic landscape patterns are formed in smaller tributary canyons. *(B)* Local terrain in this area is seen on a specially processed narrow view photograph. (Courtesy of NASA.)

southern and central parts of the planet. Almost everywhere are broad flat-floored craters which have been pitted by a myriad of smaller impact craters. We believe that this region is the most ancient Martian landscape. A large part of this southern and central region of Mars stands higher than the average elevation of the planet.

Broad plains extending over the northern part of Mars display a more sparsely cratered terrain. Some areas appear quite smooth and free of craters, and elsewhere a moderate number of impact pits can be seen. These northern plains appear to be formed from lava flows and are partly covered by eolian sediment. Most of the region lies below the average elevation of the Martian surface.

Different from the broad plains and impact terrain are regions of central volcanic and tectonic landscape. The large area near Nix Olympica and the Tharsis Ridge displays these features. This is a region of elevated terrain, and appears to be a relatively young landscape.

Large canyons and sinuous gullys can be seen in areas near the border separating the high region of impact terrain and the lowland plains of the north. The magnificent canyon pictured in Figure 16-15 is more than 100 km wide and several kilometers deep. Dendritic patterns of gulleys are clearly seen along its margins. Although the large canyon is probably a tectonic feature, the smaller features strongly indicate stream erosion. The close view in Figure 16-15 B reveals a surface strikingly similar to landscapes on the earth which have been modified by streams and landslides. Analysis of the sinuous channel pattern seen in Figure 16-16 shows the same relationships between meander length, radius of curvature, and channel width as are found for rivers on the earth. A river of comparable channel geometry could be expected to discharge at least 2700 m^3/sec (95,000 ft^3/sec) on the earth. These examples indicate that some areas of the Martian landscape have been modified by running water. But where is the water now? We can detect none on the surface. It has been suggested that water might be supplied intermittently from melting permafrost caused by volcanism. Because of cyclic perturbations of orbital and spin motions, solar heating of the Martian surface reaches a maximum every 25,000 years. Pos-

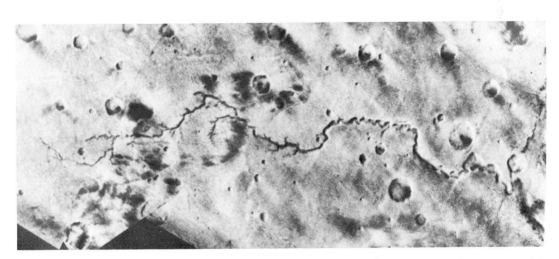

Figure 16-16 Meandering channel about 700 km long in the Mare Erythraeum region of Mars near 29° S lat., 40° W long. photographed from the Mariner 9 spacecraft at an elevation of 1666 km. (Courtesy of NASA.)

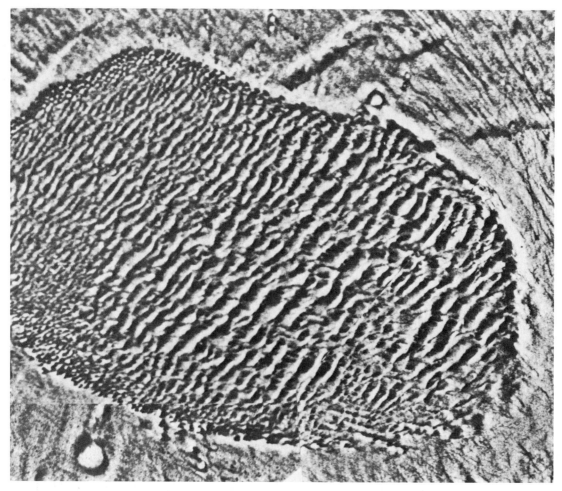

Figure 16-17 Eolian dunes on the surface of Mars in the Hellespontus region. This area is about 130 km long and 65 km wide. (Courtesy of NASA.)

sibly this could cause the icecaps to melt and rainfall to be plentiful for brief periods of time. Although the source of an intermittent water supply remains a mystery, the effects of stream erosion are evident.

Surface markings in regions of subdued relief indicate deposits of eolian sediment. Features resembling sand dunes are evident on the photographs in Figure 16-17. Atmospheric storms such as the one encountered by the Mariner 9 vehicle and the dune patterns seen in several areas indicate that eolian processes have an important effect on the geomorphology of Mars.

We have no seismic measurements to aid in discerning the internal structure of Mars. Judging from its landscape, rock near the surface is of lower density than the mean value of 4.12 gm/cm^3 for the planet. Estimates of its moments of inertia determined from paths of space vehicles indicate that density increases with depth. If the same few chemical elements that are most abundant in the earth are assumed to make up the mass of Mars, and if

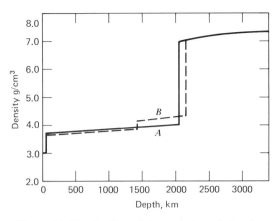

Figure 16-18 Profiles of the variation of density with depth in Mars inferred for assumed compositions and consistent with the mean density of the planet. (D. M. Johnston and others, *Journal of Geophysical Research,* September 10, 1974.)

around Mars. Observe the oblong and irregular shape of the satellite that can be seen in Figure 16-19. The largest is **Phobos** with a maximum diameter of 27 km. It follows a nearly circular orbit 6000 km above the surface of Mars. The smaller **Deimos** has a long diameter of 13 km and moves about 19,000 km above the Martian surface. Measurements of the motions of these satellites have been used to make accurate estimates of the dimensions and mass of Mars.

Mars is considerably smaller and less massive than the earth. Because of its weaker gravity field, thin atmosphere, and the scarcity of water, topographic features can stand higher and endure longer than landforms on the earth. Otherwise, its landscape and surface conditions make it more like the earth than the other terrestrial planets. The Viking Landers found no evidence of life on Mars.

similar levels of radioactivity exist there, an educated guess can be made about internal density. The estimates of density variation with depth shown in Figure 16-18 are consistent with our present limited knowledge of Mars.

There is a strong possibility that the core of Mars is liquid. This speculation is consistent with modern ideas about heating by radioactivity. The Martian magnetic field is very weak, perhaps because of the small size and low electrical conductivity of the core compared with the earth's core.

Mars must have a relatively thick and rigid outer shell. Indications of this are the large elevated regions that are not in isostatic equilibrium. This imbalance may exist because isostatic stresses caused by the gravity field are too weak to overcome the rigidity of the outer shell. Surface gravity on Mars has 39 percent of the intensity of gravity on the surface of the earth. Because of the concentration of lowland plains in the northern part of Mars, its center of mass is offset northward from the geometrical center.

Two small natural satellites follow orbits

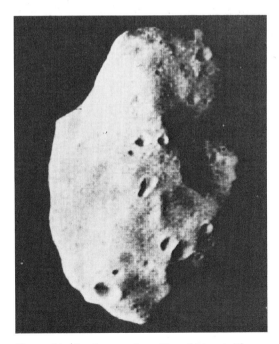

Figure 16-19 A natural satellite of Mars is Phobos. Its maximum diameter is 27 km. (Courtesy of NASA.)

Mercury

The arid dayside surface of Mercury is heated to temperatures of more than 350° C (662° F). Lacking atmosphere or oceans to transfer this heat, nightside temperatures plunge below −170° C (−274° F). The nature of Mercury has begun to unfold from information transmitted by the *Mariner 10* spacecraft. Following an orbit around the sun, this vehicle passed close to the planet on March 29 and September 21, 1974, then again on March 16 in 1975. Large areas of the surface were photographed on a scale that reveals topographic features as small as 1 km wide. Another important discovery was the relatively strong magnetic field associated with Mercury.

The cratered landscape of Mercury closely resembles the impact terrain of the moon. Figure 16-20 shows large craters with flat and terraced floors on much of the surface. We observe radial patterns of rays in some places. Elsewhere the concentric terraces and irregular mountain groups within larger crater rims such as the ones in Figure 16-21 are similar to lunar landscape features. Smaller bowl-like craters are evident almost everywhere.

Three general kinds of terrain seen in Figure 16-22 are typical of Mercury. The gently rolling landscape marked by numerous small

Figure 16-20 Mosaic of the surface of Mercury from Mariner 10 photographs taken at a distance of 50,000 km. The largest crater is about 170 km in diameter. (Courtesy of NASA.)

pits less than 10 km wide forms the **intercrater plains.** These plains surround or lie between the large flat-floored craters. Landscape displaying large craters densely spaced

Figure 16-21 These large craters on the surface of Mercury are 30 km to 35 km wide. Crater floors are terraced and central mountain groups are seen near their centers. They compare with the lunar crater Tycho illustrated in Figure 16-5 *B*.

Figure 16-22 Three landscapes common to the surface of Mercury include intercrater plains marked by numerous small pits; heavily cratered terrain characterized by closely spaced and overlapping, large flat-floored craters; and smooth plains that are areas of subdued relief sparsely pitted by craters. (Courtesy of NASA.)

and overlapping one upon another is called **heavily cratered terrain.** We believe that it developed after the more ancient intercrater plains. Broad areas of subdued relief and devoid of large craters make up the **smooth plains** of Mercury. These plains appear to be the youngest regional terrains on the planet. Surface features similar to those seen on lunar maria suggest that they may have been formed from wide-spreading lava flows.

Comparison of the dimensions of craters of Mercury with those on the moon reveal some differences. A logarithmic graph (Figure 16-23) illustrates empirical relationships between crater depth and rim diameter. The depth/diameter ratios vary similarly on both planets for craters less than 7 to 8 km wide. But we find that larger craters on the moon are proportionally deeper than those on Mercury. We also observe that for craters of the same size, rays extend to greater distances on the moon than on Mercury. These differences in crater features are caused principally by the difference in surface gravity on the two planets. The surface gravity of about 376 gals on Mercury is considerably stronger than the 162

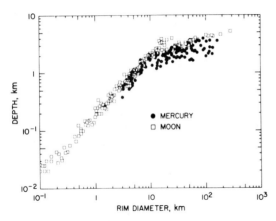

Figure 16-23 Graph showing the relationships between crater depth and rim diameter for Mercury and for the moon. (After D. E. Gault, *Journal of Geophysical Research,* page 2456, June 10, 1975.)

the principal solar planets, its mean density of 5.58 gm/cm^3 is largest. Its surface terrain is so similar to the lunar landscape that both would appear to have almost the same kind of crustal rock. Therefore, the density of the deep interior of Mercury must be larger than the value of its mean density.

The Mariner 10 spacecraft measured magnetic field intensity as strong as 400 gammas near Mercury. These measurements indicate planetary magnetism considerably weaker than is known for the earth, but much stronger than has been detected for Venus, Mars, and the moon. The magnetism of Mercury appears much too strong to be simply a remanent field. The most satisfactory hypothesis is that Mercury possesses a conductive liquid core acting as the dynamo that gener-

gal field on the moon. This apparently restricts the original heights of large crater rims formed at the time of meteoroid impact and causes the rims to become more worn down by mass wastage. Similarly, Mercury's stronger gravity reduces the distance of scattering of the ejecta that form rays which radiate from craters. Except for these differences the impact terrains of Mercury and the moon are similar.

Distinctive scarps can be traced for long distances over the surface of Mercury, but similar features have not been found on the other terrestrial planets. One such scarp is evident in Figure 16-24. It marks a vertical offset on the intercrater plain and along the floors of large craters. Some geologists believe that scarps such as this are surface projections of thrust faults in the crust of Mercury. They argue that these scarps extend in a global pattern produced by compression in the crust. According to one hypothesis this pattern of apparent thrust faults could have developed by interior contraction of the planet, amounting to a decrease of 1 to 2 km in the length of its radius.

Although Mercury is the least massive of

Figure 16-24 Curved scarp more than 300 km long can be observed on the surface of Mercury. Craters along the scarp display rim distortion and offset floors. Maximum relief on the scarp is about 3 km. (Courtesy of NASA.)

ates the magnetic field. Iron and perhaps nickel are probably the abundant constituents of this relatively high density and conductive liquid core of Mercury. Like other terrestrial planets Mercury appears to consist of concentric interior zones.

Pluto

Much of the nature of the most distant solar planet remains a mystery. It is so far from the sun that its surface temperature is probably lower than $-200°$ C ($-382°$ F). A gaseous atmosphere is improbable under such frigid conditions.

Modern measurements of orbital perturbations of the outer planets were used to estimate the value of 0.18 for the ratio of Pluto's mass and the mass of the earth. Recent measurements of the angle subtended by its illuminated disc (made with the 200-in. reflector telescope at the Hale Observatory), indicate a radius of approximately 2980 km. These estimates of mass and size suggest a mean density of about 4 gm/cm^3 for the planet. Because of its size and mean density, Pluto seems more like the terrestrial planets than the Jovian planets. No further information is available for making geological comparisons of Pluto with the other terrestrial planets.

Development of the Terrestrial Planets

The masses of the separate terrestrial planets are presumed by most planetologists to have accumulated at about the same time more than 4.6 billion years ago. No completely satisfactory hypothesis has been proposed to explain the original accumulation of these masses. We do not know the extent to which the planets formed by condensation of a gaseous nebula or by accretion of solid fragments of celestial debris. If the planets were not originally hot, it seems likely that internal heating occurred shortly after the separate masses accumulated. Heat could be produced in the insulated interior by radioactivity and mechanical compaction. Partial melting temperatures of iron silicate compounds were reached perhaps within a few hundred million years. A brief and violent episode of plantary differentiation then may have begun with the inward migration of heavy iron-rich compounds derived from early partial melting. Additional heat generated by gravitational sinking of these heavy fluids raised the temperature further. During this eposode the planet became almost totally melted and separated into concentric zones of different density. A heavy iron-rich core formed, and a thin crust of lighter silicates congealed at the surface. Planetary cooling has followed this primary differentiation stage. Long after cooling began, the planets experienced episodes of partial melting and continued magmatic differentiation within an otherwise solid outer shell.

The cooling history of a planet appears to be related principally to its size. The smallest planets, Mercury and the moon, have apparently experienced no surface volcanism for more than three billion years. The larger of the two is Mercury which retains its liquid core, whereas the smaller moon is almost, if not entirely, solid. Mars is still larger. It displays evidence of recent volcanism and very probably has a liquid core. Earth, the largest terrestrial planet, experiences volcanic activity at the present time and has a large liquid outer core. Radar images indicate that Venus may have experienced volcanism, but we have no information about whether or not it has active volcanoes now.

The landscapes of Mercury and the moon appear to have existed almost unchanged for more than three billion years. But surface processes are active on Mars where eolian and fluvial processes slowly modify the landscape. Here also an ancient terrain, densely pitted by impact craters, points to an early episode of especially intensive bombardment by meteoroids. These impact terrains are estimated to be four billion years old. The early Precambrian surface of the earth probably looked the same, but these features were long ago obliterated by the geomorphic processes that are much more active than on Mars, Mercury, and the moon. It will be interesting to see the surface of Venus unfold in future observations and to learn about the effect of its dense atmosphere on ancient terrains.

Fracture patterns and tectonic escarpments on the moon and Mars appear to be features caused principally by tension or vertical doming. In contrast, Mercury displays a pattern suggestive of planetary contraction and crustal compression. The great linear mountain chains, oceanic ridges, and submarine trenches which distinguish the earth's global tectonic system are not evident on these other planets. Will similar features eventually be recognized on Venus, which is almost the earth's twin in size and mass? Were the global tectonic mechanisms indicated by features in the ancient landscapes of Mercury and Mars once active in the earth? When we find answers to these questions, we will have learned much more about the geological past and perhaps the geological future of our own planet.

THE JOVIAN PLANETS

The giant outer planets of the solar system differ markedly from the terrestrial planets. Two space vehicles, *Pioneer 10* in 1973 and *Pioneer 11* in 1974, have passed close to Jupiter. Then Jupiter and Saturn were visited by *Voyager 1* and *2* vehicles in 1979 and 1980. Otherwise, our limited knowledge of the Jovian planets comes from earth-based observations.

Jupiter

The largest and most massive solar planet displays distinctive features that can be seen

Figure 16-25 Voyager 2 photograph of Jupiter taken at a distance of 46 million km. (Courtesy of NASA.)

through an optical telescope. It is marked by several parallel bands which are clearly seen in Figure 16-25. Especially conspicuous is the large oval spot of a red-orange hue. The rotation rates found by observing the reappearance of particular features differ from one band to another. Clearly these are features of a gaseous atmosphere consisting of parallel bands that are not rigidly joined to one another.

The Pioneer 10 spacecraft passed Jupiter in December 1973. It followed a path close to the equatorial plane, as close as 130,000 km from the surface. A year later in December 1974, the *Pioneer 11* vehicle passed closer than 50,000 km above the surface alone a subpolar trajectory. Information transmitted from these spacecraft added much to our knowledge of Jupiter. The photograph in Figure 16-26 was taken from *Voyager 2*. It reveals local features of the turbulent atmosphere as small as 170 km wide.

Jupiter was found to have a strong magnetic field. Inferred contours of surface field intensity illustrate in Figure 16-27 the princi-

The Earth in the Solar System   625

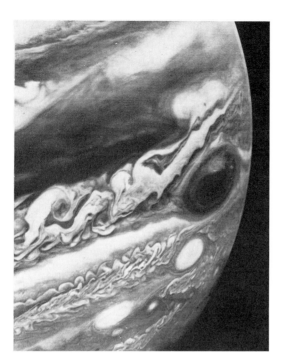

Figure 16-26 Voyager 2 photographic composite obtained at a distance of 9 million km. It shows the great Red Spot and the turbulent bands of the atmosphere nearby. Features as small as 170 km in width can be identified in this picture. (Courtesy of NASA.)

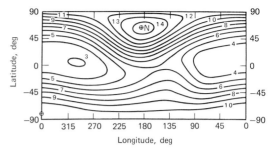

Figure 16-27 Contours depicting magnetic field intensity on the surface of Jupiter estimated from Pioneer 10 and Pioneer 11 measurements. Contour interval is 1 oerstad. (E. J. Smith and others, *Science,* May 2, 1975.)

pal form of the magnetic field. It is similar to the form of the earth's main magnetic field and more than 15 times stronger.

The atmosphere of Jupiter is a turbulent blanket more than 4000 km thick. Clouds appear to consist of frozen crystals and liquid droplets of ammonia (NH_3). Gaseous methane (CH_4) and molecular hydrogen (H_2) are important constituents. Measurements of infrared radiation indicate low temperatures of about $-140°$ C ($-252°$ F) in the atmosphere. These low values conflict with radio wave temperature measurements of $+27°$ C ($+80°$ F) at a level where atmospheric pressure was 0.5 bar. The large difference in temperature estimated by these two kinds of measurement is difficult to explain. It may be related to the effect of particles suspended in the atmosphere which could bias the infrared data.

The great red spot is an atmospheric feature that remains in about the same region, but makes subtle changes in position. It is part of the planet's atmosphere, but may be controlled by internal processes. There is no generally acceptable hypothesis to expalin this feature.

The interior of Jupiter remains unknown. Its mean density of 1.33 gm/cm^3 is quite close to the value for the sun. Hydrogen and perhaps helium appear to be the principle constituents, but hydrogen must be at least nine times more abundant. Very small amounts of carbon and nitrogen are detected. This composition is similar to the composition of the sun. The tremendous internal pressure in Jupiter is sufficient to compress hydrogen into a solid metallic state. We do not know if distinctive boundaries divide solid, liquid, and atmospheric outer shells of the planet, or if a transitional mush separates the solid planet from its atmosphere.

Jupiter has a heat balance that appears to be unique among the solar planets. Infrared radiation indicates that about twice as much heat is emitted from the planet than is received from solar radiation. We can only speculate about the source of this internal heat. If differentiation processes are still active so that heavier material is migrating toward the core, gravitational energy could be converting to heat. A mass 10 times larger than Jupiter would probably generate sufficient internal heat and pressure to ignite thermonuclear reactions, thereby becoming a star. Jupiter is too small, and therefore remains a planet.

Twelve natural satellites orbit Jupiter. The nearest passes about 110,000 km above the surface, and the most distant is 23,500,000 km away. The largest of these satellites, **Ganymede** and **Callisto,** have about the same dimensions as Mercury. During its journey in 1979 the Voyager 1 spacecraft provided us with some truly surprising observations of Jupiter's satellites. Scientists watched with amazement the transmission of photographs showing a volcanic eruption in progress on **Io,** one of the innermost satellites. Seven stupendous exlosions sent solid fragments as high as 160 km above the erupting crater. Io is believed to consist mostly of silicate rock, and is about the size of our moon. Farther from Jupiter is Ganymede, which is thought to be a mixture of silicate rock and ice. Still farther away is Callisto, which is mostly ice. We can see that, like the solar planets, Jupiter's satellites tend to be smaller, but more dense closer to the parent planet. In several ways Jupiter and its satellites are a "mini solar system."

Saturn

The planet Saturn is known from 1979, 1980, and 1981 space craft observations and from earth-based observations. It is circled by conspicuous rings, and its surface is distinctly banded (Figure 16-28). After its encounter with Jupiter the Pioneer 11 spacecraft was directed on a path that carried it close to Saturn in 1979. Then *Voyager 2* visited Saturn in 1981.

Saturn maintains 15 satellites which orbit at

Figure 16-28 Voyager 1 photograph of Saturn taken at a distance of 76 million km. Several of Saturn's satellites can be seen in the nearby space. (Courtesy of NASA.)

distances of from 98,000 to 12,000,000 km above its surface. The largest is **Titan,** which is about the same size as Mercury.

The rings of Saturn are circular, and they apparently consist of small ice particles or other ice-coated fragments. The outermost ring extends 132,000 km from the surface of the planet. The rings are probably less than a few meters thick.

The bands on the surface of Saturn mark zones in its turbulent atmosphere. There is more methane and less ammonia here than in the atmosphere of Jupiter. Molecular hydrogen and helium are also present. Atmospheric temperatures of $-180°$ C ($-292°$ F) have been detected.

Nothing is known about the transition zone between the atmosphere and the solid shell of Saturn. Its low mean density of 0.7 gm/cm^3 indicates that hydrogen must be the most abundant interior consitituent.

Uranus

The greenish disc of Uranus appears almost featureless through an optical telescope. An atmosphere containing ammonia and molecular hydrogen has been detected from earth-based observations. Temperatures on the cloud top are estimated to be $-180°$ C ($-292°$ F). The mean density of 1.6 gm/cm^3 indicates larger proportions of heavier elements than would be expected for Jupiter and Saturn. Five satellites are known to orbit Uranus, the largest being about 1200 km in diameter.

An interesting feature of Uranus is the orientation of its axis of rotation close to the plane of its orbit. This causes its north pole to be continually illuminated whereas its south pole remains in perpetual darkness for one half of its 84-year orbital period. The situation is reversed during the other half of the period.

Neptune

The disc of Neptune appears blue-green in color through an optical telescope. Methane, ethane, and acetylene have been detected in its atmosphere. The temperature is probably so low that a large portion of whatever ammonia exists is likely to crystallize. Two satellites have been discovered in orbit around Neptune. The largest is **Triton,** which has a diameter of almost 4200 km and an estimated mean density of 3.5 gm/cm^3. It is larger and more massive than the moon.

The Jovian planets and the terrestrial planets contrast not only in size and mass but also in composition. Too little is yet known about the internal structure of the Jovian planets to attempt hypotheses about their development.

IS THE EARTH UNIQUE?

We have now examined the geology of our planet, and we have looked at the most obvious features of our nearest celestial neighbors. Four nearby terrestrial planets are similar to the extent that four elements, iron, oxygen, magnesium, and silicon, account for more than 90 percent of their individual masses. Rock-forming minerals found in the earth's crust are probably quite similar to the crustal constituents of these other planets. Comparable chemical compounds most likely exist in one or another of their interior zones. Some landforms appear to have been produced by the same processes on all of these celestial bodies. Certainly the terrestrial planets are similar in some ways. To what extent is the earth unique among these planets?

The plate tectonic system that produces our sinuous mountain belts and oceanic ridges and submarine trenches has not been revealed elsewhere in the solar system. The ancient landscapes of the moon, Mercury, and Mars provide no indication that a similar tectonic system ever operated in these planets. Our estimates of their internal temperature conditions and the apparent rigidity of their outer shells suggest that this kind of tectonic system is unlikely to develop in the future. The earth appears to be a distinctively different tectonic machine. Until the landscape of Venus can be mapped we cannot tell the extent to which it might undergo similar tectonism. Even if the internal processes of that planet proved to be comparable, we know that the surface conditions are markedly different. In some ways the earth is truly a unique solar planet.

Comparisons of the solar planets inevitably lead to questions of the existence of life elsewhere in the solar system. The recent voyages of space vehicles have so far revealed no indication of life of any kind. Indeed, the biosphere of living creatures near the earth's surface appears to be unique in this corner of the universe. Probably we had the most hope of discovering some kind of organism on Mars. But the Viking Lander experiments that were designed to detect primitive life forms did not reveal any living thing. Of all the solar planets Mars offers the most hospitable conditions for human exploration.

Not only is the earth's biosphere unique among the other solar planets; it is experiencing a unique time interval when a biosphere can be maintained. A delicate balance of surface conditions has established an environment in which water is neither boiled away nor frozen. The atmosphere is not so opaque as to excessively trap solar radiation. Yet it is dense enough to effectively shield the life zone from lethal ultraviolet waves. Perhaps this environment has persisted since late Precambrian time. But it has not always been so, and there is no compelling reason to believe that a life-sustaining environment will continue indefinitely.

It would be absurd to claim that the earth is unique in the infinite expanse of the universe. However, it does provide a seemingly uncommon combination of resources and environmental conditions.

STUDY EXERCISES

1. What evidence indicates that craters on the moon were produced by meteoroid impact rather than by central volcanic eruptive processes?

2. The impact craters on Mars have depth/diameter ratios that are different from those measured for craters on the moon and for craters on Mercury. What are two reasons that would lead us to expect this difference in crater dimensions?

3. Is it possible that some processes which influence the landscape of Venus are similar to processes that influence the earth's landscape? Discuss!

4. Jupiter and its satellites have sometimes been described as a "mini solar system." Discuss some of the similarities and some of the differences between these two systems.

5. What kind of evidence indicates that a planet may have a liquid core?

SELECTED READINGS

Bowker, D. C., and J. K. Hughes, *Lunar Orbiter Photographic Atlas of the Moon,* National Aeronautics and Space Administration, Washington, D.C., 1971.

Journal of Geophysical Research, vol. 78, July 10, 1973 (entire issue devoted to geology of Mars).

Journal of Geophysical Research, vol. 80, June 10, 1975 (entire issue devoted to geology of Mercury).

Scientific American, vol. 233, September 1975 (entire issue devoted to science of the solar system).

Short, N. M., *Planetary Geology,* Prentice-Hall, Inc., Englewood Cliffs, N. J., 1975.

APPENDIX 1
MINERAL IDENTIFICATION

Minerals that are abundant in rocks or common as ores number only a few dozens, and many can be identified without special equipment if sizable pieces are available. The techniques to be described consist of direct observations and simple tests. These suffice to recognize groups of rock-making silicate minerals. More exact determinations can be made with laboratory instruments outside the scope of our introductory study.

For convenience of reference the most important minerals likely to be encountered in an introductory study of geology are arranged alphabetically in Table 1 of this appendix. The chemical formulas in the second column are for reference only, not as an aid to making identifications. In complex silicates, only generalized formulas are included; in these formulas various ion groups are designated by the capital letters A, B, and C, as follows:

A = elements with large ionic radius, such as K^+, Ca^{2+}, Na^+
B = elements with intermediate radius, such as Mg^{2+}, Fe^{2+}, Fe^{3+}, Al^{3+}
C = elements with small radius, mainly Si^{4+}, Al^{3+}

Valences greater than +1 are shown by the standard symbol. In this notation 2, 3, and 4 with a plus sign are written as raised postscripts. If the name of a laboratory specimen is known, locate the mineral in the table and be sure that you become familiar with all its physical properties. If the name of a specimen is not known, determine its physical properties and scan the appropriate columns in the table to locate the correct name by elimination.

Table 1
Properties of Some Important Minerals
(For explanation of A, B, C in chemical formulas, see text.)

Mineral	Chemical Composition	Luster	Form	Cleavage	Hardness	Specific Gravity	Other Properties; Remarks
Albite (see Plagioclase feldspars)							
Amphibole (complex group of minerals, hornblende most common)	Silicates of Ca, Mg, Fe, Al, Na. General formula: $A_{2\text{-}3}B_5C_8O_{22}(OH)_2$	Vitreous on cleavage surfaces	In long, 6-sided crystals; also in fibers and irregular grains	Two, intersecting at 56 and 124°	5 to 6	2.9 to 3.8	Commonly black, dark and light green, rarely white; some varieties used as asbestos

Appendix 1

Table 1
Properties of Some Important Minerals (Continued)

Mineral	Chemical Composition	Luster	Form	Cleavage	Hardness	Specific Gravity	Other Properties; Remarks
Apatite	$Ca_5(PO_4)_3(F,Cl,OH)$	Vitreous	In granular masses or as large crystals	Poor, in one direction	5	3.1 to 3.2	Green, brown, blue, violet; accessory mineral in many kinds of rocks, especially in calcite marbles
Aragonite	$CaCO_3$	Vitreous	In slender, needlelike crystals; granular masses	Poor, in two directions	3.5 to 4	2.94	Colorless, white, pale tints; harder and slightly heavier than calcite; effervesces in dilute HCl
Augite (see Pyroxene)							
Azurite	$Cu_3(CO_3)_2(OH)_2$	Vitreous	In crystals, in form of stalactites, in formless masses, or earthy	One perfect, another imperfect	3.5 to 4	3.7 to 3.8	Blue color and streak are diagnostic; occurs with malachite and limonite in copper deposits
Barite	$BaSO_4$	Vitreous to pearly	Crystals tabular or prismatic	Two perfect, one imperfect	2.5 to 3.5	4.3 to 4.6	Translucent to opaque; very heavy for a mineral with nonmetallic luster
Bauxite	Mixture of hydrous aluminum oxides	Dull, earthy	In earthy, clay-like masses; also in small spherical forms	None; uneven fracture	1 to 3	2.0 to 2.55	Emits strong clayish odor
Biotite (black mica)	Complex silicate of K, Mg, Fe, Al. General formula: $AB_3C_4O_{10}(OH)_2$	Pearly to nearly vitreous	In perfect thin flakes; 6-sided crystals	One cleavage direction; uniform flakes or sheets	2.5 to 3	2.7 to 3.3	Black, dark brown, or green; nearly or quite opaque; flakes are both flexible and *elastic*
Calcite	$CaCO_3$	Vitreous to dull	In tapering crystals, or granular aggregates	Three perfect, at oblique angles	3	2.71	Colorless or white; effervesces in dilute HCl
Carnotite	$K_2(UO_2)(VO_4)_2 \cdot 3H_2O$	Earthy	Earthy powder	Not visible	Very soft	4 to 4.7	Powdery form, with brilliant canary-yellow color; ore of vanadium and uranium

Table 1
Properties of Some Important Minerals (Continued)

Mineral	Chemical Composition	Luster	Form	Cleavage	Hardness	Specific Gravity	Other Properties; Remarks
Cassiterite (tinstone)	SnO_2	Adamantine to dull	Granular masses; well-formed crystals common; rounded pebbles in stream gravels	Two, indistinct; conchoidal fracture	6 to 7	6.8 to 7	Yellow to red-brown; principal ore of tin; very heavy
Chalcedony (cryptocrystalline quartz)	SiO_2	Dull	No visible crystals; commonly banded or in formless masses	None; conchoidal fracture	6 to 6.5	2.57 to 2.64	White if pure; variously colored by impurities
Chalcocite	Cu_2S	Metallic	Usually massive, fine-grained; crystals rare	None, conchoidal fracture	2.5 to 3	5.5 to 5.8	Steel-gray to black; dark gray streak; an ore of copper
Chalcopyrite	$CuFeS_2$	Metallic	Massive or granular	None, uneven fracture	3.5 to 4	4.1 to 4.3	Golden yellow to brassy color; streak dark green to black; an ore of copper
Chlorite	Variable silicate of Mg, Fe, Al. General formula: $B_3C_4O_{10}(OH)_2 \cdot B_3(OH)_6$	Greasy to vitreous	In flaky masses or 6-sided crystals	One perfect cleavage	2 to 2.5	2.6 to 2.9	Light to dark green; flakes are weak, *inelastic*, easily separated
Cinnabar	HgS	Adamantine to full	In veins; also in disseminated grains	One perfect cleavage	2 to 2.5	8.1	Red to red-brown; scarlet streak; chief ore mineral of mercury
Copper (native)	Cu	Dull-metallic	Forms of twisted leaves and wires; also irregular nodules	None; hackly fracture	2.5 to 3	8.9	Copper color, but commonly stained green; ductile and malleable; not now common as an ore
Corundum (ruby, sapphire)	Al_2O_3	Adamantine to vitreous	In separate crystals or in granular masses	Two good cleavages, with striations on planes	9	4.0 to 4.1	Blue, red, yellow-brown, green-violet; valuable as an abrasive (emery) and as gems
Diamond	C	Adamantine to greasy	In octahedral or cubic crystals; faces commonly curved	Good, in four directions	10	3.5	Used as abrasive; now a synthetic as well as natural product; high-quality stones used as gems. High heat conductivity may cause crystals to feel cold momentarily

Appendix 1

Table 1

Properties of Some Important Minerals (Continued)

Mineral	Chemical Composition	Luster	Form	Cleavage	Hardness	Specific Gravity	Other Properties; Remarks
Dolomite	$CaMg(CO_3)_2$	Vitreous to pearly	In crystals with rhomb-shaped faces; also in granular masses	Perfect in three directions, as in calcite	3.5 to 4	2.85	White, gray, or flesh-colored; some crystals have curved faces; must be scratched or powdered to effervesce in cold dilute HCl
Epidote	Variable silicate of Ca, Fe, Al. General formula: $Ca_2B_3^{3+}(SiO_4)_3(OH)$	Vitreous	In small prismatic crystals; also fibrous	One perfect cleavage, another imperfect	6 to 7	About 3.4	Yellow-green to blackish-green; commonly associated with chlorite; distinguished from olivine by cleavage and by form

Feldspars (see Plagioclase feldspars and Potassium feldspars)

Mineral	Chemical Composition	Luster	Form	Cleavage	Hardness	Specific Gravity	Other Properties; Remarks
Fluorite	CaF_2	Vitreous	In well-formed crystals and in granular masses	Good cleavage in four directions	4	3.2	Colorless, green, blue, or nearly black; commonly in veins with lead and silver ores; also in cavities in limestone and dolostone; does not effervesce with dilute HCl
Galena	PbS	Bright metallic	In cubic crystals and granular masses, coarse- or fine-grained	Perfect in three directions at right angles	2.5	7.6	Lead-gray color; streak gray to gray-black; common ore of lead; in many deposits it contains silver
Garnet (complex group)	Isomorphous silicates of Ca, Mg, Fe, Mn, Al. General formula: $(A,B^{2+})_3B_2^{3+}(SiO_4)_3$	Vitreous to resinous	Commonly in perfect crystals with 12 or 24 sides, also granular masses	None; fracture conchoidal or uneven	6.5 to 7.5	3.5 to 4.3	Color varies with composition; red, brown, yellow, green to almost black
Gold (native)	Au	Metallic	Massive or in thin irregular scales	None; hackly fracture	2.5 to 3	19.3	Yellow "golden" color; quite malleable; commonly scattered in quartz veins; nuggets occur in stream gravels

Table 1
Properties of Some Important Minerals (Continued)

Mineral	Chemical Composition	Luster	Form	Cleavage	Hardness	Specific Gravity	Other Properties; Remarks
Graphite	C	Metallic to dull	In scaly masses	Perfect, in flakes	1 to 2	2.02 to 2.23	Gray to nearly black; black streak; greasy feel; high melting point
Gypsum	$CaSO_4 \cdot 2\,H_2O$	Vitreous to pearly	In tabular, diamond-shaped crystals; also granular, fibrous, or earthly	One perfect cleavage, two imperfect	2	2.3	Usually white or colorless, transparent or translucent; cleavage plates flexible
Halite (rock salt, common salt)	NaCl	Vitreous	In cubic crystals or granular masses	Perfect in three directions at right angles	2.5	2.16	Colorless or white when pure, transparent to translucent; strong salty taste
Hematite	Fe_2O_3	Metallic or earthy	Varied: massive, granular, micaceous, earthy	None, uneven fracture	5 to 6	4.9 to 5.3	Red-brown, gray to black; red-brown streak; most important ore of iron

Hornblende (see *Amphibole*)
Hypersthene (see *Pyroxene*)

Mineral	Chemical Composition	Luster	Form	Cleavage	Hardness	Specific Gravity	Other Properties; Remarks
Kaolinite (common clay mineral)	$Al_2Si_2O_5(OH)_4$	Dull	Soft, earthy masses.	One perfect cleavage, submicroscopic	2 to 2.5	About 2.6	White if pure, usually stained yellow or other colors; plastic; emits clay odor
Kyanite	Al_2SiO_5	Vitreous to pearly	In groups of blade-like crystals	One good, another imperfect	4 to 5 parallel to crystal, 7 across crystal	3.53 to 3.65	White, pale blue, or green; occurs in metamorphic rocks; compare with sillimanite
Limonite (impure oxide)	Mixture of several hydrous oxides of iron	Dull to vitreous	Compact to earthy masses; irregular nodules	None, irregular fracture	1 to 5.5	3.5 to 4	Yellow, brown, black; streak yellow-brown; an ore of iron
Magnetite	Fe_3O_4	Metallic	Varied: massive, granular	None, uneven fracture	5.5 to 6.5	5 to 5.2	Black and opaque; black streak; strongly attracted to a magnet; important ore of iron

Table 1
Properties of Some Important Minerals (Continued)

Mineral	Chemical Composition	Luster	Form	Cleavage	Hardness	Specific Gravity	Other Properties; Remarks
Malachite	$Cu_2(CO_3)(OH)_2$	Silky to dull	Rarely in crystals; massive, with mammillary forms on surface	One perfect, another fair	3.5 to 4	3.6 to 4	Green color and streak; an ore mineral of copper that occurs in oxidized parts of copper deposits, commonly with limonite and generally with azurite

Mica (see Muscovite, "white mica," and Biotite, "black mica")
Microcline (see Potassium feldspar)

Mineral	Chemical Composition	Luster	Form	Cleavage	Hardness	Specific Gravity	Other Properties; Remarks
Muscovite (white mica)	Variable silicate of K, Al. General formula: $AB_2C_4O_{10}(OH)_2$	Vitreous to pearly	In uniform thin flakes; rarely in 6-sided crystals	One cleavage direction; perfect flakes or sheets	2 to 2.5	2.77 to 2.88	Colorless and transparent when pure but commonly greenish and mottled; flakes are flexible *and* elastic
Nepheline	$Na_3KAl_4(Si_2O_8)_2$	Crystals vitreous; cleavage surfaces greasy	Varied; massive or as scattered grains in silica-deficient rocks	Present but poorly developed	5.5 to 6	2.55 to 2.65	Colorless and translucent to white, gray, yellowish, brownish; not so hard as quartz; poorer cleavage than feldspars; gelatinizes in dilute acids
Olivine	Varied proportions of Fe, Mg. General formula: $B_2^{2+}SiO_4$	Vitreous	In small grains or granular masses	None; conchoidal fracture	6.5 to 7	3.2 to 4.3	Olive green to yellow-green; transparent to translucent
Opal (hydrous silica)	$SiO_2 \cdot nH_2O$	Waxy to vitreous	Amorphous; in irregular masses	None; conchoidal fracture	5 to 6.5	2.0 to 2.2	Various colors; translucent to opaque

Note: Other common varieties of silica, most of them containing impurities, are agate, flint, chert, and jasper.

Orthoclase (see Potassium feldspars)

Mineral	Chemical Composition	Luster	Form	Cleavage	Hardness	Specific Gravity	Other Properties; Remarks
Plagioclase feldspars (Na-Ca feldspars)	$NaAlSi_3O_8$(albite) to $CaAl_2Si_2O_8$ (anorthite) General formula: AC_4O_8	Vitreous to pearly	Commonly as irregular grains or cleavable masses; some varieties in thin plates	Two good cleavages, not quite at right angles	6 to 6.5	2.62 to 2.76	White to dark gray and also other colors; cleavage planes may show fine parallel striations; play of colors in some varieties

Table 1
Properties of Some Important Minerals (Continued)

Mineral	Chemical Composition	Luster	Form	Cleavage	Hardness	Specific Gravity	Other Properties; Remarks
Potassium feldspars (orthoclase, microcline, and sanidine)	$KAlSi_3O_8$	Vitreous	In prismatic crystals or grains with cleavage	Two good cleavages, at right angles	6	2.56 to 2.59	Commonly flesh-colored, pink, or gray; one variety green
Pyrite ("fool's gold")	FeS_2	Metallic	In cubic crystals with striated faces, commonly massive	None, uneven fracture	6 to 6.5	4.9 to 5.2	Pale brass-yellow, darker if tarnished; streak greenish-black; widely distributed; used in manufacture of sulfuric acid
Pyrolusite	MnO_2	Metallic to dull	Rarely in crystals; in coatings on fracture surfaces; commonly in concretions	Crystals have one perfect cleavage	2 to 6.5	4.5 to 5	Dark gray, black or bluish; black streak; important ore mineral of manganese
Pyroxene (complex group, *augite* and *hypersthene* most common)	Silicates of Ca, Fe, Mg, Na, Al. General formula: ABC_2O_6	Vitreous	In 8-sided, stubby crystals; also in granular masses	Two cleavages, nearly at right angles	5 to 6	3.2 to 3.9	Light to dark green, or black; alternate crystals faces at right angles (fit into corner of a box)
Quartz	SiO_2	Vitreous to greasy	Six-sided crystals, pyramids at ends, also in irregular grains and masses	None; conchoidal fracture	7	2.65	Varies from colorless and transparent to opaque with wide range of colors
Rutile	TiO_2	Adamantine to metallic	In slender prismatic crystals or granular masses	Good in one direction; has conchoidal fracture	6 to 6.5	4.2	Red-brown to black; streak brownish to gray-black; abundant in some beach sands; ore of titanium

Sanidine (see Potassium feldspars)

Mineral	Chemical Composition	Luster	Form	Cleavage	Hardness	Specific Gravity	Other Properties; Remarks
Serpentine (fibrous variety is *asbestos*)	Variable silicate of Mg, with (OH). General formula: $B_6C_4O_{10}(OH)_8$	Greasy or resinous	Massive or fibrous	Usually breaks irregularly, except in the fibrous variety	2.5 to 5	2.2 to 2.6	Light to dark green; smooth, greasy feel; translucent to opaque

Table 1
Properties of Some Important Minerals (Continued)

Mineral	Chemical Composition	Luster	Form	Cleavage	Hardness	Specific Gravity	Other Properties; Remarks
Sillimanite	Al_2SiO_5	Vitreous	In long crystals or in fibers	Perfect in one direction	6 to 7	3.2	White to greenish-gray; found in high-grade metamorphic rocks; compare with kyanite
Silver (native)	Ag	Bright metallic to dull	In flakes and irregular grains	None; hackly fracture	2.5 to 3	10 to 11	Generally tarnished to dark gray or black; cleans to silvery-white; ductile and malleable
Sphalerite (zinc blende)	ZnS	Resinous to adamantine	Fine- to coarse-granular masses; crystals common	Six directions	3.5 to 4	3.9 to 4.1	Color yellow-brown to black; streak white to yellow or brown; principal ore of zinc
Staurolite	$Al_4Fe^{2+}(SiO_4)_2(OH)_2O_2$	Vitreous to resinous	Stubby crystals, commonly twinned in form of cross	Distinct in one direction	7 to 7.5	3.7 to 3.8	Red-brown to nearly black; associated with sillimanite, kyanite, garnet
Talc	$Mg_3(OH)_2Si_4O_{10}$	Greasy, pearly	In small scales and compact masses	One perfect cleavage	1	2.58 to 2.83	White to greenish; has greasy feel
Topaz	$Al_2SiO_4(OH,F)_2$	Vitreous	In prismatic crystals; also in granular masses	One perfect cleavage	8	3.49 to 3.57	Colorless to shades of blue, yellow, or brown; found in some pegmatites and quartz veins
Uraninite (pitchblende)	UO_2 to U_3O_8	Submetallic to dull	Massive, with botryoidal forms; crystals cubic or 8-sided	None, uneven fracture	5 to 6	6.5 to 10	Black to dark brown; streak has similar color; a source of uranium and radium
Wollastonite	$CaSiO_3$	Vitreous to silky; pearly on cleavage surfaces	Fibrous or bladed aggregates of elongated crystals	Perfect in two directions	4.5 to 5	2.8 to 2.9	Colorless, white, gray, pinkish, yellowish translucent; soluble in HCl; distinguished from fibrous amphiboles by cleavage and by solubility in HCl

APPENDIX 2

The Modern Periodic Table of the Elements

Period	IA	IIA	IIIB	IVB	VB	VIB	VIIB	VIII			IB	IIB	IIIA	IVA	VA	VIA	VIIA	Noble Gases (O)
1	1 **H** 1.00797																	2 **He** 4.00260
2	3 **Li** 6.941	4 **Be** 9.01218											5 **B** 10.81	6 **C** 12.01115	7 **N** 14.0067	8 **O** 15.9994	9 **F** 18.99840	10 **Ne** 20.179
3	11 **Na** 22.98977	12 **Mg** 24.305											13 **Al** 26.98154	14 **Si** 28.086†	15 **P** 30.97376	16 **S** 32.06	17 **Cl** 35.453	18 **Ar** 39.948
4	19 **K** 39.098	20 **Ca** 40.08	21 **Sc** 44.9559	22 **Ti** 47.90	23 **V** 50.9414	24 **Cr** 51.996	25 **Mn** 54.9380	26 **Fe** 55.847	27 **Co** 58.9332	28 **Ni** 58.71	29 **Cu** 63.546	30 **Zn** 65.38	31 **Ga** 69.72	32 **Ge** 72.59	33 **As** 74.9216	34 **Se** 78.96	35 **Br** 79.904	36 **Kr** 83.80
5	37 **Rb** 85.4678	38 **Sr** 87.62	39 **Y** 88.9059	40 **Zr** 91.22	41 **Nb** 92.9064	42 **Mo** 95.94	43 **Tc** 98.9062	44 **Ru** 101.07	45 **Rh** 102.9055	46 **Pd** 106.4	47 **Ag** 107.868	48 **Cd** 112.40	49 **In** 114.82	50 **Sn** 118.69	51 **Sb** 121.75	52 **Te** 127.60	53 **I** 126.9045	54 **Xe** 131.30
6	55 **Cs** 132.9054	56 **Ba** 137.34	57 ***La** 138.9055	72 **Hf** 178.49	73 **Ta** 180.9479	74 **W** 183.85	75 **Re** 186.2	76 **Os** 190.2	77 **Ir** 192.22	78 **Pt** 195.09	79 **Au** 196.9665	80 **Hg** 200.59	81 **Tl** 204.37	82 **Pb** 207.19	83 **Bi** 208.9804	84 **Po** (210)	85 **At** (210)	86 **Rn** (222)
7	87 **Fr** (223)	88 **Ra** 226.0254	89 †**Ac** (227)	104 **Ku** (261)	105 **Ha** (260)													

atomic number → 1 **H** 1.0079 ← atomic mass

*| 58 **Ce** 140.12	59 **Pr** 140.9077	60 **Nd** 144.24	61 **Pm** (147)	62 **Sm** 150.4	63 **Eu** 151.96	64 **Gd** 157.25	65 **Tb** 158.9254	66 **Dy** 162.50	67 **Ho** 164.9304	68 **Er** 167.26	69 **Tm** 168.9342	70 **Yb** 173.04	71 **Lu** 174.97

†| 90 **Th** 232.0381	91 **Pa** 231.0359	92 **U** 238.029	93 **Np** 237.0482	94 **Pu** (244)	95 **Am** (243)	96 **Cm** (247)	97 **Bk** (247)	98 **Cf** (251)	99 **Es** (254)	100 **Fm** (257)	101 **Md** (258)	102 **No** (255)	103 **Lr** (256)

(List of Elements and Their Symbols appears on page 638)

Appendix 2

Elements and Their Symbols

Element	Symbol	Element	Symbol	Element	Symbol
Actinium	Ac	Hafnium	Hf	Potassium	K
Aluminum	Al	Hahnium	Ha	Praseodymium	Pr
Americium	Am	Helium	He	Promethium	Pm
Antimony	Sb	Holmium	Ho	Protactinium	Pa
Argon	Ar	Hydrogen	H	Radium	Ra
Arsenic	As	Indium	In	Radon	Rn
Astatine	At	Iodine	I	Rhenium	Re
Barium	Ba	Iridium	Ir	Rhodium	Rh
Berkelium	Bk	Iron	Fe	Rubidium	Rb
Beryllium	Be	Krypton	Kr	Ruthenium	Ru
Bismuth	Bi	Kurchatovium	Ku	Samarium	Sm
Boron	B	Lanthanum	La	Scandium	Sc
Bromine	Br	Lawrencium	Lr	Selenium	Se
Cadmium	Cd	Lead	Pb	Silicon	Si
Calcium	Ca	Lithium	Li	Silver	Ag
Californium	Cf	Lutetium	Lu	Sodium	Na
Carbon	C	Magnesium	Mg	Strontium	Sr
Cerium	Ce	Manganese	Mn	Sulfur	S
Cesium	Cs	Mendelevium	Md	Tantalum	Ta
Chlorine	Cl	Mercury	Hg	Technetium	Tc
Chromium	Cr	Molybdenum	Mo	Tellurium	Te
Cobalt	Co	Neodymium	Nd	Terbium	Tb
Copper	Cu	Neon	Ne	Thallium	Tl
Curium	Cm	Neptunium	Np	Thorium	Th
Dysprosium	Dy	Nickel	Ni	Thulium	Tm
Einsteinium	Es	Niobium	Nb	Tin	Sn
Erbium	Er	Nitrogen	N	Titanium	Ti
Europium	Eu	Nobelium	No	Tungsten	W
Fermium	Fm	Osmium	Os	Uranium	U
Fluorine	F	Oxygen	O	Vanadium	V
Francium	Fr	Palladium	Pd	Xenon	Xe
Gadolinium	Gd	Phosphorus	P	Ytterbium	Yb
Gallium	Ga	Platinum	Pt	Yttrium	Y
Germanium	Ge	Plutonium	Pu	Zinc	Zn
Gold	Au	Polonium	Po	Zirconium	Zr

GLOSSARY[1]

Abrasion. The mechanical wear of rock on rock.

Abyssal fan. A fanlike accumulation of sediment at the mouth of a submarine canyon.

Abyssal floor. The low areas of ocean basins.

Abyssal-hills province. A part of the ocean basin floor consisting almost completely of irregular rocky hills a few hundred meters to a few kilometers wide and having relief of 50 to 100 m.

Abyssal plain. A flat part of the ocean floor, underlain by sediment, having an imperceptible slope of less than 1:1,000.

Accessory mineral. A minor constituent of a rock that does not enter into classifying and naming of the rock.

Aeration, zone of. See **Zone of aeration.**

Agglomerate. A coarse-grained pyroclastic rock consisting largely of bombs and blocks.

Alluvial fan. See **Fan.**

Alluvial fill. A body of alluvium, occupying a stream valley, and conspicuously thicker than the depth of the stream.

Alluvium. The general name for all sediment deposited in land environments by streams.

Alpha-particle. One of the kinds of particles involved in radioactivity; identical with the nucleus of an atom of helium, consisting of 2 protons and 2 neutrons. The mass number of an α-particle is 4 and its atomic number is 2.

Amorphous solid. A solid in which the molecules, atoms, or ions are randomly arranged in an irregular network.

Amphibolite. A coarse-grained mafic metamorphic rock containing abundant amphibole.

[1]The definitions of rocks are intended only for the use of beginning students in naming hand specimens. Many rock terms are approximations that do not coincide with the more precise definitions of professional geologists based on data obtained with a petrographic microscope.

Amygdale. A vesicle that has been filled by one or more minerals.

Amygdaloidal. An adjective describing a rock containing amygdales.

Anaerobic bacteria. Bacteria that require no free oxygen but that derive oxygen from organic matter or from sulfate radicals in solution.

Andesite. A fine-grained, felsic igneous rock of intermediate composition. Porphyritic varieties contain phenocrysts of feldspar but none of quartz and have aphanitic groundmass. Andesite is a fine-grained equivalent of diorite.

Angle of respose. Also known as critical slope. As used in mass-wasting, it describes the steepest angle, measured from the horizontal, at which a material remains stable. The minimum angle along which coarse particles in an aggregate of particles begin to fall under the influence of gravity.

Angular unconformity. See *Unconformity:* **angular.**

Anorthosite. An igneous rock composed almost entirely of plagioclase.

Anticline. An upfold of layered rocks in the form of an arch and having the oldest strata in the center. The reverse of a syncline.

Aphanitic texture. A texture of equigranular igneous rocks with individual particles not visible to the unaided eye.

Aquiclude A zone in the earth that is saturated with ground water, but not permeable enough to supply water to wells or springs.

Aquifer. A body of permeable rock or regolith through which ground water moves.

Aréte. A jagged, knife-edge ridge created where two groups of glaciers have eaten into the ridge from both sides.

Arkose. A sandstone containing at least 25 per cent feldspar as well as quartz.

Artesian spring. See **Spring: artesian.**

Artesian well. A well in which water rises above the aquifer.

Ash (volcanic). Tephra ranging in size between $1/16$ and 4mm.

Asthenosphere A zone of ductile rock in the earth's mantle situated beneath the lithosphere.

Atoll. A reef that forms a nearly closed figure within which there is no land mass.

Atom. The smallest electrically balanced particle displaying the properties of an element.

Atomic mass. The mass of an atom, nearly all of which is contained in the nucleus. The mass of an atomic nucleus is expressed by the *mass number* (q. v.).

Atomic number. The number of protons in an atomic nucleus, represented by the letter Z. Also the number of electrons in a neutral atom.

Attitude. The orientation, or position, of a layer or of a surface. Measured by determining strike and dip.

Authigenic sediment. A sedimentary deposit formed in place, not from physically transported materials, and consisting of minerals that crystallized out of seawater. An **authigenic mineral** can be precipitated from interstitial water, not necessarily seawater, in a sediment.

Axial surface. An imaginary surface through the middle of the fold that passes through the axis of the fold.

Axial trough A trough in the ocean floor that extends along the axis of an oceanic ridge.

Backwash. The return sheet flow, down a sloping shore, of water from the spent swash.

Badlands. A system of closely spaced narrow ravines with little or no vegetation.

Bar. An elongate ridge of sediment, built offshore by waves and currents, whose top is always submerged.

Barbed tributary. See **Tributary: barbed.**

Barchan. See **Dune: barchan.**

Barrier. An elongate island of sand or gravel parallel with the coast.

Barrier reef. An elongate reef not connected with a land mass.

Basalt. An aphanitic igneous rock having the composition of gabbro. Specimens of basalt are opaque even on thin edges.

Basalt: oceanic. Basalt, discharged by volcanoes within ocean basins (or within continental masses), containing more than 1.75 per cent TiO_2 and generally less than 15 per cent Al_2O_3.

Base level. The limiting level below which a stream can not erode the land.

Base level: local. The level of a lake or any other base level that stands above sea level.

Base level: ultimate. Sea level, projected inland as an imaginary surface underneath a stream.

Basement complex. An assemblage of igneous and metamorphic rocks lying beneath the oldest stratified rocks of a region. Many, but not all basement rocks are of Precambrian age.

Batholith. A large pluton having an exposed area of more than 40 square miles (100km^2).

Bauxite. A mixture of hydrated oxides, containing large volumes of hydrous aluminium oxide and widely used as an ore of metallic aluminum. It is generally expressed by the chemical formula $Al_2O_3 \cdot nH_2O$.

Bay barrier. A ridge of sand or gravel that completely blocks the mouth of a bay.

Beach. A body of wave-washed sediment extending along a coast between the landward limit of wave action and the outermost breakers.

Beach drift. The movement of particles obliquely up the slope of a beach by the swash and directly down this slope by the backwash.

Bed. A stratum (q. v.) 1cm or more thick. Also, the floor of a stream channel.

Bedding plane. The top or bottom surface of a bed, or stratum.

Bedding-plane parting. A surface of separation between adjacent strata.

Bed load. See **Load: bed.**

Bedrock. Continuous solid rock that everywhere underlies regolith and locally forms the Earth's surface.

Beheaded stream. The remaining part of a stream course that has lost its upper part by stream capture (q. v.).

Beta-particle. One of the kinds of particles involved in radioactivity, consisting of either an electron or a positron (q. v.).

Bituminous coal. Black and firm soft coal that breaks into blocks and contains alternating layers having dull and bright luster.

Blowout. A deflation basin excavated in shifting sand or other easily eroded regolith.

Body wave. A seismic wave that travels entirely beneath the Earth's surface.

Botryoidal. An adjective describing minerals having rounded forms resembling grapes closely bunched.

Bottomset bed. The gently sloping, fine, thin part

of each layer in a delta.

Bouguer gravity anomaly. A value obtained by adjusting the measured value of gravity at a point on the earth's surface to account for effects of latitude, elevation, and the mass of land extending above sea level or the mass of water extending below sea level.

Boulder size. Particles of sediment having diameters greater than 256mm (approximately the size of a volleyball).

Boulder train. A group of erratics spread out fanwise.

Bowen's reaction series. See **Reaction series.**

Braided stream. See *Stream:* **braided.**

Breaker. A wave that is collapsing.

Breaker zone. The zone where waves collapse.

Breccia. A general term for a rock of any origin containing angular particles.

Brittle solid. A solid that fractures readily.

Brown clay. Formerly called "red clay." Pelagic sediment containing less than 30 percent skeletal remains of microorganisms.

Brownian movement. The movement, as a result of impacts from moving water molecules, of tiny solid particles (diameters less than 2 microns) suspended in water.

Bulk modulus. The stress required to cause a unit change of volume.

Calcareous. Containing calcium carbonate.

Calcareous tufa. A light spongy limestone precipitated by algae or bacteria or by the evaporation of water from springs and small streams.

Calcrete. See *Caliche.*

Caldera. A volcanic crater enlarged to a diameter of several miles.

Caliche. A whitish accumulation of calcium carbonate developed in a soil profile. Also known as a *calcrete.*

Calorie. The amount of heat energy required to raise the temperature of 1g of water from 15° to 16°C.

Capillary. A small opening having a diameter less than that of a human hair.

Carbonate rocks. Sedimentary rocks consisting chiefly of carbonate minerals.

Cataclastic texture. A texture of rocks resulting from breakage and pulverization of mineral particles.

Cavern. A large, roofed-over cavity in any kind of rock.

Cavitation. The formation and collapse of bubbles in a turbulent liquid.

Cement. Materials, precipitated from solution, which bind together the framework particles of a sedimentary rock.

Cementation. The binding together of particles of framework and matrix of a sediment by precipitation of mineral cement in former pore spaces. Calcite, quartz, and iron oxides are common cements.

Chalk. Limestone that is weakly cohesive.

Chert. A sedimentary rock composed of silica which is either an original precipitate or a replacement product of calcium-carbonate minerals. Commonly the original textures of the carbonate minerals are preserved.

Chute cutoff. A new channel cut across a point bar, resulting in abandonment of part of a meander.

Cinder cone. See **Tephra cone.**

Cinder, volcanic. Obsolete name for medium-grained **Tephra** (q. v.).

Circumoceanic basalt. See **Basalt: oceanic.**

Cirque. A steep-walled niche, shaped like a half bowl, in a mountain side, excavated mainly by ice plucking and frost action.

Cirque glacier. A very small glacier that occupies a cirque.

Clastic rock. A general term for a sedimentary rock having clastic texture.

Clastic sediment. See **Detritus.**

Clastic texture. A texture of sediments and sedimentary rocks resulting from physical transport and deposition of broken particles of older rocks, of older sediments, and of organic skeletal remains.

Clay size. Dimension of sedimentary particles having diameters less than 1/256 mm (4 microns).

Claystone. A clastic rock consisting predominantly of clay-size particles.

Cleavage (as applied to minerals). The capacity of a mineral to break in preferred directions along surfaces parallel to lattice planes.

Cleavage, rock. See **Rock cleavage.**

Clinometer. An instrument for measuring degree of inclination, or *dip* (q. v.).

Closed system. A reaction in which no material escapes from the scene of reaction.

Coal. A black sedimentary rock consisting chiefly of partly decomposed plant matter and containing less than 40 per cent inorganic matter.

Cobble size. Sediment particles having diameters

greater than 64 mm (about the size of a tennis ball) and less than 256 mm (about the size of a volleyball).

Cohesion. An electrostatic force of attraction among fine particles.

Col. A gap or pass in a mountain crest at a place where the headwalls of two cirques intersect.

Colluvium. A body of sediment that has been deposited by any process of mass-wasting or by overland flow.

Column. (1) A stalactite connected with a stalagmite. (2) Sometimes used as an abbreviation for **geologic column.**

Columnar joints. Joints that split rocks into long prisms, or columns. Columnar joints are a common feature in tabular bodies of igneous rock.

Compaction. The reduction in pore space within a body of fine-grained sediments in response to the weight of overlying material or to pressures within the Earth's crust.

Component (chemical). One of the minimum number of chemical constituents that must be specified to describe the chemical composition of a phase. This term is commonly used to describe the oxides that collectively represent the composition of a mineral.

Composite cone. A volcanic cone consisting partly of tephra and partly of igneous rock both extrusive and intrusive.

Composite dike. A dike composed of more than one generation of igneous rock.

Compressive stress. See **Stress:compressive.**

Conchoidal fracture. Breakage resulting in smooth curved surfaces.

Concordant contact. A contact surface of a pluton that is parallel to the layers of the intruded rocks.

Concordant pluton. A pluton having concordant contacts (q. v.).

Concretion. A localized rock body having distinct boundaries, inclosed in sedimentary rock, and consisting of substances precipitated from solution, commonly around a nucleus.

Cone of depression. A conical depression in the water table immediately surrounding a well.

Confined aquifer. An aquifer in which flow of ground water in all directions, particularly upward, is prevented by impermeable material.

Confined percolation. See **Percolation: confined.**

Conglomerate. A clastic sedimentary rock containing numerous rounded pebbles or larger particles.

Congruent melting. Melting without change of composition.

Connate water. Seawater that was trapped in a sedimentary deposit when the sediment lay on the sea floor.

Consequent stream. See **Stream: consequent.**

Contact-metamorphic aureole. A zone of altered rocks, surrounding a pluton, and originating from the effects of the magma.

Contact metamorphism. Metamorphism (q. v.) in the vicinity of a pluton and resulting from the effects of the magma on its surrounding rocks.

Continent. A major land area that stands above sea level. Compare **continental mass.**

Continental crust. See **Crust: continental.**

Continental drift. A concept that continental masses, composed largely of sialic rock, have moved widely and differentially over and through the denser mafic rock that underlies continental blocks and ocean floors.

Continental glacier. See **Glacier:** *continental.*

Continental mass. A major high-standing part of the lithosphere.

Continental rise. The gentle slope with gradient between 1:100 and 1:700 that lies seaward of the continental slope.

Continental shelf. A submerged marginal zone of a continental mass forming a shallow platform of variable width extending from the shoreline to the first prominent break in slope at a depth of 600 m or less. On most shelves the depth of this break in slope is 200 m or less.

Continental shield. An extensive area in which the Precambrian foundation rocks of a continental mass are exposed.

Continental slope. A relatively steep (3° to 6°) slope that lies seaward of a continental shelf.

Continuous reaction relationship. The exchange of materials, during cooling, between a silicate melt and a continuously growing crystal of one mineral species. An example is the exchange of Fe^{2+} and Mg^{2+} that takes place during the crystallization of olivine from a mafic magma.

Continuous seismic profiler. An acoustic device for making profiles of the thickness and internal structure of sediments beneath a body of water.

Contour. See **Contour line.**

Contour current. A subsurface density current flowing parallel to the submarine slopes at the margin of an ocean basin.

Contour interval. The vertical distance between two successive contour lines.

Contour line. A line passing through points having the same altitude above sea level; often simply called a contour.

Convection current. The movement of material, within a closed system, as a result of thermal convection arising from the unequal distribution of heat.

Coquina. An aggregation of shells and large shell fragments, cemented with calcium carbonate.

Cordillera. One of the great mountain belts of the Earth.

Cordillera, North American. See **North American Cordillera.**

Core. (1) A cylindrical sample of sediment or rock recovered from a drilled hole. (2) The central part of the Earth. See **Earth's core.**

Coriolis force. The inertial force that must be added to motions where the frame of reference rotates. The Coriolis force acts to the right where the frame of reference rotates counterclockwise (as in the Northern Hemisphere) and to the left where the frame of reference rotates clockwise (as in the Southern Hemisphere).

Correlation. In stratigraphy, this term is used to mean determination of equivalence, in geologic age and position in the sequence, of strata in different areas.

Covalent bond. A kind of bond in which atoms are held together by sharing electrons.

Crater. See **Volcanic crater.**

Creep. The imperceptibly slow downslope movement of regolith.

Crevasse. A deep crack in the upper surface of a glacier.

Critical temperature. The temperature at which the distinction between the liquid and the gaseous states of aggregation disappears.

Cross-strata. See **Strata: cross-strata.**

Crust (Earth's). The outer part of the lithosphere.

Crust: continental. The Earth's crust beneath continents, 30 to 60 k m thick, consisting of an upper part having the same elastic properties as sialic rocks and a lower part having the same elastic properties as mafic rocks.

Crust:oceanic. The Earth's crust beneath ocean basins, averaging about 5 km thick, and consisting of material having the same elastic properties as mafic rocks.

Crustal warping. Gentle bending of the crust upward or downward.

Cryptocrystalline. A texture of rocks in which the crystalline particles are so small that they can not be resolved with an ordinary microscope.

Crystal. A solid bounded by natural, regular plane surfaces formed by growth of a crystal lattice.

Crystal face. A smooth plane face of a crystal.

Crystal lattice. A systematic, regular, symmetrical network of particles within a crystal.

Crystalline particle. A mineral particle of any size lacking well-developed crystal faces. If crystal faces are present the term *crystal* is used whatever its size.

Crystalline solid. A solid in which the pattern of the particles (molecules, atoms, or ions) is repeated and symmetrical.

Crystalline texture. A texture resulting from simultaneous growth of associated crystalline particles.

Crystallization. The process of development of crystals, by condensation of materials in a gaseous state, by precipitation of materials in a solution, or by solidification of materials in a melt.

Curie point. The temperature above which a substance loses its magnetism.

Cycle of erosion. The sequence of forms, essentially valleys and hills, through which a landmass is thought to evolve from the time it begins to be eroded until it is reduced to near base level.

Darcy's law. The equation for the velocity of flow of ground water which states that in material of given permeability, velocity of flow increases as the pressure along flow lines increases.

Daughter isotope. In radioactivity, the isotope, created by decay of a parent isotope, that continually increases in amount with time.

Debris flow. The rapid downslope plastic flow of a mass of debris.

Debris flow, variety mudflow. A debris flow in which the consistency of the substance is that of mud.

Decay constant. The proportion of radioactive atoms of each isotope that decay in a unit of time; represented by the Greek letter, λ.

Declination (magnetic). The clockwise angle in the horizontal plane between true north and a magnetic line of force. Also the angle between the direction of true north and magnetic north.

Decomposition. The chemical alteration of rock materials.

Deep-focus earthquake. See **Earthquake: deep-focus.**

Deep-water wave. A wave, on the surface of a

body of water, beneath which particles of water move in circular orbits, within vertical planes, and are not influenced by the bottom.

Deflation. The picking up and removal of loose rock particles by the wind.

Deflation armor. A surface layer of coarse particles concentrated chiefly by deflation.

Deflation basin. A depression excavated by deflation.

Deformation. A change of volume, of shape, or of both volume and shape of a rock body, or a change of its original position within the Earth's crust.

Delta. A body of sediment deposited by a stream flowing into standing water of a lake or the sea. The name comes from the similarity of the plan view to the shape of the Greek letter Δ.

Dendritic stream pattern. See **Stream pattern: dendritic.**

Density. Mass per unit volume.

Density current. A localized current flowing because it consists of fluid denser than that of the body of the fluid through which it moves. The excess density may result from low temperature, high salinity, sediment held in suspension, or from various combinations of these factors.

Desert. Arid land.

Detrital sedimentary rock. A sedimentary rock composed of lithified detritus (q. v.).

Detritus. A collective term referring to broken pieces of older rocks, of minerals, or of skeletal remains of organisms. Also known as **clastic sediment.**

Diastrophism. The processes of large-scale deformation, metamorphism, and intrusion that occur in orogenic belts.

Differential weathering. The result of variations in the rate of weathering on different parts of a rock body. As a result of differential weathering, resistant parts stand in relief whereas less-resistant parts form recesses.

Diffusion. The transport of particles in the absence of bulk flow.

Dike. A tabular pluton having discordant surfaces of contact.

Dike, composite. See **Composite dike.**

Dike swarm. A group of associated dikes.

Diorite. A coarse-grained felsic igneous rock that lacks quartz, and in which the chief feldspar is plagioclase with ferromagnesian minerals (chiefly hornblende and biotite) constituting less than 50 percent.

Dip. The angle in degrees between a horizontal plane and an inclined plane, measured down from horizontal in a plane perpendicular to the strike. Dip is measured with a *clinometer*.

Dipole. An object that contains opposite charges at two points or a magnetic field consisting of both north-seeking and south-seeking components.

Dip-slip fault. A normal fault or a reverse fault on which the only component of movement lies in the vertical plane normal to the strike of the fault.

Discharge. The quantity of water passing a given point in a given unit of time.

Disconformity. A lack of continuity between two groups of parallel strata in contact but separated by an irregular surface of erosion corresponding to a gap in the geologic record.

Discontinuous reaction relationship. The exchange of materials, during cooling, between a silicate melt and crystals, which results in the dissolution of one mineral species and the simultaneous growth of a different mineral species. An example is the dissolution of olivine and the growth of pyroxene that occurs during the cooling of a mafic magma.

Discordant contact. A contact surface of a pluton that is not parallel with the layers or other boundaries within the intruded rocks.

Discordant pluton. A pluton having discordant contacts (q. v.).

Disintegration. The mechanical breakup of rocks.

Disintegration, granular. See **Granular disintegration.**

Divide. The line that separates adjacent drainage basins.

Dolerite. Medium-grained mafic igneous rock.

Dolostone. A sedimentary rock consisting chiefly of the mineral dolomite.

Doppler shift. Changes in frequency of electromagnetic waves resulting from relative motion between source and receiver.

Dormant volcano. See **Volcano: dormant.**

Drainage basin. The total area that contributes water to a stream.

Drainage:interior. Drainage that does not persist to the sea.

Drift, continental. See **Continental drift.**

Drift, glacial. See **Glacial drift.**

Drift, stratified. See **Glacial drift: ice-contact stratified.**

Dripstone. Calcite chemically precipitated from

dripping water in an air-filled cavity.

Drumlin. A streamlined hill consisting of drift, generally till, and elongated parallel with the direction of glacier movement.

Dune. A mound or ridge of sand deposited by the wind.

Dune: barchan. A crescent-shaped dune with horns pointing downwind.

Dune: longitudinal. A long, straight, ridge-shaped dune parallel with wind direction.

Dune: transverse. A dune forming a wavelike ridge transverse to wind direction.

Dune: U-shaped. A dune of U-shape with the open end of the U facing upwind.

Dunite. An ultramafic igneous rock consisting almost wholly of olivine (pronounced *dŭn'ite*).

Dust (volcanic). Tephra smaller than $1/16$ mm.

Dyne. The force required to accelerate a mass of 1g by 1 cm/sec^2.

Earth's core. The inner spherical mass below a depth of 2,900 km, consisting of a fluid outer part, approximately 6,920 km thick, and a solid inner part, about 2,720 km in diameter.

Earth's mantle. The zone, about 2,900 km thick, between the crust and the central core of the Earth. The mantle occupies about 80 per cent of the total volume of the Earth.

Earthquake: deep-focus. An earthquake with focus deeper than 300 km.

Earthquake focus. The locus of first release of the elastic energy of an earthquake.

Earthquake intensity. A relative measure of the strength of an earthquake based on observed destruction or disturbance and on human sensations.

Earthquake: intermediate-focus. An earthquake with depth of focus between 70 and 300 km.

Earthquake: shallow-focus. An earthquake with focus at depths less than 70 km.

Echo sounder. An instrument that employs sound energy to determine depth of water.

Ecliptic. The plane of the Earth's orbit around the sun.

Ecliptic, plane of. See **Plane of the ecliptic.**

Economy. The input and consumption of energy within a stream or other system and the changes that result.

Eh. The oxidation-reduction potential; measured in volts or millivolts.

Elastic deformation. Nonpermanent deformation of a body in which the stresses do not exceed its elastic limit (q. v.).

Elastic limit. The upper limit of strength of a body; the stress to which it can be subjected and still recover its original shape or volume when the forces tending to cause deformation have been removed.

Elastic rebound. The springing back of a deformed solid to its original shape as a result of rupturing under continued deforming forces or by elastic movement when the deforming forces have been removed. Elastic rebound of a body of deformed rock as a result of rupture under continued deforming forces releases great amounts of energy into the Earth's crust, creating earthquakes.

Electron. Unit particle of negative electrical charge.

Electron volt. The energy possessed by one electron which has fallen through a potential difference of 1 volt.

Emergence (of land). A fall of sea level relative to the land.

Energy. Loosely defined as (1) the capacity to do work; or more precisely as (2) the equivalent of mass according to the Einstein equation, $E = mc^2$.

Epeirogeny. Broad movements of uplift and subsidence affecting large portions of continental areas or of ocean floors.

Epicenter. The part of the Earth's surface vertically above an earthquake focus.

Equigranular texture. A texture of igneous rocks with all particles about the same size.

Erg. The mechanical energy required for a force of 1 dyne to act through 1 cm.

Erosion. A general term that describes the physical breaking down, chemical solution, and movement of broken-down and dissolved rock materials from place to place on the Earth's surface.

Erosion: sheet. The erosion performed by overland flow (q. v.).

Erratic. A transported rock fragment different from the bedrock beneath it. The agent of transport was commonly glacier ice or floating ice.

Escarpment. A steep slope or cliff.

Esker. A body of ice-contact stratified drift shaped into a long narrow ridge, commonly sinuous.

Essential mineral. A major constituent of a rock used in classifying and naming the rock.

Eustatic change of sea level. A worldwide change in the level of the sea resulting from a change in the amount of water within or in the capacity of the ocean basins.

Evaporite. A nonclastic sedimentary rock whose constituent minerals were precipitated from water solution as a result of evaporation.

Evaporite mineral. A mineral precipitated as a result of evaporation. A few examples are aragonite, dolomite, gypsum, halite, and anhydrite.

Exfoliation. The separation, during weathering, of successive shells from massive rocks. The resulting sheets of rock resemble the "skins" of an onion.

Explosion pit. A vent, drilled to the surface by volcanic gases, from which no lava issued.

Exposure. A body of bedrock not covered by regolith and forming part of the Earth's surface.

Extrusive igneous rock. A rock that originated from solidification of lava.

Fabric. The orientation of particles in a rock or a sediment.

Fabric: preferred orientation. A fabric in which the particles are aligned.

Fabric: random. A fabric in which the particles are oriented in random directions.

Facies. A distinctive group of characteristics within a rock unit, that differ as a group from those elsewhere in the same unit.

Facies, metamorphic. See **Metamorphic facies.**

Facies, mineral. See **Mineral facies.**

Fan. A fan-shaped body of alluvium built at the base of a steep slope. Known also as *alluvial fan*.

Fan, abyssal. See **Abyssal fan.**

Fault. A fracture along which the opposite sides have been relatively displaced.

Fault-block mountain. A mountain bounded by one or more faults.

Fault breccia. A breccia (q. v.) consisting of irregular pieces of rock broken as a result of faulting.

Fault drag. The bending of layers next to a fault as a result of fault movement.

Fault: footwall. The wall on (or boundary of) the block below an inclined fault.

Fault: hanging wall. The wall on (or boundary of) the block above an inclined fault.

Fault: thrust fault (or simply **thrust**). A low-angle reverse fault, with dip generally less than 45°.

Faunal succession, law of. This law, discovered by William Smith, states that fossil faunas and floras succeed one another in a definite, recognizable order.

Felsic. An adjective used to describe a rock in which light-colored minerals, chiefly feldspars, predominate.

Felsite. An aphanitic igneous rock of granitic composition, whose thin edges transmit some light.

Ferromagnesian silicate mineral. A silicate mineral containing abundant iron and magnesium.

Fiord. Also *fjord*. A glaciated trough partly submerged by the sea.

Fissile. An adjective describing rocks that split along closely spaced parting planes. In sedimentary rocks such planes parallel the stratification.

Fissure. A fracture in rocks along which the opposite walls have been pulled apart.

Fissure eruption. Extrusion of volcanic materials along an extensive fracture.

Fissure vein. A fissure in rocks, filled with mineral matter.

Flint. A popular name for dark-colored chert (q. v.).

Floodplain. That part of any stream valley which is inundated during floods.

Flow breccia. A volcanic breccia created by the breaking up of a hardened crust of extrusive rock as a result of further flow of liquid lava.

Flow: laminar. Flow in which the fluid particles move in straight, parallel paths and slip over one another along parallel plane surfaces. Also known as streamline flow.

Flow law. A statement of the quantitative relationship between rate of strain of a body of flowing ice and the shearing stress within it.

Flow layering. Layers resulting from flow of magma or lava.

Flow: overland. The movement of runoff in broad sheets or groups of small interconnecting rills.

Flow: plastic. See **Plastic flow.**

Flowstone. Material chemically precipitated from flowing water in the open air or in an air-filled cavity.

Flow: stream. The flow of surface water between well-defined banks.

Flow: turbulent. Fluid flow characterized by eddies.

Fluid. A state of aggregation of liquid or gaseous matter in which shear waves do not propagate.

Fluidized. An adjective describing a body of solid particles in a dilatant state; that is, the particles are no longer in continuous contact with one another.

Focus, earthquake. See **Earthquake focus.**

Fold. A pronounced bend in layers of rock.
Fold axis. The median line between the limbs of the fold, along the apex of an anticline or the lowest part of a syncline.
Fold: closed. A fold with an acute angle between the limbs.
Fold: isoclinal. A fold having essentially parallel limbs.
Fold: open. A fold with limbs that diverge at an obtuse angle.
Fold/ overturned. A fold having a limb in which the strata have been tilted beyond the vertical.
Fold: plunge. The angle a fold axis makes with the horizontal.
Fold: plunging. A fold with an inclined axis.
Fold: recumbent. A fold in which the axial surface is essentially horizontal.
Foliate. A general term for a metamorphic rock possessing foliation on the scale of hand specimens.
Foliation. A parallel or nearly parallel structure in metamorphic rocks along which the rock tends to split into flakes or thin slabs.
Foliation: gneissic. Foliation in which the layers are thicker and parting takes place along rougher surfaces than in the other two types (schistose and slaty).
Foliation: schistose. Foliation in which the layers are thin and parting occurs along generally smooth surfaces.
Foliation: slaty. Foliation characterized by its extremely smooth and plane parting surfaces.
Footwall. See **Fault: footwall.**
Foreset bed. The coarse, thick, steeply sloping part of each layer in a delta.
Fossil. The naturally preserved remains or traces of an animal or a plant.
Fossil fuel. A fuel that contains solar energy, locked up securely in chemical compounds by the plants or animals of former ages.
Fractional melting. Melting accompanied by a change in composition.
Fracture (as applied to minerals). The capacity of a mineral to break along irregular surfaces.
Fracture zone. A great linear system of breaks in the Earth's crust.
Fragment. A mineral or rock particle larger than a grain.
Fragmental texture. A texture of sediments and sedimentary rocks, resulting from physical transport and deposition of particles.
Framework. The rigid arrangement created by the particles of a sediment or a sedimentary rock that support one another at their points of contact.
Free air gravity anomaly. A value obtained by adjusting the measured value of gravity at a point on the earth's surface to account for effects of latitude and elevation.
Frequency (wave). The number of oscillations occurring during a specified length of time, for example, cycles per second.
Freshwater limestone. A limestone that formed in a lake, stream, or cave.
Fringing reef. A reef that is attached to the shore of a land mass.
Frost heaving. The lifting of rock waste by expansion during freezing of contained water.
Frost wedging. The mechanism involving the pushing up or apart of rock particles by the action of ice.
Fumarole. A volcano discharging gas nonexplosively.
Fusion, latent heat of. See **Latent heat of fusion.**

Gabbro. Coarse-grained mafic igneous rock.
Gaging station. A point of measurement of various attributes of a stream, such as level of the water, discharge, velocity of flow, and characteristics of the mechanical and chemical loads.
Gangue. The nonvaluable minerals of an ore.
Geochemical cycle. The cyclic chemical changes that accompany the operations of many natural processes.
Geochemistry. The study of the chemistry of natural reactions.
Geode. A hollow rounded body having a lining of crystals pointing inward.
Geologic column. A composite diagram combining in a single column the succession of all known strata, fitted together on the basis of their fossils or of other evidence of relative age.
Geologic cross section. A diagram showing the arrangement of rocks in a vertical plane.
Geologic cycle. The sum total of all internal and external processes acting on the materials of the Earth's crust.
Geologic map. A map that shows the distribution, at the surface, of rocks of various kinds or of various ages.
Geologic record. The archive of Earth history represented by bedrock, regolith, and the Earth's morphology.

Geologic time scale. The time relationships established for the geologic column (q. v.).

Geology. The science of the Earth.

Geology: physical. The study of the composition and configuration of the Earth's physical features and rock masses, their relationships to one another, and the surficial and subsurface processes that operate on and in the Earth.

Geology: historical. The systematic study of geologic history, based on relationships of rock layers, surficial sediments, morphology, measurements of radioactive minerals, and measurements of magnetic properties of rocks; history of life on Earth, including both evolutionary development and relationships of organisms to their environments as expressed in the sedimentary record; and ancient geographic relationships.

Geophysical exploration (also called **geophysical prospecting**). Exploration to infer subsurface conditions based on the distribution within rocks of some physical property, such as specific gravity, magnetic susceptibility, electrical conductivity, and elastic or other properties. In geophysical prospecting these methods are applied in the search for economically valuable substances.

Geosyncline. A huge linear or arcuate segment of the Earth's crust, which by subsidence below sea level and sediment accumulation incubates a mountain chain.

Geothermal gradient. The rate of increase of temperature downward in the Earth.

Geyser. An orifice that erupts steam and boiling water intermittently.

Glacial drift. Or, simply, **drift.** The sediments deposited directly by glaciers (**till,** q. v.), or indirectly in glacial streams, lakes, and the sea (**stratified drift,** q. v.).

Glacial drift: ice-contact stratified. Stratified drift deposited in contact with its supporting ice.

Glacial-marine sediment. See **Sediment: glacial-marine.**

Glacial plucking. The lifting out and removal of fragments of bedrock by a glacier.

Glacial polish. A smooth surface, on bedrock, abraded by a glacier.

Glacial striation. A linear scratch or groove, on bedrock, created by movement of a glacier.

Glaciated valley. A valley that has been modified by a glacier. A glaciated valley may have a troughlike (U-shaped) cross profile, hanging tributaries, steplike irregularities in its long profile, and a cirque or group of cirques at its head.

Glaciation. The alteration of a land surface by the massive movement over it of glacier ice.

Glacier. A body of ice, consisting mainly of recrystallized snow, flowing on a land surface.

Glassy texture. A texture of igneous rocks containing only glass in which crystalline particles are not present.

Gneiss. A coarse-grained foliate breaking along irregular surfaces and commonly containing prominently alternating layers of light-colored and dark-colored minerals.

Graded layer. A layer of sediment in which the particles grade upward from coarse to fine.

Gradient. Applied to a stream, it is the slope measured along the stream, on the water surface or on the bottom.

Grain. A mineral or rock particle having a diameter of less than a few millimeters and generally lacking well-developed crystal faces.

Granite. A coarse-grained igneous rock consisting in major part of potassium feldspar, some sodic plagioclase (albite), and quartz, with minor amounts of ferromagnesian minerals.

Granite gneiss. A distinctly foliated metamorphic rock with the mineral composition of granite.

Granitic rock. A general term for a coarse-grained felsic igneous rock containing quartz and minor amounts of ferromagnesian minerals.

Granitization. The transformation, without fusion, of older rocks into granitic rocks.

Granodiorite. A coarse-grained flesic igneous rock, with composition intermediate between granite and diorite, containing quartz, and in which plagioclase is the chief feldspar.

Granular disintegration. The mechanical loosening, during chemical weathering, of the individual mineral particles of bedrock.

Granular texture. A texture of equigranular igneous rock with particles ranging in size from 0.05 to 10 mm.

Gravitation, the universal law of. Sir Isaac Newton's universal law of gravitation states that every particle in the universe attracts every other particle with a force directly proportional to the product of their masses and inversely as the square of the distance between their centers.

Gravitational acceleration, Earth's (g). The inward acting force with which the Earth tends to pull all objects toward its center and which tends to make the Earth a sphere; also called the *Earth's gravity*. The Earth's gravitational acceleration, represented by g, is measured in gals (for Galileo; 1 gal = 1 cm/sec^2 of acceleration).

Gravity anomaly. A difference between the computed and the measured values of gravity at a given location. The anomaly is positive where measured values exceed computed values, and negative where measured values are less than computed values.

Gravity meter (or **gravimeter**). A sensitive measuring device for determining the value of g at any locality.

Gravity prospecting. The attempt to use sensitive measurements of g to find valuable substances associated with slight variations in the densities of rocks.

Gravity spring. See **Spring: gravity.**

Graywacke. A poorly sorted sandstone, generally dark, containing rock fragments as well as quartz.

Groin. A low wall on a beach crossing the shoreline at a right angle.

Groundmass. The particles surrounding the phenocrysts of a porphyritic igneous rock.

Ground water. The water, beneath the Earth's solid surface, contained in pore spaces within regolith and bedrock.

Guyot. A seamount having a conspicuously flat top.

Half-life (of a radioactive isotope). The time required to reduce the number of parent atoms by one-half.

Hanging valley (also **hanging tributary**). A tributary stream whose valley floor lies ("hangs") above that of the valley of a main stream. Hanging valleys are common where main valleys have been glaciated.

Hanging wall. See **Fault: hanging wall.**

Hardness. Resistance to scratching.

Hertz. A unit for expressing frequency; 1 Hertz = 1 cycle/sec (abbreviated Hz).

Hinge fault. A fault on which displacement dies out perceptibly along strike and ends at a definite point.

Historical geology. See **Geology: historical.**

Horn. A bare, pyramidal-shaped peak left standing where glacial action in cirques has eaten into it from three or more sides.

Hornfels. A tough, generally massive nonfoliate containing scattered crystals of high-temperature minerals.

Hot spring. A spring (q. v.) from which hot water flows freely. If the water is at its boiling temperature the spring is a **boiling spring.**

Humus. The decomposed residue of plant and animal tissues.

Hydraulic conductivity. The quantity of fluid that passes through material of a given cross section per unit of time, when driven by a given pressure and at a stated temperature.

Hydraulic gradient. The slope of a water table, found by determining the difference in height between two points and dividing by the horizontal distance between them.

Hydraulic plucking. The lifting out, by turbulent water, of blocks of bedrock bounded by joints and other partings.

Hydrocarbon. An organic compound consisting of hydrogen and carbon. Petroleum and natural gas are examples of natural hydrocarbons.

Hydrogeology. The study of ground water.

Hydrologic cycle. The system driven by Solar energy of water circulation from oceans to atmosphere and back to oceans either directly or via the lands.

Hydrolysis. The combination of water with other molecules.

Hydrosphere. The Earth's discontinuous water envelope.

Ice cap. An ice sheet.

Ice-contact stratified drift. See **Glacial drift: ice contact stratified.**

Ice sheet. A broad glacier of irregular shape, generally blanketing a large land surface.

Igneous rock. Rock formed by solidification of molten silicate materials.

Imbrication. The slanting, overlapping arrangement of flat particles, like shingles on a roof. Imbrication is likewise applied to inclined, overlapping thrust sheets.

Inactive volcano. See **Volcano: inactive.**

Inclination (magnetic). The angle with the horizontal measured in the vertical plane passing through the magnetic line of force.

Inclusion. An impurity, solid or liquid, within a crystal, resulting from encirclement by growth of a lattice that is larger than the impurity.

Incongruent melting. Melting accompanied by a change of composition.

Intensity, earthquake. See **Earthquake intensity.**

Interface. A boundary between bodies of matter having different physical states of aggregation, or between bodies of matter having the same physical state of aggregation but different physical properties.

Interfacial tension. The force, created parallel to an interface between a liquid and an unlike substance, which results from unequal distribution of the particles of the liquid on opposite sides of the interface.

Interior drainage. See **Drainage: interior.**

Intermediate-focus earthquake. See **Earthquake: intermediate-focus.**

Interstices. See **Pore spaces.**

Interstitial spaces. See **Pore spaces.**

Intrusive igneous rock. Rock that originated from solidification of magma emplaced in older bedrock.

Ion. A charged particle formed from an atom by adding or subtracting one or more electrons.

Ionic bond. A kind of bond in which ions are held together by strong nondirectional electrostatic forces of attraction among oppositely charged particles.

Island arc. A group of islands situated in a linear but slightly curved zone. Typically, such island chains are bordered by a submarine trench.

Isograd. A line connecting points of first occurrence of a given mineral in metamorphic rocks.

Isoseismal line. A line on a map through points of equal earthquake intensity.

Isostasy. The ideal condition of flotational balance among segments of the lithosphere.

Isotopes. Configurations of the same element, all having the same number of protons and generally similar chemical behavior but different mass numbers, A, resulting from different numbers of neutrons in the nucleus.

Isotopic date. A determination of the age of a sample by a calculation based on a measurement of its content of a suitable radioactive isotope.

Joint. A fracture, on which no appreciable movement parallel with the fracture has occurred.

Joint set. A widespread group of parallel joints.

Joint system. A combination of two or more intersecting joint sets.

Joule. A unit of mechanical energy equal to 10^7 ergs.

Juvenile water. Water, formerly dissolved in magma, which rises from deep within the Earth to become part of the hydrosphere for the first time.

Kame. A body of ice-contact stratified drift shaped as a short, steep-sided knoll or hummock.

Kame terrace. A body of ice-contact stratified drift shaped into a terracelike form along the sides of a valley.

Karst topography. An assemblage of topographic forms consisting primarily of closely spaced sinks; strikingly developed in the Karst region of Yugoslavia.

Kettle. A closed depression in drift, created by the melting out of a mass of underlying ice.

Laccolith. A concordant lenticular pluton, circular or elliptical in plan, having an essentially plane floor and a distinctly domed roof.

Lamina (plural, **laminae**). A stratum less than 1 cm thick.

Laminar flow. See **Flow: laminar.**

Lapilli tuff. A medium-grained pyroclastic rock consisting of lapilli.

Lapilli (volcanic). Tephra ranging in size between 4 and 32 mm.

Latent heat of fusion. The amount of energy added or released in changing state from a solid to a liquid.

Latent heat of sublimation. The amount of energy involved in moving particles across the solid/gas interface.

Latent heat of vaporization. The amount of energy added or released when particles move across the liquid/gas interface.

Laterite. A reddish residual product of tropical weathering, rich in oxides of iron and aluminum.

Lava. Molten silicate materials reaching the Earth's surface.

Lava dome. A dome created by the upward movement of a lava spine (q. v.).

Lava flow. A hot stream or sheet of molten material that is flowing or has flowed over the ground.

Lava fountain. Lava clots sprayed into the air from a volcano.

Lava spine. An upright, cylindrical feature created by the upward squeezing of a mass of sluggish, pasty lava.

Leaching. The continued removal, by water, of

soluble matter from bedrock or regolith.

Levee: natural. A broad, low ridge of fine alluvium built along the side of a stream channel by water spreading out of the channel during floods.

Level surface. A surface that is everywhere perpendicular to the direction in which gravity acts. Such a surface is defined to be a horizontal surface.

Lignite. A brownish-black coal, intermediate in composition between peat and bituminous coal.

Limb. The side of a fold.

Limb: overturned. The limb of a fold in which the strata have been tilted beyond the vertical.

Limestone. A sedimentary rock consisting predominantly of calcium carbonate.

Lithification. Literally, rock making. In practice, it is a general term for the conversion of sediments into sedimentary rocks.

Lithosphere. The outer zone of the solid Earth. The lithosphere includes the crust and the upper part of the mantle lying above the low-velocity zone.

Load. The material carried at a given time, by a stream, by a current of water, by the wind, or by a glacier.

Load: bed. The coarse solid particles, within a body of flowing fluid, moving along or close above the bed.

Load: suspended. The fine solid particles turbulently suspended within a body of flowing fluid.

Local base level. See **Base level: local.**

Loess. Wind-deposited silt, usually accompanied by some clay and some fine sand.

Longitudinal dune. See **Dune: longitudinal.**

Longitudinal wave. A seismic body wave that causes particles to oscillate along lines in the direction of wave travel. In seismology designated by the letter P; also called P-wave.

Long profile. A line connecting points on the surface of a stream.

Longshore current. A current in the swash zone flowing parallel to the shore.

Longshore drift. The net movement of sediment, parallel to the shore, by waves and wave-induced currents.

Love wave. A seismic surface wave that causes particles to oscillate at right angles to direction of wave travel along lines lying in horizontal planes.

Luster. The quality and intensity of light reflected from a mineral.

M-discontinuity. See **Mohorovičić discontinuity.**

Mafic rock. A rock in which ferromagnesian minerals exceed 50 percent.

Magma. Molten silicate materials beneath the Earth's surface including crystals derived from them and gases dissolved in them.

Magmatic differentiation. A collective name for the processess by which one magma generates more than one variety of igneous rock.

Magnetic anomaly. An irregularity in the earth's magnetic field that is a departure from the form and intensity of a perfect dipole field. Irregularly distributed zones of magnetized materials in the earth's crust produce the magnetic anomalies of greatest geological interest.

Magnetic declination. See **Declination, magnetic.**

Magnetic field (Earth's). The magnetic lines of force surrounding the Earth.

Magnetic pole. A point on the Earth's surface where the magnetic inclination is 90°. A north-seeking needle mounted on a horizontal axis points directly down at the North Magnetic Pole and directly up at the South Magnetic Pole.

Magnetic reversal. A reversal of the polarity of the Earth's magnetic field. The mechanism of reversal is not known.

Magnetism, remanent. The magnetism resulting from permanently magnetized ferromagnetic minerals.

Magnetometer. One of a variety of instruments for measuring the intensity or direction of the earth's magnetic field, or the magnetism of a rock specimen.

Magnitude (earthquake). See Richter Scale.

Mantle. See **Earth's mantle.**

Marble. A nonfoliate consisting chiefly of calcite or dolomite.

Marine sediment. Sediment deposited in the sea.

Marl. A mixture of calcium carbonate and clay; a common lake deposit.

Mass number. Represented by the letter-symbol A; the number of protons (Z) plus the number of neutrons (N) in an atomic nucleus.

Mass-wasting. The gravitative movement of rock debris downslope, without the aid of a flowing medium of transport such as air at ordinary pressure; water, or glacier ice.

Matrix. The small particles of a sediment or a sedimentary rock, which occupy the spaces between the larger particles that form the framework.

Meander. A looplike bend of a stream channel.

Meltwater. Water resulting from the melting of snow and glacier ice.

Mercalli scale of earthquake intensity. A scale of earthquake intensity (q. v.) with divisions, based on human sensations and on damage to man-made structures, ranging from I to XII.

Metallic bond. A kind of bond in metals in which positive ions form a fixed framework on which is superimposed a network of electrons that can move freely.

Metamorphic belt. A linear zone of the land surface extending several tens or hundreds of kilometers where an abundance of rocks produced by regional metamorphism are exposed.

Metamorphic faces. Rocks that reached equilibrium during metamorphism within a single range of environmental conditions.

Metamorphic rock. A rock formed within the Earth's crust by the transformation of a preexisting rock in the solid state without fusion and with or without addition of new material, as a result of high temperature, high pressure, or both.

Metamorphism. The changes, in mineral composition, arrangement of minerals, or both, that take place in the solid state within the Earth's crust at high temperatures, high pressures, or both.

Metamorphosed rock: contact. A rock that has been altered near a pluton. Two varieties are: (1) rocks metamorphosed by high temperature, and (2) rocks metamorphosed by high temperature and by the addition of new material.

Metamorphosed rocks: regional. Metamorphic rocks that extend through great distances in mountain chains and continental shields.

Metastable mineral. A mineral that persists under environmental conditions in which it is not the most stable form.

Meteorite. A particle of solid matter from outer space that has fallen to the ground through the atmosphere.

Microseism. A small deflection, on a seismogram, created by a seismic wave having a period ranging from 1 to 9 seconds, and not generated by an earthquake. Storms, surf, and various other motions create microseisms.

Migmatite. A rock in which thin stringers and threads of granitic material are intertwined with dark schistose layers.

Mineral. A naturally occurring substance, most but not all of which are crystalline solids and whose exteriors may or may not consist of crystal faces; whose atoms or ions of one or more elements or molecules of compounds are arranged regularly in a definite lattice, and whose chemical compositions, though constant or variable within limits, bear fixed relationships to certain physical properties.

Mineral, accessory. See **Accessory mineral.**

Mineral aggregate. A body consisting of more than one grain of the same or of different mineral species. Mineral aggregates contain more than one crystal lattice; they can occur as regolith or sediment if loosely bound, or as rock if they are tightly bound.

Mineral, essential. See **Essential mineral.**

Mineral facies. Mineral assemblages of any origin or composition that have reached chemical equilibrium under similar environmental conditions.

Mineralogy. The study of minerals.

Modulus of rigidity. The stress required to cause a unit change of shape.

Mohorovičić discontinuity. The seismic discontinuity marking the base of the Earth's crust. Abbreviated *M-discontinuity* (or, in the vernacular, "Moho").

Molecular bond. A kind of bond in which molecules having electrically balanced internal atomic charges are held together by forces of attraction between unlike partial charges distributed unevenly over the surfaces of the molecules.

Molecule. The fundamental entity of a chemical compound.

Monadnock. A conspicuous residual hill on a peneplain. The name was taken from Mount Monadnock, in New Hampshire, in 1893.

Monocline. A one-limb flexure, on either side of which the strata are horizontal or dip uniformly at low angles.

Moraine: end. A ridgelike accumulation of drift, deposited by a glacier along its front margin.

Moraine: ground. Widespread thin drift with a smooth surface consisting of gently sloping knolls and shallow closed depressions.

Moraine: lateral. An accumulation of drift along the side of a valley glacier.

Moraine: medial. A strip of drift formed by coalescence of lateral moraines at the junction of two valley glaciers.

Morphology (Earth's). The shape of the Earth's surface.

Mountain. In a general sense, any land mass that

projects conspicuously higher than its surroundings. Geologically it refers to parts of the Earth's crust having thick, crumpled strata, regionally metamorphosed rocks, and granitic batholiths.

Mountain chain. An elongate unit consisting of numerous ranges or groups, regardless of similarity in form or of equivalence in age.

Mountain making. The creation of elongate highlands by large-scale deformation of rocks in the Earth's crust.

Mountain range. A single large complex ridge or series of clearly related ridges that constitute a fairly continuous and compact unit.

Mountain, residual. See **Residual mountain.**

Mountain system. A group of ranges similar in general form, alignment, and structure, which presumably originated from the same general causes.

Mountain, volcanic. See **Volcanic mountain.**

Mud crack. A crack caused by the shrinkage of wet mud as its surface becomes dry.

Mudflow. See **Debris flow, variety mudflow.**

Natural gas. Gaseous hydrocarbons, predominantly methane, that occur in rocks underground.

Natural levee. See **Levee: natural.**

Neck cutoff. The intersection of a meander bend by the bend next upstream causing the stream to bypass the loop between the bends.

Nepheline syenite. A light-colored, coarse-grained felsic igneous rock lacking quartz, and containing nepheline.

Neptunists. A name applied to nineteenth-century geologists who believed that granite and basalt crystallized out of sea water. Compare **Plutonists.**

Neutron. An electrically neutral particle having a mass of 1.6752×10^{-24} g.

Nonfoliate. A general term for metamorphic rocks lacking foliation on the scale of hand specimens.

Normal component of gravity (g_n). The component of gravity, acting at right angles to a slope, which tends to hold particles in place. The magnitude of the normal component of gravity is determined by the cosine of the angle of slope.

Normal fault. A fault, generally steeply inclined, along which the hanging-wall block has moved relatively downward.

North American Cordillera. All the mountain units in western North America, from the eastern border of the Rocky Mountains to the Pacific coast.

Nuée ardente. An incandescent cloud consisting of superheated gases and of hot, fine-grained tephra.

Oblique-slip fault. A fault on which movement includes both horizontal and vertical components.

Obsidian. Felsic glass that transmits light along thin edges.

Ocean basin. A low part of the lithosphere that lies between continental masses and is covered by seawater.

Oceanic basalt. See **Basalt: oceanic.**

Oceanic crust. See **Crust: oceanic.**

Oceanic rise and ridge. A continuous rocky ridge, on the ocean floor, many hundreds of kilometers wide, whose relief is 600 m or more.

Oil field. A group of oil pools, usually of similar type, or a single pool in an isolated position.

Oil pool. An underground accumulation of oil or gas in a reservoir limited by geologic barriers.

Oil shale. A body of fine-grained sediment rich in hydrocarbon derivatives. The chief hydrocarbon derivative in the Green River oil shales of the Rocky Mountain region is kerogen.

Oölite. A limestone consisting predominantly of **oöids,** spherical particles of calcium carbonate, having many concentric shells, that enlarged at the site of deposition by radial growth.

Ooze. Pelagic sediment consisting of more than 30 percent of skeletal remains of microorganisms.

Ooze: calcareous. Ooze consisting of the hard parts of Foraminifera, pteropods, and coccoliths.

Ooze: siliceous. Ooze consisting of the hard parts of radiolarians, diatoms, and sponges.

Open system. A reaction in which some material can escape from the scene of reaction.

Ophiolites. Mafic and ultramafic rocks rich in chlorite, epidote, and serpentine, and which may contain some traces of deep ocean floor sediments.

Order of magnitude scale. A $\log_{10}$ scale.

Ore. An aggregate of minerals from which one or more materials can be extracted profitably.

Organic material. Material composed of hydrocarbon compounds.

Organic texture. A texture of sedimentary rocks resulting from secretion of skeletal material or from other activity by organisms.

Original horizontality, principle of. The principle

which states that most strata are nearly horizontal when originally deposited.

Orogenic belt. A mountain chain or a region where a mountain chain is in the process of formation.

Orogeny. Another name for mountain making. The deformation of the crust in the development of mountains.

Outcrop area. The area, on a geologic map, shown as occupied by a particular rock unit.

Outer ridge. A broad, smooth ridge of sediment, generally parallel to the margin of an ocean basin, and standing 200 to 2,000 m above the adjacent sea floor.

Outwash. Stratified drift deposited by streams of meltwater as they flow away from a glacier.

Outwash plain. A body of outwash that forms a broad plain.

Overland flow. See **Flow: overland.**

Oxbow lake. A curved lake occupying a cutoff meander loop.

Packing. The arrangement of particles of a sediment or sedimentary rock.

Paleomagnetism. The study of the Earth's magnetic field in former times.

Parallel strata. See **Strata: parallel.**

Parent isotope. In radioactivity, the isotope that decays and continually decreases in amount with time.

Particle. A general term used in this book; it refers to anything from electrons to larger sizes without restriction as to shape, composition, or internal structure.

Parting. A surface of separation within a rock body.

Parting, bedding-plane. See **Bedding-plane parting.**

Peat. A brownish, lightweight mixture of partly decomposed plant tissues in which the parts of plants are easily recognized.

Pebble size. Sediment particles having diameters greater than 2 mm (about the size of the head of a small wooden match) and less than 64 mm (about the size of a tennis ball).

Pedalfer. A soil in which much clay and iron have been added to the B horizon.

Pediment. A sloping surface, cut across bedrock, adjacent to the base of a highland in an arid climate.

Pedocal. A soil with calcium-rich upper horizons.

Pedology. The science of soils, their origin, use, and protection.

Pegmatite. Exceptionally coarse-grained granite with individual particles and crystals ranging in length from 1cm to many meters.

Pegmatitic texture. A texture of equigranular igneous rocks with particles larger than 10 mm.

Pelagic sediment. An open-sea deposit containing predominantly skeletal remains of microorganisms and clays or products derived from clays.

Peneplain. "Almost a plain." A land surface worn down to very low relief by streams and mass-wasting.

Perched water body. A water body that occupies a basin in impermeable material, perched in a position higher than the main water table.

Percolation. Laminar flow through interconnected spaces in saturated material.

Percolation: confined. Percolation in an aquifer, between impermeable strata, wherein ground water moves past other ground water held immobile in those strata.

Peridotite. An ultramafic igneous rock consisting of pyroxene and considerable olivine; plagioclase, if present, is only a minor constituent.

Permafrost. Ground that is frozen perenially. Permafrost occurs generally at high latitudes and locally at high altitudes.

Permeability. The capacity of a material for transmitting fluids; expressed as hydraulic conductivity (q. v.).

Petroleum. Gaseous, liquid, or solid substances, occurring naturally and consisting chiefly of chemical compounds of carbon and hydrogen. Petroleum includes both oil and natural gas.

Petrology. The study of rocks.

pH. A measure of acidity or alkalinity of solutions; the negative reciprocal of the logarithm to base 10 of the hydrogen-ion concentration.

Phase. A homogeneous and physically distinct unit of matter in a heterogeneous system. In a mixture of ice and water the ice represents one phase, the water is another phase. A rock consisting of several minerals is a heterogeneous system in which each mineral is a distinct phase.

Phenocrysts. The particles in porphyritic igneous rocks that are conspicuously larger than the remaining particles.

Phreatic zone. The subsurface zone below the

water table that is completely saturated by ground water.

Phyllite. An exceptionally lustrous fine-grained foliate parting along smooth or irregular surfaces.

Physical geology. See **Geology: physical.**

Physiographic province. A region having unities of bedrock, morphology, and morphologic history.

Piedmont glacier. A glacier on a lowland at the base of a mountain, fed by one or more valley glaciers.

Pillow. An ellipsoidal mass of lava or of extrusive igneous rock, having a fine-grained margin, and formed by the extrusion of lava under water.

Pitchstone. Felsic glass having a dull, greasy luster.

Placer. A deposit of heavy minerals concentrated mechanically. Common methods of mechanical concentration are sorting resulting from stream flow, wind action, and wave action.

Plane of the ecliptic. The plane in which the planets of the Solar System orbit around the Sun.

Plastic flow. A continuous and permanent change of shape in any direction without breakage.

Plastic solid. A solid that flows readily.

Plate (tectonic). One of the several large fragments that collectively make up the lithosphere, and which moves as a rigid unit relative to other fragments.

Plateau. An extensive upland, underlain by essentially horizontal strata, and having large areas where the surface is nearly flat.

Plateau basalt. Widespread nearly horizontal layers of extrusive basalt on a continental mass.

Playa. A dry bed of a playa lake (q. v.); a nearly level area on the floor of an intermontane basin. Occasional floods deposit silt and clay that mantle a playa.

Playa lake. An ephemeral shallow lake in a desert basin.

Plumb line. A string with a weight attached to its lower end, and fastened at its upper end to a stationary support. The string aligns in the direction that gravity is acting, and indicates the vertical direction at that location.

Plunge (of a fold). See **Fold: plunge.**

Pluton. Any body of intrusive igneous rock.

Plutonic igneous rock. A coarse-grained igneous rock that cooled slowly at great depth (generally) or at shallow depth (less commonly) within the Earth's crust.

Plutonism. A general term referring to the behavior of magmas, including their movement, internal and external reactions, and emplacement.

Plutonists. A name applied to a group of nineteenth-century geologists who believed that granite originated by igneous processes within the Earth's crust and that basalt was a volcanic product. Compare **Neptunists.**

Pluvial lake. A lake that existed under a former climate, when rainfall in the region concerned was greater than at present.

Podzol. A pedalfer soil that has been intensely leached by solutions rich in humic acids.

Point bar. A crescent-shaped bar built out from each convex ("inside") bank of a stream channel.

Poised stream. A stream that is adjusted to the landscape in such a way that it has the capacity to transport all loose debris in its channel only when the amount of water in the channel reaches flood stage.

Polar easterlies. The winds common in the polar regions.

Pool. See **Oil pool.**

Poorly sorted sediment. A sediment consisting of particles of many sizes.

Pore spaces. The spaces in a body of rock or of sediment that are unoccupied by solid materials.

Porosity. The proportion, in per cent, of the total volume of a given body of bedrock or of regolith that consists of pore spaces (q. v.).

Porphyritic. An adjective describing an igneous rock in which phenocrysts constitute less than 25 percent of the total volume.

Porphyritic texture. A texture of igneous rocks with some particles conspicuously larger than the rest of the particles.

Porphyroblasts. Particles of metamorphic rocks that are conspicuously larger than their neighbors.

Porphyry. An igneous rock with phenocrysts constituting more than 25 percent of the volume.

Positron. A particle having the same mass as an electron but with a unit positive charge.

Pothole. A cylindrical hole drilled in bedrock by a turbulent stream.

Pre-geologic time. The part of Earth history that antedates the oldest rocks.

Primary wave. The first-arriving body wave from an earthquake; a longitudinal wave (q. v.).

Principle of original horizontality. See **Original horizontality, principle of.**

Principle of stratigraphic superposition. See **Stratigraphic superposition, principle of.**

Principle of uniformity. See **Uniformity, principle of.**

Progradation. The process of outward or of forward building by a body of sediment, such as a beach, a delta, or a fan, resulting from addition of successive layers of new sediment.

Proton. Particle of unit positive electrical charge having a mass of 1.6730×10^{-24}g.

Pumice. Extremely vesicular, frothy natural glass with a high content of silica. Pumice is so light it will float on water.

Pyroclastic material. See **Tephra.**

Proclastic rock. Lithified tephra (q. v.).

Pyroclastic texture. A texture of sediments and sedimentary rocks resulting from physical transport and deposition of particles broken by volcanic activity.

Pyroxenite. An ultramafic igneous rock composed almost entirely of pyroxene.

Quartzite. A quartz-rich nonfoliate.

Radiation belts. Inner and outer zones within the Earth's magnetic field having high concentrations of charged particles.

Radioactivity. The spontaneous decay of the atoms of certain isotopes into new isotopes, which may be stable or undergo further decay until a stable isotope is finally created. Radioactivity is accompanied by the emission of α-particles, β-particles, and γ-rays and by the generation of large quantities of heat.

Radiocarbon. The radioactive isotope of carbon, C^{14}, which is created in the upper atmosphere and circulates throughout the biosphere.

Radiocarbon dating. A determination of the age of a sample by a calculation based on a measurement of its content of natural radiocarbon.

Rayleigh wave. A seismic surface wave that causes particles to oscillate in ellipses lying in a vertical plane parallel to the direction of wave travel. The particles move in the same direction as the wave advance under troughs but in the opposite direction under crests.

Reaction series. The interaction between crystals and a silicate melt in which material is exchanged as cooling occurs. This relationship was first stated by N. L. Bowen and commonly is called "Bowen's reaction series." See also **Continuous reaction relationship** and **Discontinuous reaction relationship.**

Recharge. The addition of water to the zone of saturation (q. v.).

Recumbent fold. See **Fold: recumbent.**

Reef. A massive wave-resistant structure built by the secretions of marine organisms.

Refraction (wave). A change in the direction of propagation of a wave that occurs when the wave moves from one medium into another medium where it travels with a different speed.

Regolith. The noncemented rock fragments, and mineral grains derived from rocks, which overlie the bedrock in most places. Regolith is of two kinds, residual and transported.

Rejuvenation. The development of youthful morphologic features in a land mass further advanced in the cycle of erosion. Regional uplift is the common cause of rejuvenation.

Relative humidity. The ratio of the amount of water vapor present to the maximum amount that the air mass can contain without condensation or precipitation.

Relief. The difference in altitude between the high and low parts of a land surface.

Remanent magnetism. See **Magnetism, remanent.**

Replacement. The process by which one mineral takes the place of another mineral. During replacement a fluid dissolves matter already present and at the same time deposits from solution an equal volume of a different substance.

Reservoir rock. A porous and permeable rock containing petroleum, natural gas, or both.

Residual concentration. The natural concentration of a mineral substance by removal of a different substance with which it was associated.

Residual mountain. A resistant remnant standing high as a result of long-continued erosion.

Reverse fault. A fault, generally steeply inclined, along which the hanging-wall block has moved relatively upward.

Rhyolite. A fine-grained igneous rock having the composition of granite; if porphyritic, contains quartz phenocrysts and has aphanitic groundmass.

Richter scale of earthquake magnitude. A scale of earthquake mangitude based on the logarithm (base 10) of the amplitudes of the deflections, on a seismogram, created by seismic waves and recorded on a calibrated seismograph.

Rigidity. Resistance to change of shape.

Rip current. A current, flowing seaward through the line of breakers, which returns to the open sea the water of longshore currents.

Ripple mark. A small-scale wave, in sand, created by the effects of the drag of moving water or wind. Two principle kinds are symmetrical and asymmetrical.

Roche moutonnée (plural, **roches moutonées**). A knob of bedrock, smoothed and generally striated by an overriding glacier, elongate in the direction of ice movement.

Rock avalanche. The rapid downslope flow of a mass of dry rock particles.

Rock cleavage. Closely spaced partings in rocks controlled by platy particles that have been aligned in response to pressures within the Earth's crust.

Rock cycle. That part of the geologic cycle (q. v.) concerned with the creation, destruction, and alteration of rocks during erosion, transport, deposition, metamorphism, plutonism, and volcanism.

Rockfall (and **debris fall**). The rapid descent of a rock mass, vertically from a cliff or by leaps down a slope.

Rock flour. Fine sand and silt produced by crushing and grinding in a glacier.

Rock flow. The slow movement of a rock body that is in a plastic condition.

Rock glacier. A lobate, steep-fronted mass of coarse, angular regolith, extending from the front of cliffs in a mountainous area. Downslope movement of the mass is aided by interstitial water and ice.

Rockslide. The rapid descent of a rock mass down a slope.

Roof rock. In petroleum geology, an impermeable stratum that prevents the upward escape of petroleum or of natural gas. Also, the layer of rock overlying a coal bed or a confined aquifer.

Runoff. Water that flows over the lands.

Rupture. The mechanical breaking of rocks.

Salinity. The proportion of dissolved solids in a solution.

Saltation. The jumping movement of rock particles in a current of water or air.

Salt dome. A dome, in sedimentary strata, created by the upward movement of a salt plug (q. v.).

Salt plug. A cylindrical body of rock salt that has risen upward from an underlying source bed.

Sand size. Particles of sediment having diameters larger than $1/16$ mm (about the lower limit of visibility of the individual particles with the unaided eye) and smaller than 2 mm (about the size of the head of a small wooden match).

Sandstone. A clastic sedimentary rock consisting predominantly of sand-size particles.

Sand wave. A large wave, in sand or gravel, created by the effects of the drag of a thick and swift-moving current of air or water. Sand wave is a general term that includes *dune*. In general appearance, a sand wave resembles some varieties of ripple mark, but a sand wave is many times larger than a ripple mark and can consist of coarse sand and gravel. The upcurrent sides of some sand waves are covered with ripple marks.

Saturation, zone of. See **Zone of saturation**.

Scale. The proportion between a unit of distance on a map and the unit it represents on the Earth's surface.

Schist. A well-foliated metamorphic rock in which the component flaky minerals are distinctly visible.

Scoria. An igneous rock containing abundant vesicles many of which are open at the surface.

Scoriaceous. An adjective describing an extremely vesicular igneous rock containing many vesicles open at the surface.

Sea arch. A roofed-over opening through a narrow headland, resulting from attack by waves on both sides.

Sea cave. A chasm in a sea cliff, excavated in weak rock by waves and currents.

Sea-floor trench (also **deep-sea trench**). An elongate, narrow, steep-sided depression, generally deeper than the adjacent sea floor by 2,000 m or more, extending parallel to the margin of an ocean basin.

Sea-floor spreading. The process by which a lithospheric plate moves horizontally outward from the axis of an oceanic ridge, and toward a submarine trench. Material rising from the asthenosphere is accreted to the edges of outward moving plates along the ridge axis so that new lithosphere is continuously created as the adjacent plates spread apart.

Sea level. The level continuous with that of the oceans at mean tide. Also, the surface of the sea.

Seamount. An isolated conical mound projecting more than 1,000 m, above the deep-sea floor.

Sea wall. A wall paralleling the shore, built to prevent coastal erosion by waves.

Secondary enrichment. Natural enrichment of an ore body by later addition of mineral matter, generally from percolating solutions.

Secondary mineral. A mineral that was not present in the original rock or sediment.

Secondary wave. The second-arriving body wave from an earthquake; a shear wave (q. v.).

Secular equilibrium. In radioactivity, the condition of equilibrium in which the rate of decay of the parent isotope is exactly matched by the rate of decay of every intermediate daughter isotope. When secular equilibrium has been established the concentrations of intermediate daughters remain virtually constant.

Sediment. Regolith that has been transported; mineral or organic matter deposited by water, air, or ice.

Sediment, clastic. See **Detritus.**

Sediment: glacial-marine. Marine terrigenous sediment including nonsorted mixtures of particles of all sizes dropped from floating ice.

Sediment: poorly sorted. See **Poorly sorted sediment.**

Sediment, terrigenous. See **Terrigenous sediment.**

Sediment: well-sorted. See **Well-sorted sediment.**

Sedimentary breccia. A clastic sedimentary rock containing numerous angular particles of pebble size or larger.

Sedimentary rock. A rock formed by cementation of sediment or by other processes acting at ordinary temperatures at or close beneath the Earth's surface.

Seismic. Pertaining to shock waves, natural or artificial, within the Earth.

Seismic belt. A tract subject to frequent earthquake shocks, both strong and weak. The Earth's principal seismic belts are the Circum-Pacific belt, the Mediterranean and Trans-Asiatic belt, the Mid-Atlantic belt, and the Mid-Indian belt.

Seismic discontinuity. An interface within the Earth where physical properties change. Along the discontinuity some seismic waves are reflected, others are refracted.

Seismic prospecting. The search for valuable substances underground by using seismic waves from artificial explosions to infer the subsurface structure.

Seismic seawave. Another name for **tsunami** (q. v.).

Seismic wave. A wave traveling within the Earth.

Seismogram. A record made by a seismograph.

Seismograph. An instrument for accurate recording of seismic waves.

Seismology. The study of seismic waves and the science of earthquakes.

Settling velocity. The constant velocity attained by a particle falling through a still fluid.

Shale. A fine-grained, fissile sedimentary rock composed of clay-size and silt-size particles of unspecified mineral composition.

Shallow-focus earthquake. See **Earthquake: shallow-focus.**

Shallow-water wave. A wave, on the surface of a body of water, beneath which particles of water move in elliptical orbits, within vertical planes, because the bottom influences their motion.

Shearing stress. See **Stress: shearing.**

Shear wave. A seismic wave that causes particles to oscillate along lines at right angles to the direction of wave travel. Shear waves propagate only through solids. In seismology designated by the letter S; also called S-wave.

Sheet erosion. See **Erosion: sheet.**

Shield, continental. See **Continental shield.**

Shield, volcanic. See **Volcanic shield.**

Shield volcano. See **Volcanic shield.**

Sial. The collective term for silica-rich rocks of continental masses. Derived from Si (symbol for silicon) and Al (symbol for aluminum).

Sialic rock. A rock rich in silicon and aluminum; also a felsic rock. See **Sial.**

Silicate One of a large group of minerals having the common feature of silicon and oxygen ions arranged geometrically so that each silicon ion is at the center of a tetrahedron of oxygen ions. This unit is called the SiO_4 tetrahedron. Each silicate compound has a distinct arrangement of SiO_4 tetrahedra with other ions.

Sill. A tabular pluton having concordant surfaces of contact.

Silt size. Scale of sediment particles having diameters larger than 4 microns and smaller than $1/16$ mm (about the lower limit of visibility of individual particles with the unaided eye).

Siltstone. A clastic sedimentary rock consisting predominantly of silt-size particles.

Sima. A collective term, coined from *Si* (symbol

for silicon) and *ma* (abbreviated form of magnesium), designating the mafic rocks beneath the ocean basins.

Simatic rock. See **Mafic rock.**

Sink. A large solution cavity open to the sky, generally created by collapse of a cavern roof.

Slate. A fine-grained foliate that splits along smooth planes into very thin plates.

Slaty cleavage. A closely spaced, plane foliation that divides rock into thin plates.

Slickensides. Striated and polished surfaces on rocks abraded by movement along a fault.

Sliderock. The material composing a talus.

Slip face. The straight, lee slope of a dune.

Slope, angle of. The angle, measured in the vertical plane, between an inclined portion of the Earth's surface and the horizontal. Slope is expressed in degrees or in feet per mile, the number of feet of vertical descent in a mile of horizontal distance.

Slump. The downward slipping of a coherent body of rock or regolith along a curved surface of rupture.

Snowfield. A wide cover, bank, or patch of snow, above the snowline, that persists throughout the summer season.

Snowline. The lower limit of perennial snow.

Soil. That part of the regolith which can support rooted plants.

Soil horizon. One of the subdivisions of a layered soil. Pedologists (q. v.) usually recognize three chief subdivisions, designated from top to bottom by the letters A, B, and C.

Soil, mature. A soil having a fully developed profile.

Soil profile. The succession of distinctive horizons in a soil and the unchanged parent material beneath it.

Solar energy. The energy derived from the Sun.

Solifluction. The imperceptibly slow downslope flow of water-saturated regolith.

Solifluction sediment. A sediment that has resulted from solifluction (q. v.).

Sorting. The selection, by natural processes during transport, of rock particles or other particles according to size, specific gravity, shape, durability, or other characteristics.

Source rock. In petroleum geology, a rock within which petroleum or natural gas originates.

Specific gravity. A number stating the ratio of the weight of the substance to the weight of an equal volume of pure water at 4°C.

Spectrum, wave. A collection of waves of different lengths, usually covering a considerable range.

Spit. An elongate ridge of sand or gravel projecting from the mainland and ending in open water.

Spring. A flow of ground water emerging naturally onto the surface.

Spring: artesian. A natural outflow, at the Earth's surface, of water from a confined aquifer, usually through a fissure or along a fault.

Spring: gravity. Also known as an *ordinary spring*. A spring whose flow results directly from the force of gravity.

Stack. A small, prominent island of bedrock, remnant of a former narrow promontory destroyed by wave erosion.

Stalactite. An iciclelike form of dripstone and flowstone that hangs from the ceiling of a cavern.

Stalagmite. A blunt, iciclelike form of flowstone projecting upward from the floor of a cavern.

Steady state. A condition of dynamic equilibrium in which the rate of arrival of some materials equals the rate of escape of other materials.

Stock. A pluton, roughly circular or elliptical in plan, with an exposed area of less than 40 square miles (100 km^2).

Stoping. A mechanism of emplacement of large plutons, involving repeated breaking off and engulfment of xenoliths (q. v.).

Strain. A change of shape or of volume, or of both shape and volume created by deforming forces. Strain is defined by the ratio

$$\frac{a-b}{a}$$

where *a* is the original shape or volume and *b* is the changed shape or volume.

Strata: cross-strata. Strata inclined with respect to a thicker stratum within which they occur.

Strata: parallel. Strata whose individual layers are parallel.

Stratification. The layered arrangement of the constituent particles of a rock body.

Stratified drift. Sorted and stratified glacial drift.

Stratigraphic superposition, principle of. The principle which states that in any sequence of strata, not later disturbed, the order in which they were deposited is from bottom to top.

Stratigraphic trap. An oil trap resulting from variations in permeability controlled by stratigraphic relationships.

Stratigraphy. The systematic study of stratified rocks.

Stratovolcano A high and more-or-less conically shaped volcano consisting of a mixture of congealed lava flows and pyroclastic debris.

Stratum (plural, **strata**). A definite layer of sedimentary or igneous rock consisting of material that has been spread out upon the Earth's surface.

Streak. A thin layer of powdered mineral made by rubbing a specimen on a nonglazed porcelain plate.

Stream. A body of water carrying suspended and dissolved substances and flowing down a slope along a definite path, the stream's channel.

Stream: antecedent. A stream that has maintained its course across an area of the crust that was raised across its path by folding or faulting.

Stream: braided. A stream that flows in two or more interconnected channels around islands of bed-load alluvium.

Stream capture. (Known also as **stream piracy**.) The diversion of a stream by the headward growth of another stream.

Stream: consequent. A stream whose pattern is determined solely by the direction of the slope of the land.

Stream flow. The flow of surface water between well-defined banks.

Stream pattern: dendritic. A stream pattern characterized by irregular branching in all directions.

Stream pattern: rectangular. A stream pattern characterized by right-angle bends in the stream.

Stream pattern: trellis. A rectangular stream pattern in which tributary streams are parallel and very long. Known also as grapevine pattern.

Stream piracy. See **Stream capture.**

Stream: subsequent. A stream whose course has become adjusted so that it occupies belts of weak rock.

Stream: superposed. A stream that was let down, or superposed, from overlying strata onto buried bedrock having composition or structure unlike that of the covering strata.

Stream system: well adjusted. A stream system in which most of the streams occupy weak-rock positions.

Stream terrace. A bench along the side of a valley, the upper surface of which was formerly the alluvial floor of the valley.

Strength. The ability of a rock body to resist stresses created by forces that tend to cause changes of volume or of shape, or of both volume and shape.

Stress. Force per unit area.

Stress: compressive. A stress paralleling the direction in which a body subjected to deforming forces tends to be shortened.

Stress: shearing. A stress causing parts of a solid to slip past one another, like cards in a pack.

Stress: tensile. A stress paralleling the direction in which a body subject to deforming forces tends to be elongated or pulled apart.

Striations. Scratches and grooves on bedrock surfaces, caused by grinding of rock against rock during movement of glacier ice. See also **Glacial striation.**

Strike. The compass direction of the horizontal line in an inclined plane.

Strike-slip fault. A fault on which displacement has been horizontal.

Structural geology. The study of rock deformation and the delineation of geologic structural features.

Subduction. The process by which the edge of a lithospheric plate is drawn downward into the earth and reassimilatted into the asthenosphere. Submarine trenches mark the zones where subduction is occurring.

Sublimation. The process by which particles of a solid pass directly into the gaseous state.

Sublimation, latent heat of. See **Latent heat of sublimation.**

Submarine canyon. A sinuous, V-shaped valley, having a variable number of tributaries, that crosses part or all of a continental shelf and extends down a continental slope.

Submarine Trench: See **Sea Floor Trench**

Submergence. Rise of sea level relative to the land.

Subsequent stream. See **Stream: subsequent.**

Surf. Waves of translation landward of the breakers and seaward of the backwash.

Surface wave. (1) A seismic wave traveling along the surface of the Earth. (2) A wave traveling along the surface of a body of water.

Surf zone. The zone between the outermost breaker and the outermost (lower) limit of the backwash.

Surge (glacial). A process of very rapid glacier

movement. A surging glacier can advance several meters a day. A surge may continue for several months.

Suspended load. See **Load: suspended.**

Swash. The surge of water, from a breaking wave, that flows as a thin sheet up the sloping shore.

Syncline. A downfold with troughlike form and having the youngest strata in the center.

Talus. An apron of rock waste sloping outward from the cliff that supplies it.

Tangential component of gravity (g_t). The component of gravity, acting along a slope, which tends to pull particles downslope. The magnitude of the tangential component of gravity is determined by the sine of the angle of slope.

Tar sand. A sand or sandstone whose pores have been filled with hydrocarbons in the solid state, generally known as asphalt.

Tectonic cycle. The cycle which relates the larger structural features of the Earth's crust to the kinds of rocks that form in the various stages of development of these features and to gross crustal movements.

Tensile stress. See **Stress: tensile.**

Tephra. A collective term designating all particles ejected from volcanoes, irrespective of size, shape, or composition.

Tephra cone. A small- to moderate-size cone composed of tephra.

Tephra flow ("Ash flow"). A fluidized mass of tephra, whose particles may be red hot, that flows like a liquid. A tephra flow may create a welded tuff (q. v.).

Terminal moraine. The end moraine deposited by a glacier along its line of greatest advance.

Terrace. A relatively flat, elongate surface, bounded by a steeper ascending slope on one side and a steep descending slope on the other. See also **Stream terrace.**

Terrigenous sediment. Solid particles of sediment derived from erosion of the lands.

Tetrahedron (plural, **tetrahedra**). A solid with four sides, each of which is a triangle. The arrangement of the unit clusters of oxygen and silicon atoms in silicate minerals describes a tetrahedron.

Texture. The sizes and shapes of the particles, in a rock or a sediment, and the mutual relationships among them.

Thin section. A paper-thin, transparent slice, about 30 microns thick, of a rock, mounted on a glass microscope slide.

Thrust (or **thrust fault**). See **Fault: thrust fault.**

Tidal marsh. A low, flat, coastal area thickly grown over with saltwater grasses, in large part submerged at high tide.

Tidal wave. Popular and erroneous name for **tsunami** (q. v.).

Tide. The rhythmic rise and fall of the surface of the sea caused by lunar and solar gravity.

Till. Nonsorted glacial drift.

Tillite. Till converted to solid rock.

Tombolo. A beach that connects an island with the mainland or with another island.

Topographic map. A map that delineates surface forms.

Topography. The relief and form of a land surface.

Topset bed. Stream sediment that overlies the foreset beds in a delta.

Trade winds. The prevailing winds in tropical regions.

Transform fault. A fault in the lithosphere that connects offset segments of an oceanic ridge. Horizontal, or strike slip movement occurs along transform faults. Some especially long transform faults mark boundaries between two lithospheric plates.

Transpiration. The passing of water vapor into the atmosphere from pores of plant tissues.

Transverse dune. See **Dune: transverse.**

Trap. In petroleum geology, an arrangement of reservoir rock and roof rock that localizes a body of petroleum or natural gas.

Travertine. A collective variety of limestone, including dripstone, flowstone, and calcareous tufa.

Trellis pattern. See **Stream pattern: trellis.**

Tributary: barbed. A tributary that forms an angle acute in the downstream direction at the point where it enters the main stream.

Triple point. The point on a pressure-temperature graph at which solid, liquid, and gaseous states coexist.

Tritium. A radioactive isotope of hydrogen, H^3, which is created in the upper atmosphere and circulates throughout the hydrosphere.

Tsunami. Long, low wave(s) generated by abrupt displacement of the sea floor or by landslides entering the sea. (From the Japanese *tsu*, "harbor" and *nami* "waves," pronounced tsoo-náhmĕ, and spelled the same in the singular and plural.)

Tuff. A fine-grained pyroclastic rock consisting of ash and dust.

Turbidite. A sedimentary deposit, typically displaying graded bedding, produced from sediment transported by a turbidity current.

Turbidity current. A density current whose excess density results from suspended sediment.

Turbulent flow. See **Flow: turbulent.**

Ultramafic rock. Granular igneous rock consisting almost entirely of ferromagnesian minerals.

Unconformity. A lack of continuity between units of rock in contact, corresponding to a gap in the geologic record.

Unconformity: angular. Unconformity marked by angular divergence between older and younger sedimentary strata.

Unconformity: surface of. The contact of burial between two groups of unconformable strata. The plural form is *surfaces of unconformity.*

Uniformity, principle of. The concept that relationships established between processes and materials in the modern world can be applied as a basis for interpreting the geologic record and for reconstructing Earth history.

Unit cell. The fundamental building block of a crystal: the smallest part of a crystal lattice that displays the systematic pattern of the particles.

U-shaped dune. See **Dune: U-shaped.**

Vadose zone. Subsurface zone not completely saturated by ground water, and extending from the land surface down to the water table.

Valley glacier. A glacier that flows downward through a valley.

Valley train. A body of outwash that partly fills a valley.

Vaporization, latent heat of. See **Latent heat of vaporization.**

Vapor pressure. The pressure at which both the liquid and the gaseous states of aggregation exist.

Varve. A pair of laminae deposited during the cycle of the year. *Varve* is the Swedish word of cycle.

Vein. A tabular deposit of minerals, occupying a fracture, in which the particles grew away from the walls toward the middle.

Ventifact. A rock fragment with facets that have been cut by wind action.

Vesicles. Openings, generally ellipsoidal or cylindrical, formed in molten rock material by the expansion of gas escaping from solution.

Vesicular. An adjective describing rocks containing vesicles.

Viscosity. The tendency within a flowing body to oppose flow in any direction by acting in the opposite direction.

Volcanic ash. See **Ash** (volcanic).

Volcanic block. A large fragment of volcanic rock that was solid when blasted from the vent.

Volcanic bomb. A rounded, spindle-shaped mass of volcanic rock that was molten when blasted from the vent.

Volcanic crater. A funnel-shaped depression from which gases, tephra, and some lava are ejected.

Volcanic edifice. A feature built by a volcano.

Volcanic mountain. A conical accumulation of volcanic materials.

Volcanic mudflow. A mudflow of water-saturated, predominantly fine-grained tephra.

Volcanic neck. A cylindrical filling of an ancient volcano.

Volcanic shield. A broad convex mound of extrusive igneous rock, having surface slopes of only a few degrees.

Volcanism. A term designating the aggregate of processes associated with the transfer of materials from the Earth's interior to its surface.

Volcano. A vent or a fissure through which molten and solid materials and hot gases pass upward to the Earth's surface.

Volcano: active. A volcano that is erupting or that has erupted within the previous 50 years.

Volcano: dormant. A volcano that has erupted within historic time but not within the previous 50 years.

Volcano: inactive. A volcano that has not erupted within historic time.

Water balance. Also known as **water economy.** A quantitative statement of the amounts of water circulating through various paths of the hydrologic cycle.

Water gap. A pass, in a ridge or mountain, through which a stream flows.

Water table. The upper surface of the zone of saturation.

Wave base. The depth (equal to half the wavelength) at which the bottom begins to interfere conspicuously with the motions of the water particles beneath a shoaling wave. At wave base, deep-water waves become shallow-water waves.

Wave-built terrace. A body of wave-washed sediment that lies seaward of a wave-cut terrace.

Wave-cut terrace. A bench or platform cut across bedrock by waves.

Wave-cut cliff. A coastal cliff whose base has been undermined by waves and other marine agencies.

Wave height. The vertical distance from the bottom of a wave trough to the top of a wave crest.

Wavelength. The horizontal distance between successive wave crests or between successive troughs.

Wave of oscillation. A wave that causes water particles to oscillate but not to undergo appreciable net displacement.

Wave of translation. wave that displaces the water particles within the moving crest.

Wave period. The time required for a wave to advance the distance of one wavelength.

Wave refraction. The process by which the direction of a series of waves, moving in shallow water at an angle to the shoreline, is changed so that waves become more nearly, but rarely exactly, parallel to the shore.

Wave steepness. The ratio of wave height to wavelength.

Wave velocity. The distance traveled by a wave in a unit of time.

Weathering. The chemical alteration and mechanical breakdown of rock materials during exposure to air, moisture, and organic matter.

Welded tuff. A fine-grained volcanic rock whose particles were so hot when deposited that they fused together.

Well-adjusted stream system. See **Stream system: well-adjusted.**

Well-sorted sediment. A sediment consisting of particles all having about the same size.

Wind gap. A former water gap through which a stream no longer flows.

Xenolith. A block, formerly part of the wall rock of a pluton, that has been broken loose and completely surrounded by igneous rock.

Zone of aeration. The zone in which the open spaces in regolith or bedrock are normally filled mainly with air.

Zone of saturation. The subsurface zone in which all openings are filled with water.

OPENER CREDITS

PART ONE
Department of Geological Sciences, Virginia Polytechnic Institute.

PART TWO
NASA.

PART THREE
Department of Geological Sciences, Virginia Polytechnic Institute.

PART FOUR
USDA.

PART FIVE
NASA.

INDEX

Aa lava, 241
Ablation on glaciers, 520
Ablation zone, 520
Abyssal hills, 50
Abyssal plains, 50
Accumulation on glaciers, 519
Accumulation zone, 520
Active flow of ice, 522
Adams, L. H., 113
Adams and Williamson Equation, 113
Aeromagnetic analysis, 252
Aeromagnetic anomalies in ore exploration, 276
Aftershocks, 89
Agassiz, Louis, 506
Age of earth:
 estimates of, table, 28
 radiometric dating, 29–33
Agricultural uses of water, 415
Air circulation, global, 493
Airy, G. B., 69, 70, 180
Airy isostatic model, 70, 111
Alaskan earthquake, 75–76
Albite, 205, 263
 mixture with fayalite, 263–264
Aldrin, Edwin, 599
Algae, relationship to coral, 327–328
Alkali feldspar, 223
Allen, D. W., 140
Alluvial fans, 471
Alluvial plains, meandering streams on, 471–479
Alluvium, 471
Alpha particles, 30
Altimeters, 61
Amorphous substances, 206
Amphiboles, 201, 203
Amphibolite facies, 291–292
Amplitude of seismic waves, 78
Amygdaloidal texture, 221
Ancient magnetic fields, 133–138
Andalusite, 205, 291
Anderson, Don, 103
Angular drainage patterns, 456
Angular unconformities, 371
Anhydrite, 209, 309, 332
 hydration, 439
 origin of, 334
Anhydrous silicates, 203
Animal behavior, before earthquakes, 94
Anions, 199
Anomalous magnetic field, 123, 130–133

Anorthite, 205, 222
Antarctica, 505–512
 topography, 507, 510, 512
Antarctic Ice Sheet, 507–512
Anthracite coal, 335
Anticline, 357
Apennine Mountains, Moon, 608
Apollo flights, 607
Appalachian Orogenic Belt, 373
Appalachian Plateau, 365
Aqueous fluid, 203
Aquicludes, 413
Aquifers, 413
 artesian, 414
Aragonite, 206, 326
 durability of, 343
Arenites, 321
Arete, 531
Argentiere Glacier, 515–516
Arkose, 322
 durability, 344
Armstrong, Neil, 599
Artesian aquifers, 414
Artesian springs, 414
Artesian wells, 414
Asama Volcano, SiO_2 content of, 248
Asbestos, 203, 302
Asphaltic base oil, 346, 391
Asteroids, 601
Asthenosphere, 147
 composition, 269–271
 physical properties of, 161
Atolls, 558
Atomic structure, 29–30
 of minerals, 193–200
Atoms, 193
 electrical neutrality of, 195
 electrons in shells of, 195
 model of, 193, 194
Aureole, 286
Avalanches, 491
Axial surface of a fold, 357
Azonal soils, 448

Bacon, Francis, 143
Bahama Platform, 330
Banded Precambrian ironstones, 441
Barchan dunes, 496
Barite, 209
Bars, sand, 585–586

Basalt, 8
  and extrusive rock, 249
  texture of, 219
  thermoremanent magnetism of, 133
Basaltic magma, 267
Base levels of streams, 480–481
Basins, structural, 359, 361
Batholiths, 251
Baume Scale, oil, 391
Bayous, 481
Beach dunes, 497
Bedding:
  sedimentary layers, 310–315
    table, 311
  surfaces, structure of, 315–316
Bed load, of stream, 468
Bedrock, 6
Bedrock geologic maps, 12
Bedrock topographic features, inselbergs, 476
Benchmarks, 60
Benioff, Hugo, 90
Berm, 588
Beryl, 210
Beta particles, 30
Bicarbonate ion, 441
  in sea water, 565
Binary eutectic system, 260, 263
Bingham stock, 253, 274
Biogenic sediment, 550
Birdfoot deltas, 481
Bituminous coal, 335
Block-faulted regions, 361
Blocking temperature, 32
Blowout areas, 497
Blue schist, 292
Bluffs, 582
Bog iron ores, 441
Bohr, Niels, 193
Bombs (volcanic), 243
Bonds, *see* Chemical bonds
Bottom-set bed of streams, 481
Bouguer, Pierre, 67
Bouguer gravity anomalies, 67, 111, 130
  variations, 67–68
Bouguer mass effects, 67
Bowen, N. L., 265
Bowen Reaction Series, 265
  relation to weathering, 435–436
Bragg, W. H., 189
Bragg, W. L., 189
Bragg Equation, 189
Braided channels, 460
Breakers, 578
Breakwaters, 592
Breccias, 307, 325–326
  collapse, 331

Bridging oxygen, 201
Brilliant cut, gems, 210
Brown clay, 553
Bulk modulus, 95, 377
Bullard, E. C., 139, 154, 155
Bullen, K. E., 113
Butte, 479

Cabochon cut, gems, 210
Calcareous ooze, 550
Calcite, 11, 187, 206, 326, 440
  durability of, 343
  high magnesium, 326
  low magnesium, 326
  natural, 326
Calcium carbonate, 326–328, 550
  in sea water, 565
Calcium ions, 326
  in sea water, 565
Calcium phosphate, 335
Calcium sulfate, formation of, from sea water, 334
Caldera, 235
Callisto, 626
Calving of icebergs, 520
Canadian Shield, 295–296
Canyons:
  midocean, 50
  submarine, 556
Capillarity, relation to water table, 408
Capillary fringe, 406
Cap rock, 345
Carbon 14, use in dating, 32–33
Carbonate ion, 206
Carbonate minerals, 206–207, 343
  table, 207
Carbonate mud, 326
Carbonate rocks, 307
Carbonates, 326–332
Carbonate sand, 327
Carbonation, 439–440
Carbonic acid, 439–440
Cathode tubes, hot, 188
Cations, 199
Caves, 442–446
  formation of, 442, 443
Cavity porosity, 405
Celerity of waves, 576
Cement, Portland, 345
  of sandstone, 321
Cementation, 331
Cenezoic era, 26, 27
Cenozoic Ice Age, late, 538–541
Central eruptions of volcanoes, 226–236
  Hawaiian type, 233
  Mediterranean types, 233, 235

Pelian type, 235–236
Plinian type, 235
Strombolian types, 235
Vesuvian type, 235
Vulcanian type, 235
Centrifugal force:
　on earth, 44–45
　role in tides, 573
Ceres, 601
Channel geometry, of stream, 460–462
Cheluviation, 447
Chemical bonds:
　nature of, 196
　strength of, 197–198
Chemical composition:
　of igneous rocks, 225
　of rocks, 221
Chemical elements, most abundant on earth, table, 201
Chemical phases, 222
Chemical weathering, 431–441
　sample process of, 431–433
Chlorides:
　in ores, 275
　in sea water, 564
Chlorite, 291, 297, 553
Chondrites, 268
Chondritic meteorites, 268
Chromite, 274
Cinder cones, 229
Cirques, 531
Clastic texture, 306
Clay, brown, 553
Clay minerals, 343–344, 436–438
　composition of, table, 320
Claystones, 307
Cleavage:
　fracture, 367
　of minerals, 179
Climates:
　role in chemical weathering, 433–434
　role in landscapes, 484
Clinton-type ironstone formations, 441
Coal, 307, 309, 335–336
　composition of, table, 336
　formation of, 337
　uses of, 345
Coalification, 337
Coarse-grained texture, 219
Coastal landforms, 581–594
Coasts:
　evolution of, 589–592
　maintenance of, 592–594
　submergent, 590–592
Coccolithophores, 550
Coefficient of thermal expansion, 427
Coesite, 205

Collapse breccia, 331
Collapse processes of unstable masses, 428–431
Collins, Michael, 599
Colorado Plateau, 49, 365
　weathering of sandstone in, 428, 431
Color of minerals, 179
Columns, geologic, 12–13
Composite volcanoes, 229
Composition of rocks, 221, 222
　igneous, 219
Compounds as source of ions, table, 565
Compressive stresses, 377
Concrete, 345
Cone of depression, 418–419
Confined ground water systems, 413–414
Confining stresses, 378
Conglomerates, 307, 325–326
Conjugate joint system, 367
Conservative plate margins, 147
Constructive interference of X-rays, 189
Contact metamorphism, 280, 286–289
　host rock, 300
Contacts, 354
Continental cratons, 371–376
Continental drift, 20
　hypothesis of, 143–145
　paleomagnetic pole evidence, 155
　rates of, 157, 158
Continental glaciers, 507–515
Continental margins, 53
Continental platforms, 4, 46–49
Continental rise, 54
Continental shelf, 54
Continental shields, 293–295, 373
Continental slope, 54
Continuous series of minerals, 260
　olivine, 201
Contour lines, 5–6, 46
Control of earthquakes, 94
Convection:
　in earth's core, 141
　in fluids, 167
　shapes of cells, 168
Coordination polyhedron, 199
Copernicus, Nicolaus, 608
Copper mines and stock, 253–254
Coral, 558
　relationship to algae, 327–328
Coral reefs, 327
Cordierite, 291
Cordilleran-type belt, 386
Core of earth:
　composition of, 114, 268
　convection currents in, 141
　inner zone (solid), 103
　outer zone (liquid), 103

Core of earth (continued)
  physical properties of, 113–114
  seismic analysis of, 102–103
  temperature of, 269
Cores:
  of ocean floor sediment, 163
  retrieval of, 547
Coring of deep sea sediments, 547–548
Corundum, 210, 302
Coulomb, Charles, 118
Coulomb's law, 118
Cratons, 371–376
Creep, 485
Crevasses in glaciers, 516
Cristobalite, 205
Critical refraction, 97
Cross bedding, 312–313
Cross-cutting relationships, principles of, 22
Cross sections, geologic, 12–13
Crust of earth, 6
  architecture of, 349–400
  composition of, 267
  seismic analysis of, 102–103
  seismic surveying of, 108
  structure determination of, 109
  thickness determination of, 109
Crystal faces, 182
Crystal growth, role in weathering, 426–427
Crystalline texture, 307
Crystallization, 219
  equilibrium, 260
  of igneous rocks, 259–267
Crystallographic axes, 183
Crystallography, 181–185
Crystals, 181–185
  geometric properties of, 183, 197
  symmetry of, 183
  zoned, 264
Crystal structure:
  and energy, 213
  X-ray diffraction analysis of, 189–190, 192–193
Crystal systems, classification system for identifying, table, 185
Curie temperature, 129
Current meters, 566–567
Currents, see Ocean currents; Surface currents; Tidal currents
Cutoff meander, 474
Cuvier, George, 24–25
Cycle of erosion, 500

Darcy, Henri, 412
Darcy (unit), 406
Darcy's law, 412–413
Darwin, Charles, 28, 558
Davis, William Morris, 484–485, 500, 502, 558

Debris load, 469
  of streams, 468–470
Decay constant, 31
Deep focus earthquakes, 85
Deep sea circulation, 569
Deep Sea Drilling Project, 163
Deep sea sediments, 546–563
Deflation basins, 497
Deformation, repeated episodes of, 371
Deimos, 619
Deltas, 341, 481–484
  formation of, 18, 481–484
Dendritic patterns, 456
Density:
  of earth, 42, 113
    in crust and mantle, 70, 71
    irregularities in, 63
    mean, for earth, 72–73
    values of, 114
  of minerals, 179, 200
  of rocks, 6
  of sea water, 568–569
Depressions, 49
Desert pavement, 499
Dessication cracks in bedding surfaces, 316
Destructive margins of lithospheric plates, 147
Devil's Tower, 232
Diagenesis, 325
  of coal, 335–336
  of evaporation, 334
Diagenetic processes, 331
Diamond, 7–8, 178
  crystal structure, 198
  in kimberlite mining in Africa, 232
  optical dispersion of, 186
  uses of, 216
Diatoms, 550
Dielectric constant of minerals, 179
Differential thermal expansion, role of in weathering, 427–428
Differentiation of sedimentary rocks, 341–344
Diffraction of X-rays, 188
Dikes, 256, 480
Diluvian theory, 505–506
Dimension stone, 344
Diopside, 203, 222, 291
Diorite, 223
  in lower crust, 268
  origin, 265
Dip, 352
Dip needles, 120
Dip slip faults, 361
Directed stress, 280, 377–378
Discharge of streams, 465

Discontinuities (first order), in earth, 103
Discontinuities (second order), in earth, 103
Dissolution, of carbonate rocks, 331
Dissolved ions:
    in ground water, table, 421
    role of in weathering, 436
Distributaries, 481
Dolomite, 206, 307, 326
Dolomitization, 331
    role of temperature in, 332
Domes, structural, 358–359
Donn, W. L., 532
"Doodle bugging," 94–95
Double chain silicate structure, 203
Double tetrahedral chains, 203
Doubly refractive minerals, 187
"Dousing," 416
Drainage density, 458
Drainage divides, 456
Drainage geometry, 456–460
Drainage systems, surface, 454–460
Dredging of deep sea sediments, 547–548
Drilling:
    methods used at sea, 163
    of wells, 416–417
Dripstone, 442
Drumlins, 531
Dunes, 493
Durability, 341–344
Dust clouds, 493
Dynamic Equilibrium Concept of Landscape Evolution, 502–503
Dynamo theory, 140

Earth:
    age of, 28–29
        estimates, 28
    biosphere of, 627–628
    circumference of, 43
    composition of, 267–269
    core of, see Core of earth
    crust of, see Crust of earth
    density of, see Density of earth
    elasticity of, 113–114
        values of, 114
    face of, 401
    form of, 4–6
    history of, 20, 22–34
    interior zones of, see Interior zones of earth
    life on versus extraterrestrial, 627
    magnetic field of, see Magnetic field of earth
    mantle of, see Mantle of earth
    mass of, 73
    mean density of, 72–73
    pressure conditions of, 16–17, 113–114
        values of, 114
    radius of, 43
    shape of, 40–41, 42, 46–56
    size of, controversy about, 41
        measurement of, 42–43
    in solar system, 599–628
    temperature of, 16–17, 268–269
    unique features of, 627–628
    volume of, 73
Earth flows, 487
Earth inductors, 120–121
Earthquake damage, 75–76, 86, 89, 90, 92
Earthquake predictions, 93–94
Earthquakes, 18–20, 75–94
    annual occurrence of, 87
    control of, 94
    deep focus, 85
    definition, 75
    near Denver, 94
    depth distribution of, 85
    destructive, 76
    energy of, 86–87
    epicenter of, 81
    eyewitness account, 75–76
    and faults, 89–92
    focal depths of, 82
    foci of, 79, 85
    generation of tsunami by, 580
    intensity of, 87–88
    location of, 79–86
    magnitude, 86–88
    and plate tectonics, 159–160
    and seismic waves, 75–94
    shallow, 85
    size of, 86–87
    sources of, 160–162
    and volcanic eruptions, 232
Earth structure:
    analysis of with seismic waves, 99–101
    and seismic waves, 94–115
Echo sounders, 49–50, 61, 63
Eclogite, 267–269
    in upper mantle, 268
Eddy currents, 468
Elasticity, 95
    of earth, 113–114
    values of, 114
Elasticity coefficients, 95
Elastic strain, 95, 378
Electrical conductivity of minerals, 179
Electricity, production by geothermal power, 277–278

Electromagnetic waves, wavelengths of, 186
Electromagnetism, 120
Electron microprobe, 193
Electron orbits, 195
Electrons, 29, 193
   orbital, 196
   properties of, 193
   role of, in oxidation and reduction, 440
Elevation, measurement of, 58, 60–61
Ellipsoid, reference, 43
Elsassar, W. M., 139
Eluviation, 447
Emergent coastal landscape, 590
End moraines, 532
Energy barriers, of mineral phases, 214
Energy of earthquakes, 86–87
Enstatite, 203, 222, 264
Eolian landscapes, 493–499
Eolian processes, 454
Epeirogeny, 373
   and tectonism, 382
Epicenter determination method, P-O, 82
   S-P, 82
Epicenter of earthquakes, 81
Epicontinental seas, 54, 347
Epochs, 26, 27
   time span determination of, 33–34
Equilibrium crystallization, 260
Equilibrium fusion, 260
Equilibrium line, of glacier, 520
Equipotential lines, 411
Eras, 25–26, 27
   time span determination of, 33–34
Eratosthenes, 43
Erosion, 428
   of coasts, 583
   cycle of, 500
   effect of rivers on, 17
   due to glaciers, 528–532
   relation to turbulence, of streams, 469
   by wind, 497
Eruptions of volcanoes, *see* Central eruptions of volcanoes; Fissure eruptions of volcanoes
Escarpments, ocean floor, 148
Eskers, 533
Eskola, P. E., 289
Euler, Leonhard, 151
Euler's Theorem, 151, 155
Eustatic change in mean sea level, 581
Eutectic point, 264
Eutectic system, binary, 260
Evaporation of sea water, 334
Evaporites, 307, 309, 332–335
   delicate balance needed for, 334–335
Everest, George, 69
Evolution, 28
   of coasts, 589–592

Ewing, Maurice, 542
Explosives, use of, in seismic surveying, 105, 108
Extrusive igneous rocks, 218
Extrusive sheets, 237–240

Fabrics, tectonic, 368, 370, 371
Face poles of crystals, 183
Falling mass measurement of gravity, 65–66
Faraday, Michael, 140
Fault movements, location of, 162
Faults, 19, 76, 350, 361–365
   earthquake motion related to, 92
   and earthquakes, 89–92
   in models, 382
   movement on, 89
   normal, 361
Fayalite, 201, 260
   mixture with albite, 263–264
Feldspar, 201, 205, 256
   durability, 343
   role of, in weathering, 434, 438
Feldspathic arenites, 322
Feldspathic greywackes, 344
Fennoscandianisostasy, 169–170
Ferromagnesium group, 224
Ferromagnetic minerals, 128, 207
Ferromagnetism, 128
Fine-grained texture, 219
Fire, role of in weathering, 431
Firn, 512
First order streams, 457
Fissility, 319
Fissure eruptions of volcanoes, 237–238, 240–241
   oceanic, 238–240
Fletcher's Ice Island (T3), 515
Flood-control practices, 480
Floodplains, 480
Floods, 480
Flow, turbulent, *see* Turbulent flow
Flow law for ice, 520
Flow nets, 411
Fluvial landscape features, 470–485
Fluvial processes, 454
Foci of earthquakes, 79, 85
Folding in models, 382
Folds, 19–20, 350, 354–361
Foliated metamorphic rocks, 281
Foraminifera, 550
Foreset beds of streams, 481
Formations, 354
Forsterite, 201, 260
Fossil fuels, occurrences of, 345
Fossils, 132, 318
   changes in geologic processes indicated by, 35
   guide, 24
   limitations on time scale estimates with, 33
Fra Mauro, 608

Fractional crystallization, 260
  olivine, 262—263
Fractional fusion, 260
  olivine, 262—263
Fracture cleavage, 367
Fracture porosity, 405
Fractures:
  in models, 382
  shear, 379
  and stress, 379
  tension, 379
Framework silicates, 205
Free air effect on gravity, 67
Free oscillation, 114
Freezing, role in weathering, 426
Frequency of seismic waves, 78
Frozen ground, 537
Fundamental level net, 63
Fusion:
  equilibrium, 260
  fractional, 260
Fusion processes, 259

Gabbro, 223, 265
Gal, unit of gravity, 65
Galena, 7, 179
Gamma (unit), 119
Ganges River, effect of floods, 479
Gangue minerals, 273
Ganymede, 626
Garnet, 178, 201, 286, 291, 302
Gases, volcanic, chemical composition (table), 246
Gems, 210—211
Gemstones, important, table, 211
Geocentric distances, 86
Geochemical exploration for ore, 276
Geode, 211
Geodesy, 42
Geodetic Reference System Formula (GRS 67), 66
Geoid, 56—58
Geological cycle of James Hutton, 219
Geologic columns, 12—13
Geologic cross sections, 12—13
Geologic maps, 12, 353—354
Geologic processes, present evidence of, 14
Geologic time scale, 20, 22—34
  calibration of, 33—34
  controversy about, 28—29
  perspective of, 34—37
  subdivisions of, 25—28
Geology, reason for study of, 3—4
Geomagnetic poles, 124
  variations in, 137
Geomorphology, 454
Geophones, 105
Geophysical exploration for ore, 276
Geothermal power, 276, 278

Geysers, 422—423
Gilbert, William, 117
Glacial drift, 528
Glacial-marine sediments, 553
Glacial plucking, 527
Glacial surges, 518
Glaciated landscape, 526
Glaciation, 526—537
  depositional features of, 532—537
  effect on sea level, 581
  erosional features of, 528—532
Glaciers, 18, 505—543
  advance of, 524—526
  continental, 507—515
  debris load, 526—528
  isostatic adjustment, 168—170
  landscape effects of, 505—506
  mass balance in, 519—526
  movement of, 516
  retreat of, 524—526
  worldwide location and elevation of, 525
Glaciofluvial features, 532—537
Glassy texture, 219
Global tectonic processes, 143—174
Glomar Challenger, 163
Glossary, 639—663
Gneiss, 11, 283
Goethite, 442
Gold:
  deposits of, 397
  distribution of, 271
Goldich, S. S., 434
Gold rush in California, 271
Goldschmidt, V. M., 440
Goniometers, reflecting, 182—183
Graben, 361
Graded bedding of sedimentary layers, 311—312
Graded streams, 484—485
Grade of ores, 274
Grain-supported texture, 307
Grand Canyon, 49
  erosion caused by Yellowstone River, 471
Granite, 9, 16, 223
  origin of, 265, 266
  texture of, 219
  uses of, 345
Granodiorite, 223
Graphite, 8, 178, 302
  crystal structure of, 198
  uses of, 216
Gravimeters, 66
Gravity:
  contour lines of, 67
  measurement of, 65
  measurement of petroleum in terms of, 391
  role of, in tides, 572—575
  units of, 65

Gravity anomalies, 67–69
Gravity corer, 547
Gravity surveying, 65–66
Great Diamond Hoax, 177–178
Great Lakes region isostasy, 170
Greenland, 505
    topography of, 514
Greenland Ice Sheet, 512–514
Greenschist facies, 290–293
Greywackes, 321
    origin of, 324
    properties of, 324
Griggs, David, 427
Groins, 592
Ground moraines, 532–533
Ground motion from seismic waves, 92
Ground vibrations, 76
Ground water, 403–424
    dissolved ions in, table, 423
    pressure on, 408
    role of, in dissolving limestone, 442
Ground water drainage basin, 410
Ground water drainage divides, 410
Ground water profile, 406
Ground water quality, 420–422
Ground water systems:
    confined, 413–414
    unconfined, 408–413
Ground water zones, 406–408
Guano, 335
Guide fossil, 24
Gulf Stream, 568
Gutenberg, Beno, 86
Gutenberg-Wiechert discontinuity, 102
Guyots, 51
Gypsum, 209, 309, 332
    formation of, 439
    formation of, from sea water, 334
    occurrence of, 334
    origin of, 334
    role in weathering, 426
    uses of, 345

Hack, John T., 502
Haefeli, Richard, 486
Half-life, 30
    of carbon 14, 32–33
Halite, 199, 209, 332
    formation of, from sea water, 334
    uses of, 345
Halite beds, 332
Hanging valleys, 531
Hardness:
    of minerals, 179
    of water, 421
Hawaiian basalts, comparison with oceanic basalts, 249

Hawaiian volcanic eruptions, 233, 235
Heat, radioactive sources for, 167–168
Heavily cratered terrain, on Mercury, 621
Heezan, B. C., 49–50
Hematite, 128, 179, 207, 442
Hematite group, 207
Hess, Harry, 145, 146
Hida-Sangun belt, 297
High magnesium calcite, 326
Hills, 47, 49
Hinge faults, 361
Hogbacks, 479
Holmes, Arthur, 145
Horizontal strata, 365–366
Horizontal surfaces, 56
Horn, glaciated, 531
Hornfels, 285
Hornfels facies, 292
Horsts, 361
Horton, R. E., 457
Host rocks, 251
    contact metamorphism, 300
Hot cathode tubes, 188
Hot springs, 422–423
Humbolt, Alexander von, 143
Hummocks, 497
Humus, 447
Hutton, James, 14
    great geological cycle of, 219
Hydration, 439
    silicate minerals, 203
Hydraulic geometry, 463–466
Hydraulic head, 411
Hydrocarbon compounds, presence of, in gas and oil, 346
Hydrogenic precipitates, 550, 556
Hydrolysis, 437–438
Hydrophones, 111
Hydrothermal fluids, 274
Hydrous silicates, 203
Hypotheses, nature of, 148

Ice:
    nature of, in glaciers, 512
    role of, in weathering, 426
Ice Ages, 506, 507, 537
    causes of, 542–543
    relation to climate, 542–543
Icebergs, 514
Ice-contact glacial features, 553
Iceland, volcanic activity of, 14
Ice movement, 520–524
Ice sheets:
    advance and retreat of, 542
    Antarctica, 507–512
    modern, 514–515
    motion of, 520, 523–524

Ice shelves, 510
Ice streams, mountainous terrain, 515–519
Igneous processes, 17, 217–278
 and metamorphism, 383–384
Igneous rocks, 6, 8–9
 average compositions of, 226
 classification of, 219–225
 composition of, 219
 description of, 219–225
 extrusive, 218
 formation of, 8–9, 217–218
 intrusive, 219, 250
 ores of, 271–276
 plutonic, 250
 processes involving, 217–278
 radiometric dating of, 32
 textures of, 219
Ignimbrites, 240
IGU, 223
Illite, 553
 and hydrolysis, 437
 weathering of, 437–438
Illuviation, 447
Ilmenite, 179
Imbrication, 325
Inclined seismic zones, 86
Inclined surfaces, 352
Index of refraction, 186
Induced magnetism, 129
Industrial development, relation to tectonic structures, 398–399
Industrial materials, occurrence of, 344–347
Inorganic detrital sediment, 550
Inorganic detritus, 553
In-phase X-ray beams, 189
Inselbergs, 476
Intensity of earthquakes, 87–88
Intercrater plains, 620
Interglacial stages, 539
Intergranular porosity, 404
Interior zones of earth, seismic analysis of, 101–105, 111, 113
Intermediate earthquake foci, 85
International Geophysical Year (IGY), 507
International Union of Geological Societies (IUGS), 223
Intertidal zone, 571
Intrazonal soils, 448
Intrusions, 250
Intrusive rocks:
 igneous, 219, 250
 SiO_2-rich, 258
Io, 626
Ion exchange reactions, 438–439
 use of, 439
Ions, 197
 in sea water, 563–566
  table, 563
Iron-manganese nodules, 556
Iron ore, formation of, 440
Iron oxides, 440, 441
Irrigation, effects of, on ground water levels, 415
Island arcs, 53
 composition of volcanic rock in, 248
 and orogeny, 386–387
Islands, volcanic, 239–240
Isograds, 295
Isoseismal contour lines, 87
Isoseismal maps, 87
Isostasy, 168–172
 effect on subsurface materials, 172
 and plate tectonics, 171
 role in glaciation, 543
Isostatic adjustment of glaciers, 168–170
Isotopes:
 age determination with, 32
 radioactive, 30–31
 use of, in age determination, 29–33

Japan Island Arc, 299
Japan Trench, 297
Jeffreys, Sir Harold, 103
Jetties, 594
Jodrell Bank, telescope, 602
Joint Oceanic Institutions for Deep Earth Sampling (JOIDES), 163
Joints, 350, 366–368
Joints (lava), 244
Joint set, 366
Jovian planets, 602, 624–627
Jupiter, 602, 624–626
 satellites of, 626

Kames, 533
Kame terraces, 533
Kaolinite, 434, 553
 hydration of, 439
Kaolinite minerals, hydrolysis of, 437
Karst areas, construction practices in, 446
Karst landscape, 441–446
Kelvin, Lord, 28–29
Kelvin Seamount Group, 50
Kepler, Johannes, 601
Kepler's laws, 601
Kettle lakes, 533
Kettles, 533
Kilauea Iki, oxide components in lava, 247
Kimberlite, 232, 267
Krakatoa, 235
Kuroshio Current, 568
Kyanite, 205, 291

Laccoliths, 253
Laki fissure, 238
Laminar flow, 166, 408
  in ice, 520
Laminations of sedimentary layers, 310
Landscape:
  evolution of, 499–503
    dynamic equilibrium concept of, 502–503
    glacial features of, 505–543
    role of wind, 493–499
Landslides, 488
Lateral moraines, 533
Laue, M. T. F. von, 188
Laue photographs, 188
Lavas, aa, 241
  composition of, 241–248
  pahoehoe, 241
  pillow, 243
  structure of, 8
  viscosity of, 241
Law of Constancy of Interfacial Angles, 182
Law of Gravitation, 44
Layered intrusive, 256, 258
  ore concentrations, 274
Layering of sedimentary rocks, 310–311
Layers, thickness analysis of, 100
Leaching, 447
Lead isotopes, 30
Levees, natural, 480
Level of isostatic compensation, 170
Life:
  extraterrestrial, 627
  origin of, 36–37
Lignite, 335
Limestone, 11, 307
  carbonation of, 440
  formation of, 326
Limonite, 442
Lithic, 321
Lithic arenite, 322
Lithification, 320
Lithology, in unweathered rock, 353–354
Lithosphere, 147, 384–386
  composition of, 269–271
  and location of earthquakes, 159–160
  thickness of, 160
Littrow-Taurus region, 608
Lodes, 271
Lodestone, 117
Loess, 497
Long (L) waves, 79
  seismic analysis of, 103
Longitudinal dunes, 497
Longitudinal profile of streams, 465
Longshore currents, 578
Low magnesium calcite, 326

Low-velocity zone, 103
Luminescence of minerals, 179
Lunar craters, 607
Lunar highlands, 604
Lunar maria, 604
Lunar rocks, 608
Luster of minerals, 178–179

Mackin, J. Hoover, 484–485
Madison Landslide, 491
Mafic group, 224
Magma, 17–18, 146, 217
  basaltic, 267
  crystallization of, 17
  origin of, 269–271
  properties of, 264
  seismic exploration of, 232
  temperature of, 250
Magmatic differentiation, 265, 274
Magmatic stoping, 255
Magnesium, extraction from sea water, 566
Magnetic anomalies, 118, 123, 130
  marine, see Marine magnetic anomalies
Magnetic declination, 120
Magnetic dipole, 118–119
Magnetic dip poles, 124
Magnetic domains, 128
Magnetic field of earth, 117–142
  ancient, 133–138
  anomalous, 123, 130–133
  changes, 94
  direction of, 119
    normal, 135
    reverse, 135
  intensity of, 123
  lines of force in, 120
  main, 123–128
    origin of, 139–141
  measurements of, 120–121, 123
  reversals, 134–136
  secular variations in, 124
  surveying, 123–124
  total field, 124
  variations in, 124, 127
Magnetic fields, 119
Magnetic inclination, 120
Magnetic minerals, 128
Magnetic moment, 129
Magnetic north, 120
Magnetic permeability, 118
Magnetic poles, 118
  positions, and fossil magnetism, 137, 138
    geologic periods affecting, 137
  strengths, 118–119
  unit of, 118
Magnetic properties of minerals and rocks, 130
Magnetic storms, 124, 128

Magnetic susceptibility, 129
  of minerals, 179
Magnetism:
  of earth, 117–142
  induced, 129
  origin of term, 117
  and plate tectonics hypothesis, 155–159
  remanent, see Remanent magnetism
  theory of, 118–120
Magnetite, 118, 128, 179, 207, 274, 302
Magnetohydrodynamics, 140
Magnetometers:
  nuclear precession, 121, 123
  portable, 121, 123
Magnets, 117
Magnitude of earthquakes, 86–88
Main magnetic field, see Magnetic field of earth, main
Maintenance of coasts, 592–594
Manganese nodules, 556
Manometers, 408
Mantel convection, 146
Mantel of earth, 6
  composition of, 114, 267, 268
  density of, 113
  elasticity of, 113
  physical properties of, 114, 160
  seismic analysis of, 102–103
  temperature of, 269
  viscosity values in, 166
Maps:
  bedrock geologic, 12
  continental drift features of, 148, 150
  geologic, 353–354
  isoseismal, 87
  topographic, 5–6
Marble, 11, 283, 301
Mare Humorum, 605
Mare Tranquillitatis, 607
Marianas Trench, 53, 54
Marine magnetic anomalies, 132, 135, 157, 158
Mariner spacecraft, 612, 622
Mars, 602, 612–619
  internal structure, 618
  origin and development, 616–618
  surface and atmosphere of, 613–614, 617
  surface features, 619
Marysville Stock, 287–289
Mascons, 610
Mass:
  balance in glaciers, 519–526
  under continents, 69, 72
  effect of, on gravity, 69, 70, 72
  under oceans, 70
Massive bedding, 315
Mass spectrometry, 32
Mass wastage, 454, 485–491

Master streams, 454
Maturity of sediments, 341
Mean density of earth, 72–73
Meandering channels, 460
Meandering streams:
  on alluvial plains, 471–479
  deposit of sediment by, 475–476
Meanders, 460
Mean sea level, 581
Mechanical weathering, 426–431
Medial moraines, 533
Medium bedding sedimentary layers, 310
Mediterranean volcanic eruptions, 233, 235
Melting:
  of igneous rocks, crystallization, process, 259–267
  of magma, 269
    role of water, 270
Melting process, 259
Mercalli Earthquake Intensity Scale, modified, sample of, 8
Mercer, John, 540
Mercury (planet), 602, 620–623
  craters, 621–622
Mesas, 479
Mesosphere, 147
  physical properties of, 160
Mesozoic era, 26, 27
Metallic surfaces, 178–179
Metamorphic belts, 294
Metamorphic facies, 289–300
  oxide components, 289–290
  pressure-temperature environments, 289
Metamorphic grade, 286, 295
Metamorphic minerals, 286
Metamorphic processes, 16
Metamorphic reactions, 291–293
Metamorphic rocks, 6, 11, 280–303
  common types, 281–286
  nonfoliated, 281
Metamorphic terranes, economic deposits in, 300–302
Metamorphism, 11, 280–303
  and igneous activity, 383–384
  production of ores by, 300
  regional, 280, 293–300
Metaquartzite, 283–285
Metasomatism, 300
Metastable equilibrium, 214
Metastable polymorphs, 213
Meteorites:
  age of, 32
  chondritic composition, 268
Meteoroids, impact on moon, 607
Meyer, Robert, 109
Mica minerals, 201, 204–205

Michigan basin, 360
Microcline, 205
Microearthquake recording, 93
Mid-Atlantic ridge, 52, 238
Midocean canyons, 50
Migmatites, 383
Milligal, 65
Mineral groups, most important, table, 200–211
Mineralogy, 177–216
Mineraloid, 206
Mineral resources from sea water, 565–566
Minerals, 7–8, 178
   atomic structure of, 193–200
   continuous series, 00
   identification of, 178–179, 629–637
   inner structure of, 192–193
   luster of, 178–179
   magnetic properties of, 130
   mechanical properties of, 181
   miscellaneous, table, 210
   optical properties of, 185–188
   physical properties of, 178–181
   pressure for formation of, 212–213
   response to temperature and pressure, 211–216
   temperature for formation, 212–213
Mining of ores, 397
Mississippi Delta, 341
Mississippi River:
   channel shifting of, 474
   delta of, 481, 483
   sediment carried by, 470
Mississippi Valley, 47
Mixed tides, 571
Miyashiro, A., 297
Modal analysis of rocks, 222
Modified Mercalli earthquake intensity scale, 87
   sample of, 88
Moho, see Mohorovičic discontinuity
Mohorovičic discontinuity, 102
   petrological significance, 268
   quest for data on, 109, 111
Mohs Hardness Scale, 179
Molecular diffusion, 32
Molecules, 196
Monochromatic beam of light, 186
Monocline, 356
Montmorillonite, 558
   hydration of, 437, 439
Monzodiorite-Monzogabbro group, 224
Monzonite group, 223
Moon, 604–610
  age of, 32
  density of, 73, 609–610
  mass of, 73
  origin and development of, 610
  rocks from, 607
  seismicity of, 608–609
Moonquakes, 609
Moraines, 532–533
Mountain building cycles, 384
Mountains, 49
Mt. Kilauea, 229
Mt. Pelée, 236
Mount Saint Helens, 229
   eruption of, 217–218
Mount Somma, 229
Mt. Vesuvius, 235
Movement potential, ground water, 410
Mud-supported texture, 307
Mudflows, 487
Mudstones, 307, 319–321
   composition of, 319
   modal mineralogy of, 320
Murphy, R. G., 46
Murphy classification, of topography, 46
Muscovite, 205, 256, 291

Naphthenic base oil, 346
Nappe fold, 382
Natural gas:
   creation of, 345
   occurrence of, 345, 391–397
Natural levees, 480
Natural resources, relation to tectonic structure, 393–399
Neap tides, 571
Nektonic organisms, 550
Neptune, 602, 627
Neutrons, 29, 193
Névé, 512
Newton, Isaac, 43, 572, 601
Newton's Second Law of Motion, 65
Nile River, delta of, 481, 483
Nile Valley, effect of floods, 480
Noble gases, 195
Nodules, manganese-iron, 556
Nondirected stress, 280–281, 376–377
Nonfoliated metamorphic rocks, 281
Nonmetallic surfaces, 178–179
Normal faults, 361
Normal stresses, 377
Normative analysis of rocks, 223
Nuclear explosions, detection of, 92
Nuclear precession magnetometers, 121, 123
Nuée ardentes, 235
Nye, John F., 522

Oblique slip faults, 361
Ocean basin floor, 50

Ocean basins, 4, 49–54
Ocean basin sediments, 550–559
  age of, 162–165
Ocean currents:
  deep, 569
  measurements, 566–567
  surface, 567–569
Oceanic basalts, comparison with Hawaiian basalts, 249
Oceanic ridges, 50, 52, 53, 132
  role of, in plate tectonics, 147, 152–153
Oceanic rises, 50
Oceanic seismic surveys, 111, 549–550
Oceanic volcanic eruptions, 238–240
Oceanography, 545
Ocean tides, see Tides
Ocean water, see Sea water
Ocean waves, 575–580
Oersted, Hans Christian, 119, 120
Oersted (unit), 119
Oil, see Oil shale; Petroleum
Oil shale, 11, 320
  occurrence of, 346
Old Faithful Geyser, 422
Olivines, 201
  fractional crystallization, 262–263
  fractional fusion, 262–263
  melting and crystallization, 260–262
Onawa Pluton, 286
Ooids, 327
Ophiolites, 384
Optical dispersion, 186
Optical properties of minerals, 185–188
Orbital lobes of electrons, 195
Order numbers, 457
Ores, 7
  concentration of metals in, table, 273
  of igneous origin, 271–276
  produced by metamorphism, 300
  structural settings for, 397–398
Organic accumulation and soil formation, 447
Orogenic belts, 372–373, 382
  evolution of, 384–387
  sedimentation, 383
Orogeny, 372–373, 382–391
  on constructive plate margins, 387–388
  on destructive plate margins, 384–387
Orthoclase, 205
Orthosilicates, 201
Outcrops of rocks, 6
Outwash features of glaciers, 533
Oxbow lakes, 474
Oxidation in weathering, 440–441
Oxide components, 222
  in metamorphic facies, 289–290

Oxide minerals, 207–209
  table, 209
Oxides in volcanic materials, 247
Oxygen, amount in the atmosphere, 35
  atomic structure of, 196

P-waves, reflection and refraction of, 97
  velocity of, 95
Pack ice, 514
Pahoehoe lavas, 241
Paired terraces, 479–480
Paleomagnetic fields, evidence of, 133–134
Paleomagnetic poles, 135–138
Paleomagnetism, 133–134
  and field reversals, 134–135
Paleozoic era, 26, 27
Paleozoic Ice Age, 541–542
Pangaea, 143
Parabolic dunes, 497
Paraffin base oil, 346, 391
Parallel drainage patterns, 456
Parallel laminations, 310–311
Paricutín volcano:
  birth of, 226
  oxide components in lava from, 247
Passive flow, in glaciers, 527
  of ice, 523
Pauling, Linus, 199
Pauling's Rules, 199
Peat, 337
  composition of, table, 336
Pedalfer soils, 448
Pedocals, 448
Pegmatite, 256
  ores in, 274
Pelagic sediment, 550, 554
Pelian volcanic eruptions, 235–236
Pendulums, gravity measurements with, 65–66
Peneplains, 500
Perched water bodies, 414
Percolation of petroleum and natural gas, 391
Percussion drills, 417
Peridotite, 267–269
Period of seismic waves, 78
Periods, 26, 27
  time span determination of, 33–34
Permafrost, 486–487, 537–543
Permeability, 405–406
  in sediments, table, 406
Petroleum:
  creation of, 345
  exploration of, 94–95
    and gravity, 63
  occurrences of, 345, 391–397
  in shale, see Oil shale
  viscosity of, 391–393

Phase (chemical), 222
Phase (mineral), 259
Phase (seismic), 103
Phase diagrams, polymorphic, 213
Phillipsite, 556, 558
Phobos, 619
Phosphate deposits, 551, 553
Phosphoria formation, 335
Phosphoria Limestone, 338
Phosphorites, 307, 309, 335
Phosphorus, uses of, 345
Photogrammetry, 61
Phreatic zone, 406
Phyllite, 281
Physical properties of minerals, 178—181
Piedmont glaciers, 515
Piezoelectricity of minerals, 181
Piezometric surface, 414
   fluctuations of, 414—415
Pillow lavas, 243
Pingos, 537
Pioneer spacecraft, 611, 624
Piston corer, 547
Pittsburgh Seam (coal), 335
Placer deposits, 271
Plagioclase, 205, 223
Plagioclase feldspars, 263, 264
Plains, 46—47
Planetology, 600
Planets, 597—628
   Jovian, *see* Jovian planets
   solar, *see* Solar planets
   terrestrial, *see* Terrestrial planets
Planktonic organisms, 550
Plant growth, role of in weathering, 431
Plastic deformation, 378, 379
Plastic flow of glaciers, 516
Plate tectonic hypothesis, 147
   test of, 151—152
Plate tectonics, 20
   and earthquake evidence, 160—162
   effect on ocean floor sediments, 559
   and isostasy, 171
   and magnetism, 155—159
   and origin of earthquakes, 159—160
   relationship to earthquakes, 161—162
   and seismology, 159—162
   theory of, 143—165
   and thermal convection, 165—168
   and topography, 148—155
   validity of, 165
   and volcanic activity, 248
Plinian volcanic eruptions, 235
Plumb lines, 56
Plunge direction, 357

Pluto, 16, 623
Plutonic igneous rock, 250
Plutonic processes, 16
Plutonic rock, 219
Plutons, 250—259
   associated ores, 397
Point bar, 474
Point count, 222
Poise (unit), 166
Poised streams, 485
Polar exploration, 507
Polarized light, 186
   in crystals, 187, 188
Polymorphic phase diagrams, 213
Polymorphic transformations, 292
Polymorphism, 198, 213
Polymorphs, 198
   of carbon, 215—216
Pompeii, 235
Pore space, 404
Porosity, 404—405
Porphyritic texture, 219—220
Portable magnetometers, 121, 123
Portland cement, 345
   compounds, table, 345
Post-depositional sedimentary structures, 316, 318
Potential energy:
   of crystals, 214
   of fluids, 166—167
Powder diffractometers, X-ray, 190, 192
Pratt, J. H., 69, 70, 170
Pratt isostatic earth model, 70
Precambrian era, 25—26, 27
Precambrian shields, 294
Precipitation:
   of minerals from sea water, 334
   of substances in soils, 447
Precision depth recorders, 61
Prediction of earthquakes, measurements for, 93—94
Pressure:
   in the earth, 113—114
      values of, 114
   on ground water, 408
Primary (P) waves, 79
Primitive lattices of crystals, 183
Probability of electron location:
   angular, 194
   radial, 194
Project Famous, 243
Prospecting for gold, 271
Protons, 29, 193
Pteropods, 550
Pumice, 243
Pure shear, 380
Pyrite, 209
Pyritic slate, 286

Pyroclastic debris, 229
Pyroelectricity of minerals, 181
Pyroxenes, 201, 203
Pyrrhotite, 128, 179, 209

Quartz, 7, 182, 205, 223, 256, 291
    durability of, 342–343
    weathering, resistance to, 435
Quartz arenite, 321
Quaternary Ice Age, stages in, table, 540

Radioactive decay, 30–33
Radioactive heat sources, 167–168
Radioactive isotopes, 30–33
    table, 31
Radioactivity, 30–33
    relation to estimates of earth's age, 29–33
Radiolaria, 550
Radiometric dating of rocks, 29–33
Rank, of coal, 335
Ranks of alteration (coal), 335
Reduction in weathering, 440–441
Reefs, 558
Reference ellipsoid, 43
Reflecting goniometers, 182–183
Reflection profiling, 108
Reflection of seismic waves, 96–101
Refraction:
    of seismic waves, 96–101
    of waves, 578
Refraction index, of minerals, 186
Regional metamorphism, 280, 293–300
Regolith, 6
    produced by weathering, 425
Regressive facies of sedimentary rocks, 338
Rejuvenation, 479–480
Relief, topographic, 46
Remanent magnetic moment of minerals, 179
Remanent magnetism, 129
    measurement of, 133
    oceanic crust, 158
    stability, 134
Repeated episodes of deformation, 371
Replacement, 331
    of carbonate rocks, 331
Residual stress, release of, 426–427
Reverse faults, 361
Reykjanes Ridge, 157
Rheology, 165
Rhyolite, 225
Richter, Charles F., 86
Richter magnitude scale, 86
Ridges, see Oceanic ridges
Rikitake, T., 140
Rills, 454

Rime, 519
Rio Grande rift, 250
Ripple bedding, 313, 315, 316
Rises, oceanic, 50
River channels, number and length of, table, 458
Rivers and streams, 460–470
    erosion caused by, 17
    graded, 484–485
    hydraulic geometry of, 463–466
    rejuvenation of, 479–480
    sedimentation caused by, 17
River valleys, rejuvenation effects in, 479–480
Rock falls, 488
Rock magnetism, 128–130
Rocks:
    density of, 6
    igneous, see Igneous rocks
    magnetic properties, 130
    metamorphic, see Metamorphic rocks
    normalitive analysis of, 223
    relation to minerals, 6
    relative ages of, 22, 24
    sedimentary, see Sedimentary rocks
    types of, 6
    ultramafic, see Ultramafic rocks
Rock salt, 309, 332
Rock slides, 488
Roentgen, W. C., 188
Rotary drills, 417
Runcorn, S. K., 137
Running water, 453–485
Rutherford, Ernest, 193
Ryoke-Sanbagawa belt, 297

S-waves:
    reflection and refraction of, 97
    velocity of, 95
Safe yield of wells, 419–420
Salinity of sea water, 563
Salt, uses of, 345. See also Rock salt
Saltation, 493
Salt basins, 332
Salt core structures, 395
Salt domes, 63, 333–334
    location of, 69
Salt water, contamination of wells by, 421
Salt weathering, 426
San Andreas fault, 89, 361
Sand, transport of, by wind, 493–499
Sandstone, 11, 307, 321–325
    cement of, 321
Sastrugi, 512
Saturn, 602, 626
    rings of, 626
    satellites, 626
Scarps, 361, 363

Schroedinger wave equation, 194
Schist, 282
Scoria, 243
Sea cliffs, 582
Sea floor spreading, 135, 145–148
   effect on ocean floor sediments, 559, 560
   effect on sedimentation, 164–165
   rate of, 157, 158
Sea level, 42, 56
   effects of glaciation, 540
   long-term changes in, 581
   mean, 581
Sea level surface, 57
Seamounts, 50, 558
   relation to sea floor spreading and volcanism, 250
Sea of Tranquillity, 607
Seawalls, 592
Sea water:
   contamination of wells by, 421
   density, 568–569
   dissolved constituents in, 563–566
   evaporation of, 334, 565
   precipitation of minerals from, 334
   properties of, 563–581
   salinity, 563
Secondary (S) waves, 79
Second-order streams, 457
Secular variations, in earth's magnetic field, 124
Sedimentary deposits of industrial utility, 344–347
Sedimentary differentiation, 341–344
Sedimentary facies, relationships, 336–341
Sedimentary materials, annual U. S. production, table, 344
Sedimentary particles, weight loss of in rotating barrel, 342
Sedimentary remanent magnetism, 133
Sedimentary rocks, 6, 9, 11, 304–348
   common types, 305
   economic importance of, 304
   formation of, 18
   origins of, 318–341
   perspectives on, 347
   regressive sequence of, 338
   texture, 306–307
   transgressive sequence of, 338
Sedimentary structures, 309–318
   post-depositional, 316
Sedimentation:
   along coasts, 582–583, 590
   effect of rivers on, 17
   effect of weathering on, 17
   from glaciers, 527–528
   in ocean basins, 550–579
   rates of, in ocean, 562–563
Sediment discharge, 470
Sediments:
   deep sea, 546–563
   deposit of, by meandering streams, 475–476
   ocean basin, 550–559
   in orogenic belts, 383
   particle sizes, 306
   pelagic, 550, 554
   permeability, table, 406
   transport of, in streams, 469–470
      by wind, 493
Seeps, 412
Seif dunes, 497
Seismic exploration, magma detection, 232
Seismicity, 85
   of moon, 608–609
Seismicity maps, 85
Seismic profiling, marine, 549–550
Seismic risk maps, 92–93
Seismic surveying, 95, 105–109
   field methods for, 105, 108
   marine, 111, 549–550
Seismic waves, 20, 76
   and earth structure, 94–115
   ground motion, 79
   low velocity zone of, 160
   measurement of, 77–79
   P to S conversion, 97
   reflection of, 96–101
   refraction of, 96–101
   velocity of, 95–96
   velocity of versus depth, 100
   and volcanic eruptions, 232
Seismograms, 77
   explanation of pulses, 97
   recording techniques for, 105
Seismographs, 76
Seismological observatories, 78–79
Seismologists, 81
Seismology, 75–115
   and plate tectonics, 159–162
Serpentine, 203
Shale, 11, 307, 319–321
   composition of, 319
   durability of, 344
   modal mineralogy of, 320
   petroleum content of, 10
   see also Oil shale
Shallow earthquakes, 85
Shear:
   pure, 380
   simple, 380
Shear fractures, 379
Shear modulus, 95
Shear stresses, 377
   in ice, 520

Sheeting, 427
Shields, continental, 293–295
Shield volcanoes, 227
Siderite, 300
Sierra Nevada Batholith, 251
Silica, 222
  biogenic, 551
Silicate minerals, 7, 201–206
  abundance of, 8
  composition and use, 204
  structure of, 9
Silicates, 201–206
Siliceous ooze, 551
Silicon dioxide, proportion of, in lava, 248
Sillimanite, 205, 291
Sills, 256
Siltstones, 307
Simple shear, 382
Single chain silicate structure, 203
Single crystal X-ray diffractometers, 189
Singly refractive minerals, 186
Sinkholes, 444
Sinuosity of streams, 460
Slate, 11, 281
Slump, 491
Smith, William, 24
Smooth plains, 621
Snell's law, 97, 186
Sodium chloride, 196–197
  formation of, from sea water, 334
Soil:
  varieties of, 448
  table, 450–451
Soil horizons, 447
Soil profiles, 447
Soils, 447–448
  distribution, maps, 449
Soil water, 406
Solar planets, 600–602
  dimensional and orbit geometry of, 601
  size and mass, 602
Solenhofen limestone, 379
Solifluction, 487
Sorting, 447
Space lattices:
  of crystals, 183
  X-ray determination of, 193
Space travel, 599–600
Sparker-profiler, 549
Sphalerite, 179
Spheroidal weathering, 433
Spindletop Dome and oil field, 395
Spinel group, 207
Spinner magnetometers, 133
Spirit levelling, 58, 69
Spits, 585, 586

Springs, 412
  artesian, 414
  hot, 422–423
Spring tides, 571
Stable equilibrium, 214
Stable polymorphs, 213
Stagnant regions, 00
Stalactites, 442
Stalagmites, 442
Staurolite schist, 286
Steam, use of, for electric power, 278
Steepness of waves, 589
Steno, Nicolaus, 182
Steno's law, 182
Stereographic diagrams of crystals, 183
Stishovite, 205
Stocks, 253
Stoping, magmatic, 255
Straight channels, 460
Strain, 95, 376
Strain ellipsoids, 379
Strain meter measurements, 93
Stratigraphic correlation, 24–25
Stratigraphy in oceans, 559–563
Strato-volcanoes, 229
Streak of minerals, 179
Stream flow and sediment transport, 466–470
Streamlines, 408
Streams, see Rivers and streams
Stream valleys in mountainous terrain, 471
Stream velocity, 465–466
Stress, 95, 376
  directed, 377–378
  and fractures, 379
  nondirected, 376–377
Stresses:
  compressive, 377
  confining, 378
  nondirected, 280–281, 376–377
  normal, 377
  shear, 377
    in ice, 520
  tensile, 377
Stress and strain in rock specimens, 378–380
Strike, 352
Strike slip faults, 361
Stromboli, 235
Strombolian volcanic eruptions, 235
Structural basins, 359, 361
Structural domes, 358–359
Structural settings of ores, 397–398
Structural traps, 393
Structure contour maps, 393
Subduction, 147
  and earthquake directions, 162
Submarine canyons, 54, 556

Submarine trenches, 53
Submergent coasts, 590–592
Subsidence:
    isostatic, 171
    in models, 382
Sulfide minerals, 209
Sulfide ore deposits, 397
Sulfur, occurrences of, 345
Superposed drainage systems, 456
Superposition, principle of, 22
Surface currents, 567–569
Surface drainage systems, 454–460
Surface waves, 575
Surtsey volcano, 238–240
Suspension of particles, role of, in weathering, 436
Sutter, John Augustus, 271
Swash zone, 578
Syenite group, 223
Symbiotic relationships, 327
Syncline, 357

T3 Ice Island, 515
TRM, 130
Tablelands, 47, 49
Tabular intrusions, 256
Talc, 302
Talus, 428, 431, 490
Tectonic breccia, 325
Tectonic fabrics, 368, 370–371
Tectonic plates, 147
    constructive, and orogeny, 387–388
    destructive, and orogeny, 384–387
Tectonic processes, 18–20, 147
    global, 143–174
Tectonic structures, 350–376
    and industrial development, 398–399
    mechanical analysis of, 376–382
    and natural resources, 393–399
Tectonism, 349–400
    and epeirogeny, 383
    model studies of, 380
Tellurium ores, 271
Temperature:
    in dolomitization, 332
    in earth, 268–269
Temperature-composition diagram, 261
Tensile stresses, 377
Tension fractures, 379
Terminal moraines, 532
Terminus of ice sheets and glaciers, 527
Terraces, paired, 479–480
    wave-cut, 582
Terrestrial planets, 602–624
    development of, 623–624
Terrigenous clastic rocks, 307
Terrigenous marine sediments, 556
Tetrahedral coordination, 201

Tetrahedron, SiO_4, 199
Texture:
    of igneous rocks, 219
    of sedimentary rocks, 306–307
Theory, nature of, 148
Thermal convection and plate tectonics, 165–168
Thermal expansion, role of, in weathering, 427–428
Thermoremanent magnetism (TRM), 130
Thick bedding in sedimentary layers, 310
Thin bedding in sedimentary layers, 310
Third-order streams, 457
Thrust, 361
    in models, 382
Tidal currents, 571
Tidal resonance, 574–575
Tides, 569–575
    effect of centrifugal forces on, 573
    effect of gravity on, 572–575
    mixed, 571
    neap, 571
    range of, 571
    spring, 571
Till, 528
Tillites, 538, 541
Tiltmeter measurements, 93
Time scale, geologic, 25–28
Titan, 626
Topaz, 178
Topographic maps, 5–6
Topographic profiles, 46
Topographic regions, 46
Topographic surveying, 42, 58–63
Topography, 45–46
    comparison of continental and oceanic, 54, 56
    and plate tectonic hypothesis, 148–155
    theories of development, 502
Torsion magnetometers, 121
Tourmaline, pyrroelectricity of, 181
Transantarctic Range, 389
Transform faulting, 153
Transgressive facies of sedimentary rocks, 338
Transpiration, 409
Transport of sediment, 469–470
Transverse dunes, 496
Transverse fractures, 154
Traps, petroleum and natural gas:
    structural, 393
    variable permeability, 395–396
Travel-time curves, 81, 97
    analysis with, 103
Trellis patterns, 00
Tremolite, 203, 222
Trenches, role of, in plate tectonics, 146–147
Triangular composition diagram, 223
Triangulation, 93
Tributaries, 454
Tridymite, 205

Triton, 627
Tsunami, 580
Turbidites, 554
Turbidity currents, 312, 554
Turbulence, relation of, to erosion, 469
Turbulent flow, 408
    in streams, 468–469
Tycho, 605

U-shaped valleys, 529
Ultramafic rocks, 225, 267–269
Unconfined ground water systems, 408–413
Unconformities, angular, 22, 371
Uniformitarianism, principle of, 14
Unit cells of crystals, 183
Universal Gravitation Constant, 44
Universal Law of Gravitation, 44, 573
Unstable equilibrium, 167, 214
Uplift, isostatic, 171
    in models, 382
    postglacial, 170
Upper mantle, seismic surveying of, 108. *See also* Mantle of earth
Uranium isotopes, 30
Uranus, 602, 626–627

Vadose zone, 406
Valley glaciers, 515
Valleys of mountain streams, 471
Valley of Ten Thousand Smokes, 241
Variable permeability traps, 395–396
Varved sequence of deposits, 311
Varves, 311
Vein, 273
Velocity of seismic waves, 95–96
Venus, 602, 610–612
    atmosphere of, 611
    internal properties of, 612
    radar mapping of, 611
Vertical direction, 56
Vertical movement in models, 382
Vesicular texture, 221, 243
Vesuvian volcanic eruptions, 235
Viking spacecraft, 612
Viscosity:
    of fluids, 165–166
    and isostasy, 170
    of lavas, 241
    of petroleum, 391–393
Viscous flow in mantle, 166
Volcanic activity:
    and extrusive rocks, 248–250
    intraplate, 249
    and plate tectonics, 248
Volcanic eruptions, *see* Central eruptions of volcanoes; Fissure eruptions of volcanoes
Volcanic gases, chemical composition, table, 246

Volcanic materials:
    chemical composition of, 241–248
    examination techniques for, 241
    oxide components in, 247
    physical features of, 241–248
Volcanic pipes, 232
Volcanic plateaus, 237
Volcanic processes, 14
Volcanism, 217–218
    relation to sea water, 581
Volcanoes:
    birth of, 226
    eruptions, *see* Central eruptions of volcanoes; Fissure eruptions of volcanoes
    observation of, 15
    shield type, 227
    structure of, 232
Voyager spacecraft, 624, 626
Vulcanian volcanic eruptions, 235
Vulcano, 235

Water:
    agricultures uses of, 415
    classification, 420
    consumption of, 415–416
    hardness, 421
    see also Ground water
Water level measurements, 93
Waterpocket Fold, 356
Water table, 406
    relation to capillarity, 408
    role of, in cave formation, 443
Water wells, 415–420
    drilling of, 416–417
"Water witching," 416
Wave-cut terraces, 582
Wave height, 577, 578, 580
Wave lengths, 577, 578, 580
Wave period, 577
Wave refraction, 578
Waves:
    celerity of, 576
    seismic, 76
    steepness of, 589
    see also Ocean waves; Seismic waves; Surface waves
Weathered rock, 11
Weathering, 17, 425–452. *See also* Chemical weathering; Mechanical weathering
Weathering series, susceptibility of minerals, 435
Weertman, Johannes, 518, 543
Wegener, Alfred, 143
Welded tuff, 240
Wells, artesian, 414
    contamination by sea water of, 421
    location of, 416–417
    safe yield of, 419–420
    water, 415–420

Welts, orogenic, 386
Wentworth, C. K., 306
Wentworth Scale of particle size, 306, 319
Westward drift, of earth's magnetic field, 124
West Wind Drift, 568
Williamson, E. D., 113
Wilson, J. Tuzo, 153
Wind, effect of, on landscapes, 493–499
Wollastonite, 291
Wood-Anderson seismographs, 86
World Wide Standardized Seismograph Network (WWSSN), 79, 161

Xenoliths, 255
X-ray analysis of minerals, 188–193
X-ray diffraction, crystal structure analysis with, 189–190, 192–193
X-ray diffractometers, 189–190
X-ray powder diffractometers, 190, 192
X-rays:
  discovery of, 188
  properties of, 188–189

Zeolite, 291
Zircon, 201
Zonal soils, 448
Zoned crystals, 264